Top Quark at the New Physics Frontier

Top Quark at the New Physics Frontier

Editors

Efe Yazgan
Pedro Ferreira da Silva

MDPI

Basel • Beijing • Wuhan • Barcelona • Belgrade • Novi Sad • Cluj • Manchester

Editors

Efe Yazgan
National Taiwan University
Taipei
Taiwan

Pedro Ferreira da Silva
CERN
Geneva
Switzerland

Editorial Office
MDPI
St. Alban-Anlage 66
4052 Basel, Switzerland

This is a reprint of articles from the Special Issue published online in the open access journal *Universe* (ISSN 2218-1997) (available at: https://www.mdpi.com/journal/universe/special_issues/Top_Quark_NPF).

For citation purposes, cite each article independently as indicated on the article page online and as indicated below:

Lastname, A.A.; Lastname, B.B. Article Title. *Journal Name* **Year**, *Volume Number*, Page Range.

ISBN 978-3-7258-0613-3 (Hbk)
ISBN 978-3-7258-0614-0 (PDF)
doi.org/10.3390/books978-3-7258-0614-0

Contents

About the Editors

Efe Yazgan

Efe Yazgan is a Senior Scientist specialized in high-energy physics working for National Taiwan University (NTU). His research focuses on analysing and interpreting data from the Large Hadron Collider (LHC) at CERN (Geneva, Switzerland) to understand the heaviest fundamental particles like the top quark, Higgs boson, and Z boson. After initially working on high-energy astrophysics and nanophysics in METU (Ankara, Turkey), he joined the CMS Collaboration in 2005. He completed his PhD studies at Fermi National Accelerator Laboratory (USA), with a thesis on a search for the Standard Model Higgs boson through vector boson fusion and energy reconstruction in electromagnetic and hadronic calorimeters of the CMS Experiment. Since 2007, he has been conducting his research at CERN. His main interests are top quark, Higgs boson and electroweak physics measurements, physics event modelling, and calorimetry. Some of the measurements he has made include forward–backward asymmetry of Drell–Yan pairs, spin correlations of top quarks, and underlying events accompanying top quarks. Most of these achievements were accomplished through collaborative efforts, including valuable partnerships with researchers like P. Silva. He has co-convened the Physics Event Generators groups of both CMS and the High Energy Physics Software Foundation. He has led various physics analysis and detector groups within CMS. He has numerous scientific publications across high-energy physics, astrophysics of neutron stars, and computational physics. He has given many talks in international conferences and lectures in summer schools. Currently, he focuses on searching for new Higgs bosons and top quark–Higgs boson interactions with his team at CERN from NTU.

Pedro Ferreira da Silva

Pedro Ferreira da Silva is a CERN Senior Research Physicist and a collaborator of the CMS experiment. He defended his PhD in Instituto Superior Técnico (Lisboa) on the measurement of the top quark branching ratio to b quarks in 2009. His research has focused on standard model and beyond the standard model searches using the LHC data, exploring the production of weak bosons, top quarks, and the Higgs boson. He convened the CMS Top quark group during Run 2 of the LHC and contributed to the field with different measurements of top quark production cross sections, properties, and mass. On the latter topic, Pedro has been intrigued by the experimental uncertainties affecting top quark mass measurements and has explored several different techniques, making use of different final states and reconstruction methods. These were complemented by ancillary measurements of hadronic activity and b quark fragmentation, some of which were made in close collaboration with E. Yazgan. In addition to these, Pedro has been active in searches for new physics with bosons in the final state, heavy Higgs, dark matter, as well as in the measurements of vector boson scattering processes. He's actively engaged in the detector upgrades for the High-Luminosity LHC, in particular in the design, commissioning, simulation, and performance of high granular calorimeters.

Preface

The hunt for top quarks began back in the 1970s after the proposal of the six-quark model by Kobayashi and Maskawa, and it was discovered in 1995 at the Fermilab Tevatron. It was experimentally established by five different experiments in different production modes and a variety of collision energies. At the CERN Large Hadron Collider (LHC), top quark–antiquark pairs are routinely produced at a rate of about six per minute, enabling experiments to make detailed measurements of the properties of top quarks. The analysis of the data collected at the Tevatron and the LHC experiments has revealed, so far, good agreement with the standard model (SM) predictions. The attained percentage-level precision in several measurements has been possible thanks to the excellent performance of the Tevatron and LHC accelerators and experiments. Accurate and precise theory calculations and Monte Carlo simulations also play a crucial role in the results obtained so far. In some cases, these accurate measurements are pushing the boundaries of physics and technology. The top quark is the most massive elementary particle identified to date: not only does it have a privileged Yukawa coupling to the Higgs boson, but also its mass is significantly higher than that of the Higgs boson. Owing to its large mass, the top quark decays before hadronization, making the study of "bare" quark properties possible in experimental settings. Therefore, top quark physics simultaneously pushes the frontiers of quantum chromodynamics, electroweak, and flavor physics. Of relevance is the fact that, taken together, the top quark and the Higgs boson modify the tree -level SM processes through radiative corrections. Such corrections are potentially sensitive to new physics contributions from energy scales larger than that which the current accelerators can achieve. The aim of this reprint is to provide a comprehensive review of the status and prospects of top quark physics at the LHC and possible future colliders. We have included articles that especially emphasize where the present understanding is incomplete and suggest new directions for research in this area.

Prof. Michael Tytgat has suggested E.Y. to serve as the editor of the journal issue on the top quark. We express our gratitude to him for his suggestion, as it ultimately led to the creation of the present reprint. The editors are indebted to the authors and MDPI for their support and enthusiasm for the project, as well as for their cooperation. Our special thanks go to Cici Xia for her assistance in preparing the reprint for publication. We would also like to gratefully acknowledge the CERN Scientific Information Service for their support in making this issue open access. E.Y. is grateful to the Academic Summit Program of the National Science and Technology Council of Taiwan (NSTC), Taiwan, as well as the National Taiwan University (NTU) President's Fund for their generous support.

Efe Yazgan and Pedro Ferreira da Silva
Editors

universe

Editorial

Top Quark at the New Physics Frontier

Efe Yazgan [1,*,†] and Pedro Silva [2]

1 Department of Physics, National Taiwan University, Taipei 10617, Taiwan
2 CERN, 1 Esplanade des Particules, PH-UCM Bat 40 3-A32 E24410, 1211 Geneva 23, Switzerland; pedro.silva@cern.ch
* Correspondence: efe.yazgan@cern.ch
† Current address: CERN, 1 Esplanade des Particules, EP-UCM Bat 653/1-005 C00210, 1211 Geneva 23, Switzerland.

> Whaat? You work on the top quark? Stiill?
>
> *Jack Steinberger to E.Y.*
> *in Moriond QCD, La Thuile, March 2013*

This Special Issue of *Universe* is devoted to the most massive fundamental elementary particle known, the top quark. The aim is to provide a comprehensive review of the current status and prospects of top quark physics at the Large Hadron Collider (LHC) and future colliders. We included articles that emphasize where the present understanding is incomplete and suggest new directions for research in this area. We trust that it will benefit both those seeking to learn and those seeking to review recent developments in top quark physics.

The search for top quarks began half a century ago, with the prediction of the existence of the top quark through the six-quark model by Kobayashi and Maskawa [1]. In 1977, the Υ meson was discovered by the E288 experiment at Fermilab, providing the first evidence for the existence of three families of elementary particles [2,3]. The existence of the b quark (the fifth quark) was inferred from interpreting the Υ meson as a bound state of a new heavy quark (b) and its anti-quark ($\bar{b}$) (see ref. [3] and references therein). When interpreted in a quarkonium model, the measurements of the Υ decay width to e^+e^- favored a b-quark charge of -1/3 instead of 2/3 [4,5]. With this discovery, the down-type family of quarks (d, s, and b) was established, but only two up-type quarks were observed, the u and the c quark. The measurements of $Z \to b\bar{b}$, including the forward–backward asymmetry of b quarks and the $b\bar{b}$ cross section at e^+e^- colliders, demonstrated the weak isopsin properties of the b quark [6]. With such properties, the b quark surely had to be accompanied by an upper isospin partner if the multiplet structure was present for the third quark family. Moreover, the anomaly cancellation of the electroweak (EWK) gauge theory requires that the sum of the electric charges in a family is zero. This implies that another quark with a charge of +2/3 should exist. An extra indication suggesting the existence of the top quark was the observation of fast oscillations of B–$\bar{B}$ [7,8].

Before the discovery of the top quark, using radiative corrections indirectly affecting the top quark and the measurements of EWK observables at a collision energy of $\sqrt{s} = 100$ GeV, LEP1 predicted the top quark mass (m_t) to be 173^{+13}_{-10} GeV [9]. Eventually, the top quark was discovered in 1995 at the Fermilab Tevatron with $\mathcal{O}(10)$ signal events, independently by the CDF [10] and D0 [11] experiments, using events from proton–antiproton ($p\bar{p}$) collisions at a center of mass energy of $\sqrt{s} = 1.8$ TeV. Both experiments found a signal consistent with $t\bar{t} \to W^+bW^-\bar{b}$ events and inconsistent with the background prediction, and both were also able to reconstruct mass distributions with a clear peak. The measured mass values in each experiment with central values of 176 GeV (with 7%

Citation: Yazgan, E.; Silva, P. Top Quark at the New Physics Frontier. *Universe* **2024**, *10*, 124. https://doi.org/10.3390/universe10030124

Received: 28 February 2024
Accepted: 28 February 2024
Published: 5 March 2024

relative uncertainty) and 199 GeV (with 14% relative uncertainty) were consistent with LEP1 predictions within uncertainties. This provided a critical test of the Standard Model (SM). Since then, many measurements of the top quark have been made both at the Tevatron and the LHC. Ten years before the discovery of the top quark, Lev B. Okun outlined the conditions for *reliable experimental results* [12]:

> *The physics of elementary particles is done by people. It is characteristic of man to err... Why then do physicists regard a multitude of phenomena as experimentally established, despite such mistakes?... How can it be guaranteed that these experiments are correct if so many incorrect results occurred in the past? The only guarantee is to accept a result as reliable only if it is obtained independently by several different groups employing different experimental methods. This condition is absolutely necessary but may not be sufficient, and does not provide a 100 percent guarantee. The 100 percent guarantee appears when the phenomenon recedes from the frontline of the science, when it is reproduced routinely, with the statistics of events exceeding by thousands or millions that with which the discovery was made, and when the quantities characterizing the phenomenon become known to an accuracy of several decimal places. Another way is not so much quantitative as it is qualitative: the search and discovery of a number of related phenomena that often follow the original discovery.*

It is fair to state that in present day, nearly 29 years after its discovery, the top quark satisfies the "absolute condition" of a reliable result, because it has been experimentally established by five different experiments in different production modes using different methods and a variety of collision energies. Moreover, it is observed both in $p\bar{p}$ and pp collisions. Although it would be unfair to say that the top quark has receded from the frontline of science, its existence satisfies the "100 percent guarantee" proposed by Okun. Now, at the CERN LHC, top quark–antiquark pairs are routinely produced at a rate of about six per minute, enabling experiments to make detailed measurements of the properties of top quarks. Percent-level precision in several measurements has been possible thanks to the excellent performance of the Tevatron and LHC accelerators and experiments[1]. For the top quark, the "qualitative condition" that was listed by Okun is also met. The top quark and the Higgs boson modify the tree-level SM processes through radiative corrections. Therefore, as was the case before the top quark discovery, the Higgs boson mass was predicted via an EWK fit. It was determined to be 94^{+25}_{-22} GeV, which turned out to be consistent with the measured value of the Higgs boson mass within 1.3σ [13]. This test provides a high-precision consistency check of the SM.

Significant progress has been made during the past years in improving experimental measurements and computation techniques to achieve more accurate and precise quantum chromodynamics (QCD) calculations, EWK theory calculations, and Monte Carlo simulations. The reader will comprehend that a detailed review of all these developments would require extending this Special Issue to several journal issues. Therefore, our choice, as editors, was to emphasize recent LHC results and discuss future prospects in the field, and where relevant discuss the relation to cosmology (e.g., EWK baryogenesis, stability of the vacuum, dark matter, and axion-like particles) in the respective contributions.

The top quark is an extraordinary elementary particle. It is the most massive elementary particle identified to date; not only does it have a privileged Yukawa coupling to the Higgs boson, it also has a mass that is significantly higher than that of the Higgs boson. At hadron colliders, top quarks are predominantly produced via QCD interactions. They are also produced "alone" through EWK interactions and are observed in single top quark channels. Owing to its large mass, the top quark decays before it can form a bound state, e.g., it can not form a $t\bar{t}$ meson (toponium). However, it may still be possible to observe some toponium effects in the phase space where the invariant mass of the $t\bar{t}$ pair is $2m_t \approx 345$ GeV (e.g., ref. [14]). The top quark decays before hadronization, making the study of "bare" quark properties possible in experimental settings. Moreover, the spin-decorrelation timescale for $t\bar{t}$ pairs is larger than that of the hadronization time scale.

This leaves the top and anti-top quark spins correlated and allows them to stay *entangled*. This may allow for tests of the foundations of quantum theory at high energy scales.

Top quark physics simultaneously pushes the frontier of QCD, EWK, and flavor physics. Through top quark measurements in $t\bar{t}$ and single-top quark processes, the existence of many (new) physics phenomena is verified through top quark measurements. Some of these phenomena are discussed in this issue: tests of charge-parity symmetry, lepton-flavor conservation, Lorentz-invariance violation and through that a precise test of special relativity, top quark Yukawa couplings via four-top quark production or via same-sign top quark plus a c jet, triple top quark, or single-top quark plus b jets to probe low-mass extra scalar particles, dark matter, axion-like particles, additional new particles such as color-octed vector G, neutral Z' boson, or a charged W' boson through $t\bar{t}$ asymmetries, and flavor-changing neutral currents of the top quark that connect the top quark with new scalar bosons. However, the analysis of the data collected by the first three LHC runs has revealed good agreement with the SM predictions. Currently, we do not have even a single direct or indirect indication of the existence of a new particle or interaction. Therefore, we are not in the same situation that we were in prior to the discoveries of the top quark and the Higgs boson.

Now, without any direct or indirect indication of new physics from the LHC, the scale of new physics is assumed to be above the TeV scale. Therefore, our focus has shifted to quantifying the effects of heavier hypothetical particles on our measurements at the LHC using the effective field theory (EFT) approach to identify dimension 6 operators that may affect our measurements, ensuring that kinematic distributions and cross sections align with observed data. Along with the measurements of the top quark within the SM and direct model-dependent searches for new physics, results or calculations using the EFT approach are discussed in all contributions of this Special Issue, except one contribution that does not adopt this approach; instead, this contribution promotes the general two-Higgs-doublet model (g2HDM), which offers two sets of dimension 4 operators to be investigated at the LHC and flavor experiments, specifically, new Yukawa and Higgs quartic couplings [15]. We embrace both approaches, however, we would have also welcomed a completely new revolutionary approach with no event generator, EFT, model-dependent search, or a Lagrangian. In any approach, precision measurements and open discussion are greatly needed, as emphasized by Robert B. Laughlin [16]:

> *A measurement that cannot be done accurately can never be divorced from politics and must therefore generate mythologies. The more such shades of meaning there are, the less scientific the discussion becomes. Accurate measurement in this sense is scientific law, and a milieu in which accurate measurement is impossible is lawless. The need for precision, in turn, redoubles the need for that other great Greek tradition, open discussion for ideas and ruthless separation of meaningful things from meaningless ones. Precision alone does not guarantee good law...*

We can better understand what is meant by this quotation with the difficulties encountered in the interpretation of the precise measurements of the top quark mass (see, e.g., ref. [17]). Using the measured values of the top quark and Higgs boson masses, one can say something about the stability of the EWK vacuum [18–20]. Current values of the top quark and Higgs boson masses indicate that the EWK vacuum may be meta-stable and that the SM is consistent and could be valid up to the Planck scale [21]. However, to be able to understand the stability of the EWK vacuum, we need a few times better precision in top quark pole mass measurements. This requires an electron–positron collider or a much better understanding of the meaning of the Monte Carlo mass, especially its relation to the so-called pole mass [17], or most probably the combination of both.

In this Special Issue, you will find contributions covering all these topics, although often briefly, and in most cases without going into deep detail in the theoretical aspects. Where appropriate, the contributions include the prospects for top quark measurements and related new physics searches in experiments at future colliders, such as HL-LHC, HE-LHC, FCC, ILC, CLIC, and CEPC.

Funding: E. Yazgan has received generous support from the Academic Summit Program of the National Science and Technology Council of Taiwan (NSTC), Taiwan, as well as the National Taiwan University (NTU) President's Fund.

Acknowledgments: The editors are indebted to the authors and MDPI for their support and enthusiasm for the project, as well as for their cooperation. We also would like to gratefully acknowledge the CERN Scientific Information Service for their support in making this issue open access.

Conflicts of Interest: The authors declare no conflicts of interest.

List of Contributions

1. Hosseini, Y.; Najafabadi, M.M. Prospects for Probing Axionlike Particles at a Future Hadron Collider through Top Quark Production. *Universe* **2022**, *8*, 301. https://doi.org/10.3390/universe8060301.
2. Hou, G.W.-S. On Extra Top Yukawa Couplings of a Second Higgs Doublet. *Universe* **2022**, *8*, 475. https://doi.org/10.3390/universe8090475.
3. Castro, N.F.; Skovpen, K. Flavour-Changing Neutral Scalar Interactions of the Top Quark. *Universe* **2022**, *8*, 609. https://doi.org/10.3390/universe8110609.
4. Chwalek, T.; D'Eliot, F. Top Quark Asymmetries. *Universe* **2022**, *8*, 622. https://doi.org/10.3390/universe8120622.
5. Blekman, F.; D'Eliot; F., Dutta; V., Usai, E. Four-top quark physics at the LHC. *Universe* **2022**, *8*, 638. https://doi.org/10.3390/universe8120638.
6. Behr, J.K.; Grohsjean, A. Dark Matter Searches with Top Quarks. *Universe* **2023**, *9*, 16. https://doi.org/10.3390/universe9010016.
7. Thomas-Wilsker, J. Recent Cross-Section Measurements of Top-Quark Pair Production in Association with Gauge Bosons. *Universe* **2023**, *9*, 39. https://doi.org/10.3390/universe9010039.
8. Chen, K.F.; Goldouzian, R. Tests of Charge–Parity Symmetry and Lepton Flavor Conservation in the Top Quark Sector. *Universe* **2023**, *9*, 62. https://doi.org/10.3390/universe9020062.
9. D'Hondt, J.; Kim, T.J. Measurements of the Cross-Section for the $t\bar{t}$ Heavy-Flavor Production at the LHC. *Universe* **2023**, *9*, 242. https://doi.org/10.3390/universe9050242.
10. Andrea, J.; Chanon, N. Single-Top Quark Physics at the LHC: From Precision Measurements to Rare Processes and Top Quark Properties. *Universe* **2023**, *9*, 439. https://doi.org/10.3390/universe9100439.
11. Pitt, M. Diffractive and Photon-Induced Production of Top Quark. *Universe* **2023**, *9*, 483. https://doi.org/10.3390/universe9110483.
12. Jung, A. Properties of the Top Quark. *Universe* **2024**, *10*, 106. https://doi.org/10.3390/universe10030106.

Note

1 Perhaps we can add one more condition for the top quark which is for now impossible to realize: that it becomes measurable in experiments independent of accelerators as well.

References

1. Kobayashi, M.; Maskawa, T. CP-Violation in the Renormalizable Theory of Weak Interaction. *Prog. Theor. Phys.* **1973**, *49*, 652–657. [CrossRef]
2. Herb, S.W.; Hom, D.C.; Lederman, L.M.; Sens, J.C.; Snyder, H.D.; Yoh, J.K.; Appel, J.A.; Brown, B.C.; Brown, C.N.; Innes, W.R.; et al. Observation of a Dimuon Resonance at 9.5 GeV in 400-GeV Proton-Nucleus Collisions. *Phys. Rev. Lett.* **1977**, *39*, 252–255. [CrossRef]
3. Innes, W.R.; Appel, J.A.; Brown, B.C.; Brown, C.N.; Ueno, K.; Yamanouchi, T.; Herb, S.W.; Hom, D.C.; Lederman, L.M.; Sens, J.C.; et al. Observation of Structure in the Y Region. *Phys. Rev. Lett.* **1977**, *39*, 1240–1242. [CrossRef]
4. Berger, C.; Lackas, W.; Raupach, F.; Wagner, W.; Alexander, G.; Criegee, L.; Dehne, H.C.; Derikum, K.; Devenish, R.; Flügge, G.; et al. Observation of a Narrow Resonance Formed in e^+e^- Annihilation at 9.46 GeV. *Phys. Lett. B* **1978**, *76*, 243–245. [CrossRef]
5. Bienlein, J.K.; Horber, E.; Leissner, M.; Niczyporuk, B.; Rippich, C.; Schmitz, M.; Vogel, H.; Glawe, U.; Heimlich, F.H.; Lezoch, P.; et al. Observation of a Narrow Resonance at 10.02 Gev in e^+e^- Annihilations. *Phys. Lett. B* **1978**, *78*, 360–363. [CrossRef]
6. Schaile, D.; Zerwas, P.M. Measuring the Weak Isospin of b Quarks in e^+e^- Annihilation. *Phys. Rev. D* **1992**, *35*, 3262–3266. [CrossRef] [PubMed]
7. ARGUS Collaboration. Observation of B_0-$\overline{B_0}$ Mixing. *Phys. Lett. B* **1987**, *192*, 245–252. [CrossRef]
8. CLEO Collaboration. B_0-$\overline{B_0}$ Mixing at the Upsilon(4S). *Phys. Rev. Lett.* **1989**, *62*, 2233–2236.

9. ALEPH; DELPHI; L3; OPAL; LEP Electroweak Working Group. A Combination of preliminary LEP electroweak measurements and constraints on the standard model. In Proceedings of the 17th International Symposium on Lepton Photon Interactions, Beijing, China, 10–15 August 1995; CERN-PPE-95-172.
10. CDF Collaboration. Observation of Top Quark Production in $p\overline{p}$ Collisions with the Collider Detector at Fermilab. *Phys. Rev. Lett.* **1995**, *74*, 2626–2631. [CrossRef]
11. D0 Collaboration. Observation of the Top Quark Production. *Phys. Rev. Lett.* **1995**, *74*, 2632–2637. [CrossRef] [PubMed]
12. Okun, L.B. *Particle Physics, The Quest for the Substance of Substance*; Primakoff, H., Burstein, E., Lamb, W., Lederman, L., Ruderman, M., Eds.; Harwood Academic Publishers: Chur, Switzerland, 1985; p. 61.
13. Baak, M.; Goebel, M.; Haller, J.; Hoecker, A.; Kennedy, D.; Kogler, R.; Moenig, K.; Schott, M.; Stelzer, J. The Electroweak Fit of the Standard Model after the Discovery of a New Boson at the LHC. *Eur. Phys. J. C* **2012**, *72*, 2205. [CrossRef]
14. Fuks, B.; Hagiwara, K.; Ma, K.; Zheng, Y.-J. Signatures of Toponium Formation in LHC Run 2 Data. *Phys. Rev. D* **2021**, *104*, 034023. [CrossRef]
15. Hou, G.W.-S. On Extra Top Yukawa Couplings of a Second Higgs Doublet. *Universe* **2022**, *8*, 475. [CrossRef]
16. Laughlin, R.B. *A Different Universe, Reinventing Physics from the Bottom Down*; Basic Books (A Member of the Perseus Books Group): New York, NY, USA, 2005; p. 215.
17. Hoang, A.H. What Is the Top Quark Mass? *Ann. Rev. Nucl. Part. Sci.* **2020**, *70*, 225–255. [CrossRef]
18. Alekhin, S.; Djouadi, A.; Moch, S. The Top Quark and Higgs Boson Masses and the Stability of the Electroweak Vacuum. *Phys. Lett. B* **2012**, *716*, 214–219 [CrossRef]
19. Holthausen, M.; Lim, K.S.; Lindner, M. Planck Scale Boundary Conditions and the Higgs mass. *J. High Energy Phys.* **2012**, *02*, 037. [CrossRef]
20. Elias-Miro, J.; Espinosa, J.R.; Giudice, G.F.; Isidori, G.; Riotto, A.; Strumia, A. Higgs Mass Implications on the Stability of the Electroweak Vacuum. *Phys. Lett. B* **2012**, *709*, 222–228. [CrossRef]
21. Degrassi, G. The role of the top quark in the stability of the SM Higgs potential. *Nuovo Cimento* **2014**, *37*, 47–53.

universe

Review

Properties of the Top Quark

Andreas Jung

Department of Physics & Astronomy, Purdue University, West Lafayette, IN 47906, USA; andreas.werner.jung@cern.ch

Abstract: Recent measurements of the properties of the top quark at the CERN Large Hadron Collider are discussed. The results were measured for single and top quark pair production in their final states, including jets with either one or two leptons or only in hadronic final states. Top quark properties include angular correlations, top quark spin correlations, mass, and width. When looking towards the future, top quark properties open new and even interdisciplinary avenues for probing quantum information science.

Keywords: top quark properties; top precision frontier; top quark factory

Citation: Jung A. Properties of the Top Quark. *Universe* **2024**, *10*, 106. https://doi.org/10.3390/universe10030106

Academic Editors: Efe Yazgan and Pedro Ferreira da Silva

Received: 24 November 2023
Revised: 3 February 2024
Accepted: 6 February 2024
Published: 26 February 2024

1. Introduction

The top quark, denoted as t, holds a distinctive position in the Standard Model (SM) and its various extensions. As the heaviest elementary particle known, with a mass of around 173 GeV, it was first discovered in 1995 at the Tevatron $p\bar{p}$ collider through the efforts of the CDF and D0 collaborations [1,2].

The discovery of the Higgs boson in 2012 [3,4] made it even more important to measure the mass of the top quark accurately. The precise measurement of the top quark mass, when coupled with determinations of the masses of the W and Higgs bosons, now serves as a robust self-consistency check for the SM [5–9]. The top quark is particularly exceptional due to its large Yukawa coupling and its unique role in interactions with the Higgs boson. Studying these interactions is indispensable for understanding the extension of the SM and the model's behavior at higher energy scales or even the Planck scale. In particular, the relation between the masses of the top quark and the Higgs boson [10–13] is essential with regard to the stability of the electroweak vacuum.

The top quark has an exceptionally short lifetime, approximately $\tau \approx 5 \times 10^{-25}$ s [14], effectively precluding the formation of top quark hadrons under normal conditions. However, there exists a possibility for the creation of transient bound states, known as "toponium" [15]. The depolarization time frame for top quarks surpasses $\tau \approx 10^{-23}$ s [14], allowing for the direct observation of their intrinsic properties. This unique aspect facilitates the examination of various characteristics, such as the spin, spin correlations, and polarization of top quarks. By reconstructing their quantum spin state, researchers can delve into an intriguing aspect of quantum mechanics: entanglement.

This phenomenon is characterized by the quantum states of interacting particles remaining inseparable (entangled) irrespective of the distance or spacial separation between the particles. The measurement of entanglement in top quark events provides a new handle for exploring the fundamentals of quantum mechanics at higher energy scales and is connected to the realm of quantum information science in collider experiments [16,17].

The Large Hadron Collider (LHC) at CERN is a proton-proton (pp) collider operating at various energies, and it is renowned as a "top quark factory" due to the abundant production of top quarks. The predominant production mechanisms involve the creation of top quark–antiquark pairs ($t\bar{t}$) through the strong interaction (gg / $q\bar{q} \rightarrow t\bar{t}$), serving as a fundamental test of quantum chromodynamics (QCD). Additionally, single top quarks are produced via the electroweak force, offering a means to test electroweak theory and directly probe the Cabibbo–Kobayashi–Maskawa (CKM) matrix element V_{tb}. By utilizing all available production modes, measurements of top quark properties can be utilized to enhance our understanding of fundamental particles.

In the SM, the branching fraction for a top quark decaying into Wb is $\Gamma(\mathrm{Wb})/\Gamma(\mathrm{Wq}) = 0.957 \pm 0.034$, with q = b, s, d [18]. The subsequent decay of the W boson into either $\ell\nu$ or $\mathrm{q\bar{q}'}$, where ℓ represents an electron or a muon, is utilized to categorize $\mathrm{t\bar{t}}$ events into decay channels. Namely, either both W bosons decay leptonically (dilepton final state), or only one W decays leptonically while the other W decays hadronically (lepton + jets or ℓ + jets final state), or none decay leptonically and both W decay hadronically (all-hadronic final state). Leptons in this categorization can also originate from semi-leptonic τ decays, while hadronic τ decays lead to additional jet activity in the event. The highest hadronic activity occurs in the all-hadronic final state consisting entirely of jets with jet multiplicities of up to 15 jets. The identification of jets originating from a b quarks can be achieved by utilizing the decay length of b quarks, which is in the order of $c\tau$ 10 s mm [18] and is commonly known as "b-tagging" [19,20]. It can be further improved through multivariate discriminant techniques employing a variety of variables [21]. These techniques involve a combination of variables describing the properties of secondary vertices and tracks with significant impact parameters relative to the primary vertex.

The specific final state particles of a $\mathrm{t\bar{t}}$ pair can vary, and this is not only due to the decay channel and the number of quarks and gluons involved. For example, it can vary due to single or multiple leptons and one or more jets, some of which (or all) may be b-tagged, as well as combinations of the leptons and jets folded with missing transverse momentum due to the presence of neutrinos. Different decay channels, along with the number of top quarks produced, lead to a variety of final states, which provides valuable information for experimental analyses and allows for a comprehensive study of top quark properties.

This overview delves into a selection of measurements, providing an overview of the current landscape of experimental measurements of top quark properties and opens with a brief introduction in Section 2 on how to extract top quark properties. The measurements described in Section 3 encompass various aspects, such as angular correlations linked to asymmetries in top quark production and the correlation of the top quark's spin, which are detailed in Section 3.1. It is followed by a brief summary of Lorentz invariance violation in $\mathrm{t\bar{t}}$ production. In Section 3.3, the overview delineates the present status of precision measurements of the top quark's mass, incorporating combined results from both ATLAS and CMS where applicable. Additionally, Section 3.4 discusses novel measurements to test the Yukawa coupling strength in top quark events, followed by results on the width of top quarks in Section 3.5. The review closes with Section 3.6, which explores the current state of measurements of top quark properties in associated production.

2. Accessing Top Quark Properties

In this context, what is commonly referred to as top quark properties encompasses measurements of various differential top quark cross-sections that allow (in a secondary step) for the extraction of the specific characteristics or properties (mass, spin, couplings, etc.) of the top quark. Figure 1 shows differential cross-section measurements at the LHC [22] in the ℓ + jets decay channels (see Figure 1), which provide excellent statistical power and are examples of distributions that will be used later on to extract top quark properties. A recent summary of measurement techniques and the results of differential cross-sections can be found in Ref. [23].

Examples include observations related to the polarization of top quarks and $\mathrm{t\bar{t}}$ production level asymmetries at the LHC, but this is not the only way to measure top quark properties. Precise examination of these cross-sections is essential for thoroughly testing perturbative QCD (pQCD) predictions and identifying the potential signals of new physics within their theoretical frameworks. The meticulous measurement of top quark production cross-sections ensures a precise depiction of these processes, pinpointing areas for model refinement when necessary. This precision testing is vital for advancing our comprehension of top quark behavior, pushing the boundaries of the current theoretical framework, and facilitating more robust extraction of top quark properties. By scrutinizing the behavior of top quark pairs, these measurements contribute significantly to the validation and enhancement of our understanding of the fundamental physics involved, particularly in the domain of strong interactions.

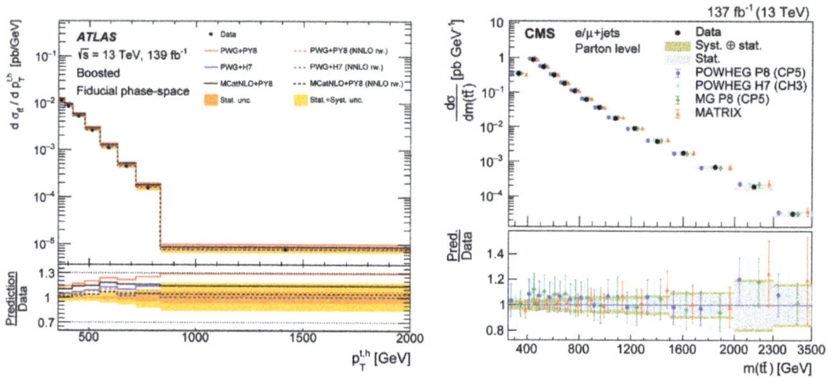

Figure 1. Differential cross-sections in the ℓ + jets channel as a function of the hadronic top quark p_T (**left**) and of m(tt̄) (**right**) [24,25].

The importance of single top quark production measurements cannot be overstated. At the LHC, the predicted cross-sections for single top quark production in the *t*- and *tW*-channels are not significantly smaller compared to the tt̄ pair production cross-section. The number of single top quark events collected at the LHC is sufficient to make similarly precise and detailed studies of top quark properties. These measurements play a crucial role as essential tests for the electroweak theory, offering a unique avenue to probe the fundamental forces and interactions involved in these processes.

Production of Top Quarks

Accessing top quark properties in both single top and top quark pair production processes provides a comprehensive understanding of the fundamental characteristics of the heaviest elementary particle. In the context of single top quark production, the measured cross-sections in different channels, such as the t-($q'g \rightarrow tq\bar{b}$) and tW-channels ($gb \rightarrow tW$), offer unique opportunities for detailed studies. Despite the cross-sections not being significantly smaller than those for tt̄ production at the LHC, single top events contain only one top quark, and the measurements serve as crucial tests for the electroweak theory. Measurements made with single top quark data have allowed for significant achievements, such as the analysis of the top quark-related components of the CKM matrix [26], the study of the polarization of individual top quarks, and leveraging single top quark events to constrain the structure of the proton, particularly by examining the ratio of the up-to-down quark content. These results are obtained via the measurements of the ratio between t and t̄ quarks.

Conversely, the production of pairs of top quarks, involving the creation of a top quark and its antiparticle, represents another crucial avenue for investigating top quark properties. The generation of tt̄ pairs occurs through QCD processes involving strongly interacting colored gluons and quarks. Present-day theoretical predictions extend to the next-to-next-to leading order (NNLO), incorporating next-to-next-to leading log (NNLL) corrections in QCD, along with electroweak corrections at the next-to-leading order (NLO).

These predictions typically have relative uncertainties of 3.5% [27], including uncertainties related to renormalization and factorization scales, the proton parton density distribution functions (PDFs), and the determination of the strong coupling constant (α_S).

Measurements in this context, often involving complex final states with multiple jets and leptons, contribute to a thorough examination of top quark characteristics, including their mass, spin, and decay properties. Measurements of single and top quark pair production cross-sections at the inclusive and multi-differential levels are described elsewhere in this issue.

3. Top Quark Properties in $t\bar{t}$ Production

The LHC produces an abundance of $t\bar{t}$ pairs, which allows for the scrutinization of their properties and behaviors with unprecedented precision. This unique capability transforms the LHC into an invaluable laboratory for unraveling the mysteries of the top quark and testing the predictions of the Standard Model. In particular, the vast amount of data allows for access to top quark properties, even at high masses for the $t\bar{t}$ pair ($m(t\bar{t})$) or large transverse momentum (p_T) scales, i.e., in the "boosted" phase space.

Measurements of $t\bar{t}$ production in the boosted regime offer a route to extracting top quark properties at high scales. This approach involves comparing measurements with predictions, leveraging the characteristics of boosted top quarks, which exhibit reduced contributions from bound state effects and Coulomb corrections, as well as lepton and trigger efficiency uncertainties that are much larger in the threshold region of $m(t\bar{t}) \approx 350$ GeV. However, studying boosted top quarks presents experimental challenges, requiring dedicated top quark taggers to maintain reconstruction efficiency.

In the analysis of boosted top quark events, the reconstruction of top quarks presents a particular challenge, especially when the lepton (either muon or electron) is non-isolated due to its close proximity to a b-jet. Hadronic decays of τ lepton decays can be exploited to study top quarks decaying into boosted τ leptons that are identified by using deep machine learning techniques [28].

Improved algorithms have been developed, resulting in better jet and lepton separation with higher efficiency for reconstructing the decay products of top quarks within the ℓ + jets channel, even when lepton isolation is not feasible. In such scenarios, the decay products of the hadronic W boson typically merge first, creating a "semi-resolved" topology. Advancing the measurements into the highly boosted phase space necessitates initial dedicated studies in this area to establish and refine the reconstruction methods [24,29]. Variables (see Figure 2) based on jet-substructure, i.e., n-subjettiness (τ_n), allow for the SM to be challenged in a new phase space with reasonable avenues for extracting top quark properties.

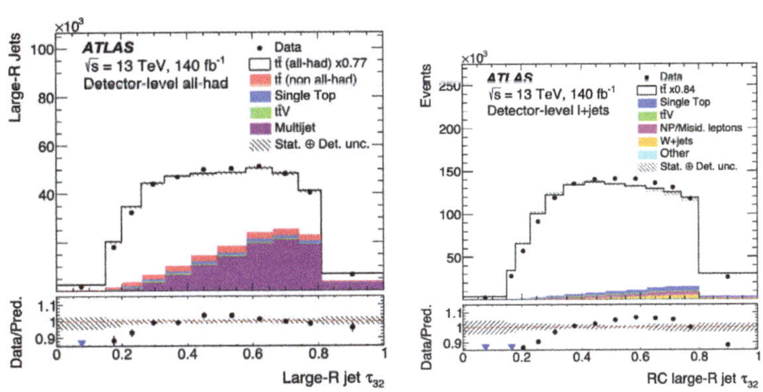

Figure 2. ATLAS measured detector level; $\tau_{32} = \tau_3/\tau_2$ distributions comparing in the **(left)** ℓ + jets and **(right)** all-hadronic channel for the data, the predicted $t\bar{t}$ signal, and the measured backgrounds [29].

3.1. Measurements of Angular Correlations

The measurement of correlations between the angles of various particles are sensitive probes with which to test the validity of the SM. Over the years, a variety of measurements have been carried out, such as production asymmetries, as well as measurements of top quark spin correlations and polarization. Typically, the angular correlations observed vary across the different center-of-mass energies during the LHC's operational phases. At the LHC, quarks tend to possess a higher average longitudinal momentum compared to anti-quarks. In the context of a pp collider, this distinction results in a broader rapidity distribution for t production as opposed to the production of $\mathrm{t\bar{t}}$. Now, this allows for the definition of production level asymmetry employing the rapidity y of a top and an anti-top quark. When using $\Delta|y| = |y_\mathrm{t}| - |y_{\bar{\mathrm{t}}}|$, the charge asymmetry A_C is given by the following equation:

$$A_\mathrm{C} = \frac{N(\Delta|y| > 0) - N(\Delta|y| < 0)}{N(\Delta|y| > 0) + N(\Delta|y| < 0)} \tag{1}$$

ATLAS and CMS have carried out measurements of charge asymmetries [30,31], including a joined analysis by ATLAS and CMS [32]. In the ensuing discussion, two of these findings are outlined. The current results from the LHC have not yet reached the level of precision of the charge asymmetry predicted by the SM at a value of $A_\mathrm{C} = 0.0064^{+0.0005}_{-0.0006}$ [33]. Notably, ATLAS has reported evidence of a non-zero charge asymmetry [30], while not yet being able to verify the predicted SM value. CMS recently published a measurement of A_C in the boosted phase space of $m(\mathrm{t\bar{t}}) \geq 750$ GeV [31]. This measurement utilizes a binned maximum likelihood fit profiling systematic uncertainties that provide improved precision. In both scenarios, the measured top quark charge asymmetry is consistent with the SM prediction at NNLO in pQCD with NLO electroweak corrections. The charge asymmetry defined in Equation (1) can be modified to rely on leptons. This minimizes the uncertainties arising from reconstructing top quarks. Recently, ATLAS presented the first search for leptonic charge asymmetry in the $\mathrm{t\bar{t}}$-associated production of W bosons [34] from initial state radiation. The ATLAS Collaboration reported an observed leptonic asymmetry of $-0.112 \pm 0.170(\mathrm{stat}) \pm 0.054(\mathrm{syst})$, which is in good agreement with theoretical predictions [33]. Although, this measurement presents a significant step forward in understanding the $\mathrm{t\bar{t}}$ + W process, the measurement is still limited by statistical uncertainties. More data from LHC Run 3 (2022–ongoing) may substantially enhance this precision. This would also be useful in searches for physics beyond the SM. Looking even further ahead, future prospects involve extending the measurements of charge asymmetry by utilizing different variables, e.g., energy asymmetry [35]. Measuring top quark asymmetries with new variables and methods would refine our understanding of top quark production asymmetries and may help us identify subtle deviations.

The angular distributions of leptons resulting from top quark decays provide a precise measurement tool that is useful for investigating variables that are sensitive to the spin correlations and polarization of top quarks. Early measurements carried out at the Tevatron [36] revealed that the spin of top quarks is correlated, aligning with the predictions made by the SM. More refined measurements conducted at the all center-of-mass energies of the LHC [37–41] offer an unprecedented opportunity to scrutinize the SM at a detailed level. Recent measurements of the difference in azimuthal angle ϕ between the decay leptons, denoted as $\Delta\phi(\ell^+\ell^-)$, conducted by both ATLAS [41] and CMS [42], exhibit noteworthy agreement, as illustrated in Figure 3. However, when comparing these findings with SM predictions, a subtle tension becomes apparent. This tension is alleviated with higher-order corrections at NNLO [43] and by employing techniques that mitigate the impact of theoretical uncertainties on acceptance corrections [44].

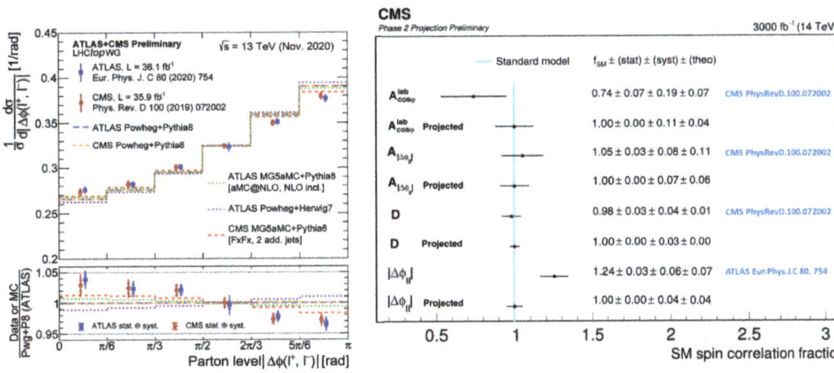

Figure 3. (**Left**): The measurements conducted by ATLAS [41] and CMS [42] on the angle of separation between the decay leptons from the $t\bar{t}$ pair are set against the projections made by the SM. (**Right**): An estimated projection concerning the accuracy of future measurements of spin correlation and polarization of top quarks at the High-Luminosity LHC (HL-LHC) [45] and compared to existing measurements [41,42].

Current and future efforts at the LHC aim to enhance differential access to spin correlation and polarization information using full Run 2 data and the eagerly anticipated first Run 3 results. This approach seeks to better understand the evident mismodeling in the distribution of the opening angle of decay leptons, $\Delta\phi(\ell\bar{\ell})$, and of the top quarks, as highlighted earlier. A recent CMS projection evaluates the precision achievable in measuring various angular distributions with the data collected during the HL-LHC [45]. With an integrated luminosity of 3000 fb^{-1} at $\sqrt{s} = 14$ TeV, providing a vast top quark data sample, statistical power becomes virtually limitless. Utilizing a generic future CMS detector simulation through DELPHES [46] offers an initial assessment of the expected precision in measuring the strength of spin correlations within the SM. The parameter D, derived from the differential helicity angle ($\cos\varphi$) distribution, is expected to have a precision of better than 3% [45]. Figure 3 (right) displays the outcome of this study, including other variables, and this is compared to existing measurements by ATLAS and CMS using partial Run 2 data. The precision of top quark spin correlation variables is leveraged to project the sensitivity to supersymmetric top quark partners (stops) in the degenerate phase space [41]. The latter is defined as the phase space where the stop mass equals the mass of its decay particles (top quark and neutralino).

3.2. Search for Lorentz Invariance Violation

A test of Lorentz invariance violation (LIV) can be achieved by a detailed examination of $t\bar{t}$ pair production with the pp collision data collected by the CMS detector at the LHC, operating at a center-of-mass energy of $\sqrt{s} = 13$ TeV. The study, motivated by the possibility of deviations from Lorentz invariance, builds upon the first of such investigations at the Tevatron [47], but it utilizes vastly more data. Data in the dilepton channel ($e\mu$ only) were selected for this study and correspond to an integrated luminosity of 77.4 fb^{-1} [48]. Specifically, the analysis explores the differential normalized cross-section for $t\bar{t}$ production as a function of sidereal time, introducing LIV as an extension of the SM—see Figure 4. In the context of an effective field theory, the predictions include the modulation of the $t\bar{t}$ cross-section with sidereal time when LIV is present. The investigation extracts bounds on LIV couplings, revealing compatibility with Lorentz invariance, with an impressive absolute precision of 0.1–0.8%.

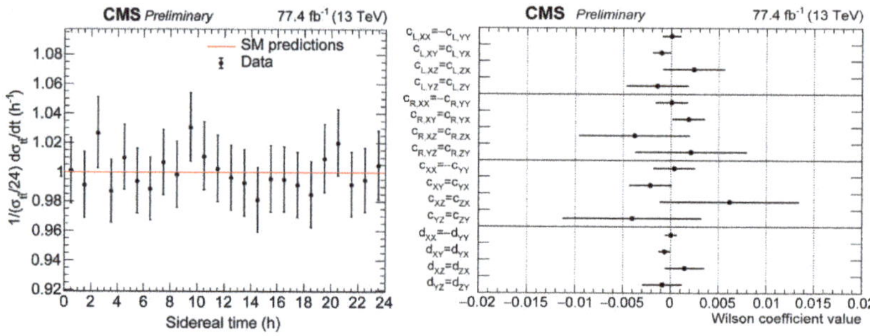

Figure 4. (**Left**): CMS data in $t\bar{t}$ production is shown as a function of siderial time and is compared to the SM expectation of no dependency at all [48]. (**Right**): Limits on EFT parameters, implementing LIV in $t\bar{t}$ production [48].

This search not only delves into potential LIV in top quark production but also serves as a precision test of special relativity with top quarks. Notably, the precision achieved in this study surpasses previous measurements [47] in this domain by two orders of magnitude, underlining the significance of this comprehensive exploration.

3.3. Top Quark Mass

Various measurements of the top quark mass, m_t, have been conducted at the LHC, and while it's impractical to cover all the details here, we focus on the latest and most relevant findings. Figure 5(left) presents a summary (as of June 2023) of top quark mass measurements utilizing kinematic or so-called "direct" methods. Both ATLAS [49] and CMS [50] experiments have measured the top quark mass with typical uncertainties well below 0.5 GeV in absolute values, which is below 0.3% in relative uncertainty. These measurements span dilepton, ℓ + jets, and all-hadronic decay channels. The Tevatron has also contributed, with the latest combination [51] and an initial world combination [52].

The world average, as depicted by the vertical grey band in Figure 5 (left), typically has a relative uncertainty of around 0.5%. This precision is somewhat diminished compared to the more recent results from the LHC, a discrepancy that is mainly due to the lack of updated LHC data in the world combination [52]. Historically, combinations have employed the BLUE method [53–55], but recent developments in measurement techniques, specifically those using profile likelihood approaches, have led to the introduction of a more precise likelihood method [56], which is now commonly adopted for LHC top quark combinations.

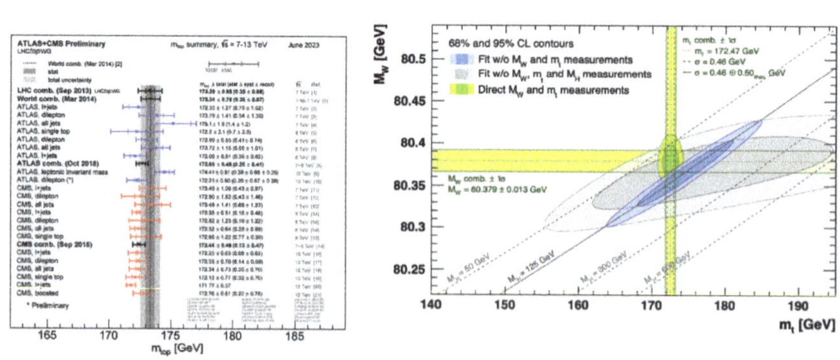

Figure 5. On the (**left**): Determinations of the top quark mass based on direct measurement techniques [57], with the following Refs. [50,52,58–73]. On the (**right**): The m_t versus m_W plane, displaying the GFitter global fit outcome derived from top quark mass results [9].

Top quark mass measurements at LHC Run 2 confirm the LHC as a top quark factory, providing ample data for precise determinations. Most recently, such a measurement was provided by CMS using profile likelihood methods with five observables [72]. Here, top quark mass determination is based on calculating the invariant mass of three jets originating from the same top quark, which includes one b jet and two light jets from the W boson. In order to reconstruct the complete $t\bar{t}$ system, a minimization of a χ^2 is carried out, taking into account the anticipated resolutions and combinatorial factors. The analysis incorporates observables, offering optimal constraints on the top quark mass and dominant uncertainties, including those impacting the jet energy scale. Figure 6 (left) illustrates the step-wise improvement in total precision by progressively adding these observables. Figure 6 (right) presents additional observables as inputs to the ML fit, along with their post-fit probability density functions. This analysis notably addresses the treatment of statistical fluctuations affecting systematic uncertainty modeling directly in relation to the likelihood, a crucial consideration for future analyses with increased data and smaller uncertainties.

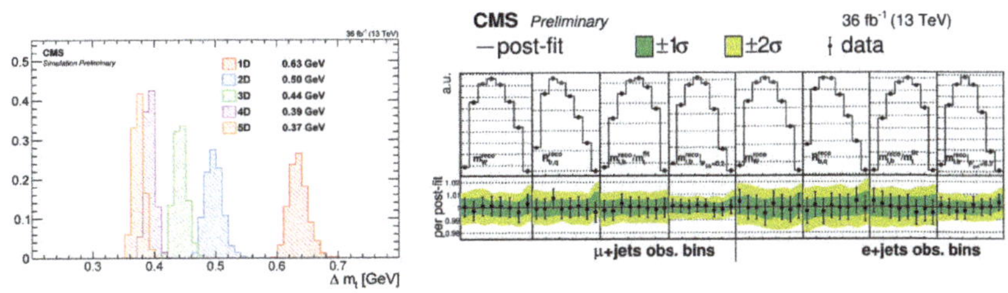

Figure 6. On the (**left**): The capability of the ML techniques to constrain uncertainties for extracting m_t is achieved by incorporating additional observables [72]. On the (**right**): The input to the ML fit comprises additional observables, along with their post-fit probability density functions [72]. The 1 sd and 2 sd uncertainty bands are depicted by the green and yellow bands, respectively.

The stability of the electroweak vacuum serves as a powerful test of the SM up to energy scales close to the Planck scale, which is also relevant for cosmology since many processes in the early universe could have triggered a decay of the electroweak vacuum [74]. In the context of particle physics, the electroweak vacuum state is closely tied to the Higgs field and its potential. A crucial question revolves around whether the universe resides in the absolute lowest energy state or if the current vacuum state is metastable, potentially capable of transitioning to a more stable state. Figure 5 (right) displays the $m_t - m_W$ plane, where m_t is determined via a combination of ATLAS and CMS measurements by the GFitter collaboration [9], and this is updated in Ref. [75]. This combination incorporates the latest measurements and is subject to an additional theoretical uncertainty of 500 MeV, addressing the "controversy" on whether the MC mass (e.g., in PYTHIA) equals the pole mass (used in fixed order calculations) in the field of top quark physics [76]. This controversy, gaining renewed attention with the precision of LHC measurements beyond 0.3%, necessitates a precise evaluation of the variances and uncertainties associated with this conversion, as outlined in a recent comprehensive review article on the subject referenced in Ref. [76].

The interplay between the masses of the top quark, the W boson, and the Higgs boson [3,4] serves as a robust self-consistency test for the SM, contributing insights into the stability or potential meta-stability of the SM vacuum [13]. Ongoing discussions, particularly regarding the W boson mass measurement, underscore the dynamic nature of our understanding in this fundamental realm. A recent revision of the W boson mass measurement by CDF [77] diverges notably from the global average. The community is actively deliberating the implications of this result [75]. Current measurements, coupled with theoretical extrapolations at NNLO in the SM, suggest a scenario where the vacuum is either meta-stable or, intriguingly, positioned precisely at the threshold between stability and meta-stability.

A substantial push towards a better understanding of how top quark mass measurements and their uncertainty relate between ATLAS and CMS is presented in Ref. [78]. A comprehensive combination of 15 top quark mass measurements from the ATLAS and CMS experiments at the LHC is presented—see Figure 7 (left). The datasets utilized cover integrated luminosities of up to 5 and 20 fb^{-1} for pp collisions at center-of-mass energies of 7 and 8 TeV, respectively. This combination encompasses measurements in top quark pair events involving both ℓ + jets and hadronic top quark decays, as well as a measurement from events enriched in single top quark production via the electroweak t-channel. The resulting combined top quark mass is $m_t = 172.52 \pm 0.14$ (stat) $\pm$ 0.30 (syst) GeV, with a total uncertainty of 0.33 GeV or better than 0.2%.

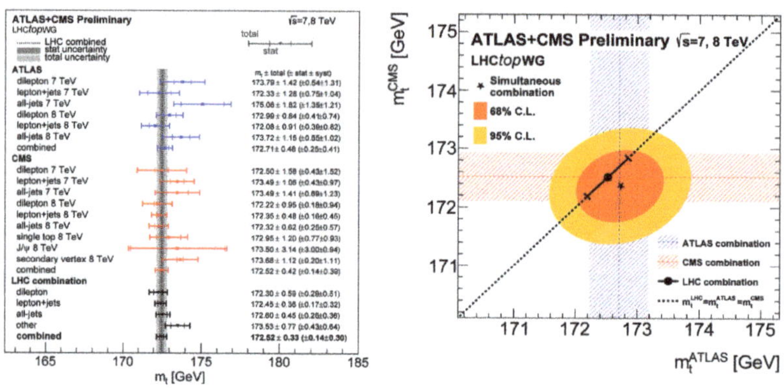

Figure 7. On the (**left**): The most recent LHC combination integrates 15 top quark mass measurements from various decay channels conducted by both ATLAS and CMS [57]. On the (**right**): The correlation between the measured m_t values in ATLAS and CMS is examined and compared to the combined value from the LHC [78].

ATLAS and CMS simultaneously extract the top quark mass based on ATLAS measurements, m_t^{ATLAS}, and based on CMS measurements, m_t^{CMS}, using a BLUE combination of 15 input measurements—see Figure 7 (right). The ellipses show the 68% and 95% confidence intervals, indicating good agreement between m_t^{ATLAS} and m_t^{CMS}. The observed correlation between them is 0.15. The central values and uncertainties for individual ATLAS and CMS combinations are represented, and the full marker denotes the central value of the LHC combination (m_t^{LHC}), assuming equality with $m_t^{ATLAS} = m_t^{CMS}$. This latest combination of top quark mass measurements has a high precision of 0.2% thanks to a substantially improved understanding of the correlations between systematic uncertainties in either set of mass measurements by ATLAS or CMS. Achieving advancements in reducing uncertainties necessitates collaborative work from the entire scientific community. Advancements are being actively pursued on both experimental and theoretical sides through the collaborative efforts of the LHC Top Working Group (LHCtopWG) [57], exemplifying a concerted drive towards scientific progress in this arena. This group serves as a hub for researchers to collectively contribute to the ongoing developments in understanding and refining our knowledge of top quark physics.

Alongside direct measurements, alternative methods for determining the top quark mass aim to enhance precision by complementing systematic uncertainties or improving the understanding of well-defined renormalization schemes [79]. A prevalent alternative technique involves extracting the top quark mass from the $t\bar{t}$ production cross-section, which measures the top quark pole mass in a direct approach [22,80]. While Tevatron had limitations, the higher LHC data statistics enable not only inclusive but also multi-differential techniques to extract the top quark mass precisely. The abundance of data allows for the simultaneous fit of α_S and the top quark mass, achieving uncertainties below 0.9 GeV, e.g., even in the boosted phase space using a "jet mass" proxy [81].

Alternative methodologies for estimating m_t focus on different features of event data. An example involves the use of the invariant mass of a muon from the W boson's leptonic decay in conjunction with a soft muon found within the b jet. This technique presents a measurable variable sensitive to the top quark mass [82]. The reliance on muons in this method reduces the vulnerability to uncertainties typically associated with the jet energy scale and the modeling of the top quark production process. Nonetheless, this approach introduces its own set of uncertainties, primarily stemming from the nuances of B hadron decay processes, including the details of B fragmentation and decay branching fractions. The effectiveness of this approach in constructing a top quark mass-sensitive observable is demonstrated in Figure 8, which showcases the post-fit distribution based on the selection of muon pairs with the same sign.

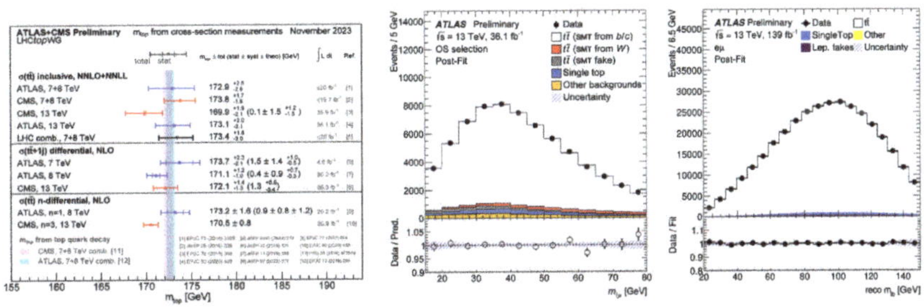

Figure 8. (**Left**): An overview of the measurements of the top quark mass by ATLAS and CMS, derived from observables in $t\bar{t}$ production [57] (**Middle**); details provided in Refs. [22,44,50,80,83–88]: The distribution after fitting, using pairs of muons with the same sign, is utilized to establish an observable sensitive to m_t [82]. (**Right**): A description of the observables employed in estimating the width of the top quark. The uncertainties at the post-fit level are determined using the correlation matrix derived from the fitting process [89].

3.4. Top Quark Yukawa Coupling

The interactions between the top quark and the Higgs boson field are governed by the Yukawa coupling strength (y_t). The top quark is the one fermion that has the highest Yukawa coupling strength. The measurements of its value can shed light on the role of the top quark in the mechanism of electroweak symmetry breaking. CMS verified the strength of the top quark Yukawa coupling by analyzing the kinematic distribution of top quark pairs [90], using dilepton events from pp collision data and the full CMS Run 2 dataset of 137 fb^{-1}. The mass of the $t\bar{t}$ system and the rapidity difference of the top quark and antiquark are particularly sensitive to y_t. The measurement yields a value of $y_t = 1.16^{+0.24}_{-0.35}$, constraining $y_t < 1.54$ at the 95% confidence level.

3.5. Top Quark Width

The precise calculation of the top quark width is achievable in the SM [91,92], and deviations could suggest new physics. Indirect measurements relying on SM couplings lack model independence since they assume SM couplings. Pioneered at the Tevatron [93,94] and refined at the LHC Run I [95], direct top quark width measurements, such as the recent one by ATLAS [89], employ width-sensitive final state distributions, utilizing the invariant masses of the lepton and b-tagged jet, $m(lb)$, and the $b\bar{b}$ pair, $m(b\bar{b})$—these allow for the constrainment of the uncertainties arising from JES uncertainties. The measured top quark width, $\Gamma_t = 1.9 \pm 0.5$ GeV, aligns with the SM prediction.

3.6. *Top Quark Properties in Associated Production*

In the initial stages of Run 2, the precise measurements of inclusive cross-sections in the associated production of vector bosons with $t\bar{t}$ pairs have provided important results to better understand the SM. These processes include $t\bar{t}$ + W, $t\bar{t}$ + Z [96], and $t\bar{t}$ + photon, offering access to enhanced charge asymmetry, Z boson coupling to $t\bar{t}$ pairs, and the determination of the top quark's electric charge. These channels also serve as crucial backgrounds for beyond the SM (BSM) searches, especially in cases involving the associated production of additional $q\bar{q}$ pairs, such as $b\bar{b}$, which contribute significantly to the $t\bar{t}$ pairs associated with the Higgs boson. Additionally, rare channels like tZq events, where a single top quark is produced with a vector boson (Z boson and a quark or jet), have been observed by ATLAS [97] and differentially measured by CMS [98]. CMS's measurements, including the Zq -to-Z$\bar{\text{q}}$ rate ratio, offer a unique avenue to explore another property of the top quark: spin asymmetry A_ℓ. It is defined as $A_\ell = 1/2Pa_\ell$ with P being the polarization of the top quark and $a_\ell \approx 1$ [99] being the spin analyzing power (see Figure 9(left)). The ratio of the Zq-to-Z$\bar{\text{q}}$ rates gives access to the polarization of the top quark [98] and Figure 9 (right) provides an overview of the associated cross-section measurement that can be utilized to extract properties once sufficient data is available. An overview of ATLAS and CMS measurements of tX with (X =Z or γ) cross-sections at 13 TeV and comparisons to NLO QCD theoretical calculations is displayed in Figure 10(left). The latest results include the observation of a single top quark with a photon by ATLAS [100] and evidence for the production of the tWZ process by CMS [101]. The different phase space regions used for the measurements are denoted as "Vis 1", "Vis 2" and "Vis 3", and they are highlighted for the ATLAS and CMS tqγ measurements. Mild tensions between the experimental results and the theoretical predictions can be seen for some of these rare processes.

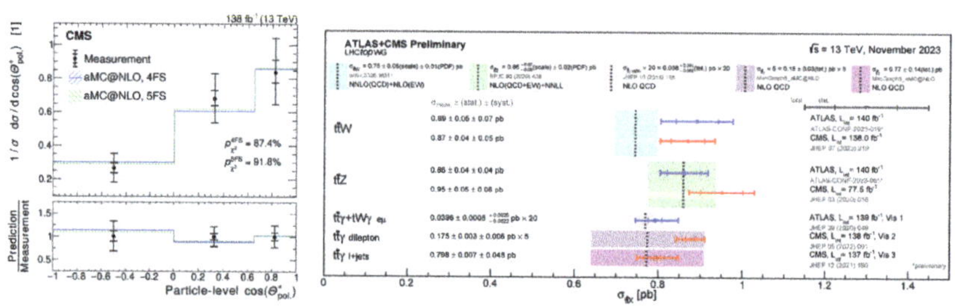

Figure 9. (**Left**): The spin asymmetry, determined through the cross-sections of top quark production in association with a Z boson, is contrasted with SM predictions [98]. (**Right**): A comprehensive overview of the associated production of vector bosons at the LHC [57] - details can be found in Refs. [41,102–107].

The top quarks produced in association with other bosons, such as photons, Higgs, or Z bosons, can be utilized to determine flavor-changing neutral currents (FCNCs). Figure 10 (right) displays the summary of LHC results for FCNCs, including comparisons with various new physics models, assuming all other FCNC processes are negligible. The limits are presented as top quark decay branching ratios, considering both FCNC top quark decay and production vertices in some cases. Both ATLAS and CMS have a variety of FCNC searches in this sector, with the most recent ones using the full Run 2 data briefly discussed here. ATLAS uses the full Run 2 data of 139 fb^{-1} to explore FCNCs involving a photon and a top quark [108]. No significant event excess is observed over the background prediction, and the 95% confidence level upper limits are placed on the strength of left- and right-handed FCNC interactions. FCNC interactions can also manifest themselves in events involving the top quark, the Higgs boson, and an up-type quark (q=c,u), where no significant excess is observed by ATLAS in the t$\to$ qH (H$\to$ $\gamma\gamma$) process [109]. Most recently, CMS submitted a search for FCNCs in events with a photon and additional jet activity [110] and new Higgs bosons [111], while ATLAS set stringent limits for FCNCs in events with heavy Higgs bosons [112]. In addition to the existence of an FCNC in the top quark sector, CMS has looked into charged lepton flavor violation involving trilepton final states. The observed data align with SM expectations, and the results have been employed to extract 95% confidence level upper limits for Wilson coefficients—more details are in Ref. [113].

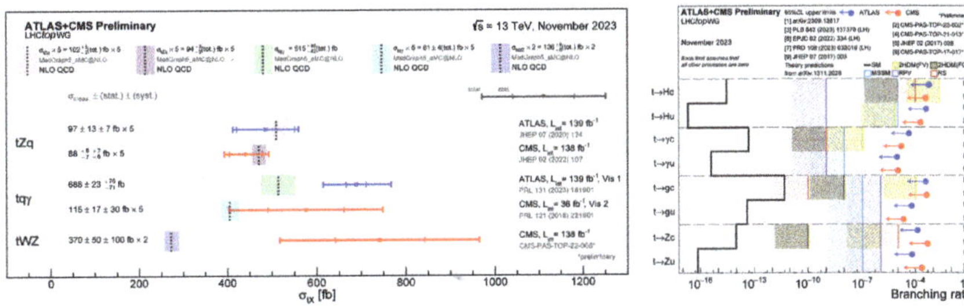

Figure 10. (**Left**): Summary of tX cross-section results by ATLAS and CMS [57] with additional information being provided in Refs. [97,98,100,101,114]. To facilitate visualization on a consistent scale, the results for tZq and the CMS tqγ measurements are multiplied by a factor of 5, while those of the CMS tWZ measurement are multiplied by a factor of 2. The theory bands encompass uncertainties arising from renormalization and factorization scales, as well as from PDFs. (**Right**): Summary of 95% confidence level observed limits on top quark decay branching ratios via FCNCs to a quark and a neutral boson (t→ Xq, where X=g, Z, γ or H; q=u or c) by the ATLAS and CMS collaborations [57]; additional information is provided in Refs. [108,109,115–122].

The exploration of four top quark production [123–125] represents one of the most intriguing and, until recently, uncharted territories at the LHC. Current limits, as illustrated in Figure 11(left), present intriguing indications of a potential cross-section enhancement in comparison to the SM prediction of $\sigma(t\bar{t} + t\bar{t})$ = 12 fb [126]. This process, involving the simultaneous production of four top quarks, serves as a valuable avenue for probing the top quark Yukawa coupling [90], as depicted in Figure 11 (right). This underscores the pivotal role of four top quark productions in advancing our understanding of top quark interactions and potentially uncovering new physics phenomena [127].

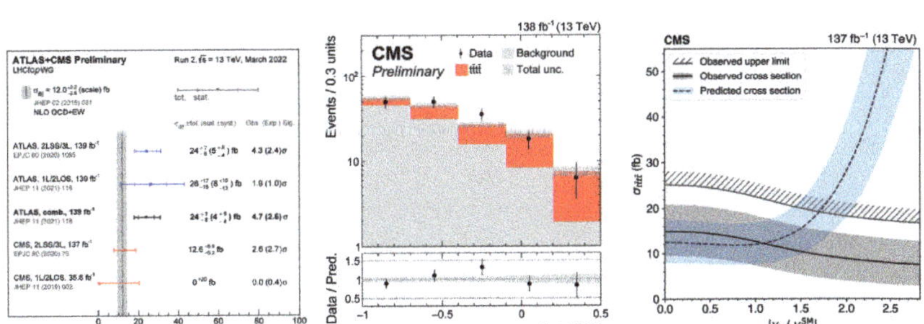

Figure 11. (**Left**): The production cross-section of $t\bar{t} + t\bar{t}$ as measured by ATLAS and CMS is compared with the theoretical prediction from the SM - further information is provided in Refs. [123,124,128,129]. (**Middle**): CMS's observation of the $t\bar{t} + t\bar{t}$ production cross-section [127]. (**Right**): The intervals of a 95% confidence level concerning the top quark Yukawa coupling, deduced from the observed $t\bar{t} + t\bar{t}$ cross-section [123–125].

The observation of the production of four top quarks in *pp* collisions marks a significant achievement, drawing from a dataset collected by the CMS experiment [127]. This milestone is based on an integrated luminosity of 138 fb^{-1}. The analysis focuses on events featuring two same-sign, three, and four charged leptons (electrons and muons), along with additional jets. By employing sophisticated multivariate discriminants to discern the signal process from predominant backgrounds, the measured signal cross-section is $17.9^{+3.7}_{-3.5}$ (stat) $^{+2.4}_{-2.1}$ (syst) fb, aligning closely with the best available theoretical predictions. The observed (expected) significance of the signal stands at 5.5 (4.9) standard deviations above the background-only hypothesis.

4. Summary

In the past decade, LHC experiments have revolutionized our understanding of top quark properties, challenging state-of-the-art theoretical predictions. Recent measurements, such as top quark spin correlations, reveal the mild tension with the Standard Model (SM), mitigated by higher-order corrections. Angular correlations, such as charge asymmetries, hint at deviations from the SM. Notably, the latest ATLAS + CMS combination of top quark mass measurements reaches a precision of 0.2%. Overcoming systematic uncertainties demands community-wide efforts, emphasizing unified signal modeling in community efforts such as the LHCtopWG. Ultimately, the HL-LHC promises a dataset exceeding 1 billion top quarks, which, in the context of top quark properties, will be a significant challenge while promising exciting new results, and this will surely also see the use of cutting-edge techniques not currently widely utilized in top quark physics.

Excitingly, top quark properties open new avenues for probing quantum information science, forming a new field of quantum observables in top quark physics and unlocking exciting opportunities for new physics discoveries.

Funding: This research was funded by Department of Energy award number DE-SC0007884.

Data Availability Statement: Data are contained within the article.

Conflicts of Interest: The author declares no conflicts of interest.

References

1. Abe, F. et al. [CDF Collaboration] Observation of Top Quark Production in $\bar{p}p$ Collisions with the Collider Detector at Fermilab. *Phys. Rev. Lett.* **1995**, *74*, 2626–2631. [CrossRef] [PubMed]
2. Abachi, S. et al. [D0 Collaboration] Observation of the Top Quark. *Phys. Rev. Lett.* **1995**, *74*, 2632–2637. [CrossRef]
3. Aad, G.; Abajyan, T.; Abbott, B.; Abdallah, J.; Abdel Khalek, S.; Abdelalim, A.A.; Abdinov, O.; Aben, R.; Abi, B.; Abolins, M.; et al. Observation of a new particle in the search for the Standard Model Higgs boson with the ATLAS detector at the LHC. *Phys. Lett. B* **2012**, *716*, 1–29. [CrossRef]
4. Chatrchyan, S.; Khachatryan, V.; Sirunyan, A.M.; Tumasyan, A.; Adam, W.; Aguilo, E.; Bergauer, T.; Dragicevic, M.; Erö, J.; Fabjan, C.; et al. Observation of a new boson at a mass of 125 GeV with the CMS experiment at the LHC. *Phys. Lett. B* **2012**, *716*, 30–61. [CrossRef]
5. de Blas, J.; Pierini, M.; Reina, L.; Silvestrini, L. Impact of the recent measurements of the top-quark and W-boson masses on electroweak precision fits. *Phys. Rev. Lett.* **2022**, *129*, 271801. [CrossRef] [PubMed]
6. de Blas, J.; Ciuchini, M.; Franco, E.; Goncalves, A.; Mishima, S.; Pierini, M.; Reina, L.; Silvestrini, L. Global analysis of electroweak data in the Standard Model. *Phys. Rev. D* **2022**, *106*, 033003. [CrossRef]
7. de Blas, J.; Chowdhury, D.; Ciuchini, M.; Coutinho, A.M.; Eberhardt, O.; Fedele, M.; Franco, E.; di Cortona, G.G.; Miralles, V.; Mishima, S.; et al. HEPfit: A code for the combination of indirect and direct constraints on high energy physics models. *Eur. Phys. J. C* **2020**, *80*, 456. [CrossRef]
8. Flächer, H.; Goebel, M.; Haller, J.; Hoecker, A.; Mönig, K.; Stelzer, J. Revisiting the global electroweak fit of the Standard Model and beyond with Gfitter. *Eur. Phys. J. C* **2009**, *60*, 543–583. [CrossRef]
9. Haller, J.; Hoecker, A.; Kogler, R.; Mönig, K.; Peiffer, T.; Stelzer, J. Update of the global electroweak fit and constraints on two-Higgs-doublet models. *Eur. Phys. J. C* **2018**, *78*. [CrossRef]
10. Holthausen, M.; Lim, K.S.; Lindner, M. Planck scale boundary conditions and the Higgs mass. *J. High Energy Phys.* **2012**, *2012*. [CrossRef]
11. Elias-Miró, J.; Espinosa, J.R.; Giudice, G.F.; Isidori, G.; Riotto, A.; Strumia, A. Higgs mass implications on the stability of the electroweak vacuum. *Phys. Lett. B* **2012**, *709*, 222–228. [CrossRef]
12. Xing, Z.; Zhang, H.; Zhou, S. Impacts of the Higgs mass on vacuum stability, running fermion masses, and two-body Higgs decays. *Phys. Rev. D* **2012**, *86*, 013013. [CrossRef]
13. Alekhin, S.; Djouadi, A.; Moch, S. The top quark and Higgs boson masses and the stability of the electroweak vacuum. *Phys. Lett. B* **2012**, *716*, 214–219. [CrossRef]

14. Bigi, I.I.; Dokshitzer, Y.L.; Khoze, V.; Kühn, J.; Zerwas, P.M. Production and decay properties of ultra-heavy quarks. *Phys. Lett. B* **1986**, *181*, 157–163. [CrossRef]
15. Fuks, B.; Hagiwara, K.; Ma, K.; Zheng, Y.J. Signatures of toponium formation in LHC run 2 data. *Phys. Rev. D* **2021**, *104*, 034023. [CrossRef]
16. Afik, Y.; de Nova, J.R.M. Entanglement and quantum tomography with top quarks at the LHC. *Eur. Phys. J. Plus* **2021**, *136*, 907. [CrossRef]
17. Abdallah, M. et al. [STAR Collaboration] Tomography of ultrarelativistic nuclei with polarized photon-gluon collisions. *Sci. Adv.* **2023**, *9*, eabq3903. [CrossRef] [PubMed]
18. Group, P.D.; Zyla, P.A.; Barnett, R.M.; Beringer, J.; Dahl, O.; Dwyer, D.A.; Groom, D.E.; Lin, C.J.; Lugovsky, K.S.; Pianori, E.; et al. Review of Particle Physics. *Prog. Theor. Exp. Phys.* **2020**, *2020*, 083C01. [CrossRef]
19. The CMS Collaboration. Identification of b-quark jets with the CMS experiment. *J. Instrum.* **2013**, *8*, P04013. [CrossRef]
20. Aad, G.; Abbott, B.; Abbott, D.C.; Abud, A.A.; Abeling, K.; Abhayasinghe, D.K.; Abidi, S.H.; AbouZeid, O.S.; Abraham, N.L.; Abramowicz, H.; et al. ATLAS b-jet identification performance and efficiency measurement with $t\bar{t}$ events in pp collisions at $\sqrt{s} = 13$ TeV. *Eur. Phys. J. C* **2019**, *79*, 970. [CrossRef]
21. Bols, E.; Kieseler, J.; Verzetti, M.; Stoye, M.; Stakia, A. Jet flavour classification using DeepJet. *J. Instrum.* **2020**, *15*, P12012. [CrossRef]
22. Sirunyan, A.M.; Tumasyan, A.; Adam, W.; Ambrogi, F.; Asilar, E.; Bergauer, T.; Brandstetter, J.; Dragicevic, M.; Erö, J.; Escalante Del Valle, A.; et al. Measurement of $t\bar{t}$ normalised multi-differential cross sections in pp collisions at $\sqrt{s} = 13$ TeV, and simultaneous determination of the strong coupling strength, top quark pole mass, and parton distribution functions. *Eur. Phys. J. C* **2020**, *80*, 658. [CrossRef]
23. Jung, A.; Kieseler, J. Top Quarks from Tevatron to the LHC. *Symmetry* **2023**, *15*, 1915. [CrossRef]
24. Aad, G. et al. [The ATLAS Collaboration] Measurements of differential cross-sections in top-quark pair events with a high transverse momentum top quark and limits on beyond the Standard Model contributions to top-quark pair production with the ATLAS detector at $\sqrt{s} = 13$ TeV. *J. High Energy Phys.* **2022**, *2022*, 63. [CrossRef]
25. Tumasyan, A. et al. [CMS Collaboration] Measurement of differential $t\bar{t}$ production cross sections in the full kinematic range using lepton+jets events from proton-proton collisions at $\sqrt{s} = 13$ TeV. *Phys. Rev. D* **2021**, *104*, 092013. [CrossRef]
26. Kobayashi, M.; Maskawa, T. CP-Violation in the Renormalizable Theory of Weak Interaction. *Prog. Theor. Phys.* **1973**, *49*, 652–657. [CrossRef]
27. Czakon, M.; Mitov, A. Top++: A program for the calculation of the top-pair cross-section at hadron colliders. *Comput. Phys. Commun.* **2014**, *185*, 2930–2938. [CrossRef]
28. Cardini, A. Tau identification exploiting deep learning techniques. *PoS* **2021**, *ICHEP2020*, 723. [CrossRef]
29. Aad, G. et al. [ATLAS Collaboration] Measurement of jet substructure in boosted $t\bar{t}$ events with the ATLAS detector using 140 fb^{-1} of 13 TeV pp collisions. *arXiv* **2023**, arXiv:2312.03797.
30. Aad, G.; Abbott, B.; Abbott, D.; Abeling, K.; Abidi, S.; Aboulhorma, A.; Abramowicz, H.; Abreu, H.; Abulaiti, Y.; Abusleme Hoffman, A.; et al. Evidence for the charge asymmetry in $pp \to t\bar{t}$ production at $\sqrt{s} = 13$ TeV with the ATLAS detector. *J. High Energy Phys.* **2023**, *2308*, 077. [CrossRef]
31. CMS Collaboration. Measurement of the $t\bar{t}$ charge asymmetry in events with highly Lorentz-boosted top quarks in pp collisions at $\sqrt{s} = 13$ TeV. *arXiv* **2022**, arXiv:2208.02751.
32. Aaboud, M.; Aad, G.; Abbott, B.; Abdinov, O.; Abeloos, B.; Abidi, S.H.; AbouZeid, O.S.; Abraham, N.L.; Abramowicz, H.; Abreu, H.; et al. Combination of inclusive and differential $t\bar{t}$ charge asymmetry measurements using ATLAS and CMS data at $\sqrt{s}$ = 7 and 8 TeV. *J. High Energy Phys.* **2018**, *2018*, 33. [CrossRef]
33. Czakon, M.; Heymes, D.; Mitov, A.; Pagani, D.; Tsinikos, I.; Zaro, M. Top-quark charge asymmetry at the LHC and Tevatron through NNLO QCD and NLO EW. *Phys. Rev. D* **2018**, *98*, 014003. [CrossRef]
34. Search for Leptonic Charge Asymmetry in $t\bar{t}W$ Production in Final States with Three Leptons at $\sqrt{s} = 13$ TeV. Technical Report, CERN, Geneva, 2022. All Figures Including Auxiliary Figures Are Available at the Link. Available online: https://atlas.web.cern.ch/Atlas/GROUPS/PHYSICS/CONFNOTES/ATLAS-CONF-2022-062 (accessed on 5 December 2023).
35. Basan, A.; Berta, P.; Masetti, L.; Vryonidou, E.; Westhoff, S. Measuring the top energy asymmetry at the LHC: QCD and SMEFT interpretations. *J. High Energy Phys.* **2020**, *2020*, 184. [CrossRef]
36. Abazov, V.; Abbott, B.; Acharya, B.; Adams, M.; Adams, T.; Agnew, J.; Alexeev, G.; Alkhazov, G.; Alton, A.; Askew, A.; et al. Measurement of spin correlation between top and antitop quarks produced in pp^{-} collisions at s=1.96 TeV. *Phys. Lett. B* **2016**, *757*, 199–206. [CrossRef]
37. Chatrchyan, S. et al. [CMS Collaboration] Measurements of $t\bar{t}$ Spin Correlations and Top-Quark Polarization Using Dilepton Final States in pp Collisions at $\sqrt{s}$ = 7 TeV. *Phys. Rev. Lett.* **2014**, *112*, 182001, [CrossRef] [PubMed]
38. Aad, G.; Abbott, B.; Abdallah, J.; Abdel Khalek, S.; Abdinov, O.; Aben, R.; Abi, B.; Abolins, M.; AbouZeid, O.S.; Abramowicz, H.; et al. Measurements of spin correlation in top-antitop quark events from proton-proton collisions at $\sqrt{s} = 7$ TeV using the ATLAS detector. *Phys. Rev. D* **2014**, *90*, 112016. [CrossRef]
39. Khachatryan, V. et al. [CMS Collaboration] Measurements of $t\bar{t}$ spin correlations and top quark polarization using dilepton final states in pp collisions at $\sqrt{s}$ = 8 TeV. *Phys. Rev. D* **2016**, *93*, 052007. [CrossRef]

40. Aaboud, M. et al. [The ATLAS Collaboration] Measurements of top quark spin observables in $t\bar{t}$ events using dilepton final states in $\sqrt{s} = 8$ TeV pp collisions with the ATLAS detector. *J. High Energy Phys.* **2017**, *3*, 113. [CrossRef]
41. Aaboud, M.; Aad, G.; Abbott, B.; Abbott, D.C.; Abdinov, O.; Abed Abud, A.; Abhayasinghe, D.K.; Abidi, S.H.; AbouZeid, O.S.; Abraham, N.L.; et al. Measurements of top-quark pair spin correlations in the $e\mu$ channel at $\sqrt{s} = 13$ TeV using pp collisions in the ATLAS detector. *Eur. Phys. J. C* **2020**, *80*, 754. [CrossRef]
42. Sirunyan, A.M.; et al. [CMS Collaboration] Measurement of the top quark polarization and $t\bar{t}$ spin correlations using dilepton final states in proton-proton collisions at $\sqrt{s} = 13$ TeV. *Phys. Rev. D* **2019**, *100*, 072002. [CrossRef]
43. Behring, A.; Czakon, M.; Mitov, A.; Papanastasiou, A.S.; Poncelet, R. Higher order corrections to spin correlations in top quark pair production at the LHC. *Phys. Rev. Lett.* **2019**, *123*, 082001. [CrossRef]
44. Aad, G. et al. [ATLAS Collaboration] Measurement of the $t\bar{t}$ production cross-section and lepton differential distributions in $e\mu$ dilepton events from pp collisions at $\sqrt{s} = 13$ TeV with the ATLAS detector. *Eur. Phys. J. C* **2020**, *80*, 528. [CrossRef]
45. CMS Collaboration. Projection of the Top Quark Spin Correlation Measurement and Search for Top Squark Pair Production at the HL-LHC. 2022. Available online: https://inspirehep.net/literature/2110182 (accessed on 5 December 2023).
46. de Favereau, J.; Delaere, C.; Demin, P.; Giammanco, A.; Lemaître, V.; Mertens, A.; Selvaggi, M. DELPHES 3: A modular framework for fast simulation of a generic collider experiment. *J. High Energy Phys.* **2014**, *2014*, 57. [CrossRef]
47. Abazov, V.M.; Abbott, B.; Acharya, B.S.; Adams, M.; Adams, T.; Alexeev, G.D.; Alkhazov, G.; Alton, A.; Alverson, G.; Aoki, M.; et al. Search for Violation of Lorentz Invariance in Top Quark Pair Production and Decay. *Phys. Rev. Lett.* **2012**, *108*, 261603. [CrossRef]
48. CERN. *Searches for Violation of Lorentz Invariance in $t\bar{t}$ Production Using Dilepton Events in Proton-Proton Collisions at $\sqrt{s} = 13$ TeV*; Technical Report; CERN: Geneva, Switzerland, 2023.
49. Aaboud, M.; Aad, G.; Abbott, B.; Abdinov, O.; Abeloos, B.; Abhayasinghe, D.K.; Abidi, S.H.; AbouZeid, O.S.; Abraham, N.L.; Abramowicz, H.; et al. Measurement of the top quark mass in the $t\bar{t} \to$ lepton+jets channel from $\sqrt{s} = 8$ TeV ATLAS data and combination with previous results. *Eur. Phys. J.* **2019**, *79*, 290. [CrossRef]
50. Khachatryan, V. et al. [CMS Collaboration] Measurement of the top quark mass using proton-proton data at $\sqrt{s}$ = 7 and 8 TeV. *Phys. Rev. D* **2016**, *93*, 072004. [CrossRef]
51. Tevatron Electroweak Working Group . Combination of CDF and D0 Results on the Mass of the Top Quark using up to 9.7 fb^{-1} at the Tevatron. *arXiv* **2014**, arXiv:1407.2682.
52. The ATLAS, CDF, CMS, D0 Collaborations . First combination of Tevatron and LHC measurements of the top-quark mass. *arXiv* **2014**, arXiv:1403.4427.
53. Lyons, L.; Gibaut, D.; Clifford, P. How to combine correlated estimates of a single physical quantity. *NIM A* **1988**, *270*, 110–117. [CrossRef]
54. Nisius, R. BLUE: A Software Package to Combine Correlated Estimates of Physics Observables within ROOT Using the Best Linear Unbiased Estimate Method—Program Manual, Version 2.1.0. 2016. Available online: http://blue.hepforge.org (accessed on 5 December 2023).
55. Nisius, R. On the combination of correlated estimates of a physics observable. *EPJC* **2014**, *74*. [CrossRef]
56. Kieseler, J. A method and tool for combining differential or inclusive measurements obtained with simultaneously constrained uncertainties. *EPJC* **2017**, *77*, 792. [CrossRef]
57. LHCtopWG Joined Experimental and Theory LHC Working Group. Available online: https://twiki.cern.ch/twiki/bin/view/LHCPhysics/LHCTopWG (accessed on 5 December 2023).
58. ATLAS Collaboration. *Combination of ATLAS and CMS Results on the Mass of the Top-Quark Using up to 4.9 fb^{-1} of $\sqrt{s} = 7$ TeV LHC Data*; Technical Report; CERN: Geneva, Switzerlands, 2013. Available online: https://atlas.web.cern.ch/Atlas/GROUPS/PHYSICS/CONFNOTES/ATLAS-CONF-2013-102 (accessed on 5 December 2023).
59. Aad, G. et al. [ATLAS Collaboration] Measurement of the top-quark mass in the fully hadronic decay channel from ATLAS data at $\sqrt{s} = 7$ TeV. *Eur. Phys. J. C* **2015**, *75*, 158. [CrossRef]
60. Aad, G.; Abbott, B.; Abdallah, J.; Abdinov, O.; Aben, R.; Abolins, M.; AbouZeid, O.S.; Abramowicz, H.; Abreu, H.; Abreu, R.; et al. Measurement of the top quark mass in the $t\bar{t} \to$ lepton + jets and $t\bar{t} \to$ dilepton channels using $\sqrt{s} = 7$ TeV ATLAS data. *Eur. Phys. J. C* **2015**, *75*, 330. [CrossRef] [PubMed]
61. The ATLAS Collaboration. *Measurement of the Top Quark Mass in Topologies Enhanced with Single Top-Quarks Produced in the t-Channel in $\sqrt{s} = 8$ TeV ATLAS Data*; Technical Report; CERN: Geneva, Switzerland, 2014. Available online: https://atlas.web.cern.ch/Atlas/GROUPS/PHYSICS/CONFNOTES/ATLAS-CONF-2014-055.(accessed on 5 December 2023).
62. Aaboud, M.; Aad, G.; Abbott, B.; Abdallah, J.; Abdinov, O.; Abeloos, B.; Aben, R.; AbouZeid, O.; Abraham, N.; Abramowicz, H.; et al. Measurement of the top quark mass in the $t\bar{t} \to$ dilepton channel from $\sqrt{s} = 8$ TeV ATLAS data. *Phys. Lett. B* **2016**, *761*, 350–371. [CrossRef]
63. Aaboud, M.; Aad, G.; Abbott, B.; Abdallah, J.; Abdinov, O.; Abeloos, B.; Aben, R.; AbouZeid, O.S.; Abraham, N.L.; Abramowicz, H.; et al. Top-quark mass measurement in the all-hadronic $t\bar{t}$ decay channel at $\sqrt{s} = 8$ TeV with the ATLAS detector. *J. High Energy Phys.* **2017**, *2017*, 118. [CrossRef]
64. Aad, G.; Abbott, B.; Abbott, D.C.; Abed Abud, A.; Abeling, K.; Abhayasinghe, D.K.; Abidi, S.H.; AbouZeid, O.S.; Abraham, N.L.; Abramowicz, H.; et al. Measurement of the top-quark mass using a leptonic invariant mass in pp collisions at $\sqrt{s} = 13$ TeV with the ATLAS detector. *J. High Energy Phys.* **2023**, *2023*, 19. [CrossRef]

65. ATLAS Collaboration. Measurement of the Top-Quark Mass in $t\bar{t} \rightarrow$ Dilepton Events with the ATLAS Experiment Using the Template Method in 13 TeV *pp* Collision Data 2022. Available online: https://cds.cern.ch/record/2826701 (accessed on 5 December 2023).
66. Chatrchyan, S.; Khachatryan, V.; Sirunyan, A.M.; Tumasyan, A.; Adam, W.; Aguilo, E.; Bergauer, T.; Dragicevic, M.; Erö, J.; Fabjan, C.; et al. Measurement of the top-quark mass in $t\bar{t}$ events with lepton+jets final states in pp collisions at $\sqrt{s} = 7$ TeV. *J. High Energy Phys.* **2012**, *2012*, 105. [CrossRef]
67. The CMS Collaboration; Chatrchyan, S.; Khachatryan, V.; Sirunyan, A.M.; Tumasyan, A.; Adam, W.; Aguilo, E.; Bergauer, T.; Dragicevic, M.; Erö, J.; et al. Measurement of the top-quark mass in $t\bar{t}$ events with dilepton final states in pp collisions at $\sqrt{s} = 7$ TeV. *Eur. Phys. J. C* **2012**, *72*, 2202. [CrossRef]
68. Chatrchyan, S.; Khachatryan, V.; Sirunyan, A.M.; Tumasyan, A.; Adam, W.; Bergauer, T.; Dragicevic, M.; Erö, J.; Fabjan, C.; Friedl, M.; et al. Measurement of the top-quark mass in all-jets $t\bar{t}$ events in pp collisions at $\sqrt{s} = 7$ TeV. *Eur. Phys. J. C* **2014**, *74*, 2758. [CrossRef]
69. Sirunyan, A.M. et al. [CMS Collaboration] Measurement of the top quark mass using single top quark events in proton-proton collisions at $\sqrt{s} = 8$ TeV. *Eur. Phys. J. C* **2017**, *77*, 354. [CrossRef]
70. Sirunyan, A.M.; Tumasyan, A.; Adam, W.; Ambrogi, F.; Asilar, E.; Bergauer, T.; Brandstetter, J.; Brondolin, E.; Dragicevic, M.; Erö, J.; et al. Measurement of the top quark mass with lepton+jets final states using pp collisions at $\sqrt{s} = 13$ TeV. *Eur. Phys. J. C* **2018**, *78*, 891. [CrossRef]
71. Tumasyan, A.; Adam, W.; Andrejkovic, J.W.; Bergauer, T.; Chatterjee, S.; Dragicevic, M.; Escalante Del Valle, A.; Fruehwirth, R.; Jeitler, M.; Krammer, N.; et al. Measurement of the top quark mass using events with a single reconstructed top quark in pp collisions at $\sqrt{s} = 13$ TeV. *J. High Energy Phys.* **2021**, *2112*, 161. [CrossRef]
72. Tumasyan, A.; Adam, W.; Andrejkovic, J.; Bergauer, T.; Chatterjee, S.; Damanakis, K.; Dragicevic, M.; Del Valle, A.; Hussain, P.; Jeitler, M.; et al. Measurement of the top quark mass using a profile likelihood approach with the lepton + jets final states in proton–proton collisions at $\sqrt{s} = 13$ TeV. *Eur. Phys. J. C* **2023**, *83*, 963. [CrossRef] [PubMed]
73. Tumasyan, A.; Adam, W.; Andrejkovic, J.W.; Bergauer, T.; Chatterjee, S.; Damanakis, K.; Dragicevic, M.; Escalante Del Valle, A.; Hussain, P.S.; Jeitler, M.; et al. Measurement of the differential $t\bar{t}$ production cross section as a function of the jet mass and extraction of the top quark mass in hadronic decays of boosted top quarks. Measurement of the jet mass distribution and top quark mass in hadronic decays of boosted top quarks in proton-proton collisions at $\sqrt{s} = 13$ TeV. *Eur. Phys. J. C* **2023**, *83*, 560. [CrossRef] [PubMed]
74. Markkanen, T.; Rajantie, A.; Stopyra, S. Cosmological Aspects of Higgs Vacuum Metastability. *Front. Astron. Space Sci.* **2018**, *5*. [CrossRef]
75. Haller, J.; Hoecker, A.; Kogler, R.; Mönig, K.; Stelzer, J. Status of the global electroweak fit with Gfitter in the light of new precision measurements. *arXiv* **2022**, arXiv.2211.07665. [CrossRef]
76. Hoang, A.H. What Is the Top Quark Mass? *Annu. Rev. Nucl. Part. Sci.* **2020**, *70*, 225–255. [CrossRef]
77. Hays, C. High precision measurement of the W-boson mass with the CDF II detector. *PoS* **2022**, *ICHEP2022*, 898. [CrossRef]
78. CERN. *Combination of Measurements of the Top Quark Mass from Data Collected by the ATLAS and CMS Experiments at $\sqrt{s} = 7$ and 8 TeV*; Technical Report; CERN: Geneva, Switzerland, 2023.
79. Sirunyan, A.M. et al. [The CMS Collaboration] Running of the top quark mass from proton-proton collisions at $\sqrt{s} = 13$TeV. *Phys. Lett. B* **2020**, *803*, 135263. [CrossRef]
80. Aaboud, M.; Aad, G.; Abbott, B.; Abdinov, O.; Abeloos, B.; Abidi, S.H.; AbouZeid, O.S.; Abraham, N.L.; Abramowicz, N.L.; Abreu, H.; et al. Measurement of lepton differential distributions and the top quark mass in $t\bar{t}$ production in *pp* collisions at $\sqrt{s} = 8$ TeV with the ATLAS detector. *Eur. Phys. J. C* **2017**, *77*, 804. [CrossRef]
81. Measurement of the jet mass distribution and top quark mass in hadronic decays of boosted top quarks in proton-proton collisions at $\sqrt{s} = 13$ TeV. *arXiv* **2022**, arXiv:2211.01456.
82. CERN . *Measurement of the Top Quark Mass Using a Leptonic Invariant Mass in pp Collisions at $\sqrt{s} = 13$ TeV with the ATLAS Detector*; Technical Report; CERN: Geneva, Switzerland, 2019. Available online: https://atlas.web.cern.ch/Atlas/GROUPS/PHYSICS/CONFNOTES/ATLAS-CONF-2019-046 (accessed on 5 December 2023).
83. Aad, G. et al. [ATLAS Collaboration] Measurement of the $t\bar{t}$ production cross-section using $e\mu$ events with b-tagged jets in pp collisions at $\sqrt{s} = 7$ and 8 TeV with the ATLAS detector. *Eur. Phys. J. C* **2014**, *74*, 3109. [CrossRef]
84. Khachatryan, V.; Sirunyan, A.M.; Tumasyan, A.; Adam, W.; Asilar, E.; Bergauer, T.; Brandstetter, J.; Brondolin, E.; Dragicevic, M.; Erö, J.; et al. Measurement of the $t\bar{t}$ production cross section in the $e\mu$ channel in proton-proton collisions at $\sqrt{s} = 7$ and 8 TeV. *J. High Energy Phys.* **2016**, *2016*, 29. [CrossRef]
85. Sirunyan, A.M.; Tumasyan, A.; Adam, W.; Ambrogi, F.; Asilar, E.; Bergauer, T.; Brandstetter, J.; Dragicevic, M.; Erö, J.; Del Valle, A.E.; et al. Measurement of the $t\bar{t}$ production cross section, the top quark mass, and the strong coupling constant using dilepton events in pp collisions at $\sqrt{s} = 13$ TeV. *Eur. Phys. J. C* **2019**, *79*, 368. [CrossRef] [PubMed]
86. Aad, G.; Abbott, B.; Abdallah, J.; Abdel Khalek, S.; Abdinov, O.; Aben, R.; Abi, B.; Abolins, M.; AbouZeid, O.S.; Abramowicz, H.; et al. Determination of the top-quark pole mass using $t\bar{t}$ + 1-jet events collected with the ATLAS experiment in 7 TeV pp collisions. *J. High Energy Phys.* **2015**, *2015*, 121. [CrossRef]

87. Aad, G.; Abbott, B.; Abbott, D.C.; Abdinov, O.; Abed Abud, A.; Abeling, K.; Abhayasinghe, D.K.; Abidi, S.H.; AbouZeid, O.S.; Abraham, N.L.; et al. Measurement of the top-quark mass in $t\bar{t}$ + 1-jet events collected with the ATLAS detector in pp collisions at $\sqrt{s} = 8$ TeV. *J. High Energy Phys.* **2019**, *2019*, 150. [CrossRef]
88. Tumasyan, A.; Adam, W.; Andrejkovic, J.W.; Bergauer, T.; Chatterjee, S.; Damanakis, K.; Dragicevic, M.; Escalante Del Valle, A.; Hussain, P.S.; Jeitler, M.; et al. Measurement of the top quark pole mass using $\mathrm{t\bar{t}}$+jet events in the dilepton final state in proton-proton collisions at $\sqrt{s} = 13$ TeV. *J. High Energy Phys.* **2023**, *2023*, 77. [CrossRef]
89. ATLAS Collaboration. *Measurement of the Top-Quark Decay width in Top-Quark Pair Events in the Dilepton Channel at $\sqrt{s} = 13$ TeV with the ATLAS Detector*; Technical Report; CERN: Geneva, Switzerland, 2019. Available online: https://atlas.web.cern.ch/Atlas/GROUPS/PHYSICS/CONFNOTES/ATLAS-CONF-2019-038 (accessed on 5 December 2023).
90. Sirunyan, A.M.; Tumasyan, A.; Adam, W.; Ambrogi, F.; Bergauer, T.; Dragicevic, M.; Erö, J.; Escalante Del Valle, A.; Fruehwirth, R.; Jeitler, M.; et al. Measurement of the top quark Yukawa coupling from $\mathrm{t\bar{t}}$ kinematic distributions in the dilepton final state in proton-proton collisions at $\sqrt{s} = 13$ TeV. *Phys. Rev. D* **2020**, *102*, 092013. [CrossRef]
91. Jezabek, M.; Kuhn, J.H. QCD Corrections to Semileptonic Decays of Heavy Quarks. *Nucl. Phys. B* **1989**, *314*, 1–6. [CrossRef]
92. Gao, J.; Li, C.S.; Zhu, H.X. Top-Quark Decay at Next-to-Next-to-Leading Order in QCD. *Phys. Rev. Lett.* **2013**, *110*, 042001. [CrossRef] [PubMed]
93. Aaltonen, T.A. et al. [CDF Collaboration] Direct Measurement of the Total Decay Width of the Top Quark. *Phys. Rev. Lett.* **2013**, *111*, 202001. [CrossRef] [PubMed]
94. Abazov, V.M. et al. [D0 Collaboration] An Improved determination of the width of the top quark. *Phys. Rev. D* **2012**, *85*, 091104. [CrossRef]
95. Aaboud, M. et al. [ATLAS Collaboration] Direct top-quark decay width measurement in the $t\bar{t}$ lepton+jets channel at $\sqrt{s} = 8$ TeV with the ATLAS experiment. *Eur. Phys. J. C* **2018**, *78*, 129. [CrossRef]
96. Aad, G. et al. [ATLAS Collaboration] Measurements of the inclusive and differential production cross sections of a top-quark–antiquark pair in association with a Z boson at $\sqrt{s} = 13$ TeV with the ATLAS detector. *Eur. Phys. J. C* **2021**, *81*, 737. [CrossRef]
97. Aad, G. et al. [The ATLAS Collaboration] Observation of the associated production of a top quark and a Z boson in pp collisions at $\sqrt{s} = 13$ TeV with the ATLAS detector. *J. High Energy Phys.* **2020**, *07*, 124. [CrossRef]
98. Tumasyan, A. et al. [The CMS Collaboration] Inclusive and differential cross section measurements of single top quark production in association with a Z boson in proton-proton collisions at $\sqrt{s} = 13$ TeV. *J. High Energy Phys.* **2022**, *02*, 107. [CrossRef]
99. Jeżabek, M.; Kühn, J. V - A tests through leptons from polarised top quarks. *Phys. Lett. B* **1994**, *329*, 317–324. [CrossRef]
100. Aad, G.; Abbott, B.; Abbott, D.C.; Abeling, K.; Abidi, S.H.; Aboulhorma, A.; Abramowicz, H.; Abreu, H.; Abulaiti, Y.; Abusleme Hoffman, A.C.; et al. Observation of Single-Top-Quark Production in Association with a Photon Using the ATLAS Detector. *Phys. Rev. Lett.* **2023**, *131*, 181901. [CrossRef]
101. Hayrapetyan H. et al. [CMS Collaboration] *Evidence for tWZ Production in Proton-Proton Collisions at $\sqrt{s} = 13$ TeV in Multilepton Final States*; Technical Report; CERN: Geneva, Switzerland, 2023.
102. The ATLAS Collaboration. Measurement of the Total and Differential Cross-Sections of $t\bar{t}W$ Production in pp Collisions at 13 TeV with the ATLAS Detector. 2023. Available onlin: https://cds.cern.ch/record/2855337?ln=en (accessed on 5 December 2023).
103. ATLAS Collaboration. Inclusive and Differential Cross Section Measurements of $t\bar{t}Z$ Production in pp Collisions at $\sqrt{s} = 13$ TeV with the ATLAS Detector, Including EFT and Spin Correlations Interpretations. 2023. Available online: http://cds.cern.ch/record/2883457 (accessed on 5 December 2023).
104. Sirunyan, A.M. et al. [The CMS Collaboration] Measurement of top quark pair production in association with a Z boson in proton-proton collisions at $\sqrt{s} = 13$ TeV. *J. High Energy Phys.* **2020**, *03*, 056. [CrossRef]
105. Tumasyan, A. et al. [The CMS Collaboration] Measurement of the inclusive and differential $\mathrm{t\bar{t}}\gamma$ cross sections in the dilepton channel and effective field theory interpretation in proton-proton collisions at $\sqrt{s}$ =13 TeV. *J. High Energy Phys.* **2022**, *05*, 091. [CrossRef]
106. Tumasyan, A. et al. [The CMS Collaboration] Measurement of the inclusive and differential $\mathrm{t\bar{t}}\gamma$ cross sections in the single-lepton channel and EFT interpretation at $\sqrt{s} = 13$ TeV. *J. High Energy Phys.* **2021**, *12*, 180. [CrossRef]
107. Tumasyan, A.; Adam, W.; Andrejkovic, J.W.; Bergauer, T.; Chatterjee, S.; Damanakis, K.; Dragicevic, M.; Escalante Del Valle, A.; Hussain, P.S.; Jeitler, M.; et al. Measurement of the cross section of top quark-antiquark pair production in association with a W boson in proton-proton collisions at $\sqrt{s} = 13$ TeV. *J. High Energy Phys.* **2023**, *2023*, 219. [CrossRef]
108. Aad, G. et al. [The ATLAS Collaboratio]. Search for flavour-changing neutral-current couplings between the top quark and the photon with the ATLAS detector at $\sqrt{s} = 13$ TeV. *Phys. Lett. B* **2023**, *842*, 137379. [CrossRef]
109. Aad, G. et al. [The ATLAS Collaboration] Search for flavor-changing neutral tqH interactions with $H \to \gamma\gamma$ in pp collisions at $\sqrt{s} = 13$ TeV using the ATLAS detector. *J. High Energy Phys.* **2023**, *12*, 195. . [CrossRef]
110. Hayrapetyan, A.; et al. [CMS Collaboration]. Search for flavor changing neutral current interactions of the top quark in final states with a photon and additional jets in proton-proton collisions at $\sqrt{s} = 13$ TeV. *arXiv* **2023**, arXiv:2312.08229.
111. Hayrapetyan, A. et al. [CMS Collaboration]. Search for new Higgs bosons via same-sign top quark pair production in association with a jet in proton-proton collisions at $\sqrt{s} = 13$ TeV. *arXiv* **2023**, arXiv:2311.03261.

112. Aad, G.; Abbott, B.; Abeling, K.; Abicht, N.J.; Abidi, S.H.; Aboulhorma, A.; Abramowicz, H.; Abreu, H.; Abulaiti, Y.; Abusleme Hoffman, A.C.; et al. Search for heavy Higgs bosons with flavour-violating couplings in multi-lepton plus b-jets final states in pp collisions at 13 TeV with the ATLAS detector. *J. High Energy Phys.* **2023**, *2023*, 81. [CrossRef]
113. Hayrapetyan, A. et al. [CMS Collaboration]. Search for charged-lepton flavor violation in the production and decay of top quarks using trilepton final states in proton-proton collisions at $\sqrt{s}$ = 13 TeV. *arXiv* **2023**, arXiv:2312.03199.
114. Sirunyan, A.M.; Tumasyan, A.; Adam, W.; Ambrogi, F.; Asilar, E.; Bergauer, T.; Brandstetter, J.; Dragicevic, M.; Erö, J.; Escalante Del Valle, A.; et al. Evidence for the Associated Production of a Single Top Quark and a Photon in Proton-Proton Collisions at $\sqrt{s} = 13$ TeV. *Phys. Rev. Lett.* **2018**, *121*, 221802. [CrossRef] [PubMed]
115. CMS Collaboration. *Search for Flavor-Changing Neutral Current Interactions of the Top Quark and Higgs Boson in Proton-Proton Collisions at* $\sqrt{s} = 13$ *TeV*. Technical Report; CERN: Geneva, Switzerland, 2023.
116. CMS Collaboration. *Search for Flavor-Changing Neutral Current Interactions of the Top Quark in Final States with a Photon and Additional Jets in Proton-Proton Collisions at* $\sqrt{s}$*=13 TeV*; Technical Report; CERN: Geneva, Switzerland, 2023.
117. Aad, G. et al. [ATLAS Collaboration] Search for flavour-changing neutral-current interactions of a top quark and a gluon in pp collisions at $\sqrt{s} = 13$ TeV with the ATLAS detector. *Eur. Phys. J. C* **2022**, *82*, 334. [CrossRef]
118. Khachatryan, V. et al. [The CMS Collaboration] Search for anomalous Wtb couplings and flavour-changing neutral currents in t-channel single top quark production in pp collisions at $\sqrt{s} = 7$ and 8 TeV. *J. High Energy Phys.* **2017**, *02*, 028. [CrossRef]
119. Aad, G. et al. [ATLAS Collaboration] Search for flavor-changing neutral-current couplings between the top quark and the Z boson with proton-proton collisions at s=13 TeV with the ATLAS detector. *Phys. Rev. D* **2023**, *108*, 032019. [CrossRef]
120. CMS Collaboration. *Search for Flavour Changing Neutral Currents in Top Quark Production and Decays with Three-Lepton Final State Using the DATA collected at* $\sqrt{s}$ *= 13 TeV*; Technical Report; CERN: Geneva, Switzerland, 2017.
121. Sirunyan, A.M. et al. [The CMS Collaboration] Search for associated production of a Z boson with a single top quark and for tZ flavour-changing interactions in pp collisions at $\sqrt{s} = 8$ TeV. *J. High Energy Phys.* **2017**, *07*, 003. [CrossRef]
122. Erbacher, R.; Gerber, C.E.; Melnikov, K.; Schwienhorst, R.; Mitov, A.; Vos, M.; Wimpenny, S.; Adelman, J.; Baumgart, M. Working Group Report: Top Quark. In Proceedings of the Snowmass 2013: Snowmass on the Mississippi, Minneapolis, MN, USA, 29 July–6 August 2013. https://arxiv.org/abs/1311.2028.
123. Aad, G. et al. [The ATLAS Collaboration] Measurement of the $t\bar{t}t\bar{t}$ production cross section in pp collisions at $\sqrt{s}$ = 13 TeV with the ATLAS detector. *J. High Energy Phys* **2021**, *11*, 118. [CrossRef]
124. Sirunyan, A.M. et al. [CMS Collaboration] Search for production of four top quarks in final states with same-sign or multiple leptons in proton-proton collisions at $\sqrt{s} = 13$ TeV. *Eur. Phys. J. C* **2020**, *80*, 75. [CrossRef] [PubMed]
125. Tumasyan, A.; Adam, W.; Andrejkovic, J.; Bergauer, T.; Chatterjee, S.; Damanakis, K.; Dragicevic, M.; Escalante Del Valle, A.; Hussain, P.; Jeitler, M.; et al. Evidence for four-top quark production in proton-proton collisions at s=13TeV. *Phys. Lett. B* **2023**, *844*, 138076. [CrossRef]
126. Frederix, R.; Pagani, D.; Zaro, M. Large NLO corrections in $t\bar{t}W^{\pm}$ and $t\bar{t}t\bar{t}$ hadroproduction from supposedly subleading EW contributions. *J. High Energy Phys.* **2018**, *02*, 031. [CrossRef]
127. Hayrapetyan, A.; Tumasyan, A.; Adam, W.; Andrejkovic, J.; Bergauer, T.; Chatterjee, S.; Damanakis, K.; Dragicevic, M.; Escalante Del Valle, A.; Hussain, P.; et al. Observation of four top quark production in proton-proton collisions at $\sqrt{s}$ = 13 TeV. *Phys. Lett. B* **2023**, *847*, 138290. [CrossRef]
128. Aad, G. et al. [ATLAS Collaboration] Evidence for $t\bar{t}t\bar{t}$ production in the multilepton final state in proton–proton collisions at $\sqrt{s} = 13$ TeV with the ATLAS detector. *Eur. Phys. J. C* **2020**, *80*, 1085. [CrossRef]
129. Sirunyan, A.M. et al. [The CMS Collaboration] Search for the production of four top quarks in the single-lepton and opposite-sign dilepton final states in proton-proton collisions at $\sqrt{s}$ = 13 TeV. *J. High Energy Phys.* **2019**, *11*, 082. [CrossRef]

universe MDPI

Review

Single-Top Quark Physics at the LHC: From Precision Measurements to Rare Processes and Top Quark Properties

Jérémy Andrea [1] and Nicolas Chanon [2,*]

[1] CNRS, IPHC UMR7178, Université de Strasbourg, 67000 Strasbourg, France; jeremy.andrea@cern.ch
[2] CNRS/IN2P3, IP2I Lyon, University Claude Bernard Lyon 1, 69622 Villeurbanne, France
* Correspondence: nicolas.pierre.chanon@cern.ch

Abstract: Since the initial measurements of single-top quark production at the Tevatron in 2009, tremendous progress has been made at the LHC. While LHC Run 1 marked the beginning of a precision era for the single-top quark measurements in some of the main production mechanisms, LHC Run 2 witnessed the emergence and exploration of new processes associating top quark production with a neutral boson. In this paper, we review the measurements of the three main production mechanisms (t-channel, s-channel, and tW production), and of the associated production with a photon, a Z boson, or a Higgs boson. Differential cross-sections are measured for several of these processes and compared with theoretical predictions. The top quark properties that can be measured in single-top quark processes are scrutinized, such as Wtb couplings and top quark couplings with neutral bosons, and the polarizations of both the W boson and top quark. The effective field theory framework is emerging as a standard for interpreting property measurements. Perspectives for LHC Run 3 and the HL-LHC are discussed in the conclusions.

Keywords: top quark; single-top quark production; top quark properties; ATLAS; CMS; LHC

Citation: Andrea, J.; Chanon, N. Single-Top Quark Physics at the LHC: From Precision Measurements to Rare Processes and Top Quark Properties. *Universe* **2023**, *9*, 439. https://doi.org/10.3390/universe9100439

Academic Editors: Efe Yazgan, Pedro Ferreira da Silva and Máté Csanád

Received: 25 July 2023
Revised: 14 September 2023
Accepted: 17 September 2023
Published: 30 September 2023

1. Introduction

After the discovery of the top quark [1,2] in 1995 at the Fermilab Tevatron, the CERN LHC era opened up many opportunities to investigate top quark processes. Both at the LHC and the Tevatron, the processes with the largest cross-sections for producing top quarks in proton–proton or proton–antiproton collisions are the $t\bar{t}$ production modes. In addition to the $t\bar{t}$ production, which arises from quantum chromodynamics (QCD) interactions, top quarks can be singly produced through electroweak interactions. This leads to the so-called single-top quark channels. The single-top quark production features many interesting properties owing to the V–A structure of the electroweak interaction. It shows specific sensitivities to parton density functions (PDFs), the V_{tb} matrix element of the CKM matrix, Wtb coupling beyond the standard model (SM), and top quark polarization, to name a few examples. Measuring inclusive cross-sections and differential cross-sections for single-top quark processes serves as an interesting test of perturbative QCD (pQCD). The associated production of a single-top quark with a boson offers insights into the coupling between the top quark and bosons, complementing the associated production of a boson with a $t\bar{t}$ pair.

Three main production modes for single-top quark processes can be distinguished: production via the exchange of a virtual W boson in the t- and s-channels, and the associated production with a W boson (tW production). The corresponding diagrams in the leading order (LO) in pQCD are presented in Figure 1.

The first observation of single-top electroweak production (t- and s-channels combined) made at the Tevatron [3,4] in 2009, followed by the observation of the t-channel [5]. The CDF and D0 collaborations performed simultaneous measurements of the s- and t-channel processes [6,7]. The p-$\bar{\text{p}}$ collisions at the Tevatron provided a unique setting for measuring the s-channel since the initial state of this process predominantly involves a light quark and a light antiquark, taken from the valence partons in the proton and antiproton. To date, the s-channel has been observed solely at the Tevatron [8]; it remains to be observed at the LHC, although there have been reports suggesting evidence of this process at both 8 TeV [9] and 13 TeV [10]. At the LHC, the largest cross-sections at $\sqrt{s} = 13$ TeV (the center-of-mass energy of Run 2) are obtained for the t-channel ($214.2^{+4.1}_{-2.6}$ pb at NNLO with MCFM [11]), followed by the tW production ($79.3^{+2.9}_{-2.8}$ pb at NLO+NNLL [12]), and the s-channel ($10.3^{+0.4}_{-0.4}$ pb at NLO with Hathor v2.1 [13,14]).

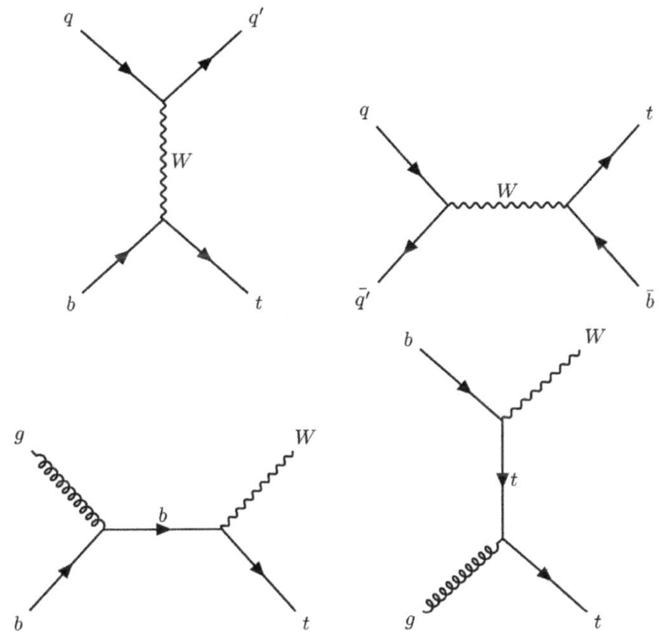

Figure 1. Examples of Feynman diagrams for single-top production at LO: t-channel (**top left**), s-channel (**top right**), and tW production (**bottom**).

At the LHC, many new processes involving single-top quarks were measured in p–p collisions. The t-channel production has been frequently measured and has been thoroughly investigated by evaluating differential cross-sections. Most of the top quark properties probed with single-top production channels are measured using the t-channel production since it yields the largest cross-section at the LHC among all the production mechanisms. The LHC is able to observe the associated tW production, for which the differential cross-sections are even measured. This channel is of particular interest because at next-to-LO (NLO) in pQCD, it features interference with the $t\bar{t}$ process. Understanding the nuances of this interference is still a focal point in the field. As noted in an earlier review (Ref. [15]), Run 1 marked the start of a precision epoch in single-top quark measurements for those main production mechanisms. This is ongoing, with remarkable scrutiny focused on the t-channel and tW production.

In addition, single-top quarks can be produced in tandem with neutral bosons. Those processes yield relatively low cross-sections; however, the additional boson in the lepton channel offers invaluable experimental leverage for measuring couplings or searching for new physics. This class of rare processes covers the production of a single-top quark with a photon ($t\gamma$), a Z boson (tZ), or a Higgs boson (tH). For each of these processes, the single-top quark can be produced via the t-channel, tW production, or s-channel, with the boson emitted from a quark line or a W boson exchange. The $t\gamma$ process was observed only at the end of Run 2. There was anticipation for the discovery of the tZ process at the onset of Run

2; nowadays, it is being measured differentially and is utilized for property measurements. The analysis methodology employed in tZ measurements is close to that of the searches for the tH final states. With great similarity in the tH and $t\bar{t}H$ final states and their common sensitivity to the top quark Yukawa coupling, the tH processes constitute a special case and are searched for simultaneously with the $t\bar{t}H$ production. Because of the destructive interference between processes where the Higgs boson emerges from a W boson or from a top-quark line, the cross-section for the tH production is so small that evidence for such processes remains elusive. However, it is already considered in several analyses because of its unique sensitivity to the sign of the top quark Yukawa coupling, which could lead to a large enhancement of its cross-section. When the review in Ref. [16] was published at the outset of Run 2, it signified the dawn of an era where the processes associating the production of a top quark with a neutral boson began to be measured. Run 2 saw the in-depth exploration of these processes, with a particular emphasis on the tZ production.

The cross-sections for all SM top quark processes measured by ATLAS are compared with theoretical predictions in Figure 2. The cross-sections for single-top production (t + X) are generally less than those for top pair production ($t\bar{t}$ + X). This top pair production acts as a large background in single-top quark searches.

A summary of the cross-section for single-top quark processes as measured at CMS is compared with theoretical predictions and presented as a function of the center-of-mass energy, as seen in Figure 3. It can be observed that the cross-section for the s-channel process does not grow as fast as that of the t-channel process as a function of the energy, which makes the search for the s-channel more difficult with recent LHC runs. The production of a single-top quark associated with a photon or Z boson results in cross-sections that are lower than those observed in the t-channel, s-channel, or tW production.

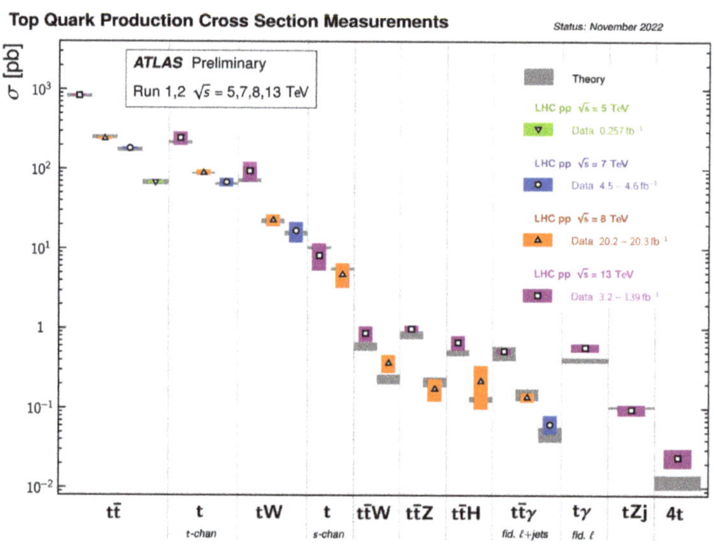

Figure 2. Summary of cross-sections for top quark processes measured by ATLAS [17] and compared with SM predictions.

We will now focus on the top quark property measurements. It is notable that the cross-sections for single-top quark production are directly proportional to the square of $|V_{tb}|$. Therefore, it is possible to determine $|V_{tb}|$ from the measurements of single-top quark cross-sections. If one assumes that $|V_{td}|, |V_{ts}| \ll |V_{tb}|$, the $|V_{tb}|$ matrix element can be extracted from the following:

$$|V_{tb}| = \sqrt{\sigma_{st}/\sigma_{st}(theo, |V_{tb}| = 1)}, \tag{1}$$

where σ_{st} is the measured cross-section and $\sigma_{st}^{theo,V_{tb}=1}$ is the expected cross-section for $|V_{tb}| = 1$. Equation (1) also assumes that no new physics effect modifies the V–A structure of the tWb interaction vertex.

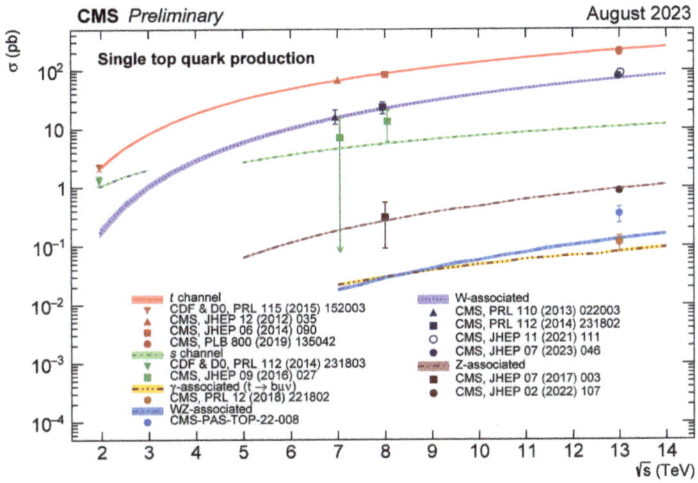

Figure 3. Summary of measured cross-sections for single-top quark production at CMS [18]. Theoretical calculations for the t-channel, s-channel, and W-associated production have been provided by N. Kidonakis to the CMS collaboration.

In single-top quark processes, the Wtb vertex appears in the top quark production and its decay, while in $t\bar{t}$ production, it appears twice in the top quark decay. Therefore, the Lorentz structure of the Wtb coupling can be investigated in detail using decay information. The single-top quark production is also sensitive to the CP property of the Wtb vertex (it is much more difficult to measure in $t\bar{t}$ production, where the CP symmetry is probed preferentially in the top quark–gluon coupling). The W boson polarization and top quark polarization can also be probed. For all of these properties, the t-channel process is usually employed as a probe because of its large cross-section. Within the t-channel, as well as in the associated production with a boson, modern tools, such as the SM effective field theory (EFT), are increasingly used to parametrize deviations from the SM in an almost model-independent way. This systematic approach of searching for signs of new physics is a novelty of Run 2.

The single-top quark production is indeed a sensitive probe in physics beyond the SM. The t-channel signature can occur via the exchange of a supersymmetric particle [19], resulting from the decay of a new heavy resonance, like a color-octet scalar [20] or a new resonance in technicolor models [21]. The s-channel shares the same final state as the possible decay of a W' boson, which is predicted in many models beyond the SM, such as supersymmetric models with R-parity violation [22], or within the paradigm of universal extra dimensions [23]. The tW final state, along with the tZ and tH states, are typical products of vector-like quark decays [24]. Excited top quarks, predicted in Randall–Sundrum models, can decay in the $t\gamma$ final state [25]. The tZ, $t\gamma$, and tH processes can also be modified by flavor-changing neutral currents (FCNCs) in the top quark production or decay, as predicted in several extensions of the SM, like the two-Higgs doublet model [26]

(2HDM), supersymmetry [27], or in warped extra dimensions [28]. In the top quark property domain, for example, CP violation is predicted in the top-Higgs boson coupling within the complex 2HDM [29], and in Wtb coupling within supersymmetric models [30]. The so-called mono-top quark signatures, designating the associated production of a single-top quark with a dark matter candidate, are areas that are being focused on in the quest for new physics (for a review, see [31]).

The outline of this review is as follows. In Section 2, after a brief note on the generation of each single-top quark process, the measurements of the three main single-top quark production mechanisms are presented: t-channel, tW production, and s-channel. The Section 3 will discuss the measurements of single-top quark production in association with a neutral boson (a photon, a Z boson, or a Higgs boson). Top quark property measurements with single-top quark production will be reviewed in Section 4, with a focus on V_{tb}, the W boson, top quark polarization, the structure of the Wtb vertex, and the interpretation in terms of the SM EFT. The conclusions of this review will be presented in Section 5.

2. Precise and Differential Measurements of Single-Top Quark Processes

2.1. The t-Channel Process: The Production Mode With The Largest Statistics

2.1.1. Features of the t-Channel Process

The so-called t-channel production mode features the largest cross-section among all single-top quark production modes. Top quarks produced in the t-channel are accompanied by a high p_T light quark that is predominantly produced in the forward region of the detector ($|\eta| > 2.5$), and of a low p_T b-quark that often fails the minimum jet p_T requirements in the analysis; as a result, it often remains experimentally invisible. Feynman diagrams are presented in Figure 4.

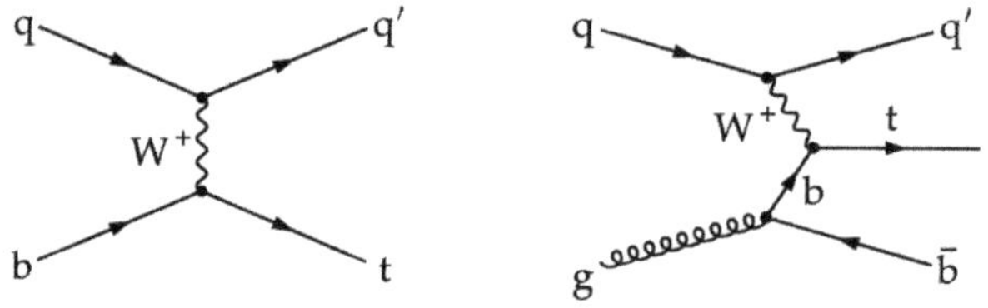

Figure 4. Feynman diagrams for single-top t-channel production at the LO in pQCD [32], in the 5-flavor scheme (**left**), and 4-flavor scheme (**right**).

The initial b quark is accounted for in the theoretical calculation according to the 5-flavor scheme (5FS) or the 4-flavor scheme (4FS), as shown in the Feynman diagrams in Figure 4. Each one considers a b quark PDF within the proton (5FS); one can consider that the proton is made of light-flavored quarks in the sea (4FS), in which case, the b-quark arises from virtual gluons. In the 5FS, the uncertainty associated with the PDF can be relatively large because b-quark PDFs are not necessarily well known. On the other hand, 4FS calculations usually suffer from higher sensitivity to QCD renormalization and factorization scales. The decision as to whether to employ the 4FS or the 5FS is particularly important for the t-channel signature, where the additional jet (the so-called recoiling or spectator jet) is relatively forward, and its pseudorapidity (η) distribution is sensitive to the PDF. It has been observed that the η distribution of the recoiling jet ($\eta(j')$) in data is actually better described using the 4FS, while inclusive cross-sections are more accurately described with the 5FS.

With single-top quark production in the t-channel (as well in the s-channel), one of the incoming light quarks can be a valence quark of the proton, depending on whether a top or an antitop quark is being produced. This leads to a larger cross-section for top quark production ($134.2^{+2.6}_{-1.7}$ pb at $\sqrt{s}$ =13 TeV, calculated at NNLO with MCFM [11]) than for antitop quark production ($80.0^{+1.8}_{-1.4}$ pb).

2.1.2. Experimental Techniques for the t-Channel Measurement

The t-channel process was the first single-top production mode observed at the LHC [33,34], thanks to its large cross-section and its manageable signal-over-background ratio. For this channel, clear discriminating observables exist between the signal and background, such as the $\eta(j')$ distribution. Most of the t-channel analyses share a lot of common features in the event selection, background estimation, separation of the signal from the background, and signal extraction. The following paragraphs provide general descriptions of the analysis methods applied in t-channel measurements, and are valid, to a large extent, to the other single-top quark measurements discussed in this paper.

The top quark decays at almost 100% to a W boson and a b-quark. Top quark decays are said to be leptonic ($t \to bW \to bl\nu$) or hadronic ($t \to bW \to bqq'$). The hadronic decay of top quarks produced in the single-top t-channel leads to a signature with several jets, and suffers from an overwhelming QCD multijet background. For this reason, only the leptonic decay of the top quark is usually studied. The experimental signature for the analysis presented here targets the leptonic decay products from the W bosons: a charged lepton (electron or muon potentially arising from tau lepton decay), and the presence of a significant missing transverse energy $\not{E}_T$ originating from a neutrino. Leptons are accompanied by a (mainly) forward light-quark jet and a b-quark jet arising from the top quark decay.

The data sample considered usually selects events with a trigger requiring at least one lepton with a large p_T isolated from hadronic activity. The usage of b-quark identification ("b-tagging") at the trigger level was investigated in earlier analyses [35], but was found to add a significant complexity for a limited gain, especially with increasing luminosity. To summarize a typical event selection, the presence of only one high p_T-charged lepton (electron or muon with $p_T > 20$ GeV) is required, with a significant missing transverse energy ($\not{E}_T > 30$ GeV) and the presence of at least two high p_T jets ($p_T > 30$ GeV), with one of them from a b-quark and the other failing this requirement, while possibly being detected in the forward region ($|\eta(j')| < 4.7$).

The backgrounds can be classified as arising from two main sources: events containing a charged lepton produced from a boson decay (referred to as prompt lepton), and events with hadronic objects misidentified as prompt leptons. Given that prompt leptons are typically distanced from significant hadronic activity while non-prompt leptons are surrounded by hadrons, a potent method to reject the prevalent QCD multijet background is to require the charged lepton to be isolated. An isolation variable is devised by accumulating the hadronic energy around charged leptons, and this needs to be small. As the modeling of a non-prompt background is hardly well-simulated, non-prompt backgrounds are usually estimated directly from the data, possibly leading to large systematic uncertainties. This estimate is performed, for instance, by inverting the lepton isolation requirement, thus enriching the events in QCD multijet processes. The shape of a distribution of interest is then used as a data-driven estimate of the non-prompt lepton background.

The major prompt lepton background events are chiefly from $t\bar{t}$ production with semi-leptonic decays, where jets are not well reconstructed or do not pass the b-tagging requirements. The $t\bar{t}$ process has been extensively studied; precise measurements have been confronted with theoretical predictions. This process is well described by the state-of-the-art Monte Carlo (MC) generators, such that single-top measurements rely on simulations to describe $t\bar{t}$ kinematics, while the normalization is usually estimated or constrained from data.

The associated production of a single W boson with additional jets, referred to as "W+jets" in the following, constitutes the second main source of background events. The W+jets processes were measured at the LHC, and the event kinematics show a good agreement between data and MC predictions. However, the kinematics of the W+jets process varies slightly depending on the flavor of the additional jets. For this reason, several analyses actually split the W+jets simulation into W+b, c, or light jets, measuring the normalization of each contribution separately.

Finally, other subdominant processes after the selection contribute to the background events, such as the Drell–Yan production, when one of the two leptons is not reconstructed or does not pass the lepton selection. These processes are usually estimated from simulations.

The *t*-channel analyses capitalized on the rise of the LHC profile likelihood method [36,37] to simultaneously estimate the background contributions and constrain the systematic uncertainties from the data. Background normalization is adjusted within the fit, possibly using control regions enriched in background events, usually defined by jet and b-tagged jet multiplicities. For instance, the $t\bar{t}$ background can be controlled by fitting events with at least three jets and two b-tagged jets $(3j, 2t)$. The W+jet background can be controlled with events containing two jets and no b-tagged jet $(2j0t)$, using the distribution in the transverse mass of the W boson $(m_T(W))$, showing a broad resonance for W bosons, as shown in Figure 5. The signal events are mainly expected in the region defined by asking for two jets, one of which is a b-tagged jet $(2j1t)$. The signal is extracted from a combined fit in $(3j, 2t)$, $(2j0t)$, and the $(2j1t)$ regions. Discriminating observables in each of these regions are fitted together with common nuisance parameters representing the systematic uncertainties.

Several distributions can be used to discriminate the signals from backgrounds. In the early versions of analyses, the most obvious observables included the pseudorapidity of the recoiling jet or the reconstructed top-quark mass. In the most precise measurements, the discriminating variables in the $(3j, 2t)$ and $(2j1t)$ regions are constructed from multivariate analyses, such as boosted decision trees (BDTs) or neural networks (NNs), using various kinematic observables as input. An example is shown in Figure 5. In the latest published cross-section measurement at 13 TeV [32], the BDTs are trained using input variables related to the absolute value of the pseudorapidity of the untagged jet, $|\eta(j')|$, the reconstructed top quark mass, the transverse W boson mass, $m_T(W)$, the distance in the $\eta - \phi$ space between the b-tagged and the untagged jet, $\Delta R(b, j')$, the absolute difference in the pseudorapidity between the b-tagged jet used to reconstruct the top quark and the selected lepton, $|\Delta\eta(b, l)|$.

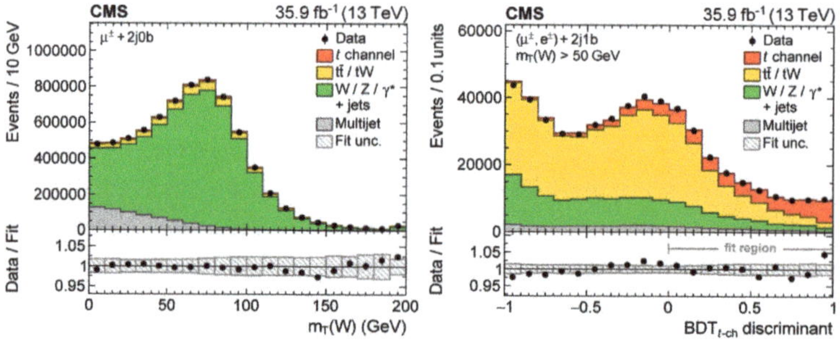

Figure 5. Distribution of the transverse mass of the W boson in the muon channel of the $2j0t$ region (**left**), the BDT discriminant in the $2j1t$ category (**right**) [32].

Thanks to the large amount of integrated luminosity collected at the LHC, the uncertainties related to the *t*-channel measurements are no longer statistically dominated. Remarkably, one can even select a relatively pure sample of *t*-channel events by applying stringent requirements on the BDT discriminants, as illustrated in Figure 6 (taken from [32]), showing the distribution in the cosine of the top quark polarization angle $\cos\theta^*$ in a background-enriched region (with the requirement of $BDT_{t-ch} < 0$) and in a signal-enriched region ($BDT_{t-ch} > 0.7$). The sample can be vastly enriched in signal events while still providing a large event yield.

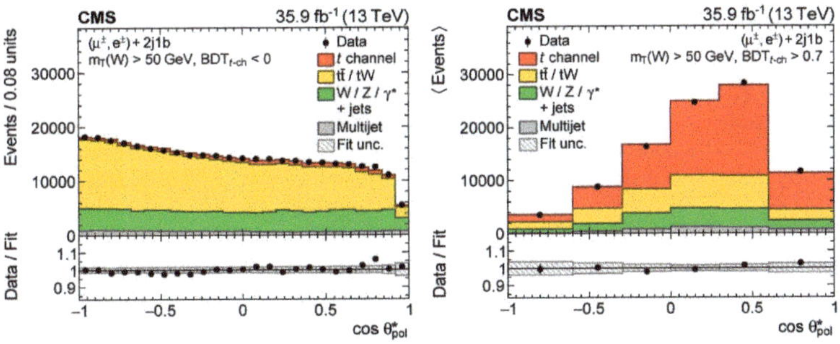

Figure 6. Distributions in the cosine of the top quark polarization angle with a background-dominated selection (**left**) and a signal-enriched selection (**right**) for events in the $2j1t$ region [32].

The main sources of systematic uncertainties impacting t-channel measurements can be summarized as follows:

- Integrated luminosity: Typically a few percent (depending on the dataset).
- Signal and background modeling from SM theoretical predictions: Uncertainties in the modeling of signal acceptance and in the modeling of distributions used as discriminant observables are usually major sources of systematic uncertainty in top quark physics. This includes renormalization and factorization scale variations (accounting for missing higher-order contributions in pQCD), parton shower and hadronization, PDFs, the choice of a matching scheme between fixed-order predictions and the parton shower, the choice of flavor scheme (4FS or 5FS), and MC statistical uncertainty. These uncertainties are treated by generating various MC samples, or including various event weights in the MC samples, with generation parameters varied up and down. The same uncertainties are also included for most backgrounds, which are estimated from simulations.
- Data-driven background estimate: Due to the inadequate representation in simulations of jets misidentified as leptons, the non-prompt lepton background is directly estimated from the data. Usually, these estimations are complicated and rather imprecise. It is rare to lower the relative systematic uncertainty below 30% before any constraints from the fit.
- Simulation-to-data corrections: Several corrections (so-called scale factors) to the reconstructed objects are applied to the simulation to improve its agreement with the data. These corrections are derived from dedicated analyses estimating the associated systematic uncertainties. The corrections are typically related to trigger and lepton selections, jet energy scale and resolution, and b-tagging.

In the most recent analyses, the statistical uncertainty provides a small contribution to the total uncertainty (less than 5%). The relative sizes of the systematic uncertainties depend on the analysis strategy; for instance, the choice of the discriminating observable matters. The use of the $|\eta_{j'}|$ distribution naturally leads to large uncertainties on the jet energy scale and resolution (up to about 5%), since controlling such corrections in the forward part of the detector is difficult. Using a multivariate discriminant can significantly reduce the jet energy scale and resolution uncertainties to a few percent, most likely due to the higher signal purity and increased constraining power. Another large source of systematic uncertainty arises from the signal modeling, which can be lowered by performing a fiducial measurement, as described in Section 2.1.4. Fiducial measurements are performed within a generator-level acceptance to avoid the extrapolation from the visible phase space to the full process phase space, thus reducing the modeling uncertainties.

2.1.3. Summary of the Latest Measurements of the t-Channel-Inclusive Cross-Section

A summary of the latest measurements of the cross-section for t-channel production at $\sqrt{s}$ =7, 8, and 13 TeV from the ATLAS and CMS collaborations can be seen in Table 1, where the combinations made by the LHC*top*WG are also shown when available. Figure 7 shows the relative total uncertainty of the *t*-channel cross-section $\Delta\sigma_{t-chan}/\sigma_{t-chan}$ as a function of the integrated luminosity accumulated at different center-of-mass energies. The most precise 13 TeV measurement was recently released, as a conference note, by the ATLAS collaboration [38]. The *t*-channel cross-section has also recently been measured at 5.02 TeV by the ATLAS collaboration [39].

In Table 1, it quickly becomes evident that the statistical uncertainty soon turns into a secondary source of uncertainty, with precision measurements primarily dominated by systematic uncertainties. Comparing the most precise result measured at 7 TeV [40] with the published measurement at 13 TeV [41], it becomes clear that while several systematic uncertainties related to detector effects and background estimates have decreased, there is an increase in all the theory uncertainties related to the signal modeling. While experimental systematic uncertainties can be reduced further, a significant improvement in the total precision of the inclusive *t*-channel cross-section requires effort in the signal modeling involving the theory community. The largest uncertainties that are common to both ATLAS and CMS are related to the parton shower used in the simulation samples of the *t*-channel and $t\bar{t}$ processes. Sources of large uncertainties can include the choice of the parton shower algorithm, the matching scheme used to interface the NLO fixed-order matrix element with the parton shower, models of hadronization, or final state radiation. Prescriptions should be refined and agreed within ATLAS and CMS (an ongoing effort), and work is needed to decrease the uncertainty based on physics arguments. For instance, improved algorithms, such as antenna-based parton showers, could be tested [42]. The developments of parton showers at NLL [43] or even higher accuracy [44] could bring about large improvements in the future. Eventually, exploring in situ constraints of these uncertainties from ancillary measurements in data [45] is another path to consider.

Table 1. Summary of the most recent and precise *t*-channel cross-sections from the ATLAS and CMS collaborations, and their combinations for 7 and 8 TeV.

	Cross-Section (pb)	$\Delta\sigma_{t-chan}/\sigma_{t-chan}$
7 TeV		
ATLAS [46]	$68 \pm 2 \pm 8 \pm 1$	0.122
CMS [35]	$67.2 \pm 3.7 \pm 4.6 \pm 1.5$	0.091
Combination [40]	$67.5 \pm 2.4 \pm 5.5 \pm 1.1$	0.090
8 TeV		
ATLAS [47]	$89.6 \pm 1.2^{+6.8}_{-5.9} \pm 1.7$	0.076
CMS. [48]	$83.6 \pm 2.3 \pm 7.1 \pm 2.2$	0.093
Combination [40]	$87.7 \pm 1.1 \pm 5.5 \pm 1.5$	0.066
13 TeV		
ATLAS [49]	$247 \pm 6 \pm 45 \pm 5$	0.185
ATLAS [38]	221 ± 13	0.059
CMS [41]	$207 \pm 2 \pm 30 \pm 5$	0.147

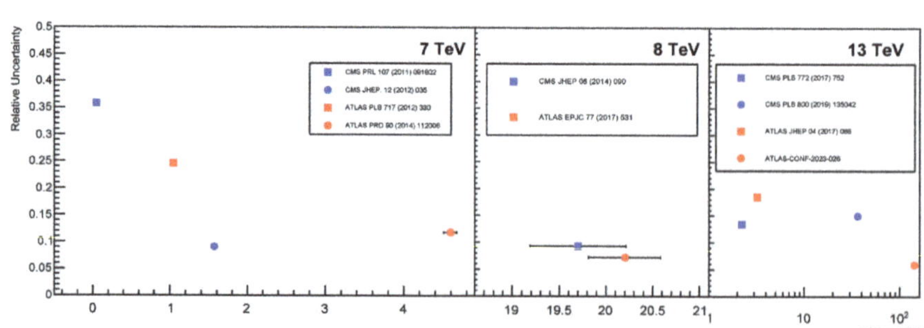

Figure 7. Evolution of the relative total uncertainty in the inclusive *t*-channel cross-section measurements, plotted as a function of the integrated luminosity at $\sqrt{s}$ =7, 8, and 13 TeV.

2.1.4. Measurements of Fiducial and Differential Cross-Sections

With the latest investigations into the t-channel process, leveraging the extensive statistics of LHC Run 2, differential cross-section distributions [32] were measured. Differential cross-sections provide precious information on the theory modeling, and can also be used to constrain the parameters of the EFT. Differential measurements are critical for a deeper understanding of the t-channel process and for identifying any deviations from SM predictions.

The distributions measured at the reconstructed level are unfolded to theoretically well-defined observables, correcting for detector and acceptance effects. The basic principle consists of determining corrections from simulations to infer the "true" top quark properties, by accounting for the signal acceptance induced by the selection for the detector resolution and the efficiencies. The unfolded distributions can be compared in a robust way with theoretical predictions. Two fundamental unfolding levels are usually defined in top quark physics:

- **Parton level:** Corresponds to the generated on-shell top quarks after QCD radiative corrections.
- **Particle level:** Corresponds to (pseudo-)top quarks reconstructed from simulated particles after QED and QCD radiation, particle decay, and hadronization, with a dedicated algorithm.

With the definitions adopted in [32], measurements unfolded to parton and particle levels are confronted with NLO theoretical predictions for various observables, like the top-quark p_T, rapidity y, $\cos\theta^*$, or W boson p_T. Beyond the differential cross-sections, the charge ratios of the cross-section σ_t to $\sigma_{t+\bar{t}}$ are also measured. This observable is sensitive to the PDFs. Figure 8 presents examples of differential cross-sections and cross-section ratios. The measurements show good agreement between data and NLO predictions, validating our understanding of the electroweak interactions in the production of single-top quarks for most of the observables that were scrutinized. However, the precision reached (even in the differential cross-sections normalized to the total cross-section, thus canceling the impacts of several uncertainties) is not yet completely sufficient to unambiguously determine which generator agrees best with the data. As noted by the authors, a few trends can still be highlighted. The predictions with the 4FS well describe the W boson p_T while the 5FS does not; neither the 4FS nor 5FS predictions are able to nicely reproduce the entire distribution of the top quark p_T in the data. This latter trend can be confirmed with deeper studies in the future since a possible mis-modeling could be of great importance for measurements of top quark properties employing the t-channel, and in measurements or searches in which the SM t-channel is an important background (for instance, the s-channel searches). It should be noted that including t channel differential distributions in PDF fits allows for reducing the gluon and light quark PDF uncertainties [50]; for this purpose, it is essential to release experimental correlation matrices, which were not made public in the latest measurements.

The so-called fiducial cross-section is defined at the particle level, and is less sensitive than the inclusive cross-section to the systematic uncertainties arising from signal modeling. In inclusive cross-section measurements, the number of signal events is measured in the visible phase space at the reconstructed level, defined by the detector acceptance and selection efficiencies. The observed number of events is then extrapolated to the full phase space based on simulations. This extrapolation induces a large systematic uncertainty related to the modeling of signal events in the simulation. In contrast, the measurement of the fiducial cross-section is performed in the visible phase space, and extrapolated to the fiducial phase space volume that is defined as close as possible to the phase space of the selected dataset. The fiducial single-top t-channel cross-section was measured [47] at $\sqrt{s} = 8$ TeV and led to a reduction of about 2% in the size of the systematic uncertainties related to the QCD scale and the NLO matching. This resulted in a significant improvement in the precision, and provided a robust method for comparing data with theoretical calculations.

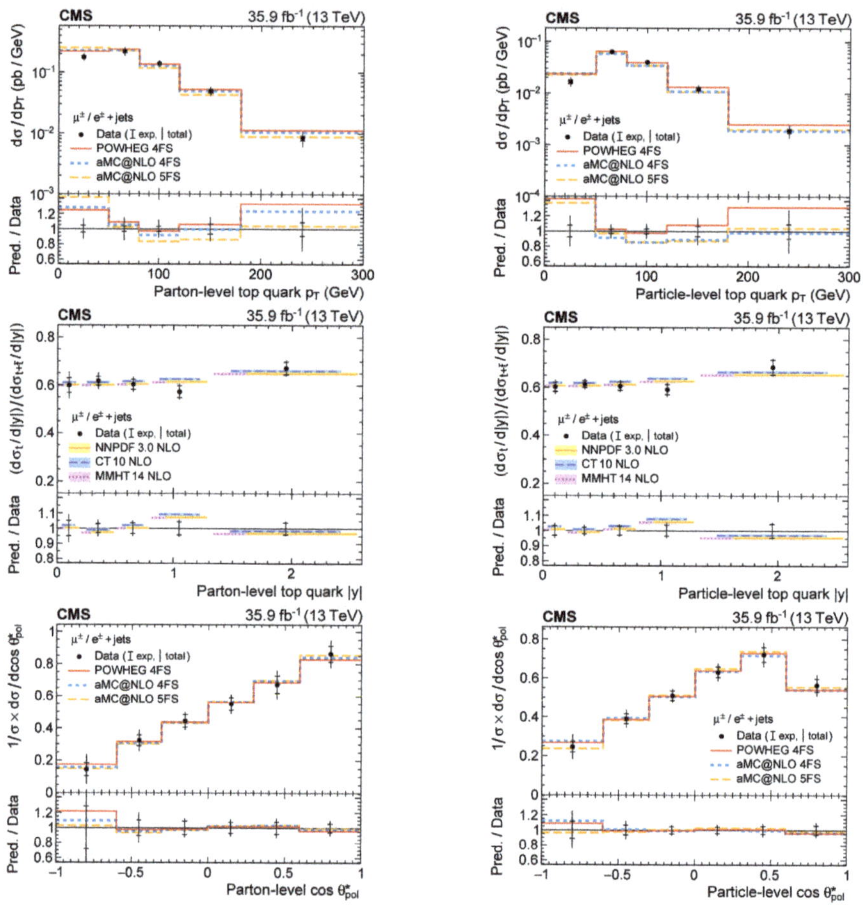

Figure 8. Unfolded differential cross-section measurements: top quark p_T (**upper row**), charge ratio as a function of the rapidity y (middle row), and $\cos\theta^*$ (**bottom row**) at the parton level (**left**) and particle level (**right**) [32].

2.2. *The tW Process, and Its Interplay with the $t\bar{t}$ Process*

2.2.1. Introduction to the tW process

The tW process features a top quark produced in association with a W boson, either initiated by a gluon and a b-quark (in the 5FS, see Figure 9), or with the b-quark produced by gluon splitting (in the 4FS). Because the PDFs for bottom and anti-bottom quarks in the proton in the 5FS are assumed to be the same, the predicted cross-section for tW^- and $\bar{t}W^+$ is identical at LO (and almost identical at a higher order) [12].

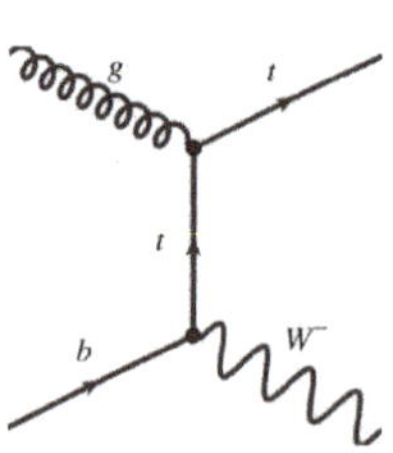

Figure 9. Example of the Feynman diagram for tW production at the LO in pQCD within the 5FS [51].

There is some degree of overlap between the tW process and the $t\bar{t}$ process, since the tW production at NLO in pQCD features resonant diagrams, which interfere with LO diagrams of $t\bar{t}$ production. The NLO corrections to the production of tW include tWb processes, where the Wb system can also arise from the decay of an on-shell top quark. Examples of LO Feynman diagrams for tWb processes are shown in Figure 10. Since the cross-section for $t\bar{t}$ production is much higher than that of tW production, these corrections are very large. As a result, there is ambiguity in the way the tW + 1 jet processes are defined. A similar situation is occurring in the FCNC processes [52].

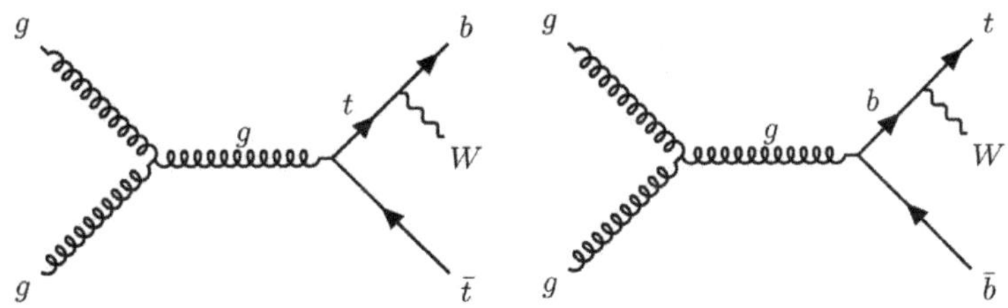

Figure 10. Examples of Feynman diagrams for the production of tWb with (**left**) and without (**right**) an on-shell top.

The definition of the tW process, therefore, relies on the treatment of this interference and presents challenges at a theoretical level, depending on the choice of suppressing the interference to define independent simulation samples for the tW process at NLO, or including it in the simulation in a consistent way between tW and the $t\bar{t}$ process. Two methods exist to suppress the interference [53]. In the diagram removal (DR) method, the resonant $t\bar{t}$ diagrams are excluded at the level of the matrix element calculation. In the diagram subtraction (DS) approach, the $t\bar{t}$ resonant contributions are removed from the cross-section calculation by means of counter terms. Thus, a comparison of the DR and DS prediction provides an estimation of the importance of the interference terms and treatment, which is small for the usual kinematic selection applied [54].

2.2.2. Measurements of the tW Process

The tW process has not been measured at the Tevatron, as its cross-section is small at the Tevatron center-of-mass energy in $p-\bar{p}$ collisions. The measurement of the tW process is more challenging than that of the t-channel since a very large background from $t\bar{t}$ events mimics the signal with almost the same experimental signature. The ATLAS and CMS collaborations presented evidence for this process in the dilepton channel at 7 TeV [55,56], while the inclusive cross-section was measured at 8 TeV [57,58] and 13 TeV [59,60]. The measurements at 13 TeV performed with larger collected data samples allowed measuring the differential cross-sections [61,62] for the first time. The tW process was measured in the lepton+jets channel, more difficult owing to larger backgrounds, by ATLAS using 8 TeV collisions [51] and at CMS using 13 TeV collisions [63].

The dilepton decay channel for the tW process refers to processes where one lepton arises from the top quark decay through Wb and another lepton is produced by the associated W boson decay. We describe features of the ATLAS [61] and CMS [62] analyses measuring differential cross-sections at 13 TeV with the dilepton channel, where the leptons refer to electrons or muons. Nominal SM predictions for the tW process use the DR scheme. For this analysis, the main background contribution after event selection is the $t\bar{t}$ process in the dilepton decay channel, amounting to nearly 80% of the event yield after selection. The signal region is defined with exactly one reconstructed jet being tagged as a b-jet (so-called $1j1b$ region) to remove contributions from doubly resonant diagrams. In general, a selection on the transverse missing energy does not need to be applied (among recent measurements, the ATLAS 13 TeV inclusive cross-section measurement [59] is an exception); this variable is used to reconstruct kinematic quantities and provide input to machine learning techniques. Figure 11 shows the number of events after selection, sorted in bins of the number of jets and b-jets. Two (three) regions that are defined depending on the number of jets and b-jets are used to measure the inclusive cross-section by ATLAS (CMS), with dedicated BDTs.

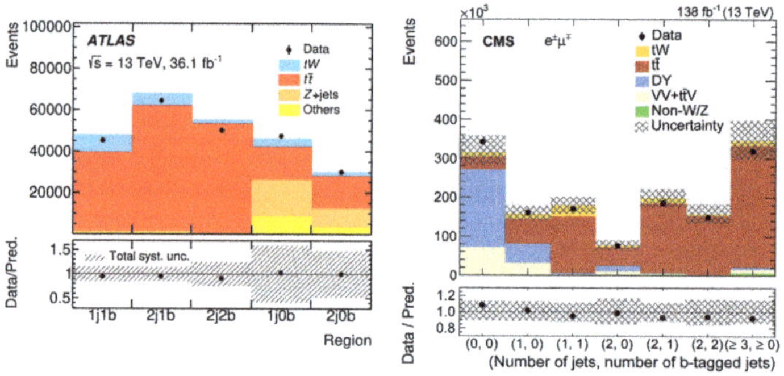

Figure 11. Categories in the number of jets and b-jets in the tW dilepton analysis by ATLAS [61] (**left**) and CMS [62] (**right**).

The inclusive cross-section at 13 TeV is measured to be $\sigma_{tW} = 79.2 \pm 0.9$ (stat) $^{+7.7}_{-8.0}$ (syst) $\pm$ 1.2 (lumi) pb at CMS using 138 fb^{-1} [62], and $\sigma_{tW} = 94 \pm 10$ (stat.) $^{+28}_{-22}$ (syst.) $\pm$ 2 (lumi.) pb by ATLAS using 3.2 fb^{-1} [59], in agreement with SM theoretical predictions. The dominant systematic uncertainty is the jet energy scale, followed by the background normalization and the theory uncertainties on tW process modeling.

The $1j1b$ region—by both ATLAS and CMS—is used to extract the differential cross-sections. In the ATLAS analysis, an additional selection is applied to the output of the BDT to increase the separation between the signal and backgrounds for the differential measurement. In CMS, a veto on additional loose jets is also applied. The data are corrected for detector effects and compared to theoretical predictions, such as the invariant mass of the dilepton and b-jet in Figure 12. In most of the measured bins, the data and simulations agree within less than 1σ; however, more data are needed to discriminate between the different ways of modeling the signal.

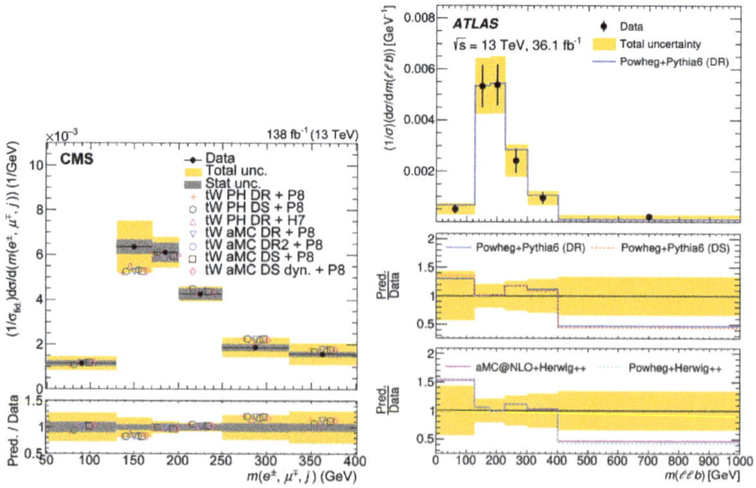

Figure 12. Invariant mass of the dilepton and b-jet in the tW dilepton analysis by ATLAS [61] (**left**) and CMS [62] (**right**), comparing data and several predictions for tW modeling.

Measuring the tW process in the lepton+jets channel, which targets a final state where one of the two W bosons decays leptonically and the other hadronically, is a challenging task, owing to the prominent $t\bar{t}$ and W+jet backgrounds arising from the selection. Machine learning methods are used to enhance the signal-over-background ratio, the NN by ATLAS [51], and the BDT at CMS [63]. ATLAS extracts the signal using a two-dimensional distribution in the NN output and the invariant mass of the hadronically decaying W in events with at least three jets (including one b-jet). CMS employs the BDT outputs in three regions, whether there are two, three, or four jets in the event (including one b-jet). The analyses lead to evidence for the tW process in the lepton+jets channel using 8 TeV data by ATLAS, and an observation using 13 TeV data at CMS. The measured inclusive cross-sections are in agreement with the SM predictions, and the precision is already dominated by systematic uncertainties. The main systematic uncertainties arise from the jet energy scale, background normalization, and $t\bar{t}$ or tW modeling.

The lepton+jets analysis shows that, nowadays, more difficult channels are used to measure the tW process. One of the next steps would be to scrutinize the tails of kinematic distributions by using boosted jet tagging, allowing to access highly boosted regions [64,65] that are sensitive to new resonances, like excited b-quarks appearing in theories beyond the SM, such as composite models [66]. Differential distributions are measured with the dilepton channel and will be investigated more differentially in the future. Despite having a smaller cross-section than the t-channel, the tW process could also be used to measure SM parameters. Similar to measurements performed in the t-channel, measuring charge ratios would be interesting since they are sensitive to PDFs; this would require separating top from antitop contributions in tW production with advanced techniques like the matrix element method [67].

2.2.3. Understanding the Interference between tW and $t\bar{t}$ Processes

While the above-mentioned measurements of tW process are designed to minimize the interference with the $t\bar{t}$ process by mostly selecting events with only one b-jet, a recent analysis by ATLAS [68] targeted a phase space with exactly two b-jets, where the interference effect was maximized. This analysis utilized a variable defined as the invariant mass of a lepton and a b-jet as a proxy for the top quark mass. Since there is ambiguity in assigning leptons and b-jets to a given top quark, a particular choice is made:

$$m_{b\ell}^{minimax} = min\Big(max(m_{b_1\ell_1}, m_{b_2\ell_2}), max(m_{b_1\ell_2}, m_{b_2\ell_1})\Big), \tag{2}$$

where particles 1 and 2 are interchangeable. This variable is defined in such a way that, at LO, $m_{b\ell}^{minimax} < \sqrt{m_t^2 - m_W^2}$. The cross-section above this value has increased sensitivity to the interference between single and double resonant contributions.

Events are selected if there are two leptons and two jets satisfying a tight b-tagging criterion, with a veto on further leptons using a loose requirement (which suppresses backgrounds arising from $t\bar{t}$ associated with heavy flavor jets). The analysis measures the normalized differential cross-section in a phase space at the generator level as close as possible to the reconstructed level, as a function of the $m_{b\ell}^{minimax}$ observable.

The data are compared to simulations at the particle level in Figure 13, after the background subtraction and correction for detector effects. The simulation sample matching the best data across the entire range of $m_{b\ell}^{minimax}$ includes both tW and $t\bar{t}$, as well as their interference with POWHEG [69]. Samples featuring interference suppression with the DR or DS scheme do not reproduce the data at large values of $m_{b\ell}^{minimax}$.

Due to the datasets expected at LHC Run 3 and the HL-LHC, one can expect new measurements to probe the nature of the interference in more depth. The lepton+jets final state could be scrutinized as well for this purpose, since the theoretical predictions at NLO, including the interference, were recently made available [70].

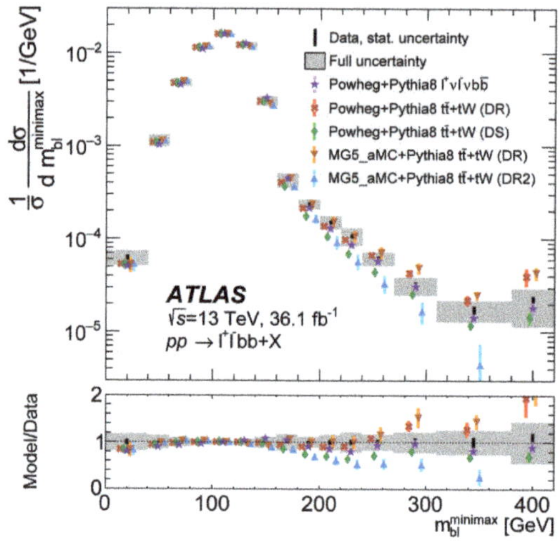

Figure 13. Data comparison of MC predictions for the normalized differential cross-section of the tW process in a region maximizing the interference, as a function of the $m_{b\ell}^{minimax}$ variable [68]. The region sensitive to the interference lies above $\sqrt{m_t^2 - m_W^2}$.

2.3. The Challenging s-Channel

The final state for top quark production in the s-channel is similar to that of the t-channel in Section 2.1, except that the top quark is now produced with a b or $\bar{b}$ quark in the final state instead of a light quark (in the 5FS). The process occurs through the exchange of a time-like W boson instead of a space-like W boson, as shown in Figure 14. The virtual W boson has to be far away from its resonant mass to produce a top quark, and this highly suppresses the corresponding cross-section, which makes the observation of the s-channel very challenging. The top quark is more likely to be produced with central b-jets than with a forward light jet.

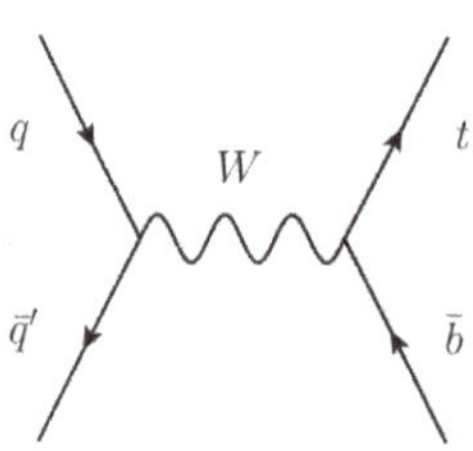

Figure 14. Feynman diagram for single-top s-channel production at the LO in pQCD in the 5FS [9].

The s-channel process is observed at the Tevatron [8], using 9.7 fb^{-1} of proton–antiproton collisions collected at D0 and CDF at $\sqrt{s} = 1.96$ TeV. This process remains to be observed at the LHC. At CDF, the lepton+jets channel and $\not{E}_T$+jets channel are used, while the lepton+jets channel is used at D0. Multivariate techniques are employed to identify the b-jets and reduce the contribution of background processes. Events are classified in categories depending on the number of jets, and the number and quality of b-jets. Multivariate discriminants are built to extract the s-channel cross-section using a Bayesian statistical technique. The result is $\sigma_s = 1.29^{+0.26}_{-0.24}$ pb, which is in agreement with the SM prediction of $\sigma = 1.05 \pm 0.06$ pb at an approximate NNLO with NNLL accuracy [71] at the Tevatron. The s-channel process was observed at 6.3 σ at the Tevatron. Figure 15 shows the measured cross-section for each channel at the Tevatron along with their combined results.

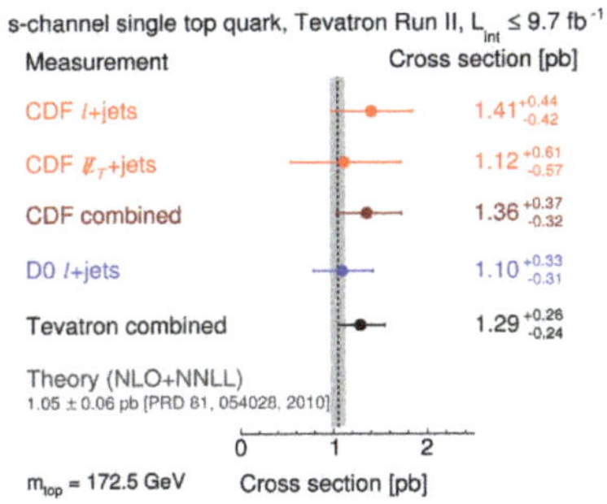

Figure 15. Measured cross-sections for the *s*-channel at the Tevatron along with their combined results [8].

Proton–antiproton collisions at the Tevatron mainly provide a quark and an antiquark in the initial state while the LHC does not. Furthermore, the b-quark content in the proton is larger at the LHC. As a result, the ratio of the *s*-channel to the *t*-channel cross-section decreases from the Tevatron to the LHC. Furthermore, the ratio of signal-to-background is quite favorable at the Tevatron relative to the LHC. When the energy in the p–p center of mass increases, this search becomes more difficult: the quark luminosity increases at a slower pace than the gluon luminosity when increasing the center-of-mass energy. As an example, the ratio of the *s*-channel to the $t\bar{t}$ cross-section changes from 2.1% at 8 TeV to 1.2% at 13 TeV [10].

For all of these reasons, searches for the *s*-channel are very challenging at the LHC. The first search by ATLAS using 8 TeV data resulted in a significance of 1.3σ [72] (with 1.4σ expected); however, a subsequent search on the same dataset employed the matrix element method (MEM [73], which has been used since the early measurements of the top quark mass at the Tevatron [74,75]), leading to an observed significance of 3.2σ [9] (with 3.9σ expected). CMS analyzed Run 1 data using the 7 and 8 TeV datasets, resulting in an observed significance of 2.5σ with an expected significance of 1.1σ [76]. Recently, ATLAS performed a search using the same analysis techniques with the MEM as in their 8 TeV paper, analyzing Run 2 data at 13 TeV [10]. Despite the unfavorable signal-to-background ratio at 13 TeV compared to 8 TeV, a similar observed (3.3σ) and expected (3.9σ) significance was achieved.

Since the ATLAS result is the latest, with the largest observed significance, and the only one published using 13 TeV data, we will provide details on this analysis. The lepton+jet channel is analyzed, with one electron or muon having $p_T > 30$ GeV and at least two jets with $p_T > 25$ GeV. Events from multijet production are reduced by requiring $\not{E}_T > 35$ GeV and $m_{T,W} > 30$ GeV. In the signal region, exactly two jets are required, and both of them must be b-tagged. A validation region targets the *W*+jets process, where one of the jets must fail the b-tag requirement. Events are also validated using two regions enriched in the $t\bar{t}$ process, with three or four jets, among which, two must be b-tagged. The normalization for multijet production is estimated from the data, while the other background processes are taken from the simulation. A dedicated method, the MEM, is employed to further reduce the backgrounds. The MEM consists of calculating a probability density, representing the compatibility of each event with signal and background hypotheses, using exact calculations at the LO in pQCD. Hypotheses for the *s*-channel, *t*-channel, $t\bar{t}$ production, and *W* boson production are considered. A likelihood is built by combining these hypotheses, and the less likely events are discarded. The shape of the likelihood distribution in the signal region is then used to extract the *s*-channel cross-section. The post-fit distribution is shown in Figure 16, left, and the signal after background subtraction is shown in Figure 16, right. The measured cross-section is $\sigma = 8.2 \pm 0.6(\text{stat})^{+3.4}_{-2.8}(\text{syst})$ pb, in agreement with the SM prediction of $\sigma_{SM} = 10.32^{+0.40}_{-0.36}$ pb at the NLO accuracy in pQCD.

As a side note, predictions at the NNLO in pQCD are available [77] and could be used by the LHC experiments.

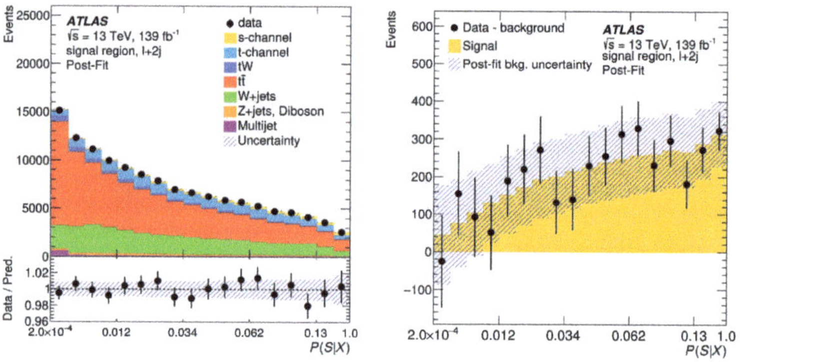

Figure 16. Results of the s-channel searches by ATLAS: (**left**) post-fit distribution comparing data and simulation for the MEM likelihood, and (**right**) signal distribution after the background subtraction [10].

As a conclusion, the observation of the s-channel remains to be achieved at the LHC. A result for a CMS analysis using Run 2 data is desired. The analysis is already systematic-dominated; therefore, new techniques should be employed to reduce the uncertainties. A simultaneous fit using signal and control regions could be used to further constrain the background contributions. An involved analysis technique beyond the MEM, like a deep NN (DNN), could also improve the significance. Despite maintaining an unfavorable signal-to-background ratio compared to the Tevatron, the searches should be pursued at the LHC with Run 3 data and at the HL-LHC to make an observation.

3. Associated Production of a Single-Top Quark with a Neutral Boson

3.1. A Newcomer: Associated Production of a Single-Top Quark with a Photon ($t\gamma$)

The production of a photon in association with a top quark ($t\gamma$) is a rare process, accessible at the LHC. The cross-section predicted at NLO in pQCD with Madgraph5_aMC@NLO [78] is $2.95 \pm 0.13(\text{scale}) \pm 0.03(\text{pdf})$ pb (as quoted by CMS [79]), requiring the photon p_T to be greater than 10 GeV before the top quark decay in the 5FS. The cross-section is dominated by t-channel diagrams with the radiation of a photon ($t\gamma q$), featuring a forward jet due to the electroweak nature of the t-channel. The cross-section for the $t\gamma q$ production is known at the approximate NNLO [80]. Measuring the $t\gamma$ process extends the landscape of the measured top quark process and is an experimental challenge, owing to its low cross-section. The $t\gamma$ final states are also powerful tools used to constrain the FCNC [52]. Together with $t\bar{t}\gamma$ processes, they can be used to constrain the top-γ coupling. Examples of Feynman diagrams are shown in Figure 17. The photon can be emitted in the initial state, final state, or in the top quark decay.

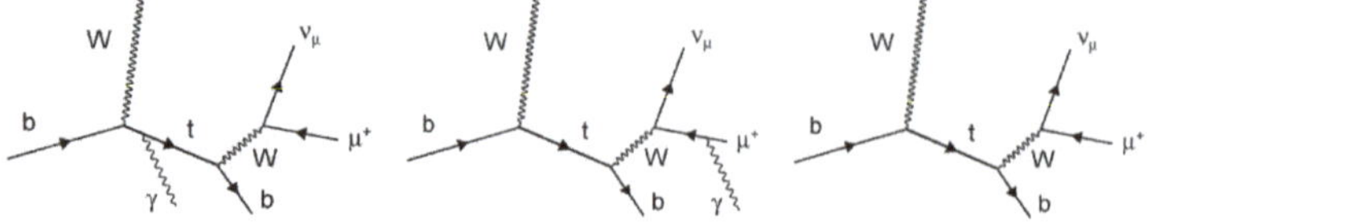

Figure 17. Examples of Feynman diagrams for $t\gamma q$ processes at the LO in pQCD [79].

The searches for the $t\gamma q$ process led to evidence at CMS [79] using 36 fb^{-1} of Run 2 LHC data, and an observation by ATLAS with the full Run 2 dataset [81]. Special care is needed in single-top t-channel MC samples to remove photons produced in the parton shower since they could be double-counted with photons produced at the matrix element level in the $t\gamma q$ signal samples. There is also some freedom in the signal definition: photons arising from the top quark decay are treated as backgrounds in the ATLAS analysis [81]. The dominant backgrounds contain prompt leptons and photons, like $t\bar{t}\gamma$ and $W+\gamma$ processes, and processes involving jets or electrons misidentified as photons (hereafter denoted as "fake photons"). A control region is defined to measure the $t\bar{t}\gamma$ background. The $W+\gamma$ process also benefits from a control region in the ATLAS analysis. The fake photon backgrounds are two-fold, either arising from the misidentification of an electron as a photon, or of a jet as a photon. In the ATLAS analysis, both are estimated with dedicated methods from the data, while in the CMS analysis, only the backgrounds made of jets misidentified as photons are estimated from the data. To maximize the sensitivity to the signal, the signal extraction is performed by constructing a discriminant with a BDT (CMS) and a DNN (ATLAS). Both analyses make use of the forward jet to discriminate the signal against the backgrounds, including the pseudorapidity as an input variable to the machine learning algorithm. The discriminants are shown in Figure 18.

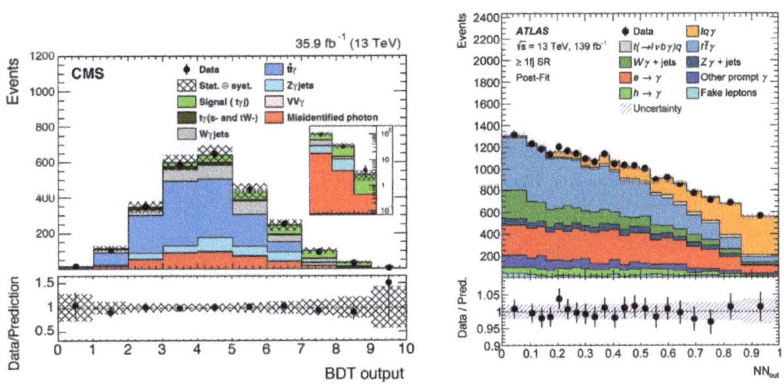

Figure 18. Discriminants used in the $t\gamma q$ signal extraction: BDT output at CMS [79] (**left**) and DNN output by ATLAS [81] (**right**) in the signal region with one forward jet.

The observed (expected) significance obtained is 4.4σ (3.0σ) at CMS [79] and 9.1σ (6.7σ) by ATLAS [81]. With such a large significance, the $t\gamma$ processes can provide sufficient statistics for a first differential cross-section measurement at LHC Run 3. Complementing the $t\bar{t}\gamma$ channel with the $t\gamma q$ channel to probe the top-photon coupling will become especially relevant at the HL-LHC [82]. As a side note, the single-top $tW\gamma$ process has also been measured simultaneously with the $t\bar{t}\gamma$ process [83] because of the interference at NLO, similar to the tW and $t\bar{t}$ processes, but in a $t\bar{t}\gamma$ phase space chosen without particular enhancement of the $tW\gamma$ process or the interference.

3.2. *A Path toward Top-Z Coupling: Single-Top Quark Production with a Z Boson (tZ)*

The first process observed for single-top quark production in association with a neutral boson is actually the single-top quark production with a Z boson (tZ), due to the datasets made available at the LHC. In general, the tZ processes refer to the production of a single-top quark in association with a Z boson, including the interferences between on-shell and off-shell γ^* or Z bosons. Similar to the $t\gamma$ process, the process with the largest cross-section is provided in the t-channel (tZq).

The Feynman diagrams for tZq production at the LO in pQCD can be seen in Figure 19. The inclusive tZq cross-section predicted at NLO in the SM, as calculated with MG5_aM@NLO, is 800 fb $^{+6.1\%}_{-7.4\%}$ [84]. Because of its clear signature and interesting signal-to-background ratio, the tZq process is measured in the three-lepton channel. The cross-section for tZq production in the three-lepton decay channel, calculated at NLO with MG5_aMC@NLO, and including a dilepton invariant mass cut of $m_{\ell\ell} > 30$ GeV, is $94.2^{+1.9}_{-1.8}$ (QCD scale) $\pm$ 2.5 (PDF) fb [85]. The cross-section for tZq production could be updated with the latest predictions at the approximate NNLO [86] in future measurements.

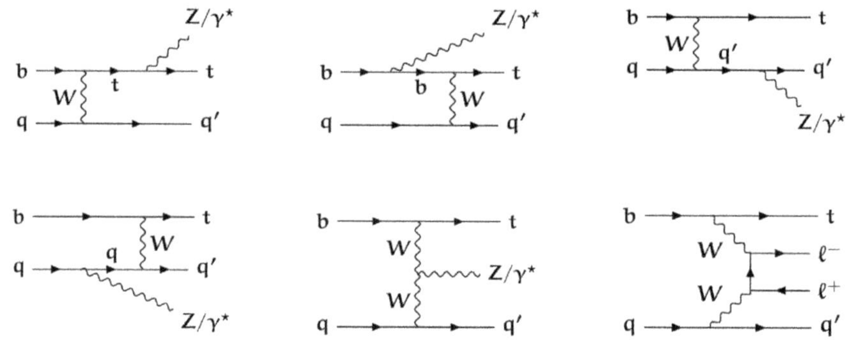

Figure 19. Representative Feynman diagrams for the tZq production at the LO in pQCD [87].

The tZq production has several interesting features. In a similar way with the t-channel process without an associated Z boson, the top and antitop quarks from the tZq production are strongly polarized, making this process an excellent probe for studying $t-Z$ couplings, particularly in the context of EFT measurements. It is also sensitive to triple-gauge couplings WWZ, in a complementary manner with the diboson production. Both are potentially sensitive to physics beyond the SM.

Data are selected with a combination of single-lepton or double-lepton triggers. Events are selected events if they contain three well-identified and isolated leptons (electrons or muons possibly arising from τ lepton decays). A pair of same-flavored opposite-charged leptons, compatible with a Z boson decay, is then required. Because the tZq process is a t-channel process, it contains a light jet preferentially produced at large $|\eta|$, a b-tagged jet arising from the top quark decay, and missing transverse energy arising from the neutrino from the W boson decay.

Similar to other analyses presented in this review, the signal is extracted from signal and control regions defined by the number of jets and b-tagged jets. The first signal region requires $N_j = 2$ or $N_j = 3$ with $N_b = 1$ (so-called $2j1b$ and $3j1b$ regions). These regions contain most of the signal with the WZ+jets process as the dominant background, and with contributions from other diboson processes. For larger jet multiplicities and b-tagged jet multiplicities ($N_j \geq 3$, $N_b \geq 2$), the dominant background source arises from $t\bar{t}Z$ events, with contamination from $tt + W, H$ processes. A control region with $N_b = 0$ allows constraining the diboson contribution.

Other background sources are from non-prompt lepton events in $t\bar{t}$ or Z+jets processes. While backgrounds presenting three prompt leptons are estimated from simulations and constrained from the data in the likelihood fit, events containing at least one non-prompt lepton are not well-described by simulations and are, therefore, more difficult to estimate. In the CMS observation paper [88], the analysis uses a fully data-driven technique, where the probabilities for measuring a non-prompt lepton are measured from a region where one lepton fails the lepton isolation. The ATLAS observation paper [89] uses a semi data-driven technique, where the normalization of the non-prompt background is estimated from the data in control regions, and the kinematic distributions are determined from simulations of $t\bar{t}$+tW and Z+jets events, by replacing b-jets with non-prompt leptons and accounting for the needed corrections.

The discriminating variables used in the fit are based on multivariate discriminants (BDT or NN), which include kinematic variables related to the reconstructed Z bosons or top quarks, the pseudorapidity of the spectator jet $|\eta_{j'}|$, dijet invariant mass, or kinematic variables related to the lepton from the W decay. Examples of NN output distributions from ATLAS [89] can be found in Figure 20.

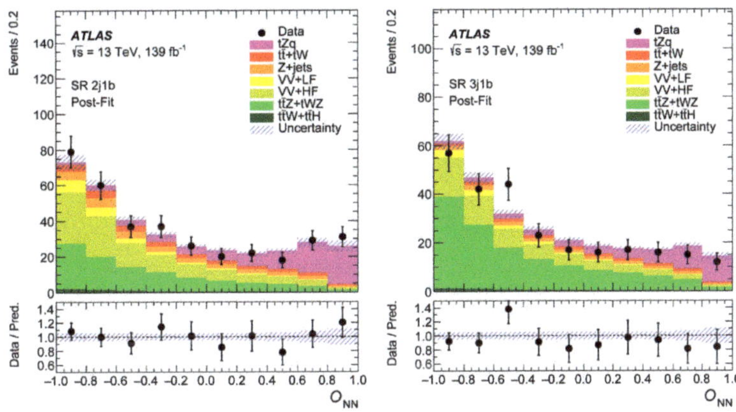

Figure 20. Distributions of the NN output in the signal regions $2j1b$ (**left**) and $3j1b$ (**right**) in the ATLAS analysis [89].

The most recent inclusive tZq cross-sections measured by ATLAS [89] and CMS [87] are compatible with the SM. The precision is still dominated by statistical uncertainties, although with the CMS results, systematic and statistical uncertainties are almost of the same level. The dominant sources of systematic uncertainties are mainly experimental, and are not identical between ATLAS and CMS (while the analysis techniques are relatively similar). For ATLAS, the dominant uncertainties arise from background modeling and normalization, the jet energy scale, and the lepton selection; for CMS, the systematic uncertainties are dominated by the QCD scale uncertainties in the signal modeling, non-prompt lepton background estimation, and WZ process normalization. All these uncertainties have similar orders of magnitude. Differences in the relative sizes of the systematic uncertainties by ATLAS and CMS are probably explained by differences in the analysis strategy, noticeably in the treatment of the background estimates.

Due to the large integrated luminosity provided by the LHC, it is now possible to measure differential cross-sections for tZq production [87]. This analysis follows a different approach to extract the signal. A multi-class NN is used to separate the tZq process from the $t\bar{t}Z$, WZ, and other $t+X$ processes. The signal region is then sub-divided based on the bins of the observables of interest, at the detector level. The NN score of the tZq node is used to extract the signal in each bin. Similarly, the NN score of the $t\bar{t}Z$ node is used to constrain the $t\bar{t}Z$ background. An unfolding procedure infers the particle- or parton-level distributions. Examples of differential cross-section measurements for $p_T(Z)$ and $p_T(t)$ at the parton level, and $|\eta(j')|$ and $\cos\theta^*_{pol.}$ at the particle level are shown in Figure 21. The $\cos\theta^*_{pol.}$ variable is the cosine of the polarization angle of the top quark, defined as:

$$\cos\theta^*_{pol.} = \frac{\vec{p}(q'^*)\cdot\vec{p}(l^*_t)}{|\vec{p}(q'^*)||\vec{p}(l^*_t)|} \quad (3)$$

with $\vec{p}(q'^*)$ and $\vec{p}(l^*_t)$, the three momenta of the light jet and the lepton from the top quark decay. A good agreement between data and predictions is observed. This very promising publication presents the first differential measurements of a rare single-top process, and can serve as the basis for future studies. In particular, it provides a clear procedure to perform a differential measurement, featuring an interesting signal extraction based on a multi-class discriminant.

Eventually, the first measurement for the tWZ process led to evidence [90] (presented as a conference note by CMS). This very rare process can be seen as a tZ production in the tW channel, where it shares similar modeling issues since it interferes at NLO with the $t\bar{t}Z$ process [91]. The analysis techniques are similar to those of the tZq analysis, using a multi-class NN, with a multi-lepton signature targeted. This result opens up a new era for measuring top quark processes associated with multi-bosons.

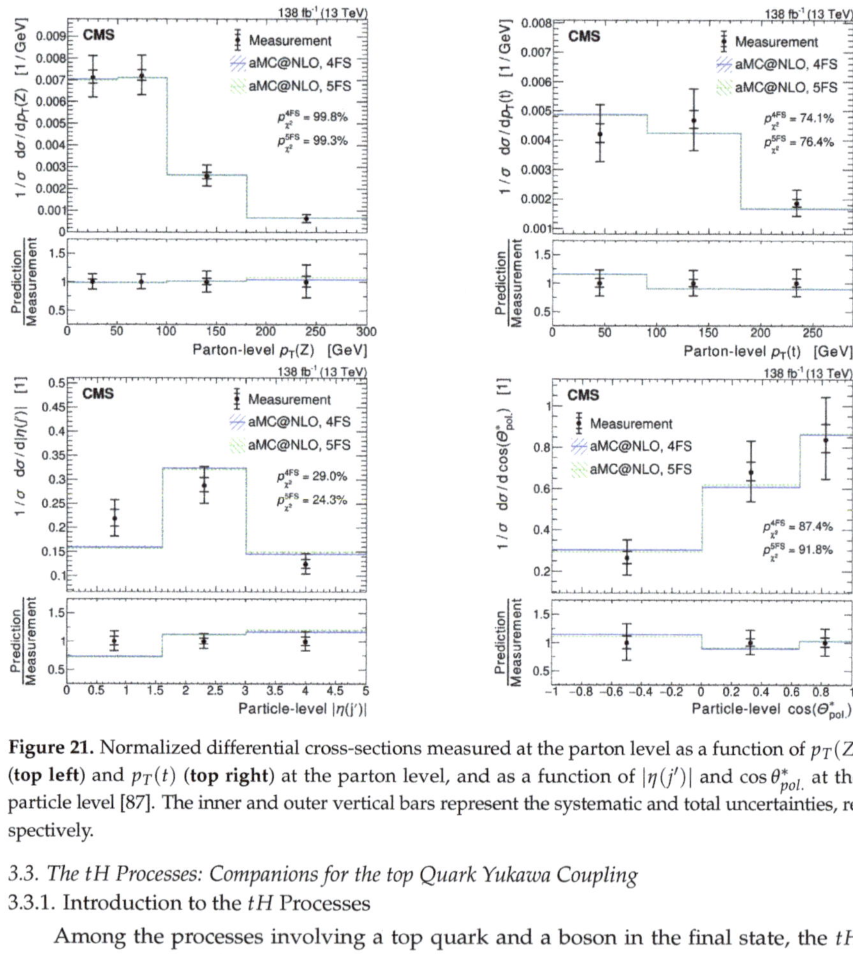

Figure 21. Normalized differential cross-sections measured at the parton level as a function of $p_T(Z)$ (**top left**) and $p_T(t)$ (**top right**) at the parton level, and as a function of $|\eta(j')|$ and $\cos\theta^*_{pol.}$ at the particle level [87]. The inner and outer vertical bars represent the systematic and total uncertainties, respectively.

3.3. *The tH Processes: Companions for the top Quark Yukawa Coupling*

3.3.1. Introduction to the tH Processes

Among the processes involving a top quark and a boson in the final state, the tH processes are produced with the lowest cross-section predicted in the SM, of approximately 71 fb and 16 fb at NLO for the t-channel (tHq process) and the tW-associated production (tHW process) with $\sqrt{s}$ =13 TeV [92]. The latest predictions for the tHq cross-section are computed at NNLO in pQCD [93]. The tHq processes share many properties with the $t\gamma q$ and the tZq processes, noticeably their modeling in the 4FS or 5FS schemes, and the production of an associated quark in the forward direction. The Feynman diagrams for the production of tHq are depicted in Figure 22. The tHW production is also considered in the analyses.

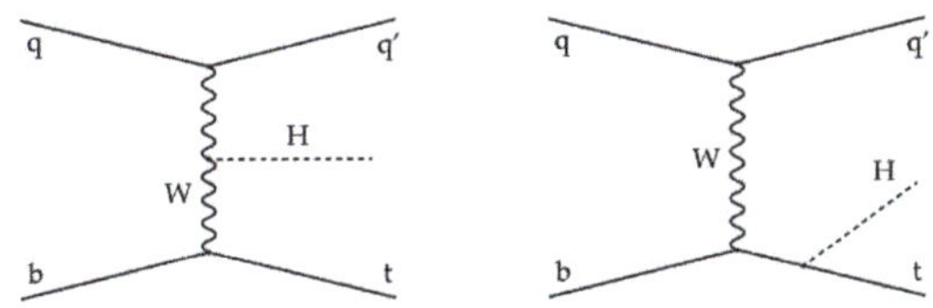

Figure 22. Examples of Feynman diagrams for tHq processes at the LO in pQCD [94].

The search for tH processes is traditionally performed in association with the search for the Higgs boson measurement in the $t\bar{t}H$ production mode, whose cross-section is larger than the cross-section of the tHq process by a factor of 10, as shown in Figure 23. The amplitude for tHq production features an interesting property, as it features interference between diagrams where the Higgs boson is emitted from a top quark line and those arising from W boson exchange. This property makes the measurement of the tHq process appealing since it provides access to the sign of the Yukawa coupling of the top quark. If the sign of the Yukawa coupling κ_t is negative, the interference becomes constructive, for instance, by increasing the cross-section by a factor of approximately 12 if $\kappa_t = -1$ [95]. The tH final states, on equal footing with the tZ and the $t\gamma$ final states, are also used in the FCNC searches [52].

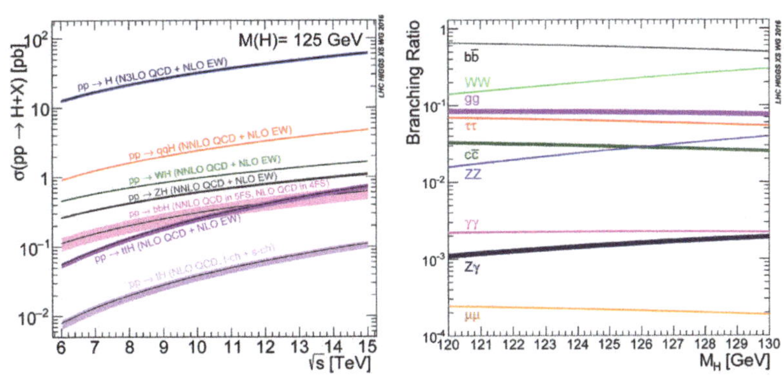

Figure 23. (**Right**) Cross-section for the Higgs boson production as a function of $\sqrt{s}$. (**Left**) branching ratio for the Higgs boson decay [92].

3.3.2. Searches for the tH Processes

The early searches for the tH processes at 8 TeV [94] attempted to directly measure the tH production, while the $t\bar{t}H$ process was treated as a background. It was, however, realized that by varying the value of the top quark Yukawa coupling, the cross-section for both the tH and $t\bar{t}H$ processes would be modified in a correlated way. Nowadays, the searches for the tHq process are performed in a combined measurement with the $t\bar{t}H$ process, either targeting the measurement of the top quark Yukawa coupling or simultaneously measuring the cross-section for the tH and $t\bar{t}H$ processes.

Branching ratios for the Higgs boson decay are shown in Figure 23. The ATLAS and CMS analyses target the main decay modes of the Higgs boson: $H \to \gamma\gamma$; $H \to WW$, $H \to ZZ$, and $H \to \tau\tau$ (grouped under the naming of the "multilepton final state" since W, Z, τ, and associated top quarks can decay leptonically), and to a lesser extent, $H \to b\bar{b}$ (suffering from a lack of available luminosity to achieve similar sensitivity as the other channels). We present here the methodology and the latest results.

The analysis of the $H \to \gamma\gamma$ decay channel with Run 2 data by ATLAS [96] and CMS [97] follows a similar strategy to the measurements of the other production mechanisms of the Higgs boson. The small $H \to \gamma\gamma$ branching fraction (close to 0.2% at $m_H = 125$ GeV) is compensated by the excellent resolution of the electromagnetic calorimeters (the effective mass resolution on the Higgs boson is close to 1.5 GeV, depending on the analysis categories). The background processes involving jets reconstructed as photons are reduced using photon isolation and information on the shape of the electromagnetic energy deposit, with sequential criteria by ATLAS and a multivariate method at CMS. Several event classes are constructed, specifically targeting a given production mechanism. For each event class targeting the tH processes, the background is reduced by the means of a BDT discriminant, which is subsequently fitted with a smoothly falling function. In the latest versions of the analysis [96,97], several subcategories are built to specifically target the $t\bar{t}H$ and tH processes in kinematic bins, and the fit is interpreted in the so-called "simplified template cross-section" framework (STXS) [98]. The STXS framework is a convention used to provide results in kinematic bins at particle levels within a defined acceptance for each Higgs boson production mechanism. In the CMS analysis, a category at the reconstructed level specifically targets tH in the leptonic channel, and a DNN discriminant is used to improve the separation between $t\bar{t}H$ and tH. Using this category, together with many reconstructed-level event classes in a simultaneous fit, the cross-section for the tH processes at the STXS level is quoted to be $6.3^{+3.4}_{-3.7}$ times the SM expectation (in the so-called "maximal merging scenario", where fewer STXS categories are used at the particle level than in the "minimal merging scenario"). In the ATLAS analysis, four reconstructed categories targeting tH processes are defined, where two categories specifically target the tHq processes with either a positive or a negative top quark Yukawa coupling (defined using the output of a NN), one category targets the tHW process, and the remaining category gathers events with low-scores of the BDT for tHq and $t\bar{t}H$. At the STXS level, the cross-section for the tH processes is $2.1^{+4.2}_{-3.1}$ times the SM expectation.

Using the multilepton channel, CMS [99] reported measurements of the cross-section for $t\bar{t}H$ and tH production simultaneously with Run 2 data. This analysis uses multiple final states. For leptonic top decay, the configurations are the same-sign $2\ell + 0\tau_h$ (where $\ell = e, \mu$ and τ_h denotes hadronically decaying τ), $3\ell + 0\tau_h$, $2\ell + 1\tau_h$ (both same-sign and opposite-sign), $1\ell + 2\tau_h$, $4\ell + 0\tau_h$, $3\ell + 1\tau_h$ and $2\ell + 2\tau_h$. For hadronic top decay, the configurations are $1\ell + 1\tau_h$ and $0\ell + 2\tau_h$. The sensitivity arises mainly from the same-sign channel $2\ell + 0\tau_h$, $3\ell + 0\tau_h$, and $1\ell + 2\tau_h$. In those main categories, the analysis employs a multi-class DNN, separately providing discriminants for $t\bar{t}H$ and tH, while using simpler BDTs in the other categories. In the same-sign channels, $2\ell + 0\tau_h$, $2\ell + 1\tau_h$, categories are further divided according to the lepton flavor and whether the b-jet number is larger or smaller than 2. The jet-faking lepton background is estimated with a data-driven method by relaxing lepton identification criteria in a region enriched in multijet events. Backgrounds resulting from mismeasuring the lepton charge are determined using $Z \to ee$ events. The dominant background arises from $t\bar{t}W$ and $t\bar{t}Z$ processes, estimated from simulation. The background arising from the conversion of leptons in the detector is estimated from the simulation. The signal is extracted using bins in the multivariate discriminants. Several control regions with 3ℓ and 4ℓ final states are also used in the fit. Two parameters of interest are measured: the signal strength μ for $t\bar{t}H$ and for tH processes. The signal strength for tH production is 5.7 ± 2.7(stat) $\pm$ 3.0(syst). Additionally, a 2-dimensional distribution of the likelihood as a function of $\mu_{t\bar{t}H}$ and μ_{tH} is measured, as shown in Figure 24.

A few results arising from combinations of final states are also reported. A dedicated CMS analysis specifically targeting tH at 13 TeV using 36 fb^{-1} of Run 2 data [100] employed the $H \to \gamma\gamma$ and multilepton final states, as well as the final state $H \to b\bar{b}$ in the VH production mode with the single-lepton decay of the associated boson. The multilepton analysis uses simpler techniques than those previously described [97,99], and trained multivariate methods with tH processes as the signal. The $H \to \gamma\gamma$ analysis reinterprets the content of the $t\bar{t}H$ categories of a previous analysis. The $H \to b\bar{b}$ analysis brings little sensitivity and will not be described here. The combined measurement results in an observed limit on the cross-section for the tH production of 1.94 pb at 95% CL in the SM hypothesis. The results for the hypotheses with negative top quark Yukawa coupling are also reported. In commemoration of the 10th anniversary of the Higgs boson discovery, grand combinations were performed by both CMS [101] and ATLAS [102], including many final states. Categories specifically targeting the tH processes are taken from the $H \to \gamma\gamma$ channel by ATLAS, and $H \to \gamma\gamma$ and the multilepton channel at CMS. The CMS combination reports a measured signal strength of $\mu_{tH} = 6.05^{+2.66}_{-2.42}$.

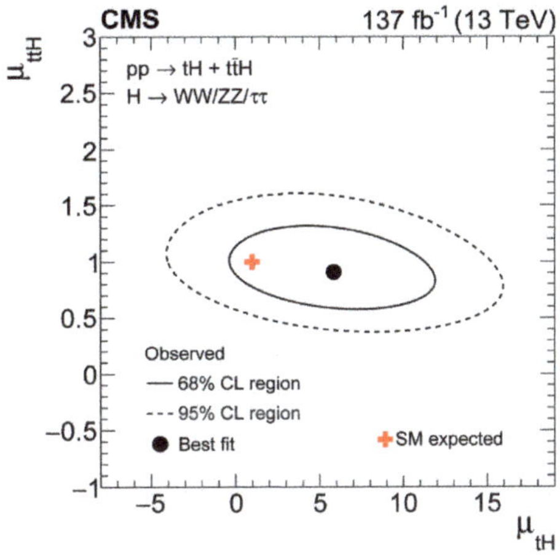

Figure 24. Likelihood contour as a function of the signal strengths for $t\bar{t}H$ and tH processes in the multilepton analysis [99].

Studies estimating the sensitivity to the tH processes at the HL-LHC were expecting a relative uncertainty of 90% on the tH signal strength in the SM hypothesis [103] (with the $t\bar{t}H$ signal strength floating); however, these studies were based on early projections and would need to be updated with the latest ATLAS and CMS results.

3.3.3. Probing the Sign of the Top Quark Yukawa Coupling

With the analyses from ATLAS and CMS for $H \to \gamma\gamma$ final state [96,97], combined with the CMS multilepton analysis [99] and the earlier CMS combination [100]—all of which include categories that specifically target the tH processes—it is now feasible to determine the sign of the top quark Yukawa coupling, thanks to the interference observed between Feynman diagrams that showcase the Higgs boson coupling to both the top quark and the W boson. The modifier κ_t of the top quark Yukawa coupling in the SM, $y_{t,SM}$, is defined as $\kappa_t = y_t/y_{t,SM}$. Furthermore, since a similar interference is also present in the $H \to \gamma\gamma$ decay between the top quark loop and the W boson loop, further sensitivity is gained in this channel. Sensitivity to the positive values remains dominated by the $t\bar{t}H$ process in direct measurements, and by the $gg \to H$ process (involving a top quark loop) in indirect measurements because of the larger cross-section.

Figure 25 shows the likelihood fit value as a function of the κ_t parameter. The best-fit value is positive and close to 1, while a second minimum of the likelihood is found at a value close to -1. As shown on the left side of Figure 25, including the parameterization of the gluon fusion mechanism as a function of κ_t in the likelihood provides more weight to the positive value of κ_t. On the contrary, when only the tH and $t\bar{t}H$ processes are included, more sensitivity is gained on the sign of κ_t. Values outside of $0.65 < \kappa_t < 1.25$ in the first case and $0.87 < \kappa_t < 1.20$ in the second case are excluded at 95% CL by the $H \to \gamma\gamma$ analysis by ATLAS. The CMS multilepton analysis results in $-0.9 < \kappa_t < -0.7$ or $0.7 < \kappa_t < 1.1$ at 95% CL.

Projections for the measurement of the top quark Yukawa coupling at the HL-LHC are reported by CMS [101] without emphasis on a possible negative coupling. A precision on the order of 3-4% on κ_t would be achievable, while a precision on the order of 10% is achieved today [101,102].

The tH processes, together with the $t\bar{t}H$ process, can also be used to set constraints on a CP-odd top quark Yukawa coupling. Such measurements were performed by ATLAS with the $H \to \gamma\gamma$ [104] and $H \to b\bar{b}$ [105] channels, and at CMS with the $H \to \gamma\gamma$ [106] and multilepton [107] channels. Since the $t\bar{t}H$ process has a larger cross-section than the tH processes, most of the sensitivity will come from the former, and these measurements will not be described here.

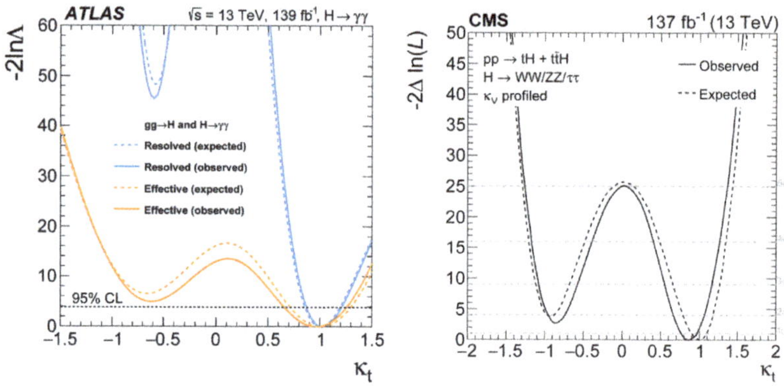

Figure 25. Log-likelihood as a function of the κ_t parameter, in the ATLAS $H \to \gamma\gamma$ analysis [96] (**left**), and in the CMS multilepton analysis [99] (**right**).

4. Discovery Potential of Property Measurements and Interpretations

The large number of single-top events produced at the LHC and the high precision obtained in single-top measurements allow for the measurements of top quark properties, which can be seen as tests of the SM or a search for physics beyond the SM. Although single-top quark production has a lower cross-section compared to $t\bar{t}$ production, the production and subsequent decay of single-top quarks to Wb engage the Wtb vertex twice, during both the top quark production and its decay. This interesting feature can be used to measure several interconnected properties: the couplings of the Wtb vertex, including the CKM matrix element $|V_{tb}|$, the W polarization, and the top quark polarization. Precision measurements of the Wtb couplings can be expressed in terms of CP-even and CP-odd effective couplings or within the EFT. Apart from the $|V_{tb}|$ measurement, which can be inferred from the single-top cross-section, the general experimental strategy for measuring all other properties consists of performing various angular analyses of the top quark decay, and choosing suitable angular distributions to measure the parameters of interest. Additional couplings can be probed within the EFT, including four-fermion couplings, and couplings between the top quark and neutral bosons.

This review will not discuss the top quark mass measurement using the single-top t-channel or the CPT symmetry tests comparing top and antitop quark masses in single-top events (for a recent result, see [108]), since the precision is not yet at the required level for competing with $t\bar{t}$ measurements. This section will cover the other above-mentioned top quark properties using the single-top quark as a probe, reaching a precision similar to or better than that achieved in $t\bar{t}$ measurements.

4.1. Measurement of the CKM Matrix Element $|V_{tb}|$

Because the V_{tb} CKM matrix element is close to unity in the SM, the measurement of V_{tb} is particularly intriguing, and its study is an excellent way to better understand the SM and search for signs of new physics. The measurement of the V_{tb} CKM matrix element is strongly related to the electroweak nature of the single-top production. The cross-section for the single-top production can be used to test the unitarity of the CKM matrix. Assuming the values of $|V_{td}|$ and $|V_{ts}|$ are much smaller than those of $|V_{tb}|$, the measured single-top cross-section can be used to determine $|V_{tb}|$ according to the following formula [40]:

$$|f_{L_V} V_{tb}| = \sqrt{\frac{\sigma^{meas}}{\sigma^{theo}}}, \tag{4}$$

with σ^{meas} denoting the measured cross-section, σ^{th} denoting the SM theoretical cross-section, assuming $|V_{tb}| = 1$, and f_{LV} denoting an anomalous form factor (of the kind vectorial left-handed, as in the SM), which can be different from 1 in new physics models. Such a method was used to reinterpret several single-top cross-sections at 7 and 8 TeV. Their combinations, including ATLAS and CMS results for t-channel, tW production, and s-channel, were performed in the context of the LHC*top*WG, leading to the most precise $|f_{LV} V_{tb}|$ measurement to date, as shown in Figure 26. One can see that the t-channel measurements dominate the combination. The latest t-channel measurement at 13 TeV [38] improves over this combination by approximately 30% in precision, with $|V_{tb}| = 1.014 \pm 0.031$ reported.

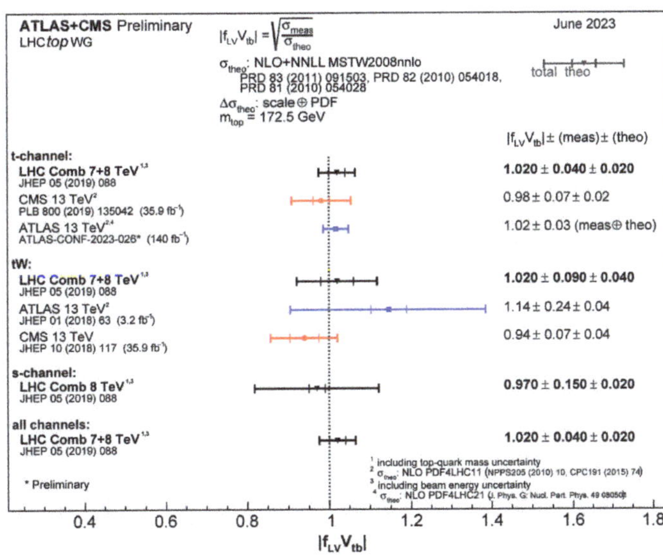

Figure 26. Summary of the ATLAS and CMS extractions of the CKM matrix element V_{tb} from single-top quark measurements [40], compared with theoretical predictions at NLO+NNLL accuracy [71,109,110].

It is possible to release the assumption that $|V_{td}|$ and $|V_{ts}|$ are negligible compared to $|V_{tb}|$. Such a method has also been pursued, consisting of measuring $|V_{tb}|$, $|V_{td}|$ and $|V_{ts}|$ in a model-independent way, using single-top t-channel-enriched events [111]. The main principle of the analysis relies on considering several single-top t-channel signals, according to the presence of a tWb vertex in single-top production ($ST_{b,q}$), in top quark decay ($ST_{q,b}$), or in both ($ST_{b,b}$). Several signal regions, based on the jet and b-tagged jet multiplicities, can be defined and fitted simultaneously. Further discrimination between $ST_{b,q}$, $ST_{q,b}$, and $ST_{b,b}$ is obtained using kinematic and angular properties of the involved processes, using the fact that (1) PDFs are different for each of them, and (2) the presence of an additional b-jet from gluon-splitting can affect top quark reconstruction. Using the constraint of CKM unitarity ($|V_{tb}|^2 + |V_{ts}|^2 + |V_{td}|^2 = 1$), a precision similar to that of the combination [40] is achieved [111] with an integrated luminosity of 35.9 fb^{-1} of 13 TeV proton–proton collisions. The method allows performing the measurements under the constraints of BSM scenarios. The results are compatible with previous measurements and the SM predictions.

4.2. W Boson Polarization Fractions

The V–A structure of the electroweak theory, together with the mass of the particles involved, determine the fractions of longitudinal, left-handed, and right-handed W boson polarization (sometimes called helicity fractions), denoted, respectively, as F_0, F_L, and F_R. Predictions for these fractions computed at NNLO in pQCD are [112] $F_0 = 0.687 \pm 0.005$, $F_L = 0.311 \pm 0.005$, and $F_R = 0.0017 \pm 0.0001$. Experimentally, the fractions can be measured within the W rest frame where the W boson arises from leptonic top decay, using the angle θ^*, defined as the angle between the direction of the charged lepton and the reversed direction of the b-quark. The differential decay rate is:

$$\frac{1}{\Gamma}\frac{d\Gamma}{dcos\theta^*} = \frac{3}{4}(1 - cos^2\theta^*)F_0 + \frac{3}{8}(1 - cos\theta^*)^2 F_L + \frac{3}{8}(1 + cos\theta^*)^2 F_R, \quad (5)$$

with $F_0 + F_L + F_R = 1$. The differential decay rate as a function of $cos\theta^*$ is illustrated in Figure 27.

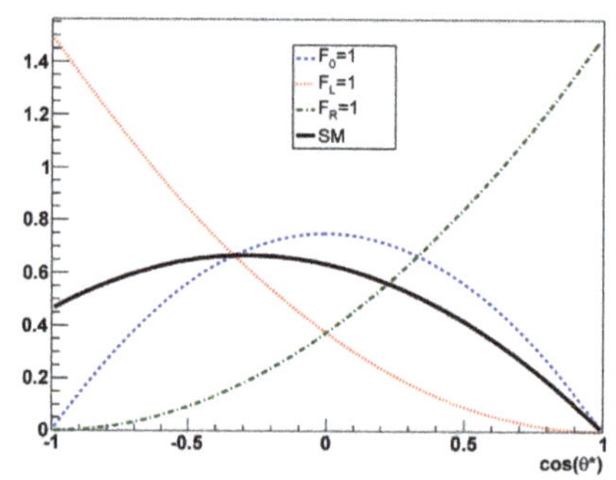

Figure 27. The differential decay rate as a function of $cos\theta^*$ in several scenarios for W boson polarization [113].

The fractions are obtained from a fit of the $cos\theta^*$ distribution to the data. The W boson polarization fractions have been measured at CDF and D0 [114] with a precision on F_0 of the order of 10-15%, using $t\bar{t}$ decay. At the LHC, the single-top production in the t-channel and its large cross-section offer the possibility of measuring the polarization fractions in single-top decay in addition to $t\bar{t}$ decay. The fractions were measured at 8 TeV with CMS [115], as $F_L = 0.298 \pm 0.028(stat) \pm 0.032(syst)$, $F_0 = 0.720 \pm 0.039(stat) \pm 0.037(syst)$, and $F_R = -0.018 \pm 0.019(stat) \pm 0.011(syst)$. The precision achieved with single-top measurements justifies its inclusion in ATLAS and CMS combination of 8 TeV results [116], leading to $F_0 = 0.693 \pm 0.014$, $F_L = 0.315 \pm 0.011$, and $F_R = -0.008 \pm 0.007$. The 7 TeV results were obtained by analyzing $t\bar{t}$ and were not considered since they were expected to bring about negligible improvement.

ATLAS also employed the "generalized helicity fractions and phases" formalism [117] by the means of amplitude decomposition in several angular distributions in the top quark rest frame. Among the parameters measured, the transverse polarization fraction using single-top, and decaying at 7 and 8 TeV [118,119], yields $F_T = F_L + F_R = 0.30 \pm 0.05$ [119] as the best result. ATLAS also measures the phase between amplitudes for longitudinally and transverse W bosons recoiling against left-handed b-quarks [118,119], providing no sign of CP violation. From this formalism, left- and right-handed fractions could in principle be calculated.

4.3. *Wtb* Effective Couplings and Interpretation in the SM EFT

The Wtb effective couplings were also measured, either at CMS (as extracted from the W boson polarization fractions [115] or measured directly [120]) or by ATLAS, by analyzing the single-top amplitudes [118,119] or measuring various angular asymmetries [121]. The Lagrangian describing the Wtb-effective couplings reads [122]:

$$L_{Wtb} = -\frac{g}{\sqrt{2}}\bar{b}\gamma^{\mu}(V_L P_L + V_R P_R)tW^{-}_{\mu} - \frac{g}{\sqrt{2}}\bar{b}\frac{i\sigma^{\mu\nu}q_{\nu}}{m_W}(g_L P_L + g_R P_R)tW^{-}_{\mu} + h.c. \quad (6)$$

where V_L, V_R denote the vectorial left-handed and right-handed Wtb couplings, and g_L, g_R denote the tensorial left-handed and right-handed Wtb couplings (sometimes called, respectively, f_V^L, f_V^R, f_T^L, f_T^R, depending on the convention [120]). In the SM at the LO in pQCD, $V_L = V_{tb}$ while $V_R = g_L = g_R = 0$. The couplings V_R, g_L, g_R are complex and can be CP-odd if their imaginary part is non-zero. The V_{tb} CKM matrix element is inferred from the single-top cross-section measurement, as discussed in Section 4.1.

The analyses have moderate sensitivity to the right-handed vectorial coupling and left-handed tensorial coupling. With a simultaneous fit of both parameters, ATLAS reports $|V_R/V_L| < 0.37$ and $|g_L/V_L| < 0.29$ at 95% CL [119], and CMS reports $f_V^R < 0.16$ and $f_T^L < 0.057$ at 95% CL [120], including f_V^L in the fit using inclusive cross-section information. The best sensitivity on the Wtb couplings is obtained on the g_R coefficient. ATLAS obtained with a simultaneous fit $-0.12 < Re(g_R/V_L) < 0.17$ and $-0.07 < Im(gR/V_L) < 0.06$ at 95% CL [119]. If using single-top cross-section information and assuming a null imaginary part, CMS obtains $|Re(f_T^R)| < 0.046$. These results can be compared with the combination of an 8 TeV W boson polarization fraction (including $t\bar{t}$ channels) [116]: $-0.11 < Re(V_R) < 0.15$, $-0.08 < Re(g_L) < 0.05$, and $-0.04 < Re(g_R) < 0.02$. Since the imaginary part of g_R cannot be accessed easily from the $t\bar{t}$ process and would need a dedicated analysis [122], the single-top measurements, such as [118,119], are irreplaceable.

The results obtained in the effective coupling formalism can be translated into the modern framework of the SM EFT [123], adding all operators to the SM Lagrangian and respecting gauge invariance. The Wtb couplings considered in Equation (6) (V_L, V_R, g_L, g_R) are, respectively, related to the following four dimension-6 operators:

$$O^{(3)}_{\phi q} = \frac{c^{(3)}_{\phi q}}{\Lambda^2} i(\phi^{+}\overleftrightarrow{D}^{I}_{\mu}\phi)(\bar{q}\gamma^{\mu}\tau^{I}q), \quad (7)$$

$$O_{\phi tb} = \frac{c_{\phi tb}}{\Lambda^2}(\phi^{+}\overleftrightarrow{D}^{I}_{\mu}\phi)(\bar{t}\gamma^{\mu}\tau^{I}b), \quad (8)$$

$$O_{tW} = \frac{c_{tW}}{\Lambda^2}(\bar{q}\sigma^{\mu\nu}\tau^{I}t)\tilde{\phi}W^{I}_{\mu\nu}, \quad (9)$$

$$O_{bW} = \frac{c_{bW}}{\Lambda^2}(\bar{q}\sigma^{\mu\nu}\tau^{I}b)\phi W^{I}_{\mu\nu}, \quad (10)$$

using notations from [124]. Results from the combination of the W boson polarization at 8 TeV are [116]: $-3.48 < Re(c_{\phi tb}) < 5.16$, $-0.48 < Re(c_{tW}) < 0.29$, and $-0.96 < Re(c_{bW}) < 0.67$. A translation from the best measurement of $Im(g_R)$ [119] to the EFT formalism using [123] gives $-0.82 < Im(c_{bW}) < 0.70$.

4.4. Top Quark Polarization

Recently, via an analysis of the top quark polarization, ATLAS directly measured the coefficient $Im(c_{tW})$ for the first time [125], using the full Run 2 dataset at 13 TeV. Because of parity conservation in QCD, top quarks in $t\bar{t}$ production are unpolarized, while top quarks are mostly polarized in single-top production. The polarization vector $\vec{P}$ is defined with components $P_i = 2 < S_i >$, where S_i is the top quark spin along the i direction [126], in the top quark rest frame, where the z' direction is defined as the W boson direction, the x' direction is defined as the spectator quark direction projected on the transverse plane, and the y' axis completes the direct basis. On this basis, the values of the polarization vectors are close to $(-0024, 0, 0.965)$ for the top quark and $(-0.073, 0, -0.957)$ for top antiquark produced in the t-channel at NNLO in pQCD [125]. The top quark polarization can be extracted from angular distributions of top decay products defined in the top quark rest frame, given by the following general formula:

$$\frac{1}{\Gamma}\frac{d\Gamma}{dcos\theta_X} = \frac{1}{2}(1 + \alpha_X P_X cos\theta_X), \tag{11}$$

where θ_X is the angle between the top quark spin axis and the direction of motion of the chosen decay X, α_X is the spin analyzing power associated with the X, and P_X is the top quark degree of polarization along the direction of X. The measurement of top quark polarization in [125] is performed using the angular distributions related to the charged lepton (shown to have the largest spin analyzing power, close to 1) arising from the top decay and projected in the previously defined directions. If the top quark polarization had previously been measured at the LHC along the z direction (for instance in [121]), the measurement [125] is the most precise and includes x' and y' directions. For top and antitop quarks, it leads, respectively, to $P_x' = 0.01 \pm 0.18$, $P_y' = -0.029 \pm 0.027$, $P_z' = 0.91 \pm 0.10$, and $P_x' = -0.02 \pm 0.20$, $P_y' = -0.007 \pm 0.051$, $P_z' = -0.79 \pm 0.16$. The polarizations along directions x' and z' are also reported in Figure 28. Using the same angular distributions, ATLAS reports $-0.9 < Re(c_{tW}) < 1.4$ and $-0.8 < Im(c_{tW}) < 0.2$ at 95% CL.

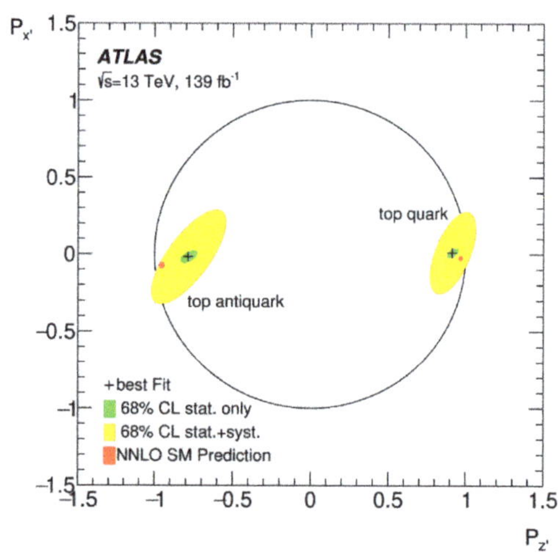

Figure 28. Top quark polarization in the single-top t-channel production along x' and z' directions for the top and antitop quarks [125].

4.5. Discussion on other Couplings with Single-Top Quark Measurements in the SM EFT

If anomalous coupling measurements in single-top quark processes are primarily interesting for Wtb couplings, other couplings are also actively measured, e.g., the coupling between heavy quarks and light quarks, the coupling between heavy quarks and neutral bosons, and the coupling between heavy quarks and leptons. The discussion in this section excludes the FCNC (for a review, see [52]).

In general, single-top production with a boson can help constrain the coupling between top quarks and neutral bosons. The top-Z (resp. top-Higgs) coupling impacts the single-top quark produced in association with a Z boson (resp., a Higgs boson). The top-gluon coupling impacts the tW channel (since tW channel LO diagrams feature one gluon in the initial state) and any production channel considered at NLO, where gluons can be emitted from top quarks. The process of single-top production accompanied by a photon has just been observed and could be used in the near future for measuring the top-γ coupling [82]. It has also been emphasized that the tZq and tHq processes can be greatly impacted by some of these couplings [127]. However, the cross-sections for processes of single-top production in association with bosons ($t + V$) are lower than those of top pairs produced in association with bosons ($t\bar{t} + V$); therefore, analyses of $t + V$ final states are generally swamped by $t\bar{t} + V$ backgrounds. As a consequence, measuring the tZ, tH, and tg couplings requires, for consistency, the modeling of the anomalous couplings in $t + V$ and $t\bar{t} + V$ simulation samples, which will help in constraining the couplings. It is difficult to disentangle what is the exact contribution of single-top production to the sensitivity in these couplings. We will, therefore, limit ourselves to providing some examples, where the contributions of single-top processes are explicitly included. Generic searches for measuring top quark couplings in the multi-lepton lepton final state define many event classes, targeting a great number of EFT operators that impact tZq and tHq processes [128], e.g., nine operators involving two quarks and one or more bosons (with some impacting the Wtb vertex considered at the production level only), as well as seven operators involving two heavy quarks and two leptons. An updated analysis [129] involving more operators needs to be published. The top gluon coupling was considered in [128] by including its impact on gluon radiation at LO. Measurements of EFT operators in $t\bar{t}Z + tZq$ final states [130] include five operators involving two quarks and one or more bosons (including Wtb vertex) and uses machine learning to maximize sensitivity. The Yukawa coupling is measured in $t\bar{t}H$ analyses by including its impact on tHq, as discussed in Section 3.3.

A recent measurement of the t-channel process using full Run 2 data by ATLAS [38] (to be published) sets constraints on the coupling between light and heavy quarks (the $C_{q,Q}^{(1,3)}$ coefficient within the SMEFT framework), in a competitive manner with global fits reinterpreting LHC data.

The EFT is a consistent framework-preserving gauge invariance in a model-independent way [131] (as long as new physics appears at a high energy scale); therefore, there is a tendency to employ the EFT framework more widely, replacing previous anomalous coupling frameworks. The LHC*top*WG, together with the LHC EFT WG, are working on prescriptions toward the combination of direct top quark EFT measurements.

5. Conclusions

After more than 10 years of data-taking with the LHC, the understanding of the physics involving single-top quark processes has undergone a spectacular change. Prior to the LHC, a single-top production was discovered, singling out the t- and s-channels. Nowadays, 14 years on, the differential cross-sections for the t-channel and tW production modes are measured in great detail. The t-channel is routinely used for top quark property measurements. This ranges from the structure of the Wtb vertex to the W boson and top quark polarization, not to mention the top quark mass measurement. The tW production is employed to probe delicate interference effects with the $t\bar{t}$ process. The s-channel process remains to be observed at the LHC, but initial evidence suggests that such an observation is on the horizon. The cross-sections measured for single-top quark production in the t-channel, tW channel, and s-channel by ATLAS and CMS are compared with theoretical predictions in Figure 29.

Run 2 of the LHC offered a new opportunity for observing and exploring the associated production of top quarks and neutral bosons. After its observation, the production of a single-top quark with an associated Z boson was measured differentially for the first time. It is now employed as a probe of various couplings within the SM EFT framework. The associated production with a photon has also been observed with the full Run 2 dataset. The tH processes are used to probe the sign of the top quark Yukawa coupling; however, they have not been observed as yet. The cross-sections measured by ATLAS and CMS for single-top quark production associated with a γ or Z boson are compared with theoretical predictions in Figure 30.

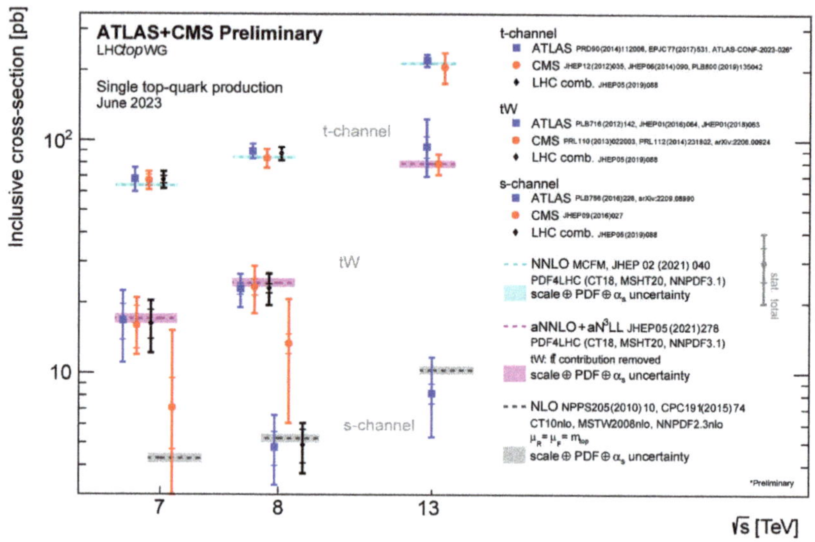

Figure 29. Summary of single-top quark cross-section measurements by ATLAS and CMS in the t-channel, tW production, and s-channel, as functions of the center of mass energy, compared with theoretical predictions at NNLO [11], approximate NNLO+N^{3}LL [12], and NLO [14,132] accuracy, provided by the LHC*top*WG [133].

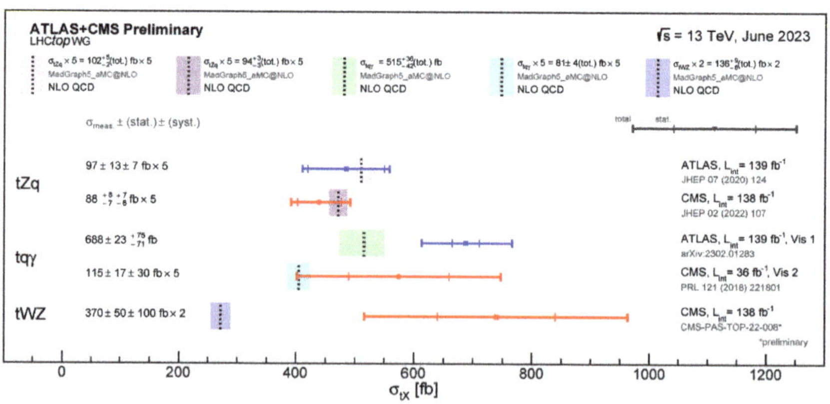

Figure 30. Summary of single-top quark cross-section measurements by ATLAS and CMS in the associated production with a Z boson or a photon compared with theoretical predictions at NLO [78] accuracy; provided by the LHC*top*WG [133].

The Run 3 of the LHC is ongoing, with a center-of-mass energy of 13.6 TeV, resulting in a predicted increase of about 11% in the inclusive cross-section relative to 13 TeV for the three main production modes [11,12,77,134] (and a similar increase for the $t\bar{t}$ process [135,136]). One can expect measurements for all of the processes discussed in this review to be performed at this unprecedented energy, verifying if the data still agree with the SM predictions. The luminosity is expected to accumulate during Run 3 in a way that is at least comparable to Run 2, producing a new dataset that is larger by a factor of 1.4 (around 140 fb^{-1}, collected separately by ATLAS and CMS during Run 2, and 200 fb^{-1} during Run 3). Statistically dominated measurements will profit from this step in center-of-mass energy and expected luminosity, such as the measurement of the $t\gamma q$ process, where a first differential cross-section can be targeted. Hopefully, some of the limitations of the previous measurements will be lifted to improve the precision, provided that additional work on the systematic uncertainties is carried out. For instance, more work is needed on the topic of the parton shower modeling since it is now the largest source of uncertainties in the t-channel measurements and an important source in the measurement of the tW production. Global efforts will also be needed to reduce the systematic uncertainties in the tZq measurement, which is dominated by several sources of large experimental uncertainties. Most of the measurements of the top quark properties in the single-top quark area employ either the t-channel or the tZ production and are impacted by the above uncertainty sources. For the top quark coupling measurements (Wtb vertex and top-boson couplings), it is expected that the movement toward the generalized usage of the EFT will be pursued, allowing the search for new physics in precision measurements in a unified way and allowing for the combination of complementary measurements. Finally, there is hope that the s-channel process could be observed at LHC Run 3 by reducing the uncertainties and refining the analysis techniques. On the other hand, the search for the tH processes will continue, although their observation will have to be postponed to the HL-LHC, where it will remain a challenge [103].

Beyond these extensions of the already engaged single-top quark program, new possibilities can be explored at the HL-LHC. Using boosted top quarks with a jet substructure is one of them (already used for $t\bar{t}$ measurements [137] or in Ref. [90]), since more events will be available in the tails of the distributions to search for new physics [138]. The process of producing three top quarks is occasionally categorized within single-top quark physics. It stands as a minor background in the four top quark process measurements [139,140], and deserves a direct search [141]. The measurement of the tWZ production at Run 2 was the first of its kind, featuring a single-top quark accompanied by two bosons; its observation could be within reach, likely at the HL-LHC, where it could be used to probe the top-boson couplings [142]. And even rarer processes can be reached, i.e., single-top quark production with a combination of two W,Z bosons or photons could be measured beyond tWZ; some studies suggest that the production of single-top quarks through vector boson fusion is another rare process to explore, offering high sensitivity to new top quark couplings [143]. In general, the program of measuring the top quark couplings within the EFT is still in its infancy. One can foresee that the couplings in which the single-top quark area is relevant will be measured systematically at the HL-LHC [144]. For instance, searching for a possible CP violation in the top-Higgs boson coupling will be conducted, where separating tH from $t\bar{t}H$ will be crucial [145], or measuring the top quark couplings to the gauge bosons, such as the top-γ coupling [82]. Combining measurements from other top quark production modes, along with insights from B physics, electroweak, and Higgs boson measurements, will certainly be essential, and could lead to the observation of statistical deviations indicative of physics beyond the SM.

Author Contributions: Writing—review and editing, J.A. and N.C. All authors have read and agreed to the published version of the manuscript.

Funding: This research received no external funding.

Data Availability Statement: No new data were created or analyzed in this study. Data sharing is not applicable to this review.

Conflicts of Interest: The authors declare no conflict of interest.

References

1. CDF Collaboration. Observation of top quark production in $\bar{p}p$ collisions. *Phys. Rev. Lett.* **1995**, *74*, 2626–2631. [CrossRef] [PubMed]
2. D0 Collaboration. Observation of the top quark. *Phys. Rev. Lett.* **1995**, *74*, 2632–2637. [CrossRef] [PubMed]
3. D0 Collaboration. Observation of Single Top Quark Production. *Phys. Rev. Lett.* **2009**, *103*, 092001. [CrossRef] [PubMed]
4. CDF Collaboration. First Observation of Electroweak Single Top Quark Production. *Phys. Rev. Lett.* **2009**, *103*, 092002. [CrossRef]
5. D0 Collaboration. Model-independent measurement of t-channel single top quark production in $p\bar{p}$ collisions at $\sqrt{s} = 1.96$ TeV. *Phys. Lett. B* **2011**, *705*, 313–319. [CrossRef]

6. D0 Collaboration. Evidence for S-Channel Single Top Quark Production in $p\bar{p}$ Collisions at $\sqrt{s}$ = 1.96 TeV. *Phys. Lett. B* **2013**, *726*, 656–664. [CrossRef]
7. CDF Collaboration. Updated Measurement of the Single Top Quark Production Cross Section and Vtb in the Missing Transverse Energy Plus Jets Topology in $p\bar{p}$ Collisions at $\sqrt{s} = 1.96$ TeV. *Phys. Rev. D* **2016**, *93*, 032011. [CrossRef]
8. CDF Collaboration; D0 Collaboration. Observation of s-channel production of single top quarks at the Tevatron. *Phys. Rev. Lett.* **2014**, *112*, 231803. [CrossRef]
9. ATLAS Collaboration. Evidence for single top-quark production in the s-channel in proton-proton collisions at $\sqrt{s}$ =8 TeV with the ATLAS detector using the Matrix Element Method. *Phys. Lett. B* **2016**, *756*, 228–246. [CrossRef]
10. ATLAS Collaboration. Measurement of single top-quark production in the s-channel in proton–proton collisions at $\sqrt{s}$ = 13 TeV with the ATLAS detector. *J. High Energy Phys.* **2023**, *6*, 191. [CrossRef]
11. Campbell, J.; Neumann, T.; Sullivan, Z. Single-top-quark production in the t-channel at NNLO. *J. High Energy Phys.* **2021**, *2*, 40. [CrossRef]
12. Kidonakis, N.; Yamanaka, N. Higher-order corrections for tW production at high-energy hadron colliders. *J. High Energy Phys.* **2021**, *5*, 278. [CrossRef]
13. Aliev, M.; Lacker, H.; Langenfeld, U.; Moch, S.; Uwer, P.; Wiedermann, M. HATHOR: HAdronic Top and Heavy quarks crOss section calculatoR. *Comput. Phys. Commun.* **2011**, *182*, 1034–1046. [CrossRef]
14. Kant, P.; Kind, O.M.; Kintscher, T.; Lohse, T.; Martini, T.; Mölbitz, S.; Rieck, P.; Uwer, P. HatHor for single top-quark production: Updated predictions and uncertainty estimates for single top-quark production in hadronic collisions. *Comput. Phys. Commun.* **2015**, *191*, 74–89. [CrossRef]
15. Giammanco, A. Single top quark production at the LHC. *Rev. Phys.* **2016**, *1*, 1–12. [CrossRef]
16. Giammanco, A.; Schwienhorst, R. Single top-quark production at the Tevatron and the LHC. *Rev. Mod. Phys.* **2018**, *90*, 035001. [CrossRef]
17. ATLAS Collaboration. Top Cross Section Summary Plots—June 2023. 2023. Available online: https://atlas.web.cern.ch/Atlas/GROUPS/PHYSICS/PUBNOTES/ATL-PHYS-PUB-2023-014 (accessed on 24 July 2023).
18. CMS Collaboration. CMS Top Quark Physics Summary Figures. Available online: https://twiki.cern.ch/twiki/bin/view/CMSPublic/PhysicsResultsTOPSummaryFigures (accessed on 24 July 2023).
19. Tait, T.M.P.; Yuan, C.P. Single top quark production as a window to physics beyond the standard model. *Phys. Rev. D* **2000**, *63*, 014018.
20. Drueke, E.; Nutter, J.; Schwienhorst, R.; Vignaroli, N.; Walker, D.G.E.; Yu, J.H. Single Top Production as a Probe of Heavy Resonances. *Phys. Rev. D* **2015**, *91*, 054020. [CrossRef]
21. Burdman, G. Scalars from top condensation models at hadron colliders. *Phys. Rev. Lett.* **1999**, *83*, 2888–2891.
22. Oakes, R.J.; Whisnant, K.; Yang, J.M.; Young, B.L.; Zhang, X. Single top quark production as a probe of R-parity violating SUSY at p p and p anti-p colliders. *Phys. Rev. D* **1998**, *57*, 534–540.
23. Burdman, G.; Dobrescu, B.A.; Ponton, E. Resonances from two universal extra dimensions. *Phys. Rev. D* **2006**, *74*, 075008.
24. Aguilar-Saavedra, J.A. Identifying top partners at LHC. *J. High Energy Phys.* **2009**, *11*, 30. [CrossRef]
25. Hassanain, B.; March-Russell, J.; Rosa, J.G. On the possibility of light string resonances at the LHC and Tevatron from Randall-Sundrum throats. *J. High Energy Phys.* **2009**, *7*, 77. [CrossRef]
26. Atwood, D.; Reina, L.; Soni, A. Phenomenology of two Higgs doublet models with flavor changing neutral currents. *Phys. Rev. D* **1997**, *55*, 3156–3176.
27. Cao, J.J.; Eilam, G.; Frank, M.; Hikasa, K.; Liu, G.L.; Turan, I.; Yang, J.M. SUSY-induced FCNC top-quark processes at the large hadron collider. *Phys. Rev. D* **2007**, *75*, 075021.
28. Agashe, K.; Perez, G.; Soni, A. Collider Signals of Top Quark Flavor Violation from a Warped Extra Dimension. *Phys. Rev. D* **2007**, *75*, 015002.
29. Fontes, D.; Mühlleitner, M.; Romão, J.C.; Santos, R.; Silva, J.a.P.; Wittbrodt, J. The C2HDM revisited. *J. High Energy Phys.* **2018**, *2*, 73. [CrossRef]
30. Bi, X.J.; Dai, Y.B. CP violation in semileptonic decays of the top quark within MSSM. *Eur. Phys. J. C* **2000**, *12*, 125–135.
31. Behr, J.K.; Grohsjean, A. Dark Matter Searches with Top Quarks. *Universe* **2023**, *9*, 16. [CrossRef]
32. CMS Collaboration. Measurement of differential cross sections and charge ratios for t-channel single top quark production in proton–proton collisions at $\sqrt{s} = 13$ TeV. *Eur. Phys. J. C* **2020**, *80*, 370. [CrossRef]
33. CMS Collaboration. Measurement of the t-channel single top quark production cross section in pp collisions at $\sqrt{s} = 7$ TeV. *Phys. Rev. Lett.* **2011**, *107*, 091802. [CrossRef] [PubMed]
34. ATLAS Collaboration. Measurement of the t-channel single top-quark production cross section in pp collisions at $\sqrt{s} = 7$ TeV with the ATLAS detector. *Phys. Lett. B* **2012**, *717*, 330–350. [CrossRef]
35. CMS Collaboration. Measurement of the Single-Top-Quark t-Channel Cross Section in pp Collisions at $\sqrt{s} = 7$ TeV. *J. High Energy Phys.* **2012**, *12*, 35. [CrossRef]
36. Cowan, G.; Cranmer, K.; Gross, E.; Vitells, O. Asymptotic formulae for likelihood-based tests of new physics. *Eur. Phys. J. C* **2011**, *71*, 1554; Erratumin *Eur. Phys. J.C* **2013**, *73*, 2501. [CrossRef]

37. ATLAS Collaboration, CMS Collaboration, and the LHC Higgs Combination Group. Procedure for the LHC Higgs Boson Search Combination in Summer 2011. *CMS-NOTE-2011-005, ATL-PHYS-PUB-2011-11*. Available online: https://cds.cern.ch/record/1379837 (accessed on 24 July 2023).
38. ATLAS Collaboration. Measurement of t-Channel Production of Single Top Quarks and Antiquarks in pp Collisions at 13 TeV Using the Full ATLAS Run 2 Dataset. *ATLAS-CONF-2023-026*. Available online: https://atlas.web.cern.ch/Atlas/GROUPS/PHYSICS/CONFNOTES/ATLAS-CONF-2023-026 (accessed on 24 July 2023).
39. ATLAS Collaboration. Measurement of t-Channel Single-Top-Quark Production in pp Collisions at $\sqrt{s} = 5.02$ TeV with the ATLAS Detector. *ATLAS-CONF-2023-033*. Available online: https://atlas.web.cern.ch/Atlas/GROUPS/PHYSICS/CONFNOTES/ATLAS-CONF-2023-033 (accessed on 24 July 2023).
40. ATLAS Collaboration; CMS Collaboration. Combinations of Single-Top-Quark Production Cross-Section Measurements and $|f_{LV}V_{tb}|$ Determinations at $\sqrt{s} = 7$ and 8 TeV with the ATLAS and CMS Experiments. *J. High Energy Phys.* **2019**, *5*, 88. [CrossRef]
41. CMS Collaboration. Measurement of the single top quark and antiquark production cross sections in the t channel and their ratio in proton-proton collisions at $\sqrt{s} = 13$ TeV. *Phys. Lett. B* **2020**, *800*, 135042. [CrossRef]
42. Fischer, N.; Prestel, S.; Ritzmann, M.; Skands, P. Vincia for Hadron Colliders. *Eur. Phys. J. C* **2016**, *76*, 589. [CrossRef] [PubMed]
43. Dasgupta, M.; Dreyer, F.A.; Hamilton, K.; Monni, P.F.; Salam, G.P.; Soyez, G. Parton showers beyond leading logarithmic accuracy. *Phys. Rev. Lett.* **2020**, *125*, 052002. [CrossRef] [PubMed]
44. Ferrario Ravasio, S.; Hamilton, K.; Karlberg, A.; Salam, G.P.; Scyboz, L.; Soyez, G. A parton shower with higher-logarithmic accuracy for soft emissions. *arXiv* **2023**, arXiv:2307.11142 https://doi.org/10.48550/arXiv.2307.1114.
45. Corcella, G.; Franceschini, R.; Kim, D. Fragmentation Uncertainties in Hadronic Observables for Top-quark Mass Measurements. *Nucl. Phys. B* **2018**, *929*, 485–526. [CrossRef]
46. ATLAS Collaboration. Comprehensive measurements of t-channel single top-quark production cross sections at $\sqrt{s} = 7$ TeV with the ATLAS detector. *Phys. Rev. D* **2014**, *90*, 112006, [CrossRef]
47. ATLAS Collaboration. Fiducial, total and differential cross-section measurements of t-channel single top-quark production in pp collisions at 8 TeV using data collected by the ATLAS detector. *Eur. Phys. J. C* **2017**, *77*, 531. [CrossRef]
48. CMS Collaboration. Measurement of the t-channel single-top-quark production cross section and of the $|V_{tb}|$ CKM matrix element in pp collisions at $\sqrt{s}$= 8 TeV. *J. High Energy Phys.* **2014**, *6*, 90. [CrossRef]
49. ATLAS Collaboration. Measurement of the inclusive cross-sections of single top-quark and top-antiquark t-channel production in pp collisions at $\sqrt{s} = 13$ TeV with the ATLAS detector. *J. High Energy Phys.* **2017**, *4*, 86. [CrossRef]
50. Nocera, E.R.; Ubiali, M.; Voisey, C. Single Top Production in PDF fits. *J. High Energy Phys.* **2020**, *5*, 67. [CrossRef]
51. ATLAS Collaboration. Measurement of single top-quark production in association with a W boson in the single-lepton channel at $\sqrt{s} = 8$ TeV with the ATLAS detector. *Eur. Phys. J. C* **2021**, *81*, 720. [CrossRef]
52. Castro, N.F.; Skovpen, K. Flavour-Changing Neutral Scalar Interactions of the Top Quark. *Universe* **2022**, *8*, 609. [CrossRef]
53. White, C.D.; Frixione, S.; Laenen, E.; Maltoni, F. Isolating Wt production at the LHC. *J. High Energy Phys.* **2009**, *11*, 74. [CrossRef]
54. Aguilar-Saavedra, J.A. Single top quark production at LHC with anomalous Wtb couplings. *Nucl. Phys. B* **2008**, *804*, 160–192. [CrossRef]
55. ATLAS Collaboration. Evidence for the associated production of a W boson and a top quark in ATLAS at $\sqrt{s} = 7$ TeV. *Phys. Lett. B* **2012**, *716*, 142–159. [CrossRef]
56. CMS Collaboration. Evidence for Associated Production of a Single Top Quark and W Boson in pp Collisions at $\sqrt{s} = 7$ TeV. *Phys. Rev. Lett.* **2013**, *110*, 022003. [CrossRef] [PubMed]
57. ATLAS Collaboration. Measurement of the production cross-section of a single top quark in association with a W boson at 8 TeV with the ATLAS experiment. *J. High Energy Phys.* **2016**, *1*, 64. [CrossRef]
58. CMS Collaboration. Observation of the associated production of a single top quark and a W boson in pp collisions at $\sqrt{s} = 8$ TeV. *Phys. Rev. Lett.* **2014**, *112*, 231802. [CrossRef]
59. ATLAS Collaboration. Measurement of the cross-section for producing a W boson in association with a single top quark in pp collisions at $\sqrt{s} = 13$ TeV with ATLAS. *J. High Energy Phys.* **2018**, *1*, 63. [CrossRef]
60. CMS Collaboration. Measurement of the production cross section for single top quarks in association with W bosons in proton-proton collisions at $\sqrt{s} = 13$ TeV. *J. High Energy Phys.* **2018**, *10*, 117. [CrossRef]
61. ATLAS Collaboration. Measurement of differential cross-sections of a single top quark produced in association with a W boson at $\sqrt{s} = 13$ TeV with ATLAS. *Eur. Phys. J. C* **2018**, *78*, 186, [CrossRef] [PubMed]
62. CMS Collaboration. Measurement of inclusive and differential cross sections for single top quark production in association with a W boson in proton-proton collisions at $\sqrt{s} = 13$ TeV. *J. High Energy Phys.* **2023**, *7*, 46. [CrossRef]
63. CMS Collaboration. Observation of tW production in the single-lepton channel in pp collisions at $\sqrt{s} = 13$ TeV. *J. High Energy Phys.* **2021**, *11*, 111. [CrossRef]
64. ATLAS Collaboration. Search for the production of single vector-like and excited quarks in the Wt final state in pp collisions at $\sqrt{s} = 8$ TeV with the ATLAS detector. *J. High Energy Phys.* **2016**, *2*, 110. [CrossRef]
65. CMS Collaboration. Search for a heavy resonance decaying into a top quark and a W boson in the lepton+jets final state at $\sqrt{s} = 13$ TeV. *J. High Energy Phys.* **2022**, *4*, 48. [CrossRef]
66. Baur, U.; Spira, M.; Zerwas, P.M. Excited Quark and Lepton Production at Hadron Colliders. *Phys. Rev. D* **1990**, *42*, 815–824. [CrossRef] [PubMed]

67. Brochet, S.; Delaere, C.; François, B.; Lemaître, V.; Mertens, A.; Saggio, A.; Vidal Marono, M.; Wertz, S. MoMEMta, a modular toolkit for the Matrix Element Method at the LHC. *Eur. Phys. J. C* **2019**, *79*, 126. [CrossRef]
68. ATLAS Collaboration. Probing the quantum interference between singly and doubly resonant top-quark production in pp collisions at $\sqrt{s} = 13$ TeV with the ATLAS detector. *Phys. Rev. Lett.* **2018**, *121*, 152002. [CrossRef] [PubMed]
69. Ježo, T.; Lindert, J.M.; Nason, P.; Oleari, C.; Pozzorini, S. An NLO+PS generator for $t\bar{t}$ and Wt production and decay including non-resonant and interference effects. *Eur. Phys. J. C* **2016**, *76*, 691. [CrossRef]
70. Ježo, T.; Lindert, J.M.; Pozzorini, S. Resonance-aware NLOPS matching for off-shell $t\bar{t} + tW$ production with semileptonic decays. *arXiv* **2023**, arXiv:2307.15653. https://doi.org/10.48550/arXiv.2307.15653.
71. Kidonakis, N. NNLL resummation for s-channel single top quark production. *Phys. Rev. D* **2010**, *81*, 054028. [CrossRef]
72. ATLAS Collaboration. Search for s-channel single top-quark production in proton–proton collisions at $\sqrt{s} = 8$ TeV with the ATLAS detector. *Phys. Lett. B* **2015**, *740*, 118–136. [CrossRef]
73. Kondo, K. Dynamical Likelihood Method for Reconstruction of Events with Missing Momentum. 1: Method and Toy Models. *J. Phys. Soc. Jap.* **1988**, *57*, 4126–4140. [CrossRef]
74. D0 Collaboration. A precision measurement of the mass of the top quark. *Nature* **2004**, *429*, 638–642. [CrossRef]
75. CDF Collaboration. Precision measurement of the top quark mass from dilepton events at CDF II. *Phys. Rev. D* **2007**, *75*, 031105. [CrossRef]
76. CMS Collaboration. Search for s channel single top quark production in pp collisions at $\sqrt{s} = 7$ and 8 TeV. *J. High Energy Phys.* **2016**, *9*, 27. [CrossRef]
77. Liu, Z.L.; Gao, J. s-channel single top quark production and decay at next-to-next-to-leading-order in QCD. *Phys. Rev. D* **2018**, *98*, 071501. [CrossRef]
78. Alwall, J.; Frederix, R.; Frixione, S.; Hirschi, V.; Maltoni, F.; Mattelaer, O.; Shao, H.S.; Stelzer, T.; Torrielli, P.; Zaro, M. The automated computation of tree-level and next-to-leading order differential cross sections, and their matching to parton shower simulations. *J. High Energy Phys.* **2014**, *7*, 79. [CrossRef]
79. CMS Collaboration. Evidence for the associated production of a single top quark and a photon in proton-proton collisions at $\sqrt{s} = 13$ TeV. *Phys. Rev. Lett.* **2018**, *121*, 221802. [CrossRef]
80. Kidonakis, N.; Yamanaka, N. QCD corrections in $tq\gamma$ production at hadron colliders. *Eur. Phys. J. C* **2022**, *82*, 670. [CrossRef]
81. ATLAS Collaboration. Observation of single-top-quark production in association with a photon using the ATLAS detector. *arXiv* **2023**, arXiv:2302.01283.
82. Fael, M.; Gehrmann, T. Probing top quark electromagnetic dipole moments in single-top-plus-photon production. *Phys. Rev. D* **2013**, *88*, 033003. [CrossRef]
83. ATLAS Collaboration. Measurements of inclusive and differential cross-sections of combined $t\bar{t}\gamma$ and $tW\gamma$ production in the eμ channel at 13 TeV with the ATLAS detector. *J. High Energy Phys.* **2020**, *9*, 49. [CrossRef]
84. ATLAS Collaboration. Measurement of the production cross-section of a single top quark in association with a Z boson in proton–proton collisions at 13 TeV with the ATLAS detector. *Phys. Lett. B* **2018**, *780*, 557–577. [CrossRef]
85. CMS Collaboration. Measurement of the associated production of a single top quark and a Z boson in pp collisions at $\sqrt{s} = 13$ TeV. *Phys. Lett. B* **2018**, *779*, 358–384. [CrossRef]
86. Kidonakis, N.; Yamanaka, N. Soft-gluon corrections for tqZ production. *Phys. Lett. B* **2023**, *838*, 137708. [CrossRef]
87. CMS Collaboration. Inclusive and differential cross section measurements of single top quark production in association with a Z boson in proton-proton collisions at $\sqrt{s} = 13$ TeV. *J. High Energy Phys.* **2022**, *2*, 107. [CrossRef]
88. CMS Collaboration. Observation of Single Top Quark Production in Association with a Z Boson in Proton-Proton Collisions at $\sqrt{s} =13$ TeV. *Phys. Rev. Lett.* **2019**, *122*, 132003. [CrossRef]
89. ATLAS Collaboration. Observation of the associated production of a top quark and a Z boson in pp collisions at $\sqrt{s} = 13$ TeV with the ATLAS detector. *J. High Energy Phys.* **2020**, *7*, 124. [CrossRef]
90. CMS Collaboration. Evidence for tWZ Production in Proton-Proton Collisions at $\sqrt{s} = 13$ TeV in Multilepton Final States. *CMS-PAS-TOP-22-008*. Available online: https://cms-results.web.cern.ch/cms-results/public-results/preliminary-results/TOP-22-008/index.html (accessed on 24 July 2023).
91. ATLAS Collaboration. Studies on Top-Quark Monte Carlo Modelling for Top2016. Available online: https://atlas.web.cern.ch/Atlas/GROUPS/PHYSICS/PUBNOTES/ATL-PHYS-PUB-2016-020/ (accessed on 24 July 2023).
92. LHC Higgs Cross Section Working Group. Handbook of LHC Higgs Cross Sections: 4. Deciphering the Nature of the Higgs Sector. *arXiv* **2016**, arXiv:1610.07922. https://doi.org/10.23731/CYRM-2017-002.
93. Forslund, M.; Kidonakis, N. Soft-gluon corrections for the associated production of a single top quark and a Higgs boson. *Phys. Rev. D* **2021**, *104*, 034024. [CrossRef]
94. CMS Collaboration. Search for the associated production of a Higgs boson with a single top quark in proton-proton collisions at $\sqrt{s} = 8$ TeV. *J. High Energy Phys.* **2016**, *6*, 177. [CrossRef]
95. Farina, M.; Grojean, C.; Maltoni, F.; Salvioni, E.; Thamm, A. Lifting degeneracies in Higgs couplings using single top production in association with a Higgs boson. *J. High Energy Phys.* **2013**, *5*, 22. [CrossRef]
96. ATLAS Collaboration. Measurement of the properties of Higgs boson production at $\sqrt{s} = 13$ TeV in the $H \to \gamma\gamma$ channel using 139 fb^{-1} of pp collision data with the ATLAS experiment. *J. High Energy Phys.* **2023**, *7*, 88. [CrossRef]

97. CMS Collaboration. Measurements of Higgs boson production cross sections and couplings in the diphoton decay channel at $\sqrt{s}$ = 13 TeV. *J. High Energy Phys.* **2021**, *7*, 27. [CrossRef]
98. Berger, N.; Bertella, C.; Calvet, T.P.; Calvetti, M.; Dao, V.; Delmastro, M.; Duehrssen-Debling, M.; Francavilla, P.; Haddad, Y.; Kivernyk, O.; et al. Simplified Template Cross Sections—Stage 1.1. *arXiv* **2019**, arXiv:1906.02754. https://doi.org/10.48550/arXiv.1906.02754.
99. CMS Collaboration. Measurement of the Higgs boson production rate in association with top quarks in final states with electrons, muons, and hadronically decaying tau leptons at $\sqrt{s} = 13$ TeV. *Eur. Phys. J. C* **2021**, *81*, 378. [CrossRef] [PubMed]
100. CMS Collaboration. Search for associated production of a Higgs boson and a single top quark in proton-proton collisions at $\sqrt{s} = 13$ TeV. *Phys. Rev. D* **2019**, *99*, 092005. [CrossRef]
101. CMS Collaboration. A portrait of the Higgs boson by the CMS experiment ten years after the discovery. *Nature* **2022**, *607*, 60–68. [CrossRef] [PubMed]
102. ATLAS Collaboration. A detailed map of Higgs boson interactions by the ATLAS experiment ten years after the discovery. *Nature* **2022**, *607*, 52–59; Erratum in *Nature* **2022**, *612*, E24. [CrossRef] [PubMed]
103. CMS Collaboration. Sensitivity Projections for Higgs Boson Properties Measurements at the HL-LHC. *CMS-PAS-FTR-18-011*. Available online: https://cms-results.web.cern.ch/cms-results/public-results/preliminary-results/FTR-18-011/index.html (accessed on 24 July 2023).
104. ATLAS Collaboration. *CP* Properties of Higgs Boson Interactions with Top Quarks in the $t\bar{t}H$ and tH Processes Using $H \to \gamma\gamma$ with the ATLAS Detector. *Phys. Rev. Lett.* **2020**, *125*, 061802, [CrossRef]
105. ATLAS Collaboration. Probing the *CP* nature of the top-Higgs Yukawa coupling in $t\bar{t}H$ and tH events with $H \to b\bar{b}$ decays using the ATLAS detector at the LHC. *arXiv* **2023**, arXiv:2303.05974. https://doi.org/10.48550/arXiv.2303.05974.
106. CMS Collaboration. Measurements of $t\bar{t}H$ Production and the CP Structure of the Yukawa Interaction between the Higgs Boson and Top Quark in the Diphoton Decay Channel. *Phys. Rev. Lett.* **2020**, *125*, 061801. [CrossRef]
107. CMS Collaboration. Search for CP violation in tt̄H and tH production in multilepton channels in proton-proton collisions at $\sqrt{s}$ = 13 TeV. *J. High Energy Phys.* **2023**, *7*, 92. [CrossRef]
108. CMS Collaboration. Measurement of the top quark mass using events with a single reconstructed top quark in pp collisions at $\sqrt{s}$ = 13 TeV. *J. High Energy Phys.* **2021**, *12*, 161. [CrossRef]
109. Kidonakis, N. Next-to-next-to-leading-order collinear and soft gluon corrections for t-channel single top quark production. *Phys. Rev. D* **2011**, *83*, 091503. [CrossRef]
110. Kidonakis, N. Two-loop soft anomalous dimensions for single top quark associated production with a W^- or H^-. *Phys. Rev. D* **2010**, *82*, 054018. [CrossRef]
111. CMS Collaboration. Measurement of CKM matrix elements in single top quark *t*-channel production in proton-proton collisions at $\sqrt{s} = 13$ TeV. *Phys. Lett. B* **2020**, *808*, 135609. [CrossRef]
112. Czarnecki, A.; Korner, J.G.; Piclum, J.H. Helicity fractions of W bosons from top quark decays at NNLO in QCD. *Phys. Rev. D* **2010**, *81*, 111503. [CrossRef]
113. CMS Collaboration. Measurement of the W boson helicity fractions in the decays of top quark pairs to lepton + jets final states produced in pp collisions at $\sqrt{s} =$ 8TeV. *Phys. Lett. B* **2016**, *762*, 512–534. [CrossRef]
114. CDF Collaboration; D0 Collaboration. Combination of CDF and D0 measurements of the *W* boson helicity in top quark decays. *Phys. Rev. D* **2012**, *85*, 071106, [CrossRef]
115. CMS Collaboration. Measurement of the W boson helicity in events with a single reconstructed top quark in pp collisions at $\sqrt{s} = 8$ TeV. *J. High Energy Phys.* **2015**, *1*, 53. [CrossRef]
116. CDF Collaboration; D0 Collaboration. Combination of the W boson polarization measurements in top quark decays using ATLAS and CMS data at $\sqrt{s} = 8$ TeV. *J. High Energy Phys.* **2020**, *8*, 51. [CrossRef]
117. Boudreau, J.; Escobar, C.; Mueller, J.; Sapp, K.; Su, J. Single top quark differential decay rate formulae including detector effects. *arXiv* **2013**, arXiv:1304.5639. https://doi.org/10.48550/arXiv.1304.5639.
118. ATLAS Collaboration. Search for anomalous couplings in the *Wtb* vertex from the measurement of double differential angular decay rates of single top quarks produced in the *t*-channel with the ATLAS detector. *J. High Energy Phys.* **2016**, *4*, 23. [CrossRef]
119. ATLAS Collaboration. Analysis of the *Wtb* vertex from the measurement of triple-differential angular decay rates of single top quarks produced in the *t*-channel at $\sqrt{s}$ = 8 TeV with the ATLAS detector. *J. High Energy Phys.* **2017**, *12*, 17. [CrossRef]
120. CMS Collaboration. Search for anomalous Wtb couplings and flavour-changing neutral currents in t-channel single top quark production in pp collisions at $\sqrt{s} = 7$ and 8 TeV. *J. High Energy Phys.* **2017**, *2*, 28. [CrossRef]
121. ATLAS Collaboration. Probing the W tb vertex structure in t-channel single-top-quark production and decay in pp collisions at $\sqrt{s} = 8$ TeV with the ATLAS detector. *J. High Energy Phys.* **2017**, *4*, 124. [CrossRef]
122. Aguilar-Saavedra, J.A.; Carvalho, J.; Castro, N.F.; Veloso, F.; Onofre, A. Probing anomalous Wtb couplings in top pair decays. *Eur. Phys. J. C* **2007**, *50*, 519–533.
123. Buckley, A.; Englert, C.; Ferrando, J.; Miller, D.J.; Moore, L.; Russell, M.; White, C.D. Constraining top quark effective theory in the LHC Run II era. *J. High Energy Phys.* **2016**, *4*, 15. [CrossRef]
124. Grzadkowski, B.; Iskrzynski, M.; Misiak, M.; Rosiek, J. Dimension-Six Terms in the Standard Model Lagrangian. *J. High Energy Phys.* **2010**, *10*, 85. [CrossRef]

125. ATLAS Collaboration. Measurement of the polarisation of single top quarks and antiquarks produced in the t-channel at $\sqrt{s}$ = 13 TeV and bounds on the tWb dipole operator from the ATLAS experiment. *J. High Energy Phys.* **2022**, *11*, 40. [CrossRef]
126. Aguilar-Saavedra, J.A.; Amor Dos Santos, S. New directions for top quark polarization in the *t*-channel process. *Phys. Rev. D* **2014**, *89*, 114009. [CrossRef]
127. Degrande, C.; Maltoni, F.; Mimasu, K.; Vryonidou, E.; Zhang, C. Single-top associated production with a *Z* or *H* boson at the LHC: The SMEFT interpretation. *J. High Energy Phys.* **2018**, *10*, 5. [CrossRef]
128. CMS Collaboration. Search for new physics in top quark production with additional leptons in proton-proton collisions at $\sqrt{s} = 13$ TeV using effective field theory. *J. High Energy Phys.* **2021**, *3*, 95. [CrossRef]
129. CMS Collaboration. Search for physics beyond the standard model in top quark production with additional leptons in the context of effective field theory. *arXiv* **2023**, arXiv:2307.15761. https://doi.org/10.48550/arXiv.2307.15761.
130. CMS Collaboration. Probing effective field theory operators in the associated production of top quarks with a Z boson in multilepton final states at $\sqrt{s}$ = 13 TeV. *J. High Energy Phys.* **2021**, *12*, 83. [CrossRef]
131. Degrande, C.; Greiner, N.; Kilian, W.; Mattelaer, O.; Mebane, H.; Stelzer, T.; Willenbrock, S.; Zhang, C. Effective Field Theory: A Modern Approach to Anomalous Couplings. *Ann. Phys.* **2013**, *335*, 21–32. [CrossRef]
132. Campbell, J.M.; Ellis, R.K. MCFM at the Tevatron and the LHC. *Nucl. Phys. B Proc. Suppl.* **2010**, *205-206*, 10–15. [CrossRef]
133. LHC Top Working Group. LHCTopWG Summary Plots. Available online: https://twiki.cern.ch/twiki/bin/view/LHCPhysics/LHCTopWGSummaryPlots (accessed on 24 July 2023).
134. LHC Top Working Group. Predictions at NNLO of Single Top-Quark Production Cross-Sections. Available online: https://twiki.cern.ch/twiki/bin/view/LHCPhysics/SingleTopNNLORef (accessed on 24 July 2023).
135. LHC Top Working Group. NNLO+NNLL Top-Quark-Pair Cross Sections. Available online: https://twiki.cern.ch/twiki/bin/view/LHCPhysics/TtbarNNLO (accessed on 24 July 2023).
136. Czakon, M.; Mitov, A. Top++: A Program for the Calculation of the Top-Pair Cross-Section at Hadron Colliders. *Comput. Phys. Commun.* **2014**, *185*, 2930. [CrossRef]
137. CMS Collaboration. Measurement of differential $t\bar{t}$ production cross sections in the full kinematic range using lepton+jets events from proton-proton collisions at $\sqrt{s}$ = 13 TeV. *Phys. Rev. D* **2021**, *104*, 092013. [CrossRef]
138. Aguilar-Saavedra, J.A.; Mangano, M.L. New physics with boosted single top production at the LHC and future colliders. *Eur. Phys. J. C* **2020**, *80*, 5. [CrossRef]
139. ATLAS Collaboration. Observation of four-top-quark production in the multilepton final state with the ATLAS detector. *Eur. Phys. J. C* **2023**, *83*, 496. [CrossRef]
140. CMS Collaboration. Observation of four top quark production in proton-proton collisions at $\sqrt{s}$ = 13 TeV. *arXiv* **2023**, arXiv:2305.13439. https://doi.org/10.48550/arXiv.2305.13439.
141. Ahmed, I.; Bi, N.; Ather, M.W.; Amjad, M.S. Observability of triple top quark signal at future hadron colliders. *Prog. Theor. Exp. Phys.* **2023**, *2023*, 053B01. [CrossRef]
142. Faham, H.E.; Maltoni, F.; Mimasu, K.; Zaro, M. Single top production in association with a WZ pair at the LHC in the SMEFT. *J. High Energy Phys.* **2022**, *1*, 100. [CrossRef]
143. Maltoni, F.; Mantani, L.; Mimasu, K. Top-quark electroweak interactions at high energy. *J. High Energy Phys.* **2019**, *10*, 4. [CrossRef]
144. Durieux, G.; Camacho, A.G.; Mantani, L.; Miralles, V.; López, M.M.; Llácer Moreno, M.; Poncelet, R.; Vryonidou, E.; Vos, M. Snowmass White Paper: Prospects for the measurement of top-quark couplings. In Proceedings of the Snowmass 2021, Washington, DC, USA, 17–27 July 2022.
145. Bahl, H.; Bechtle, P.; Heinemeyer, S.; Katzy, J.; Klingl, T.; Peters, K.; Saimpert, M.; Stefaniak, T.; Weiglein, G. Indirect $\mathcal{CP}$ probes of the Higgs-top-quark interaction: Current LHC constraints and future opportunities. *J. High Energy Phys.* **2020**, *11*, 127. [CrossRef]

universe

MDPI

Article

Recent Cross-Section Measurements of Top-Quark Pair Production in Association with Gauge Bosons

Joshuha Thomas-Wilsker

Institute of High Energy Physics, Chinese Academy of Sciences, Beijing 100049, China; jthomasw@cern.ch

Abstract: This article reviews recent cross-section measurements of $t\bar{t}$ production in association with a photon, W or Z boson at the Large Hadron Collider (LHC). All measurements reviewed use proton–proton (pp) datasets collected by the ATLAS and CMS experiments between 2016 and 2018 from collisions at a centre-of-mass energy of 13 TeV during the LHC Run 2. Differential and inclusive cross-section measurements are discussed along with the constraints on the effective field theory operators accessible through each process. Finally, we discuss the potential for measurements of these processes at future colliders.

Keywords: top quark; pair production; cross-section; EFT; ttX; LHC; CMS; ATLAS

1. Introduction

The top quark has several unique features that distinguish it from other Standard Model (SM) particles. With its electroweak (EW) scale mass of approximately 172 GeV it is by far the most massive of the fundamental SM particles. This mass, along with an associated Yukawa coupling value close to unity, suggests it may have a special role in the EW symmetry-breaking mechanism. It also has a uniquely short lifetime of $\mathcal{O}(10^{-25})$ seconds which prevents it from hadronising before it decays[1], making it the only quark for which it is possible to study bare quark properties via its decay products.

This unconventional particle provides us with a tool with which we can scrutinise predictions of SM parameters and test a plethora of Beyond the Standard Model (BSM) hypotheses. Several model-dependent searches for BSM physics look for deviations in top-pair production rates and could verify theoretical models that predict the existence of top super-partners, vector-like quarks or even Dark Matter. There are also many model-independent searches that use an effective field theory framework to search for anomalous couplings. Additionally, there are many measurements at the LHC for which SM top-quark processes are important backgrounds and therefore also benefit from improved measurements in the top sector.

In proton–proton collisions at the LHC, the dominant top-quark production mechanism produces top quarks in pairs via the QCD process $gg \rightarrow t\bar{t}$. Due to the CKM matrix element $|V_{tb}|$ being so large, the top-quark decays almost exclusively via the process $t \rightarrow bW$. Thus, most top-quark pairs are produced via the interaction $gg \rightarrow t\bar{t} \rightarrow bW^{+}bW^{-}$. The $t\bar{t}$ process is often categorised according to the decay of the two W bosons. These categories are referred to as dileptonic, semi-leptonic or full hadronic, and are often studied independently due to the varying backgrounds and final state signatures.

The focus of this article is on $t\bar{t}$ production in association with an additional gauge boson ($t\bar{t}X$), as exemplified in Figure 1. More explicitly, the latest ATLAS and CMS cross-section measurements of $t\bar{t}$ production in association with either a photon (γ), W or Z boson. These measurements typically assume SM-like processes to obtain inclusive and differential cross-sections; however, several of them also provide interpretations using the Standard Model Effective Field Theory (SMEFT) framework [2,3]. These processes provide a deep insight into the nature of the couplings in the top-quark interactions with gauge boson.

Citation: Thomas-Wilsker, J. Recent Cross-Section Measurements of Top-Quark Pair Production in Association with Gauge Boson. *Universe* **2023**, *9*, 39. https://doi.org/10.3390/universe9010039

Academic Editors: Efe Yazgan, Pedro Ferreira da Silva and Santiago Peris

Received: 23 October 2022
Revised: 19 December 2022
Accepted: 20 December 2022
Published: 6 January 2023

The publications discussed focus on cross-section measurements performed using datasets collected during the LHC Run 2, where high energy (13 TeV) collisions and huge datasets (approximately 140 fb^{-1} integrated luminosity) make it possible investigate these rare $t\bar{t}$ processes in more detail than ever before. The future of these measurements is also discussed, focusing on their potential at the HL-LHC and the main future collider candidates.

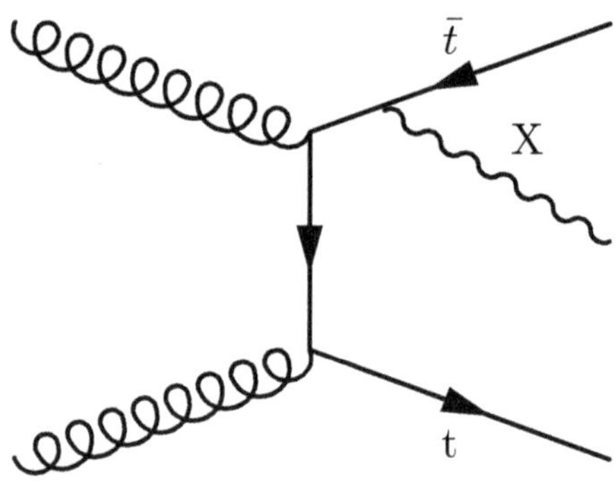

Figure 1. Leading-order Feynman diagram for gluon-gluon top-pair production ($gg \rightarrow t\bar{t}$) in association with a boson (X).

2. $t\bar{t}Z$ Measurements

Inclusive and differential measurements of the $t\bar{t}Z$ cross-section are interesting because they directly probe the coupling between the top quark and the neutral EW Z boson, also known as the t-Z coupling. Furthermore, several BSM theories [4,5] also predict anomalous neutral EW top-quark couplings that can drastically change the amplitude and subsequently the measured cross-section. Such couplings have also been interpreted using an effective field theory (EFT) approach [6]. The attraction here is that the EFT approach provides a model-independent way to interpret possible deviations in a cross-section measurement from its SM value.

This process is also an important background for several SM measurements, for example single-top production in association with a Z boson, $t\bar{t}H$ and many BSM searches [7]. A precise measurement of the process is therefore beneficial to analyses looking to minimise the uncertainties associated with this process.

The first measurements of $t\bar{t}Z$ were performed by ATLAS and CMS at 8 TeV. However, both ATLAS and CMS have also measured this process using partial Run 2 datasets of 36.1 fb^{-1} and 35.9 fb^{-1}, respectively, from 13 TeV collisions, where the production rate increases approximately by a factor of 4 [8,9].

Events were selected with two or more leptons to simultaneously extract the $t\bar{t}Z$ and $t\bar{t}W$ production cross-sections. The 3 and 4 lepton categories are the most sensitive to the $t\bar{t}Z$ process. Observed and expected significance values in both experiments for the $t\bar{t}Z$ process are well above 5σ in these measurements. ATLAS measured the cross-section to be $\sigma_{t\bar{t}Z} = 0.95 \pm 0.08(\text{stat}) \pm 0.10(\text{syst})$ pb while CMS measured a value of $\sigma_{t\bar{t}Z} = 0.99^{+0.09}_{-0.08}(\text{stat}) \pm^{0.12}_{0.10}(\text{syst})$ pb. One can see that, due to the large dataset, the statistical uncertainty is dramatically reduced and the systematic uncertainty on this result is already of a similar size. CMS also provide the first limits on anomalous t-Z couplings with $t\bar{t}Z$ data using an effective field theory (EFT) framework. Typically, this process provides the tightest constraints on this coupling.

Both collaborations [10,11] now measure $t\bar{t}Z$ separately from $t\bar{t}W$ using Run 2 datasets of 139 fb^{-1} and 77.5 fb^{-1} for ATLAS and CMS, respectively. In both analyses, events with 3 or 4 isolated leptons (electrons or muons) are selected, targeting processes where one or both top quarks decay leptonically along with leptonic decays of the Z boson. Event and object quality requirements ensure the leptons are isolated and consistent with either the decay of a W boson (from the top-quark decay) or a Z boson. B-tagging algorithms are used to distinguish jets that originate from the hadronisation of b-quarks from those originating from light (up, down , strange or charm) quarks or gluons. Events are then further categorised according to the flavour and multiplicity of the jets in the event.

The ATLAS analysis selects events at detector level (using objects reconstructed from detector signals) with a minimum of two jets along with the aforementioned 3 or 4 lepton signature. Further signal region requirements are applied to maximise the sensitivity to $t\bar{t}Z$ production while ensuring enough signal events are retained to prevent the statistical uncertainty from becoming too large in the differential measurement. Additionally, control regions are defined to estimate background contributions from processes with prompt leptons from EW boson decays. Control region definitions can be found in Figure 2 where WZ/ZZ plus light jet processes dominate. The event yields from control regions are constrained by the observed data yields in these regions, which are then extrapolated to predict their contribution in the signal regions. WZ/ZZ plus b-jet production is not included in this method and are instead predicted directly using simulated templates which are included in the signal extraction procedure.

Another significant background contribution comes from processes where the selected lepton is not from the prompt decay of a vector boson (aka non-prompt/fake-lepton). This background mostly stems from $t\bar{t}$ dilepton processes where additional non-prompt leptons can originate from leptonically decaying heavy-flavour hadrons and/or jets that 'fake' a leptonic signature and is subsequently misidentified as a lepton. The contribution from this background is estimated using the matrix-method [12,13] which relies on the different probabilities that prompt and fake leptons pass the identification, isolation and impact parameter requirements. All other background processes are estimated from simulation, normalised to the latest theoretical cross-section prediction [14–16].

In comparison, the latest CMS inclusive cross-section measurement employs a very similar detector-level event selection. The measurement selects events with 3 or 4 lepton signatures and at least one jet. Events are then categorised according to the number of leptons, light (up, down, strange and gluon) flavour jets and heavy (bottom) flavour jets. The background processes are the same and are grouped in a mostly identical manner. All background processes with prompt leptons are modelled using the state-of-the-art simulation and normalised to the latest cross-section calculation. The normalisation of the WZ/ZZ plus jets processes are not extracted in the fit but are assigned uncertainties to cover the difference between data and the simulation in a dedicated control region. Backgrounds with fake/non-prompt leptons are estimated using the "fake rate" method in which estimates are made of the rate at which fake leptons pass the lepton selection requirements in control regions, and then this is extrapolated to the signal regions.

Both analyses extract the inclusive cross-section through a simultaneous maximum likelihood fit of the predicted yields of the signal and background processes to data in the signal regions. The signal strength ($\mu = \frac{\sigma^{best\,fit}}{\sigma^{SM}}$) is a free parameter in the fit and uncertainties are included in the fit as nuisance parameters constrained by Gaussian functions. The ATLAS analysis simultaneously fits data in the control regions and the WZ/ZZ plus light jets backgrounds treated as free parameters in the fit. The yields for the fitted simulation and data in the signal regions for both analyses can been seen in Figures 3 and 4.

Variable	3ℓ-WZ-CR	4ℓ-ZZ-CR
N_ℓ ($\ell = e, \mu$)	$= 3$	$= 4$
	1 OSSF lepton pair with	2 OSSF lepton pairs with
	$\lvert m_{\ell\ell} - m_Z \rvert < 10\,\text{GeV}$	$\lvert m_{\ell\ell} - m_Z \rvert < 10\,\text{GeV}$
$p_T(\ell_1, \ell_2, \ell_3, \ell_4)$	$> 27, 20, 20\,\text{GeV}$	$> 27, 20, 10, 7\,\text{GeV}$
N_{jets}	≥ 3	–
$N_{b\text{-jets}}$@85%	$= 0$	–
E_T^{miss}	–	$20\,\text{GeV} < E_T^{\text{miss}} < 40\,\text{GeV}$

Figure 2. ATLAS control regions [11].

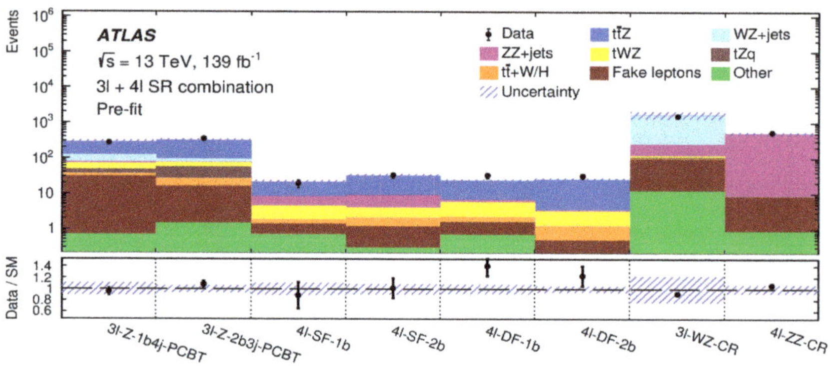

Figure 3. ATLAS signal regions [11].

The inclusive cross-section measured by ATLAS [11] from the combined fit in the 3 and 4 lepton signal regions, corresponding to a fiducial volume in which the Z-boson invariant mass lies between 70 and 110 GeV, is found to be

$$\sigma_{\text{ATLAS}}^{\text{pp}\to\text{t}\bar{\text{t}}\text{Z}} = 0.99 \pm 0.05(\text{stat.}) \pm 0.08(\text{syst.})\,\text{pb} \tag{1}$$

where the dominant systematic uncertainties originate from the $\text{t}\bar{\text{t}}\text{Z}$ parton shower modelling, tWZ background modelling and the identification.

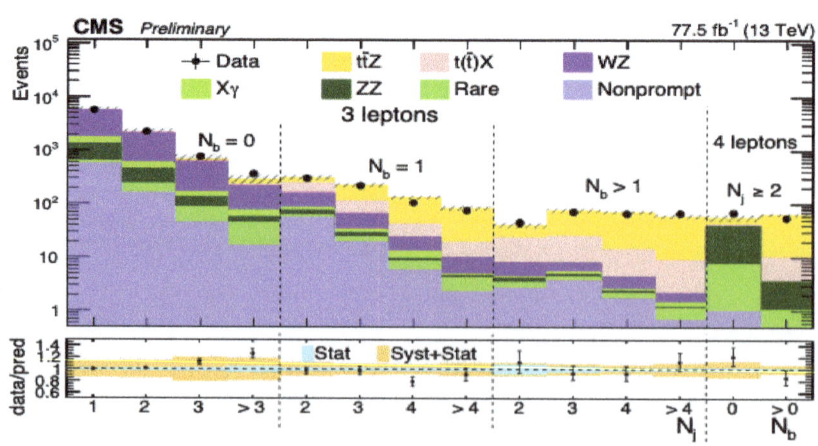

Figure 4. CMS signal regions [10].

The CMS cross-section measurement [10] yielded a value of

$$\sigma_{\text{CMS}}^{\text{pp}\to\text{t}\bar{\text{t}}\text{Z}} = 1.00^{+0.06}_{-0.05}(\text{stat.}) \pm^{+0.07}_{-0.06}(\text{syst.})\,\text{pb} \tag{2}$$

The results are evidently in excellent agreement with one another and reasonable agreement with the SM theoretical prediction [15,16] of

$$\sigma_{\text{theo.}}^{\text{pp}\to\text{t}\bar{\text{t}}\text{Z}} = 0.88^{+0.09}_{-0.15}\,\text{pb} \tag{3}$$

Several differential cross-section measurements investigate the kinematics of the $t\bar{t}Z$ system. In general, these measurements are performed by first subtracting background estimates from the data and then implementing an unfolding procedure which removes detector effects from the data so it can be compared with theoretical predictions. Migration matrices are constructed as part of the method that ensures resolution and acceptance affects are accounted for. The ATLAS measurement uses an iterative Bayesian unfolding to distributions defined using either particle or parton-level objects. Particle-level objects are defined using the collection of stable particles from the full matrix element plus parton shower simulation, i.e., baryons and mesons. Parton-level objects are defined using the unstable particles before any hadronisation effects have been simulated, i.e., quarks and gluons.

The fiducial volumes in which the measurements are made are defined using particle and parton-level objects, respectively, with a selection designed to be as close to the selection used in the inclusive measurement as possible. The background contributions are estimated in the same way as for the inclusive cross-section measurement. The WZ/ZZ plus jets background normalisation is corrected using normalisation factors obtained in a fit of the inclusive cross-section, based on the 3 and 4 lepton regions. All backgrounds are subsequently subtracted from the data. Several observables are measured, with most resulting in agreement between the background subtracted, unfolded data and the NLO simulation with which it is compared. Figure 5 shows the agreement between the unfolded particle-level data distribution of the Z-boson transverse momentum and the four theoretical predictions.

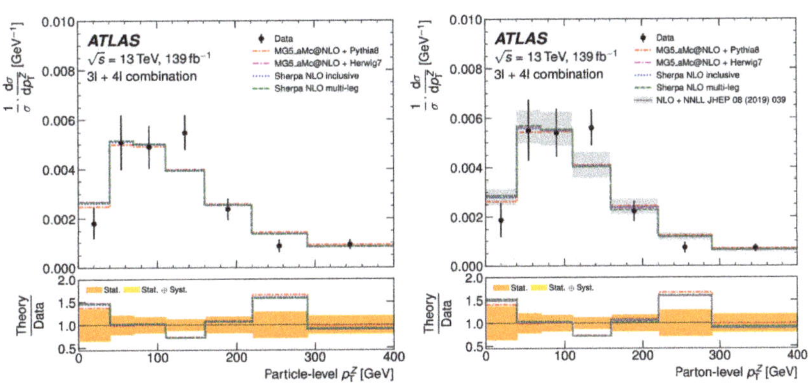

Figure 5. Comparison of normalised unfolded particle- and parton-level distribution of the transverse momentum of the Z boson in observed data from ATLAS [11] with Theoretical expectations obtained from different generators: Sherpa 2.2.1 [17] generator at NLO QCD accuracy using either multi-leg or inclusive setups and MG5_aMC@NLO [18] at NLO QCD accuracy interfaced with either the Pythia [19] or Herwig [20] parton shower models.

The differential cross-section measurement from CMS is performed in the same fiducial volume as defined for the inclusive measurement. Data are unfolded to parton level using the TUnfold package [21], which implements a least square fit with a Tikhonov regularisation. The unfolded distribution of the Z-boson transverse momentum is shown in Figure 6 along with the prediction from the MadGraph5_aMC@NLO Monte Carlo simulation.

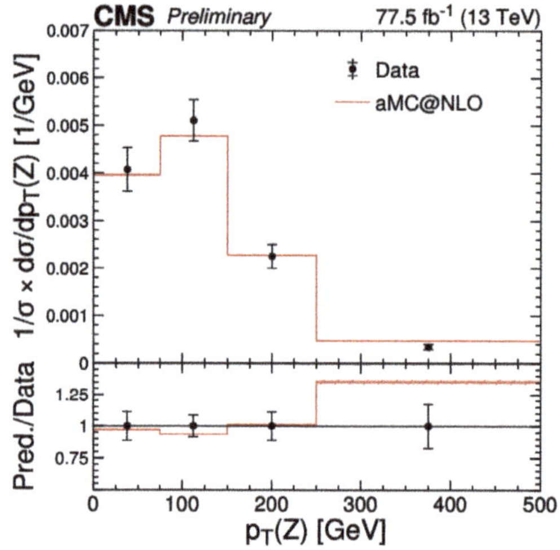

Figure 6. Comparison of normalised unfolded parton−level distribution of the transverse momentum of the Z boson in observed data from CMS [10] with Theoretical expectations obtained from different generators: Sherpa 2.2.1 generator at NLO QCD accuracy using either multi-leg or inclusive setups and MG5_aMC@NLO at NLO QCD accuracy interfaced with either the Pythia or Herwig parton shower models.

CMS also provide an interpretation of the results in the context of the Standard Model Effective Field Theory (SMEFT) in the Warsaw basis. Anomalous couplings are parameterised by 59 independent Wilson coefficients (WC's) of mass dimension 6, of which 15 are relevant for top-quark interactions. Of these 15, processes involving t-Z interactions can provide competitive constraints on four Wilson coefficients: c_{tZ}, $c_{tZ}^{[I]}$, $c_{\Phi t}$ and $c_{\Phi Q}$. The first two can induce anomalous EW dipole moments while the second two can induce anomalous neutral-current couplings. The values for these parameters will affect the kinematics and normalisation of processes with such vertices and can therefore be probed using differential distributions of the $t\bar{t}Z$ process. Signal yield predictions for non-zero (and zero = SM point) values of anomalous couplings are simulated in an independent sample at LO accuracy. Ratios of the BSM and SM points in a two-dimensional parton-level plane of the $p_T(Z)$ and $cos\theta_Z^*$ distributions are used to re-weight the nominal SM NLO $t\bar{t}Z$ sample. To validate this procedure, the distributions from the reweighted NLO SM sample and the dedicated LO BSM sample are then compared at various points in the WC parameter space after the full event reconstruction and are found to be in agreement.

A binned likelihood function $\mathcal{L}(\theta)$ is constructed from the product of Poisson probabilities and nuisance parameters from the bins in the differential distribution. The values of the nuisance parameters are maximised for each point in the BSM parameter plane to find which point maximises the likelihood. The test statistic

$$q = -2log(\frac{\mathcal{L}(\hat{\theta})}{\mathcal{L}(\hat{\theta}_{max})})$$

where $\mathcal{L}(\hat{\theta})$ is the likelihood function which maximises the nuisance parameters at a given BSM point and $\mathcal{L}(\hat{\theta}_{max})$ is the maximised likelihood function at the BSM point with the maximum likelihood. The test statistic q is shown for 1 and 2-dimensional scans of the WCs in Figure 7. For the 1-dimensional scan, all other WCs are fixed to their SM value. All results agree with the SM.

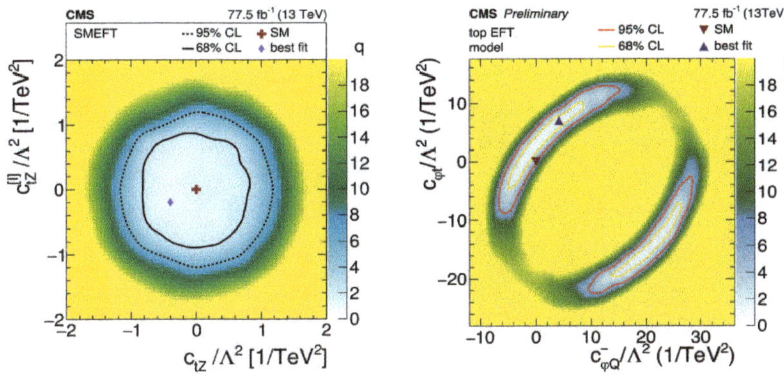

Figure 7. Two−dimensional scan of c_{tZ} with $c_{tZ}^{[I]}$ and $c_{\Phi t}$ with $c_{\Phi Q}$ Wilson coefficients [10].

3. Simultaneous t$\bar{t}$Z and tZq Measurements using Machine Learning Techniques

To probe the t-Z interaction even further, CMS has constructed a novel analysis [22] in which EFT effects on t-Z sensitive processes are targeted using bespoke machine learning algorithms. The analysis targets t$\bar{t}$Z, tZq and tWZ processes with at least three leptons and uses multivariate algorithms to exploit the EFT effects in a multi-observable phase-space, creating observables which are optimally sensitive to the effects of EFT operators.

As with the aforementioned measurements, the focus of the measurement is on operators that can affect the couplings between third generation quarks and EW vector bosons. Thus, the same operators are studied but excluding the imaginary component of the complex Wilson coefficient $c_{tZ}^{[I]}$ as it does not conserve CP. Two additional operators are studied however: c_{tW} probing the t-W EW dipole moment and $c_{\Phi Q}^{3}$ which probes the left-handed SU(2) triplet current operator.

A multi-classifier is trained to discriminate between the signals and major backgrounds. Separate binary classifiers are trained to discriminate between events generated under the SM and BSM (non-zero WC values) hypotheses. Training datasets are constructed from events randomly sampled from the SM scenario (labelled as background) and BSM scenario (labelled as signal). A novel approach that parameterises the event-weights as a 2nd order polynomial is used [23]. This makes it possible to smoothly interpolate predictions of the yields in bins of kinematic distributions, between the multitude of different combinations of WC values representing different EFT scenarios. It also allows for the interference between EFT operator amplitudes and either other EFT or the SM amplitudes to be taken into account in the simulation making it possible to exploit these kinematic differences of the various scenarios using a neural network. Separate networks are trained for t$\bar{t}$Z and tZq due to their largely different kinematics (tWZ is not explicitly targeted due to its smaller cross-section and similar kinematics to t$\bar{t}$Z). Training is also performed separately for each operator, along with one training course which targets all five operators simultaneously, allowing for a more global EFT interpretation. Post-fit distributions of the 1D and 5D EFT classifiers are shown in Figure 8. It is important to note that for larger WC values, the impact on the yield in the more signal-like bins grows stronger, demonstrating how effective these discriminators can be.

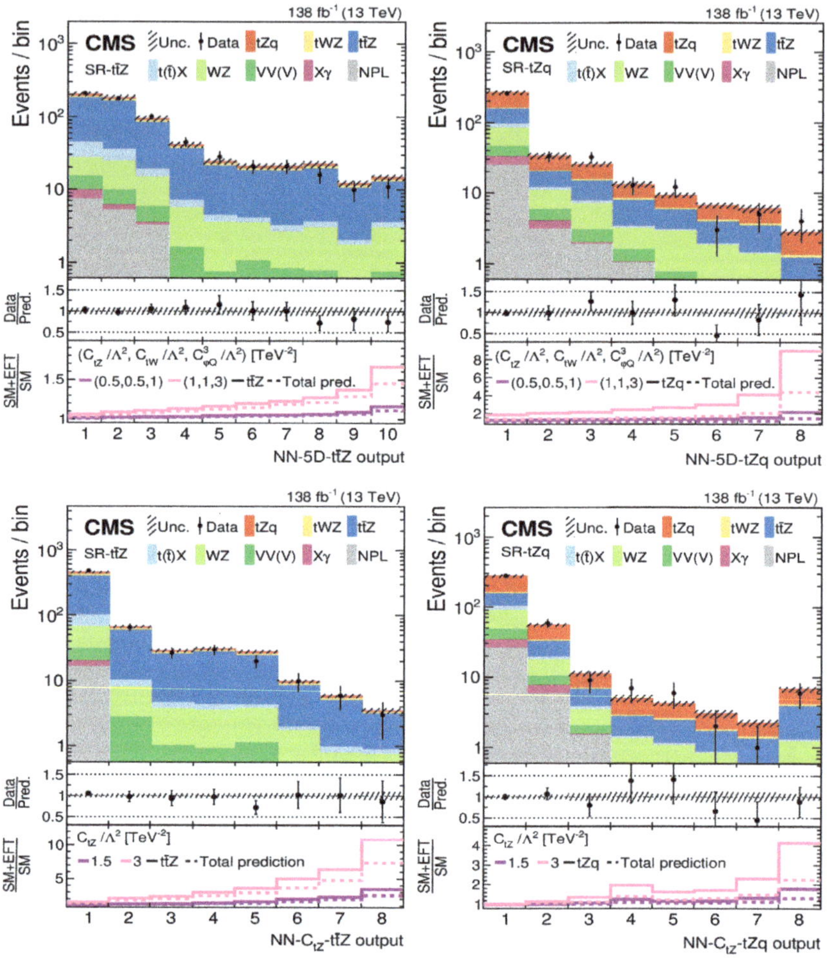

Figure 8. Post−fit distributions of the EFT neural networks in the $t\bar{t}Z$ and tZq signal regions from [22]. The top row shows the 5D discriminant while the bottom row shows the discriminant trained to target the effects of the c_{tZ} operator. The middle ratio plot demonstrates the data/MC agreement, while the lower ratio demonstrates the increasing impact on the yields in each bin from larger WC values.

The distributions of these NN's are fit to data in a maximum likelihood fit, where the likelihood is constructed in the same manner as was described in Section 2, to establish 68% and 95% CL confidence intervals on the values of the WC's. Five 1D scans (one for each operator) of the likelihood are performed, maximising the likelihood in steps of the WC value while fixing the other WCs to zero. Two-dimensional and five-dimensional scans are performed; however, the fit in this case uses the NN trained using distributions sampled from simultaneous variations of the 5 WC. The 95% CL confidence intervals for the 1D and 5D fits are shown in Figure 9.

WC/Λ^2 [TeV^{-2}]	95% CL confidence intervals			
	Other WCs fixed to SM		5D fit	
	Expected	Observed	Expected	Observed
c_{tZ}	$[-0.97, 0.96]$	$[-0.76, 0.71]$	$[-1.24, 1.17]$	$[-0.85, 0.76]$
c_{tW}	$[-0.76, 0.74]$	$[-0.52, 0.52]$	$[-0.96, 0.93]$	$[-0.69, 0.70]$
$c^{3}_{\varphi Q}$	$[-1.39, 1.25]$	$[-1.10, 1.41]$	$[-1.91, 1.36]$	$[-1.26, 1.43]$
$c^{-}_{\varphi Q}$	$[-2.86, 2.33]$	$[-3.00, 2.29]$	$[-6.06, 14.09]$	$[-7.09, 14.76]$
$c_{\varphi t}$	$[-3.70, 3.71]$	$[-21.65, -14.61] \cup [-2.06, 2.69]$	$[-16.18, 10.46]$	$[-19.15, 10.34]$

Figure 9. 95% CL confidence intervals for the 1D and 5D fits in [22].

Results of the 2D scans to compare with Figure 7 are shown in Figure 10. One can see very competitive results are obtained for common operators. All reported WC values agree with their expected SM value.

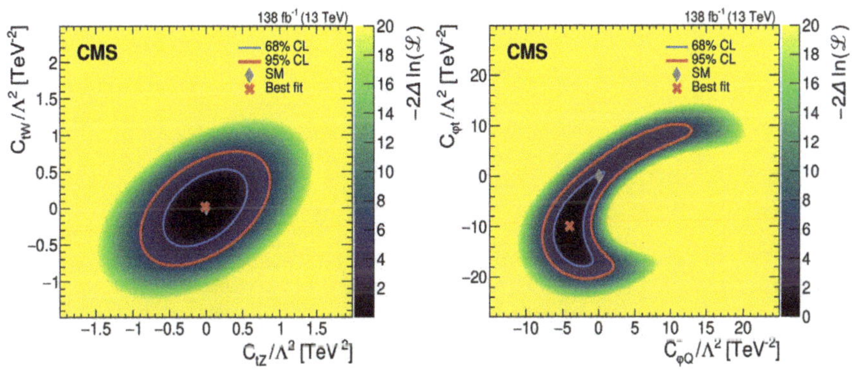

Figure 10. 95% CL confidence intervals for the 2D fits in [22].

4. $t\bar{t}W$ Measurements

The $t\bar{t}W$ process is unique among processes in which the $t\bar{t}$ system is produced with an associated boson. At leading-order the W boson can only be produced in the initial state, as is shown in Figure 11. The dominant contribution to the total amplitude is form quark-initiated processes. The W boson in fact polarises the incoming quarks and subsequently the top-quark pair leading to an enhancement in the decay product asymmetry at LO, exemplifying the need to take special care of spin correlations in any simulation [24]. Furthermore, the dominance of the quark-initiated production also leads to the $t\bar{t}W\pm$ asymmetry, in which $t\bar{t}W+$ production dominates over $t\bar{t}W-$, and is sensitive to the parton density function (PDF) of the proton.

Fixed order calculations of $t\bar{t}W$ at NLO in QCD ($\alpha_S^3\alpha$) have existed for a long time [25] and have been matched to parton shower [26,27], with NLO EW corrections ($\alpha_S^2\alpha^2$) coming later [16].

Persistent tensions between the measurements and predictions of the $t\bar{t}W$ cross-section have driven a lot of recent activity in the theory community. Calculations have become increasingly more sophisticated despite the many difficulties that arise when calculating the higher-order corrections for this process.

$t\bar{t}W$ production with an additional parton (e.g., $t\bar{t}Wj$ and $t\bar{t}Wjj$) generate large augmentations to the total cross-section with large NLO corrections as they introduce gluon-initiated production processes [28]. To merge the matrix elements of these processes with PS machinery, dedicated studies have been performed, with an improved multi-leg matching scheme [29].

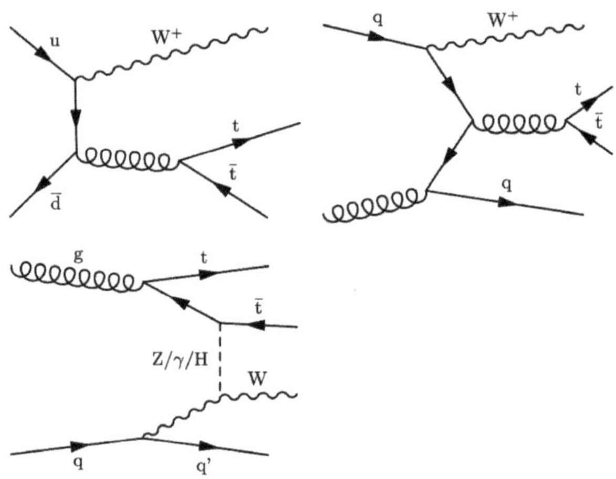

Figure 11. Leading-order (top left) and next-to-leading-order (top right and bottom) Feynman diagram for the $t\bar{t}W$ process. The last diagram is an example of the sub-leading electroweak corrections.

Calculations at NLO in QCD that account for the next-to-next-to-leading logarithmic (NNLL) [30] effects are now available as well as NLO QCD with NNLL effects with NLO EWK corrections [31,32]. Sub-leading EW corrections ($\alpha^3\alpha_S$) to $t\bar{t}W$ have in fact been found to have a larger effect than expected (approximately 10%) [33–35], primarily due to contributions from amplitudes represented by top-W-boson scattering diagrams.

Recent work has also included calculations of the full NLO cross-section including fixed order corrections and full LO spin correlations of decay products using POWHEG [35]. Some emphasis has also been put on the need for off-shell calculation which culminated in full off-shell calculations at NLO in QCD [36–38], off-shell calculations incorporating NLO EWK corrections [39] and finally the development of procedures to incorporate off-shell effects into NLO+PS procedures [40]

As mentioned in Section 2, $t\bar{t}W$ inclusive cross-section measurements have in the past been simultaneously extracted the $t\bar{t}Z$ cross-section due to the difficulties in disentangling these two rare processes. The previous measurements from CMS used data collected in 2016, selecting events with two or more leptons. Events selected with two leptons of the same sign charge provide the most sensitivity to the $t\bar{t}W$ process. The inclusive cross-section was measured to be $\sigma_{t\bar{t}W} = 0.77^{+0.12}_{-0.11}\text{stat.}^{+0.13}_{-0.12}\text{syst.}$ pb with an observed (expected) significance of 5.3 (4.5) standard deviations [9]. ATLAS made a similar measurement, extracting a cross-section value of $\sigma_{t\bar{t}W} = 0.87 \pm 0.13\text{stat.} \pm 0.14\text{syst.}$ pb and an observed (expected) significance of 4.3 (3.4) [8].

With the full Run 2 dataset available CMS has performed a new analysis that independently measures the inclusive $t\bar{t}W$ cross-section in the two lepton (same-sign) and three or more lepton channels. Although the much larger dataset significantly reduces the statistical uncertainty, new techniques have been developed to reduce the systematic uncertainty from 16% in the 2016 measurement to 6%. One of the key developments was a new multivariate analysis (MVA) algorithm designed to distinguish between leptons from the decays of W bosons (prompt leptons) and leptons originating in either the decay of heavy quarks (b or c quarks) or misidentified hadronic jets (non-prompt leptons). Although non-prompt leptons are generally easy to distinguish from prompt leptons, when background processes are large enough, they will still produce many objects with lepton-like signatures, such that further steps are needed to reduce their contribution to a signal region. The non-prompt background in this analysis primarily stems from the $t\bar{t}$ process. The new MVA algorithm brings a large improvement in the signal efficiency of the analysis compared with when a cut-based identification method was used in the previous iteration.

In the same-sign dilepton category, a multi-class deep neural network (DNN) is used to discriminate between signal and background using kinematic distributions of the jets and leptons in the event. The network is trained to distinguish between four processes: $t\bar{t}W$, non-prompt lepton backgrounds (modelled using $t\bar{t}$ simulation), $t\bar{t}Z$ or $t\bar{t}H$, and $t\bar{t}\gamma$. The distribution of the $t\bar{t}W$ output node provides an optimally discriminating variable.

A likelihood function is built from the Poisson probabilities to obtain the observed yields in bins of the discriminating variables in several event categories, with terms incorporating the various uncertainties and the correlations. A binned profile likelihood fit to the observed data is then performed using predicted signal and background distributions simultaneously in all event categories.

In the dilepton channel events are categorised according to the selected leptons' flavour and charge. The DNN $t\bar{t}W$ output node is the discriminating observable that is then used in the fit. In the tri-lepton category, events are categorised according to their number of jets, medium b-tags and the charge of the selected leptons, and the tri-lepton mass $m(3\ell)$ is used as the fit observable.

The inclusive $t\bar{t}W$ production cross-section is measured to be $\sigma_{t\bar{t}W} = 868 \pm 40(\text{stat}) ^{+52}_{-50}(\text{syst})$ fb [41], which is the most precise measurement to date. A breakdown of the cross-section measurement in the different channels is found in Figure 12 where it is compared with two theory predictions. The SM prediction at NLO+NNLL accuracy with FxFx jet merging represents the latest theory prediction [29] giving a cross-section of $\sigma^{\text{theo}}_{t\bar{t}W} = 592^{+155}_{-96}(\text{scale}) \pm 12(\text{PDF})$ fb. Measured and predicted cross-sections are within two standard deviations of one another. The central value of the measurement in data is approximately 1.5 times larger than the comparative theory prediction.

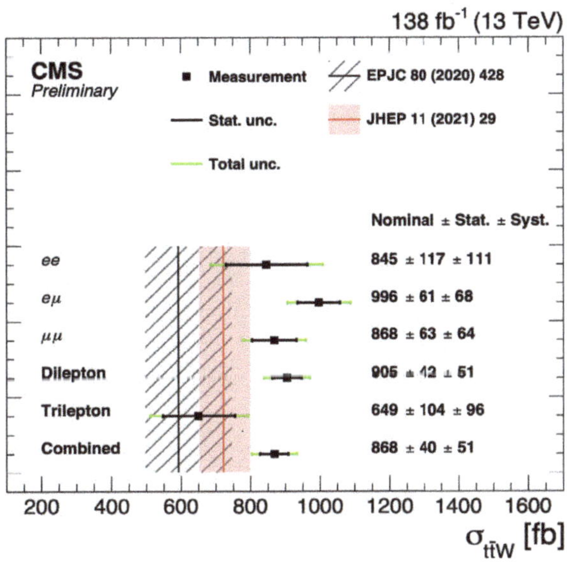

Figure 12. Measurements of the inclusive $t\bar{t}W$ cross−section [41]. The combined result is shown with a breakdown of the measurement obtained in the different dilepton and tri−lepton channels, as well as the measurement obtained in the different lepton flavour categories of the dilepton channel. The black inner error bar indicates the statistical uncertainty, while the green outer error bar represents the full systematic plus statistical uncertainty. The measurements are compared with two SM predictions. The prediction shown by the black line is from Ref. [31] while the prediction represented by the red line comes from Ref. [29] and includes FxFx predictions. The central lines of these two vertical lines represent the nominal prediction, while the band represents the combined uncertainty from the scale and PDF theory variations in the calculation.

The dominant systematic uncertainties originate from the uncertainty on the luminosity determination, the background estimation of the electron charge misidentification rate and the b-jet identification. All these uncertainties have significantly reduced with respect to the last iteration.

A simultaneous measurement of the $t\bar{t}W^{+}$ and $t\bar{t}W^{-}$ cross-sections is performed. The results in Figure 13 show that the measured cross-sections are significantly lower than the theoretical prediction. A measurement of the ratio of these two cross-sections is performed, as there are partial correlations between the systematic uncertainties of the two cross-sections that are reduced when measuring the ratio directly. This measurement is shown in Figure 14 to also be low in the theoretical prediction, but in agreement within the uncertainties.

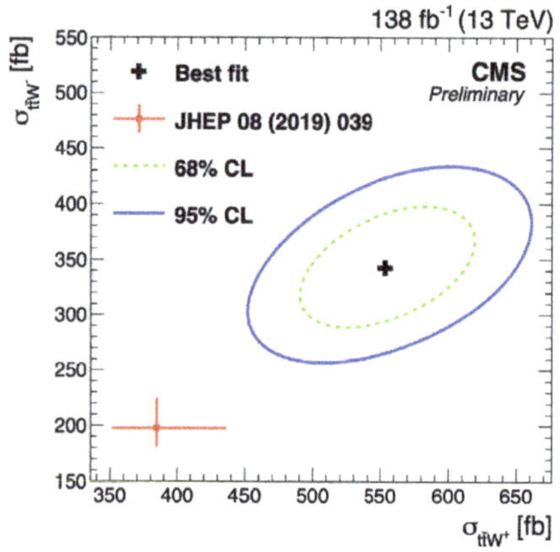

Figure 13. Contours showing the 68% and 95% CL intervals from the likelihood fit in which the $t\bar{t}W^{+}$ and $t\bar{t}W^{-}$ processes are measured simultaneously as independent parameters [41]. The best fit value of the fit is indicated by the black cross, with the theory prediction from Ref. [31] shown by the red cross. The theory prediction included is without the FxFx jet merging.

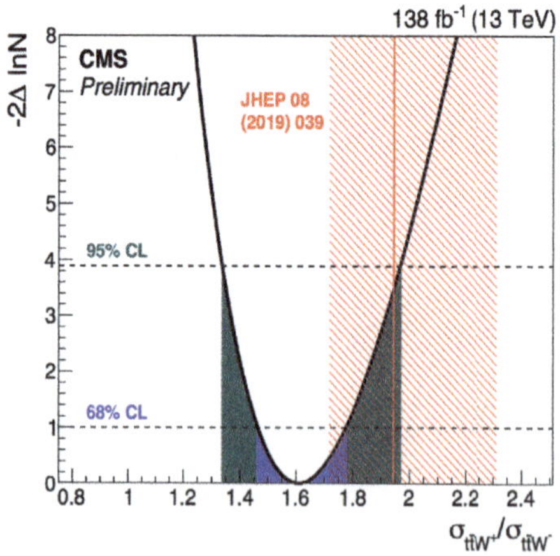

Figure 14. Negative log−likelihood scan for values of the ratio of $t\bar{t}W^{+}$ and $t\bar{t}W^{-}$ cross-sections. The best fit value is found at the minimum of the curve, while the dashed horizontal lines represent the CL limits [41]. The red line and hatched band represent the central value and total uncertainty of the theory prediction without the FxFx merging in Ref. [31].

5. t$\bar{\text{t}}\gamma$ Measurements

Studies of the t$\bar{\text{t}}\gamma$ production process probe the behaviour of the tγ electroweak coupling. The cross-section is sensitive to new physics that can occur via anomalous dipole moments of the top. Differential measurements provide additional sensitivity to said modifications that may affect spectra more or less in a particular kinematic regime. Such measurements typically compare state-of-the-art theory predictions with data to stress test the SM, and can be used to probe for BSM physics in a mode independent way.

The t$\bar{\text{t}}\gamma$ process is the rarest of the processes discussed in this review. Despite the small production cross-section, the associated production of a photon creates a very distinctive signature that manifests as an isolated energy deposit in the electromagnetic calorimeter without any associated tracks in the silicon tracker. This, along with several jets and leptons, facilitates a high purity event selection. As a result, evidence of this process was first seen by the CDF Collaboration in $\sqrt{s} = 1.96$ TeV collisions [42]. It was subsequently observed at the LHC by the ATLAS Collaboration in $\sqrt{s} = 7$ TeV proton–proton collisions [43] and has been measured by both ATLAS and CMS in $\sqrt{s} = 8$ TeV [44,45].

Both collaborations have now also measured this process using 13TeV pp collisions. The first measurement at this energy scale was performed by the ATLAS collaboration in leptonic final states [46] using a luminosity of 36.1 fb^{-1}, which accounts for a subset of the full Run 2 dataset. Subsequently, measurements using the full Run 2 dataset of 138 fb^{-1} were performed by CMS in the single-lepton [47] and dilepton [48] final states. Similarly, ATLAS uses a full Run 2 dataset of 139 fb^{-1} use targeting the dilepton (eμ) [49] final state only.

The targeted signals in all analysis includes the processes demonstrated in Figure 15, in which the photon not only originates from the top-quark decay but also the charged fermions radiated from the decay products of the top quark, and from the incoming parton. No attempt to differentiate between the sources is made, but requirements on the photon kinematics are implemented to suppress photons from the top-quark decay products.

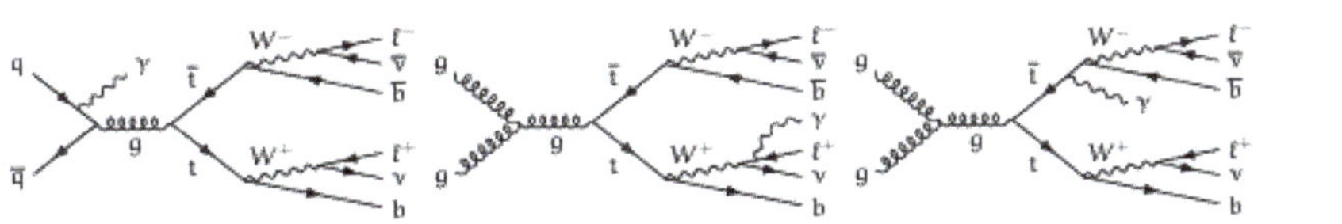

Figure 15. Leading-order Feynman diagram for the t$\bar{\text{t}}$fl process. Each diagram demonstrates a different production mechanism for the high energy photon in the process.

ATLAS performed its latest dilepton measurement in the eμ channel only, due to the clean final state and small background contribution. This enables an analysis strategy without having to implement complicated MVAs to discriminate signal from background thus simplifying the subsequent comparison with theoretical calculations. In particular, the analysis targets a comparison with the pp $\rightarrow$ bWbWfl calculation in reference [50,51]. The calculation includes all resonant and non-resonant diagrams, interference and off-shell effects of the top quarks and W bosons, meaning the signal considered combines both resonant t$\bar{\text{t}}\gamma$ and non-resonant tWfl production as demonstrated in Figure 16.

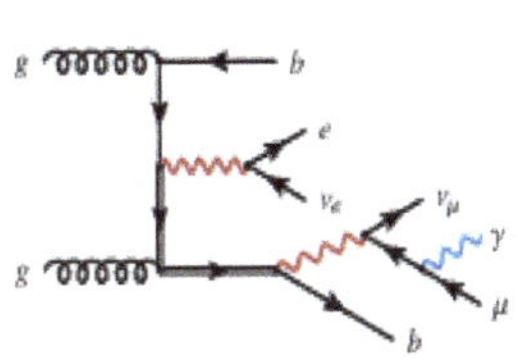

Figure 16. Leading-order Feynman diagram for the tWγ process. Red gauge boson lines represent W bosons while blue gauge boson lines represent photons [49].

Each analysis defines its own signal region at the detector-level where events are selected with exactly one photon, at least one b-tagged jet, and a channel-dependent number leptons (electrons or muons) and jets. After the full event selection, the persisting backgrounds can be broadly categorised as coming from four sources, three of which originate from events in which the photon or the lepton has been misidentified. Each measurement then defines a fiducial volume using particle-level objects, except for the ATLAS dilepton $e\mu$ measurement, which uses parton-level objects. A summary of the different fiducial volumes is shown in Table 1.

Table 1. Table outlining the fiducial selections made in each analysis. All selections are made on particle-level objects except for the ATLAS dilepton measurement. Additional requirements ensure leptons (in all cases, only electrons and muons are considered) are isolated and that the lepton energy incorporates that of radiated photons. Additional photon requirements also ensure isolation and that it does not originate from hadronic decay. Additional vetoes are applied to events in which leptons and photons are produced in close proximity. In particle-level selections, b-jets are defined using ghost-matching [52]. Leptons in the parton-level definition are required to come from the W-boson decay. Superscripts 1 and 2 refer to objects ordered by transverse momentum from highest to lowest.

Experiment	Final State	Photon	Leptons (e/μ)	Jets	b-jets
ATLAS [46]	$N_\ell = 1(=2)$, $N_\gamma = 1$, $N_j \geq 4(\geq 2)$, $N_b \geq 1$	$p_T > 20\,\text{GeV}$, $\lvert\eta\rvert < 2.37$	$p_T > 25\,\text{GeV}$, $\lvert\eta\rvert < 2.5$	$p_T > 25\,\text{GeV}$, $\lvert\eta\rvert < 2.5$	$p_T > 25\,\text{GeV}$, $\lvert\eta\rvert < 2.5$
ATLAS [49] (parton level)	$N_e = 1$, $N_\mu = 1$, $N_\gamma = 1$, $N_b = 1$	$E_T > 20\,\text{GeV}$, $\lvert\eta\rvert < 2.37$	$p_T > 25\,\text{GeV}$, $\lvert\eta\rvert < 2.5$	$p_T > 25\,\text{GeV}$, $\lvert\eta\rvert < 2.5$	$p_T > 25\,\text{GeV}$, $\lvert\eta\rvert < 2.5$
CMS [47]	$N_\ell = 1$, $N_\gamma = 1$, $N_j \geq 3$, $N_b \geq 1$	$p_T > 20\,\text{GeV}$, $\lvert\eta\rvert < 1.44$	$p_T^e > 35\,\text{GeV}$, $p_T^\mu > 35\,\text{GeV}$, $\lvert\eta\rvert < 2.37$	$p_T > 30\,\text{GeV}$, $\lvert\eta\rvert < 2.4$	$p_T > 30\,\text{GeV}$, $\lvert\eta\rvert < 2.4$
CMS [48]	$N_\ell = 2(\text{OS})$, $N_\gamma = 1$, $N_b \geq 1$	$p_T > 20\,\text{GeV}$, $\lvert\eta\rvert < 1.44$	$p_T^1 > 25\,\text{GeV}$, $p_T^2 > 15\,\text{GeV}$, $\lvert\eta\rvert < 2.4$	$p_T > 30\,\text{GeV}$, $\lvert\eta\rvert < 2.4$	$p_T > 30\,\text{GeV}$, $\lvert\eta\rvert < 2.4$

Events in which the selected photon candidate originates from a misidentified jet or non-prompt photon from the decay of a hadron make up the hadronic-fake background. The main process contributing to this background is $t\bar{t}$ where one of the jets in the final state is misidentified as a photon. All analyses use data-driven methods to derive scale factors in regions enriched with the hadronic-fake background which are then applied to the simulated hadronic-fake background prediction in the signal region.

Events in which the selected photon candidate originates from an electron make up the electron-fake background. This is the dominant background source in the dilepton channels. Electron-to-photon fake rates are measured using the tag-and-probe method in control regions using the $Z \rightarrow ee$ process. The fake rate scale factors are determined by taking the ratio between the fake rate measured in the data and simulation in bins of p_T and η.

Additionally, the backgrounds in which one or more leptons result from either a jet or a non-prompt lepton from heavy-flavour decays (fake-lepton) are estimated directly from data, contributing mainly to the single-lepton channel. The main contribution to this background comes from SM processes in which jets are produced uniquely through the strong interaction i.e., QCD events. The photon in such events can be either prompt or fake. The background contributions from events with a prompt photon, excluding signal events and fake-lepton backgrounds with a prompt photon, are estimated using simulated samples. These include $W\gamma$, $Z\gamma$, single-top+γ, diboson, and $t\bar{t}V$.

All the analyses discussed report inclusive and differential cross-sections measured in fiducial volumes defined according to the kinematics of the final state particles. Differential distributions of certain variables provide information on specific aspects of the $t\bar{t}\gamma$ process. Photon kinematics such as its p_T and η are sensitive to the coupling between the top quark and photon. Distributions of the angular separation between the photon and the top quarks decay products are sensitive to the origin of the photon. Furthermore, studying observables that do not involve the photon provide information on the $t\bar{t}$ system itself.

5.1. Inclusive Cross-Section Measurements

The latest ATLAS measurement of the inclusive cross-section in the single-lepton channel also includes a simultaneous measurement on the dilepton channel [46]. This measurement was performed using a smaller dataset collected in 2016 only consisting of 36.1 fb^{-1}, somewhat smaller than the more recent ATLAS dilepton $e\mu$ measurement [49] that will be described later. Using a neural network to discriminate the $t\bar{t}\gamma$ signal from backgrounds at detector-level, this distribution was then used as the input distribution to a profiled likelihood fit in which the fiducial cross-section is extracted. Several fits are performed, either independently fitting to the data in each channel or fitting to the data in each channel simultaneously. A correction factor for the signal efficiency and event migration into the fiducial region is also used when quoting the results. The measured inclusive fiducial cross-section measurements from [46] are found to be

$$\sigma^{SL}_{fid} = 521 \pm 9(\text{stat}) \pm 41(\text{sys})\ \text{fb}$$
$$\sigma^{DL}_{fid} = 69 \pm 3(\text{stat}) \pm 4(\text{sys})\ \text{fb}$$

A breakdown of the results, normalised to their corresponding NLO SM predictions, can be seen in Figure 17. In the single-lepton channel, the dominant uncertainties are related to the estimates of the jet energy and resolution scales as well as the background modelling, which is dominated by $t\bar{t}$ modelling, used to model the hadronic and electron-fake backgrounds. In the dilepton channel the uncertainty is still dominated by the statistical uncertainty of the data, with the largest systematic uncertainty coming from the signal and background modelling, which is dominated by $Z\gamma$ modelling.

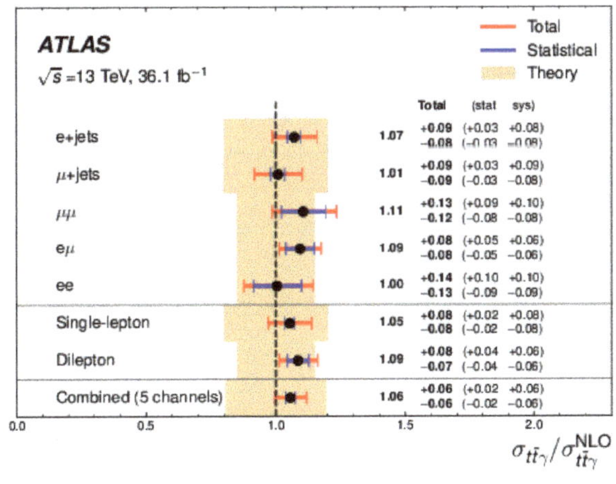

Figure 17. Inclusive $t\bar{t}\gamma$ production cross-section measurements by ATLAS in leptonic channels [46]. The NLO prediction from theory is shown in the dashed vertical line, with the uncertainty shown in the beige band. The measured values in data are represented by the black points, where the associated total and statistical uncertainties are shown in the red and blue lines, respectively. Results in each of the different lepton flavour channels are also shown.

CMS performs a similar measurement of the inclusive cross-section in the single-lepton channel. The fiducial phase-space is defined at particle level and can be found in Table 1. It is the same for both the inclusive and differential measurements. Signal regions are defined at detector-level and are designed to be as close as possible to the fiducial volume as possible. Additionally, orthogonal control regions are defined, enriched in the major backgrounds, are used in a fit to data to constrain associated uncertainties. The observed and expected yields in the signal and control regions along with the systematic uncertainties, are used to construct a binned likelihood function. The likelihood fit performed to extract the inclusive fiducial cross-section is performed separately to the one for the differential measurement. For the inclusive measurement, events in the signal and control regions are first categorised according to the flavour of the lepton. In the control regions, events are further categorised according to the photon transverse momentum, whereas in the signal regions the M_3 variable is used. This M_3 variable represents the invariant mass of the three jets that maximises their vector p_T sum. Nuisance parameters are assigned to account for the normalisation of the misidentified electron, $Z\gamma$ and $W\gamma$ backgrounds. The resulting fiducial inclusive cross-section measurement [47] is found to be

$$\sigma(t\bar{t}\gamma) = 798 \pm 7(\text{stat}) \pm 48(\text{syst})\ \text{fb}$$

A breakdown of the inclusive measurement in the different channels can be seen in Figure 18. The leading systematic uncertainties according to their post-fit impact on the measured cross-section come from the normalisation of the $W\gamma$ background, the non-prompt background estimation and the integrated luminosity estimation.

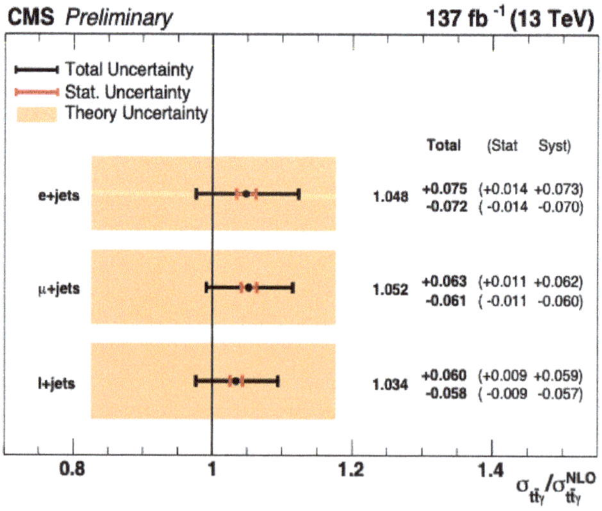

Figure 18. Inclusive $t\bar{t}\gamma$ production cross−section measurements by CMS in the single−lepton channel [47]. Results are also shown for the individual lepton flavour channels.

To extract the inclusive $t\bar{t}\gamma$ fiducial cross-sections in the dilepton channel, the CMS measurement uses a very similar strategy to the single-lepton case, making the two measurements easier to combine. The fiducial phase-space is defined at particle level, for which the full definition can be found in Table 1. A profile likelihood fit to the photon p_T distribution in data across the three data taking periods of Run 2 is performed. The resulting inclusive fiducial cross-section is found to be [48]

$$\sigma_{fid} = 175.2 \pm 2.5(\text{stat}) \pm 6.3(\text{syst})\ \text{fb}$$

This agrees with the predicted inclusive cross-section of

$$\sigma_{SM} = 155 \pm 27\,fb$$

The predicted inclusive cross-section is about 12% (0.7 standard deviations) lower than the measurement. This is shown in Figure 19 along with the breakdown of the fit in the individual channels. However, the large theory uncertainties that impact the prediction from Madgraph make it difficult to draw strong conclusions on the agreement between the prediction and the unfolded data. The predicted cross-section is scaled to the NLO $2 \rightarrow 3\,\mathrm{pp} \rightarrow \mathrm{t\bar{t}}\gamma$ process, but does not include processes in which the photon is radiated from the final state decay products of the top quark. This is one potential cause of the discrepancy between the results.

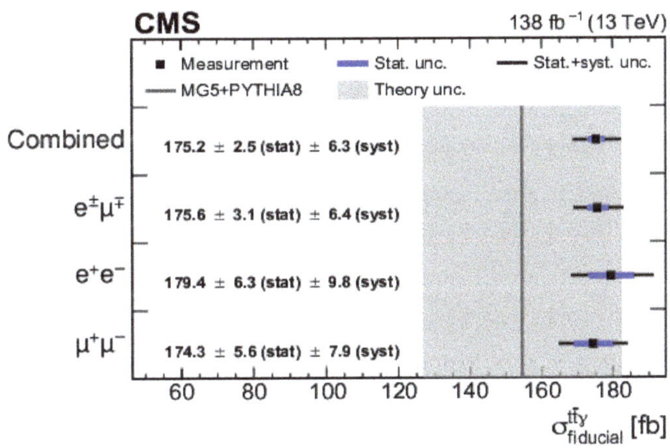

Figure 19. Inclusive $\mathrm{t\bar{t}}\gamma$ production cross-section measurements by CMS in the dilepton channel [48]. Results are also shown for both the combined measurement and the breakdown for the individual dilepton channels.

ATLAS measures dilepton fiducial cross-sections in the $e\mu$ final state using a profile likelihood fit to the S_T distribution (scalar sum of all transverse momenta in the event) in data. This variable provides good separation between the signal and backgrounds. The fiducial volume is defined in Table 1 and is the same for both the inclusive and differential cross-section measurements. The selection mimics that of the theory calculation with which the experimental results are compared [50,51]. The inclusive cross-section is measured to be

$$\sigma_{fid} = 39.6 \pm 0.8\,(\mathrm{stat})^{+2.6}_{-2.2}\,(\mathrm{syst})\,\mathrm{fb}$$

Ref. [49], which agrees with the dedicated theoretical calculation which predicts a value of

$$\sigma_{fid} = 38.50^{+0.56}_{-2.18}\,(\mathrm{scale})^{+1.04}_{-1.18}\,(\mathrm{PDF})\,\mathrm{fb}$$

Refs. [50,51]. As is shown in Table 2, the cross-section measurements all tend to agree with the predicted values at NLO within uncertainties when taking the branching ratios into consideration. Differences in the fiducial cross-sections between the experiments stem from the differences in the fiducial volumes outlined in Table 1. In particular, the CMS single-lepton fiducial cross-section is measured to be much higher than in ATLAS due to the inclusion of events with three jets and a looser dR() selection.

Table 2. Table of the inclusive $t\bar{t}\gamma$ production fiducial cross-section measurements from ATLAS and CMS.

Experiment	$t\bar{t}$ Decay Channel	$\sigma_{fid}^{t\bar{t}\gamma}$ (fb)
CMS [47]	Single lepton	798
ATLAS [46]	Single lepton	521
CMS [48]	Dilepton	175
ATLAS [49]	Dilepton ($e\mu$)	39.6

5.2. *Differential Cross-Section Measurements*

CMS has reported differential $t\bar{t}\gamma$ fiducial cross-sections in both the single lepton [47] dilepton [48] channels. The single-lepton publication reports differential fiducial cross-section measurements as a function of the photons p_T, $|\eta|$ and the difference in angle between the lepton and photon ($\Delta R(\ell, \gamma)$). Results were obtained simultaneously for the 3 and 4 jet regions, the lepton flavour channels, and the different data taking periods. The same control regions are used as in the inclusive measurement. After the profile likelihood fit, backgrounds are subtracted from the observable distribution in data and subsequently unfolded to particle level. The unfolded differential cross-section is defined in the same fiducial phase-space as the inclusive cross-section. Distributions of the unfolded observables are shown in Figure 20 where a comparison with simulations obtained using Madgraph_aMC@NLO interfaced with three different parton shower algorithms is shown. In the bulk of the distribution, the dominant uncertainties are similar to those in the inclusive cross-section measurement. For $p_T(\gamma) > 120$ GeV, the uncertainties in the jet energy scale, photon identification efficiency and colour re-connection modelling are the largest sources of uncertainty.

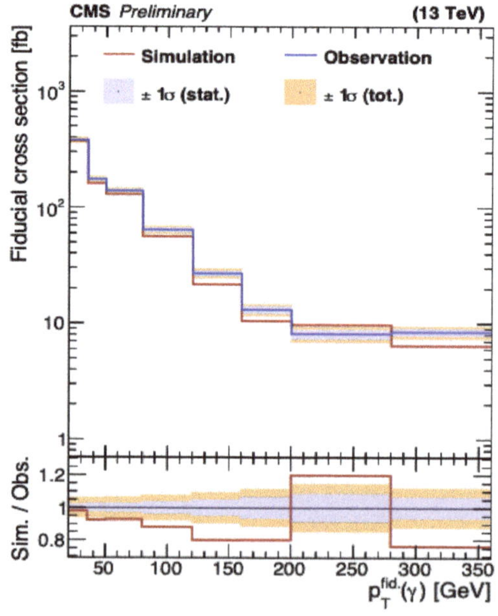

Figure 20. Differential $t\bar{t}\gamma$ production cross-section measurements by CMS in the single-lepton channel [47]. Results are also shown as a function of the transverse momentum of the photon at particle level.

In the dilepton channel, differential cross-sections are reported with respect to 12 observables that are unfolded to particle level in the same fiducial volume as the inclusive cross-section measurement. These are compared with two predictions using Madgraph_aMC@NLO event generator interfaced with two parton shower simulations: Pythia8 with the CP5 tune [53] and Herwig [54] v7.14 with the CH3 tune [55]. An example of the unfolded distribution of the transverse momentum of the photon at particle level in the dilepton channel is shown in Figure 21. No significant deviation between the measured distribution and either of the predictions is observed, but due to the size of the theory uncertainties it is once again difficult to come to a conclusion regarding their agreement.

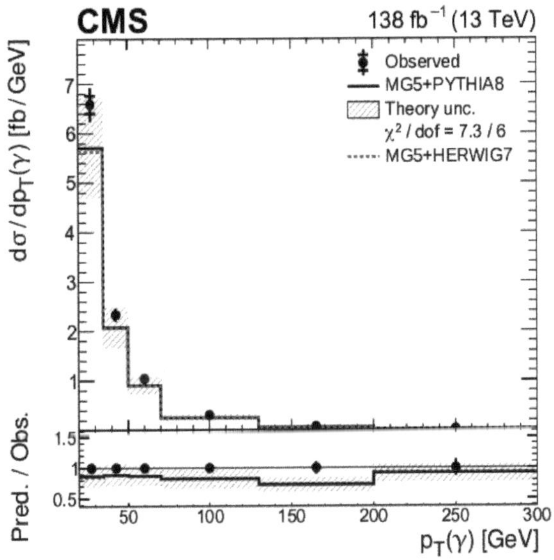

Figure 21. Distribution of the absolute production cross-section of $t\bar{t}\gamma$ in the dilepton channel as a function of the p_T of the photon, as measured by the CMS experiment [48]. Observed data unfolded to particle level is compared with the predicted distribution from the Madgraph generator with two different parton shower models. Theoretical uncertainties evaluated using the Pythia 8 prediction are shown in the shaded grey bands.

ATLAS has reported differential cross-section measurements in both the leptonic [46] and dilepton ($e\mu$) [49] channels. To extract the distributions, no fit to data is performed. The major backgrounds are subtracted from the data using the estimates outlined earlier after which detector effects are removed using an unfolding procedure which is applied to the observed detector level distributions to obtain the true distribution of the signal at particle or parton level. The differential cross-section is normalised to unity resulting in distributions shown in Figure 22 for [46]. Absolute differential distributions are also provided and can be found in the paper.

In the case of the dilepton ($e\mu$) channel, ATLAS measures differential cross-sections as a function of a similar set of variables described in the CMS dilepton measurement. Distributions are unfolded to parton level and can therefore be directly compared with the aforementioned theory prediction via both normalised and absolute differential cross-sections. Additionally, a comparison is made with two leading-order simulations using Madgraph interfaced with Pythia or Herwig. A comparison of the parton-level cross-section as a function of the photon p_T in simulation and the unfolded data are shown in Figure 23. In general, all distributions agree well; however, one trend that was recognised was that the NLO prediction tends to describe most distributions better than the LO prediction.

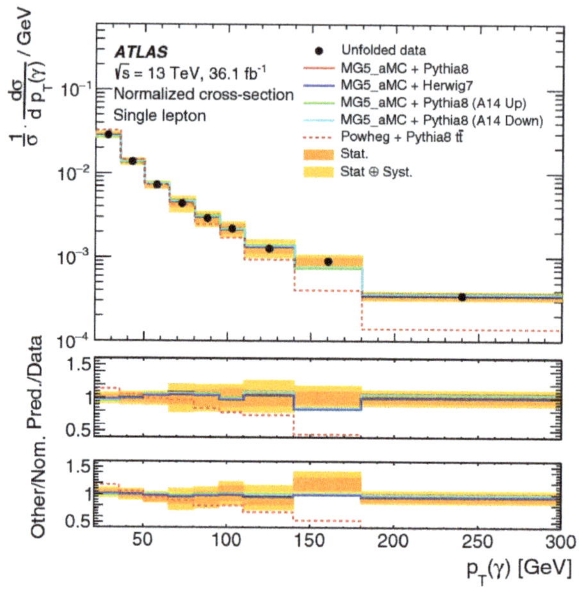

Figure 22. Normalised differential cross-section as a function of the photon transverse momentum [46]. Unfolded distributions are compared with predictions using the $MG5_aMC + Pythia8$ together with up and down variations of the Pythia8A14 tune parameters, the $MG5_aMC + Herwig7$ and POWHEG + Pythia8 $t\bar{t}$ where the photon radiation is modelled in the parton shower.

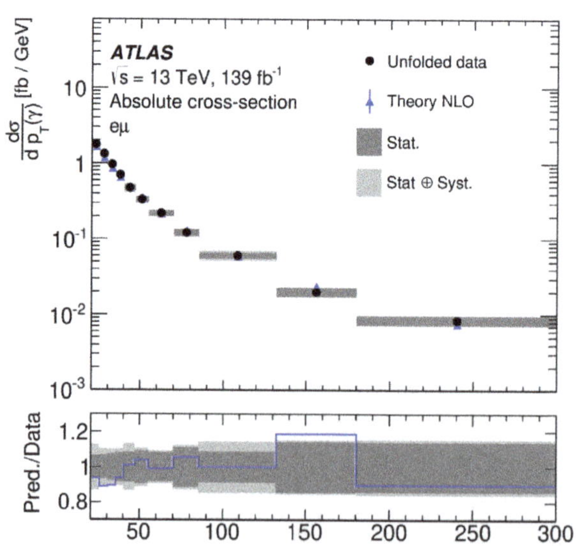

Figure 23. Distribution of the absolute production cross−section of $t\bar{t}\gamma$ in the $e\mu$ channel as a function of the p_T of the photon, as measured by the ATLAS experiment [49]. Observed data unfolded to parton level is compared with the predicted distribution from the theoretical prediction from [50,51]. The systematic and statistical uncertainties are shown in the grey bands.

5.3. EFT Interpretations

The CMS measurements also provide limits on Wilson coefficients that induce electroweak dipole moments

$$c_{tZ} = Re(-sin\theta_W C_{uB}^{33} + cos\theta_W C_{uW}^{33})$$
$$c_{tZ}^{I} = Im(-sin\theta_W C_{uB}^{33} + cos\theta_W C_{uW}^{33})$$

A maximum likelihood fit using the $p_T(\gamma)$ spectrum, which is sensitive to such modifications, is performed to obtain 68% and 95% CL intervals on the targeted coefficients. The fit is performed in the signal regions only. The intervals for a given Wilson coefficient are obtained by either fixing the other WC to its SM value (1D), or simultaneously profiling the two WCs (2D). The results of both tests are shown in Figure 24. No deviation from the SM values is observed. The 1D scans show more stringent intervals than the $t\bar{t}Z$ measurements. This is partially because models with non-zero WC values predict a harder $p_T(\gamma)$ spectrum, which is not observed in the tails of the data distribution. The precision with which CMS can reconstruct photon kinematics is a major contributing factor to this measurements ability to improve upon the latest limits.

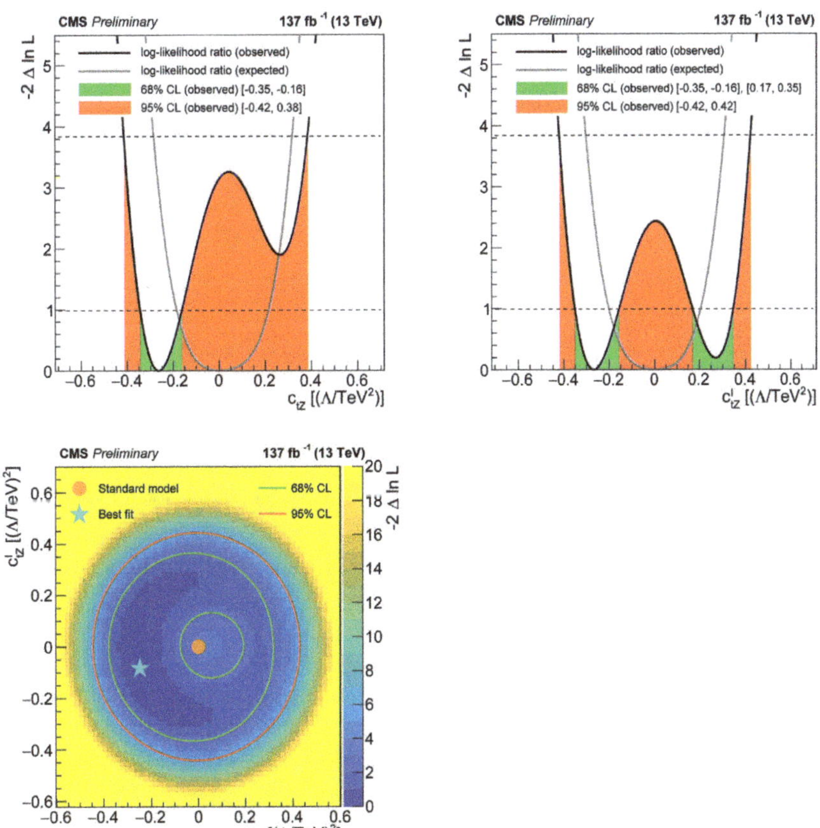

Figure 24. Best fit values for the explored EFT Wilson coefficients by CMS in the single−lepton channel [47]. Both the 1D and 2D scans are shown.

The CMS dilepton measurement performs a profile likelihood fit in the same way as the inclusive measurement, to obtain the best fit values for the Wilson coefficients probed. A combined profile likelihood fit is also performed with the single-lepton analysis. Although the dilepton channel benefits from a higher purity of signal, the single-lepton channel profits from a higher number of signal events with a high p_T photon, making it sensitive to modifications in the kinematics of the photon caused by anomalous Wilson coefficient values. The 1D and 2D scans of the Wilson coefficients in both the dilepton and combined fits can be found in Figure 25. No sign of anomalous couplings is observed. A comparison with the constraints from other measurements is also shown in Figure 25. The results in this publication provide the best limits to date on the c_{tZ} and c^I_{tZ} Wilson coefficients in Figure 26.

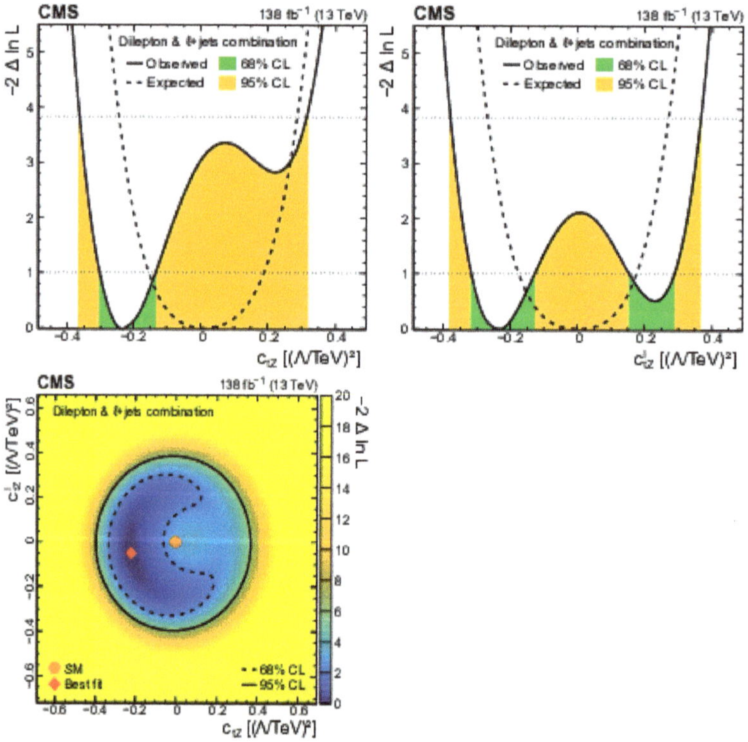

Figure 25. Distributions of the observed (solid line) and expected (dashed line) negative log-likelihood difference from the best fit value for the one-dimensional and two-dimensional scans of the studied Wilson coefficients. The results are obtained from the fit to data using the photon p_T distribution. The plots shown here are from the combination of the single lepton and dilepton analyses [48].

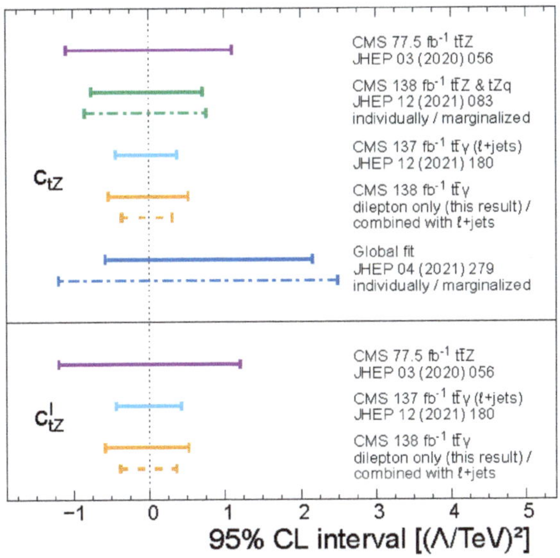

Figure 26. Comparison of observed 95% CL intervals for the two Wilson coefficients, c_{tZ} (**upper panel**) and c_{tZI} (**lower panel**) from CMS measurements of: $\mathrm{t\bar{t}Z}$, $\mathrm{t\bar{t}}\gamma$ single lepton, $\mathrm{t\bar{t}}\gamma$ dilepton. The results are shown from the one-dimensional scans, i.e., all other Wilson coefficients have values set to zero. The dashed lines indicate the results from the combination with the single−lepton channel. In the case of the global fit and the $\mathrm{t\bar{t}Z} + tZq$, the solid lines represent the result where all Wilson coefficients are fixed to zero, whereas the dashed lines show the results from the marginalised limits. The tightest constraint to date on these Wilson coefficients comes from the combination of the $\mathrm{t\bar{t}}\gamma$ single lepton and dilepton channels [48].

6. Measurements at the HL-LHC and Future Colliders

Cross-section measurements of rare $\mathrm{t\bar{t}} + X$ process are incredibly useful probes of top-quark couplings to gauge bosons and are therefore a key ingredient to furthering our knowledge at the high energy frontier. Anomalous couplings are predicted by several BSM theories including composite Higgs models, models with extra dimensions and those predicting vector-like quarks [4,56,57].

The LHC has already produced a sizeable sample of rare top-quark processes, which has been used to measure the cross-sections for several $\mathrm{t\bar{t}} + X$ processes with an uncertainty that is considered to be on the cusp of what is commonly referred to as a 'precision measurement'. The large dataset from the HL-LHC will cement these measurements in the precision regime and allow more precise probes of anomalous couplings affecting these amplitudes. Extrapolations of current measurements to future datasets and accelerators provide estimates of what might be achieved, help to establish physics goals and highlight the improvements required to achieve them.

Although the HL-LHC will provide a huge rare top dataset, enhancing the boosted regime in particular, it is also interesting to look towards machines planned even further in the future. Several machines fall into this category and are typically designed to push the precision frontier through the clean environment provided via lepton collisions, or the energy frontier through the high energies achievable at large circular hadron colliders. Results from future lepton colliders are particularly interesting in the context of this article as ultra-precise measurements of top-quark EW interactions will be achievable. Future colliders, of both hadrons and leptons, at or above the energy frontier ($\geq$10 TeV) have the potential to improve the sensitivity of Standard Model EFT (SMEFT) fits to new physics, particularly to four-fermion operators for which there is a strong increase in sensitivity at higher energies.

To fully harness the power of precision measurements in a truly model-independent search for new physics, it is best to take a global approach to SMEFT fits [58]. This requires a combination of the broadest dataset possible in a high-dimensional fit of many operators affecting several SM processes into account. Several of these operators are of particular interest given the scope of this article, namely operators affecting top EW couplings. So far, this article has only discussed measurements of SMEFT parameters using $t\bar{t}Z/\gamma$ however, operators can affect several SM processes in many different ways and hence a global fit of these operators using many processes can provide important constraints.

The outlook for measurements of EFT parameters affecting top EW couplings has in fact been studied and reported in several publications. A comparison of the expected 95% confidence interval for several EFT operators, using the LHC Run 2 dataset and the extrapolated values using the HL-LHC dataset are shown in Figure 27. The figure shows the results from a global EFT fit performed in Ref. [59]. A full list of analyses included in the global fit can be found in Table 3. It should be noted that, although the HL-LHC data are shown to bring an improvement to the global fits of almost all of the operators in question (Figure 27), the individual 95% confidence intervals on operators $C^{-}_{\varphi Q}$ and $C^{3}_{\varphi Q}$ are not enhanced. This is due to their reliance on the legacy $e^+e^- \to b\bar{b}$ measurements of R_b and A^{bb}_{FBLR} at the Z-pole from LEP and SLC. The inclusion of the Tevatron s-channel single-top measurement provides complementary constraining power on these operators and is still the most sensitive measurement of this process which, at the time of writing this article, still illudes measurements at the LHC.

Not all processes used in the global fit are relevant for this article; however, the plot highlights the importance of $t\bar{t}+X$ measurements at the present and in the future. All projections (including for lepton colliders discussed later) are based on similar approximations to the 'S2' scenario used in projections of Higgs boson measurements [60] where many statistical and experimental uncertainties scale as $\frac{1}{\sqrt{L_{int}}}$, with L_{int} representing the integrated luminosity. With respect to uncertainties at the end of the LHC Run 2, the complete HL-LHC program approximates that experimental uncertainties will reduce by a factor of 5, while theory and modelling uncertainties are reduced by a factor of two. The reduction in theory uncertainties assumes that N^{2}LO calculations will be achieved for the rare top processes and that large steps forward in Monte Carlo modelling are made in the next 10 years, ready for when the new colliders are expected to start running.

Table 3. Measurements included in the top-quark EW sector EFT fit [59]. The table includes the process, observable, centre-of-mass energy, integrated luminosity and experiment for each measurement. Where the experiment is cited as LHC, a combination of ATLAS and CMS measurements were used. Where Tevatron is cited, a combination of CDF and D0 results were used. LEP/SLD refers to different experiments from these two accelerators.

Process	Observable	$\sqrt{s}$	Luminosity	Experiment
$pp \to t\bar{t}$	$\frac{d\sigma}{dm_{t\bar{t}}}$	13 TeV	140 fb^{-1}	CMS
$pp \to t\bar{t}$	$\frac{dA_C}{dm_{t\bar{t}}}$	13 TeV	140 fb^{-1}	ATLAS
$pp \to t\bar{t}H + tHq$	σ	13 TeV	140 fb^{-1}	ATLAS
$pp \to t\bar{t}Z$	$\frac{d\sigma}{dp_T^Z}$	13 TeV	140 fb^{-1}	ATLAS
$pp \to t\bar{t}\gamma$	$\frac{d\sigma}{dp_T^\gamma}$	13 TeV	140 fb^{-1}	ATLAS
$pp \to tZq$	σ	13 TeV	77.4 fb^{-1}	CMS
$pp \to t\gamma q$	σ	13 TeV	36 fb^{-1}	CMS
$pp \to t\bar{t}W$	σ	13 TeV	36 fb^{-1}	CMS
$pp \to t\bar{b}$ (s-chan)	σ	8 TeV	20 fb^{-1}	LHC
$pp \to tW$	σ	8 TeV	20 fb^{-1}	LHC
$pp \to tq$ (t-chan)	σ	8 TeV	20 fb^{-1}	LHC
$pp \to Wb$	F_0, F_L	8 TeV	20 fb^{-1}	LHC
$p\bar{p} \to t\bar{b}$ (s-chan)	σ	1.96 TeV	9.7 fb^{-1}	Tevatron
$e^+e^- \to b\bar{b}$	R_b, A^{bb}_{FBLR}	91 GeV	202.1 pb^{-1}	LEP/SLC

This study highlights the need for further advances in theoretical calculations and modelling for HL-LHC measurements where, according to the current 'S2' model for projections, theory uncertainties will for the first time dominate over experimental and statistical sources. The current state of the art for the theory predictions of the relevant processes, as well as the desires of the experimental community for future predictions are reported here.

The latest $t\bar{t}+W$ calculations have been discussed at length in the relevant section in this article as the area is particularly active. To summarise, the latest calculations have been performed using matrix element using perturbative calculations with precision up to the NLO terms in QCD and include additional next-to-next-to-leading log (NNLL) effects [30], as well as predictions for NLO+NNLL in QCD with NLO EW corrections. Full off-shell calculations up to NLO in QCD [36–38] have also recently been developed and now with the possibility to combine NLO EW and QCD corrections to off-shell $t\bar{t}W$ [39] and a procedure to apply the full off-shell corrections within the NLO+PS setup [40]. Future NNLO calculations could bring a factor of two improvement in the precision of the calculation.

NLO QCD calculations of $t\bar{t}\gamma$ have been available for a while [50]. The most recent NLO calculation was in fact for the process $t\bar{t}\gamma+tW\gamma$ [51] in the $e\mu$ final state. The inclusion of NNLO QCD corrections in a full $t\bar{t}\gamma$ calculation will become necessary if the full potential of the data at the HL-LHC is to be exploited.

The latest $t\bar{t}Z$ cross-section calculation is of NLO QCD+EW precision. This not only takes into account the Z/γ interference, but also includes the off-shell $t\bar{t}\gamma^{\star}$ contributions. The theory uncertainty in this calculation is $^{+0.09}_{-0.10}$ [14–16]. This is mainly a result of the proton PDF, QCD scale and α_S. The measurements in Section 2 show that the total systematic uncertainty of the inclusive and differential cross-section measurement are already very close to this. A more precise theory calculation in the future would have a great impact on the achievable precision of future EFT measurements sensitive to effects from the $\mathcal{O}_{tZ}$ operator.

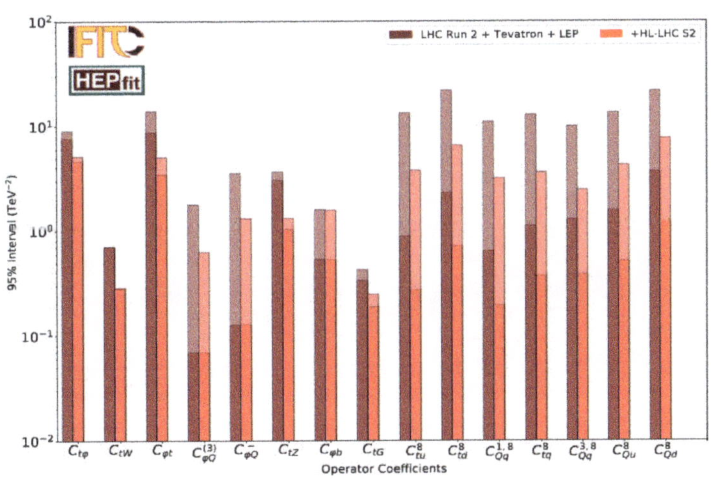

Figure 27. Comparison of expected 95% confidence intervals on Wilson coefficients for dimension-six operators affecting top-quark production and decay measurements using the LHC Run 2 dataset and the HL-LHC dataset [59]. Only linear terms proportional to Λ^{-2} are accounted for in the dependence of the observables on the Wilson coefficients. The solid bars show the constraint of from the single parameter fits, while the translucent bars show the marginalised constraints from the global fit.

Figure 27 shows the Wilson coefficients for several EFT operators along the x-axis. $t\bar{t}X$ processes are sensitive to the first six couplings from the left. The remaining couplings often affect top-pair production via QCD mechanisms and can be investigated more precisely using other $t\bar{t}$ processes. Differential measurements of $t\bar{t}Z$ and $t\bar{t}\gamma$ as a function of the Z boson or photon transverse momentum, respectively, are essential probes of the effects of the O_{tZ} operator. With increasing statistics, several measurements of rare top processes could be measured to much greater precision. Notably, precise differential measurements of $t\bar{t}W$ would provide essential information on this key background to measurements of $t\bar{t}H$ and four-top process in multi-lepton final states to name but a few.

Across all selected operators, a factor of two to four times the current Run 2 limits is expected with the HL-LHC dataset, both for the individual and marginalised bounds. The exceptions to this are the individual bounds of $C^{-}_{\varphi Q}$ and $C^{3}_{\varphi Q}$, which are very dependent on the bounds from the $Zb\bar{b}$ measurements at the Z-pole. Sensitivity to operators affecting EW couplings could be dramatically improved in the future through the harvesting and analysis of data large datasets in the boosted regime [61]. An additionally interesting insight from this reference is that, although not included in the fits performed in the document, the two-quark two-lepton ($\mathcal{O}_{qq\ell\ell}$) operators can be probed at the LHC and beyond and by including analyses targeting for instance the off-Z-peak dilepton invariant mass region in $t\bar{t}\ell^{+}\ell^{-}$, the sensitivity of EFT fits can be enhanced.

Although the HL-LHC provides a much larger dataset with which to study EW couplings, the processes that provide the most sensitivity remain $pp \rightarrow t\bar{t}Z$ and $pp \rightarrow t\bar{t}\gamma$. Future lepton colliders provide an excellent opportunity to perform high-precision tests for anomalous EW couplings affecting top-quark pair processes. One of the benefits of $e^{+}e^{-}$ machines is that once the centre-of-mass energy exceeds twice the top mass, the dominant $t\bar{t}$ production mechanism becomes $e^{+}e^{-} \rightarrow Z/\gamma \rightarrow t\bar{t}$, providing direct access to the top-quark EW couplings in a very clean environment. Furthermore, lepton colliders can distinguish the coupling between the top quark and photon from the top-quark coupling with a Z boson. At circular lepton colliders, this is facilitated via a measurement of the final state polarisation in semileptonic top-quark decays, whereas at a linear collider this can be done using different beam polarisations configurations [62–65].

Figure 28 compares expected limits on the different EFT operator coefficients using combinations of the data collected from the HL-LHC combined with data taken in the final stages of four different future lepton colliders: the CEPC, FCC, ILC and CLIC [59]. Important information on the different working configurations of each future machine is shown in Table 4. Though not as important for the processes and operators discussed here, it is worth noting that the different runs have different centre-of-mass energies above the top-quark pair production threshold, which can be used to disentangle the four-fermion $e^{+}e^{-}t\bar{t}$ operator coefficients from the two-fermion operator coefficients. This is because the four-fermion operators scale quadratically with energy whereas the two-fermion operators either remain constant or grow linearly. Given the energies above the $t\bar{t}$ threshold in the circular collider scenarios are very close, this disentanglement is more difficult with such machines.

The data from circular colliders (FCC-ee and CEPC) operating at centre-of-mass energies equal to and slightly above the $t\bar{t}$ threshold, are expected to improve in the constraints on the bottom and top operators at the HL-LHC by a factor of 2 to 5 for several two-fermion operators. The constraining power on four-fermion operators is limited by the energy reach. The data for the linear colliders (ILC and CLIC) was simulated at two centres of mass energies above the $t\bar{t}$ threshold and provides impressive constraints on all operators. As mentioned, it is due to these different collision energies that even the bounds on the four-fermion operators become competitive once the centre-of-mass energy surpasses 1 TeV.

Table 4. Table of the working configurations for several future e^+e^- colliders from Ref. [59]. The machines listed in the table are: the International Linear Collider (ILC), the Circular Electron–Positron Collider (CEPC), the Compact Linear Collider (CLIC) and the Future Circular Lepton Collider (FCC-ee). The polarisation, energy and luminosity for 3 to 4 different running stages are listed along with references to the relevant documentation.

Machine	Polarisation	Energy	Luminosity
ILC [66]	P(e^+,e^-): ($\pm$30%,$\mp$80%)	250 GeV	$2ab^{-1}$
		500 GeV	$4ab^{-1}$
		1 TeV	$8ab^{-1}$
CLIC [67]	P(e^+,e^-): ($\pm$30%,$\mp$80%)	380 GeV	$1ab^{-1}$
		1.4 TeV	$2.5ab^{-1}$
		3 TeV	$5ab^{-1}$
FCC-ee [68]		Z-pole	$150ab^{-1}$
	Unpolarised	240 GeV	$5ab^{-1}$
		350 GeV	$0.2ab^{-1}$
		365 GeV	$1.5ab^{-1}$
CEPC [68]		Z-pole	$57.5ab^{-1}$
	Unpolarised	240 GeV	$20ab^{-1}$
		350 GeV	$0.2ab^{-1}$
		360 GeV	$1ab^{-1}$

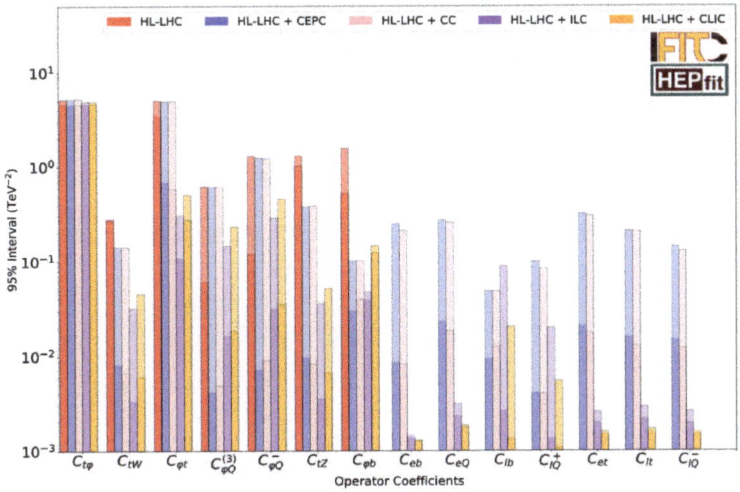

Figure 28. Comparison of expected 95% confidence intervals combining data from the HL-LHC with data from several proposed lepton collider experiments [59]. $q\bar{q}t\bar{t}$ and C_{tG} coefficients are not shown in the figure as e^+e^- collider measurements provide no additional sensitivity; however, all operators are included in the global fit. The solid bars show the constraint of from the single parameter fits, while the translucent bars show the marginalised constraints from the global fit. N.B. label HL-LHC+CC refers to the addition of FCC results.

Looking further ahead, collisions at higher centre of mass (beyond 10TeV) could be achieved with for example a 100 km hadron collider, a linear electron–positron collider or compact circular muon collider [69–71]. As was alluded to earlier, the energy-growing sensitivity of the global SMEFT fits to new physics, especially through four-fermion operators, makes measurements at such machines invaluable. Given the absence of new physics signals, model-independent searches such as this provide one of the best chances of finding deviations from the SM and guiding the future of HEP.

7. Conclusions

The top quark is a unique particle in the known universe and while there are many priorities for high energy physics research, its distinctive features suggest it may have a special role in the SM. Therefore, understanding the top quark with absolute clarity remains a top priority for high energy physics experiments. The absence of new resonant particles has driven the development of novel methods to detect the presence of new physics, including indirect searches looking for anomalous couplings involving SM particles using EFT's. Such measurements require immense precision and a wealth of data. This has been the case for several years regarding the dominant QCD top-pair production mechanism. However, over the coming years, several rare top-pair processes will enter this regime, providing essential probes of anomalous couplings and new insights into where to look for this evasive new physics.

$t\bar{t}Z/\gamma$ measurements have had a sub 10% precision for some time and have provided differential measurements and constraints on the relevant EFT operators while the most recent $t\bar{t}W$ measurements have a precision of around 7%, though unfortunately no differential or EFT measurement has been performed at the time of writing this article. With an influx of more data from Run 3 and beyond, the collected dataset of all the rare top-pair processes will be large enough to perform both differential and EFT measurements. Additionally, as we increase the dataset size the boosted regime will become more populated and, due to energy-growing effects in certain EFT operators, these regimes will become much more important and provide complementary constraints.

EFT measurements become so important going forward, allowing us to scrutinise the SM and use the power of precision measurement across diverse datasets to probe a wide range of operators in a model-independent manner to perform comprehensive searches for new physics. It is clear from the projections that as we look towards the HL-LHC, the achievable constraints on EFT parameters grow 2–4 times stronger in the top EW sector. However, these constraints will grow even stronger at future lepton colliders, which show further improvements of between a factor of 2–5.

Funding: This research received no external funding.

Institutional Review Board Statement: Not applicable.

Informed Consent Statement: Not applicable.

Data Availability Statement: Not applicable.

Acknowledgments: Not applicable.

Conflicts of Interest: The authors declare no conflict of interest.

Notes

1 There have in fact been phenomenological investigations into the possible existence of top-quark pair-bound states, also known as 'Toponium' [1], in order to explain excesses seen in LHC Run 2 data measurements of top-quark pair production with dilepton decays, where the top-quark pair is produced near threshold

References

1. Fuks, B.; Hagiwara, K.; Ma, K.; Zheng, Y.J. Signatures of toponium formation in LHC run 2 data. *Phys. Rev.* **2021**, 104, 034023. [CrossRef]
2. Buchmller, W.; Wyler, D. Effective Lagrangian analysis of new interactions and flavour conservation. *Nucl. Phys. B* **1986**, *268*, 621–653. [CrossRef]
3. Grzadkowski, B.; Iskrzyński, M.; Misiak, M.; Rosiek, J. Dimension-six terms in the Standard Model Lagrangian. *J. High Energy Phys.* **2010**, *2010*, 85. [CrossRef]
4. Richard, F. Present and future constraints on top EW couplings. *arXiv* **2014**, arXiv:1403.2893.
5. The ALEPH Collaboration; The DELPHI Collaboration; The L3 Collaboration; The OPAL Collaboration; The SLD Collaboration; The LEP Electroweak Working Group; The SLD Electroweak and Heavy Flavour Groups. Precision electroweak measurements on the Z resonance. *Phys. Rep.* **2005**, *427*, 257–454. [CrossRef]

6. Bylund, O.B.; Maltoni, F.; Tsinikos, I.; Vryonidou, E.; Zhang, C. Probing top quark neutral couplings in the Standard Model Effective Field Theory at NLO QCD. *J. High Energy Phys.* **2016**, *2016*, 52. [CrossRef]
7. CMS Collaboration. Search for physics beyond the standard model in final states with two opposite-charge same-flavor leptons, jets, and missing transverse momentum in pp collisions at 13 TeV. *J. High Energy Phys.* **2021**, *123, 4.*
8. ATLAS Collaboration. Measurement of the $t\bar{t}Z$ and $t\bar{t}W$ cross sections in proton-proton collisions at $\sqrt{s} = 13$ TeV with the ATLAS detector. *Phys. Rev. D* **2019**, *99*, 072009. [CrossRef]
9. CMS Collaboration. Measurement of the cross section for top quark pair production in association with a W or Z boson in proton-proton collisions at $\sqrt{s} = 13$ TeV. *J. High Energy Phys.* **2018**, *2018*, 11. [CrossRef]
10. CMS Collaboration. Measurement of top quark pair production in association with a Z boson in proton-proton collisions at $\sqrt{s} = 13$ TeV. *J. High Energy Phys.* **2020**, *2020*, 56. [CrossRef]
11. ATLAS Collaboration. Measurements of the inclusive and differential production cross sections of a top-quark–antiquark pair in association with a Z boson at $\sqrt{s} = 13$ TeV with the ATLAS detector. *Eur. Phys. J. C* **2021**, *81*, 737. [CrossRef]
12. The ATLAS Collaboration. Search for supersymmetry at $\sqrt{s} = 8TeV$ in final states with jets and two same-sign leptons or three leptons with the ATLAS detector. *J. High Energy Phys.* **2014**, *2014*, 35. [CrossRef]
13. The ATLAS Collaboration. Estimation of non-prompt and fake lepton backgrounds in final states with top quarks produced in proton-proton collisions at $\sqrt{s} = 8$ TeV with the ATLAS detector. In Proceedings of the 7th International Workshop on Top Quark Physics, Cannes, France, 29 September–3 October 2014.
14. The ATLAS Collaboration. Modelling of the $t\bar{t}H$ and $t\bar{t}V$ $(V = W, Z)$ processes for $\sqrt{s} = 13$ TeV ATLAS analyses. *CDS* **2016** , *5*, ATL-PHYS-PUB-2016-005.
15. CERN. CERN Yellow Reports: Monographs, Vol 2 (2017): Handbook of LHC Higgs cross sections: 4. Deciphering the nature of the Higgs sector. *arXiv* **2017**, arXiv:1610.07922.
16. Frixione, S.; Hirschi, V.; Pagani, D.; Shao, H.-S.; Zaro, M. Electroweak and QCD corrections to top-pair hadroproduction in association with heavy bosons. *J. High Energy Phys. 2015*, 184. [CrossRef]
17. Bothmann, E.; Chahal, G.S.; Höche, S.; Krause, J.; Krauss, F.; Kuttimalai, S.; Liebschner, S.; Napoletano, D.; Schönherr, M.; Schulz, H.; et al. Event generation with Sherpa 2.2. *arXiv* **2019**, arXiv:1905.09127.
18. Frederix, R.; Frixione, S.; Hirschi, V.; Pagani, D.; Shao, H.-S.; Zaro, M. The automation of next-to-leading order electroweak calculations. *J. High Energy Phys.* **2018**, *2018*, 185. [CrossRef]
19. Sj′ostrand, T.; Mrenna, S.; Skands, P. A brief introduction to PYTHIA 8.1. *Comput. Phys. Commun.* **2008**, *178*, 852–867. [CrossRef]
20. Bähr, M.; Gieseke, S.; Gigg, M.A.; Grellscheid, D.; Hamilton, K.; Latunde-Dada, O.; Plätzer, S.; Richardson, P.; Seymour, M.H.; Sherstnev, A.; et al. Herwig++ physics and manual. *Eur. Phys. J. C* **2008**, *58*, 639–707. [CrossRef]
21. Schmitt1, S. TUnfold, an algorithm for correcting migration effects in high energy physics. *J. Instrum.* **2012**, *7*, T10003. [CrossRef]
22. CMS Collaboration. Probing effective field theory operators in the associated production of top quarks with a Z boson in multilepton final states at $\sqrt{s} = 13TeV$. *J. High Energy Phys.* **2021**, *2021*, 12.
23. CMS Collaboration. Search for new physics in top quark production with additional leptons in proton-proton collisions at $$ \sqrt{s} $$ = 13 TeV using effective field theory. *J. High Energy Phys.* **2021**, *2021*, 95. [CrossRef]
24. Maltoni, F.; Mangano, M.L.; Tsinikos, I.; Zaro, M. Top-quark charge asymmetry and polarization in $t\bar{t}W\pm$ production at the LHC. *Phys. Lett. B* **2014**, *736*, 252–260. [CrossRef]
25. Campbell, J.M.; Ellis, R.K. $t\bar{t}W$ production and decay at NLO. *J. High Energy Phys.* **2012**, *2012*, 52. jhep07(2012)052. [CrossRef]
26. Garzelli, M.V.; Kardos, A.; Papadopoulos, C.G.; Trocsanyi, Z. $t\bar{t}W$ and $t\bar{t}Z$ Hadroproduction at NLO accuracy in QCD with Parton Shower and Hadronization effects. *J. High Energy Phys.* **2012**, *2012*, 56. [CrossRef]
27. Maltoni, F.; Pagani, D.; Tsinikos, I. Associated production of a top-quark pair with vector bosons at NLO in QCD: impact on $t\bar{t}H$ searches at the LHC. *J. High Energy Phys.* **2016**, *2016*, 113. [CrossRef]
28. von Buddenbrock, S.; Ruiz, R.; Mellado, B. Anatomy of inclusive $t\bar{t}W$ production at hadron colliders. *Phys. Lett. B* **2020**, *811*, 135964. [CrossRef]
29. Frederix, R.; Tsinikos, I. On improving NLO merging for $t\bar{t}W$ modelling. *J. High Energy Phys.* **2021**, *2021*, 29. /JHEP11(2021)029. [CrossRef]
30. Kulesza, A.; Motyka, L.; Schwartländer, D.; Stebel, T.; Theeuwes, V. Associated production of a top quark pair with a heavy electroweak gauge boson at NLO+NNLL accuracy. *Eur. Phys. J. C* **2019**, *79*, 1–15. [CrossRef]
31. Broggio, A.; Ferroglia, A.; Frederix, R.; Pagani, D.; Pecjak, B.D.; Tsinikos, I. Top-quark pair hadroproduction in association with a heavy boson at NLO+NNLL including EW corrections. *J. High Energy Phys.* **2019**, *2019*, 39. [CrossRef]
32. Kulesza, A.; Motyka, L.; Schwartländer, D.; Stebel, T.; Theeuwes, V. Associated top quark pair production with a heavy boson: differential cross sections at NLO + NNLL accuracy. *Eur. Phys. J. C* **2020**, *80*, 428. [CrossRef]
33. Frederix, R.; Pagani, D.; Zaro, M. Large NLO corrections in $t\bar{t}W\pm$ and $t\bar{t}t\bar{t}$ hadroproduction from supposedly subleading EW contributions. *J. High Energy Phys.* **2018**, *2018*, 31. [CrossRef]
34. Asaf, Dror, J.; Farina, M.; Salvioni, E.; Serra, J. Strong tW Scattering at the LHC. *J. High Energy Phys.* **2016**, *2016*, 71. [CrossRef]
35. Cordero, F.F.; Kraus, M.; Reina, L. Top-quark pair production in association with a $W^{\pm}$ gauge boson in the POWHEG-BOX. *Phys. Rev. D* **2021**, *103*, 094014. [CrossRef]
36. Bevilacqua, G.; Bi, H.Y.; Hartanto, H.B.; Kraus, M.; Worek, M. The simplest of them all: $t\bar{t}W^{\pm}$ at NLO accuracy in QCD. *J. High Energy Phys.* **2020**, *2020*, 43. [CrossRef]

37. Denner, A.; Pellicciolі, G. NLO QCD corrections to off-shell $t\bar{t}W^{\pm}$ production at the LHC. *J. High Energy Phys.* **2020**, *2020*, 69. [CrossRef]
38. Bevilacqua, G.; Bi, H.Y.; Hartanto, H.B.; Kraus, M.; Nasufi, J.; Worek, M. NLO QCD corrections to off-shell $t\bar{t}W^{\pm}$ production at the LHC: correlations and asymmetries. *Eur. Phys. J. C* **2021**, *81*, 675. [CrossRef]
39. Denner, A.; Pellicciolі, G. Combined NLO EW and QCD corrections to off-shell $t\bar{t}W$ production at the LHC. *Eur. Phys. J. C* **2021**, *81*, 354. [CrossRef]
40. Bevilacqua, G.; Bi, H.Y.; Febres Cordero, F.; Hartanto, H.B.; Kraus, M.; Nasufi, J.; Reina, L.; Worek, M. Modeling uncertainties of $t\bar{t}W^{\pm}$ multilepton signatures. *Phys. Rev. D* **2022**, *105*, 014018. [CrossRef]
41. The CMS Collaboration. Measurement of the cross section of top quark-antiquark pair production in association with a W boson in proton-proton collisions at $\sqrt{s}$ = 13 TeV. *arXiv* **2022**, arXiv:2208.06485.
42. CDF Collaboration. Evidence for $t\bar{t}\gamma$ Production and Measurement of $\sigma_{t}\bar{t}\gamma/\sigma_{t}\bar{t}$. *Phys. Rev. D* **2011**, *84*, 031104. 10.1103/PhysRevD.84.031104. [CrossRef]
43. ATLAS Collaboration. Observation of top-quark pair production in association with a photon and measurement of the $t\bar{t}\gamma$ production cross section in pp collisions at $\sqrt{s} = 7$ TeV using the ATLAS detector. *Phys. Rev. D* **2015**, *91*, 072007. [CrossRef]
44. ATLAS Collaboration. Measurement of the $t\bar{t}\gamma$ production cross section in proton-proton collisions at $\sqrt{s} = 8$ TeV with the ATLAS detector. *J. High Energy Phys.* **2017**, *2017*, 86. [CrossRef]
45. CMS Collaboration. Measurement of the semileptonic $t\bar{t} + \gamma$ production cross section in pp collisions at $\sqrt{s} = 8$ TeV. *J. High Energy Phys.* **2017**, *2017*, 6. [CrossRef]
46. ATLAS Collaboration. Measurements of inclusive and differential fiducial cross-sections of $t\bar{t}\gamma$ production in leptonic final states at $\sqrt{s} = 13$ TeV in ATLAS. *Eur. Phys. J. C* **2019**, *79*, 382. [CrossRef]
47. CMS Collaboration. Measurement of the inclusive and differential $t\bar{t}\gamma$ cross sections in the single-lepton channel and EFT interpretation at $\sqrt{s}$ = 13 TeV. *J. High Energy Phys.* **2021**, *2021*, 180. [CrossRef]
48. CMS Collaboration. Measurement of the inclusive and differential $t\bar{t}\gamma$ cross sections in the dilepton channel and effective field theory interpretation in proton-proton collisions at $\sqrt{s}$ =13 TeV. *J. High Energy Phys.* **2022**, *2022*, 91. [CrossRef]
49. ATLAS Collaboration. Measurements of inclusive and differential cross-sections of combined $t\bar{t}\gamma$ and $tW\gamma$ production in the $e\mu$ channel at 13 TeV with the ATLAS detector. *J. High Energy Phys.* **2020**, *2020*, 49. [CrossRef]
50. Bevilacqua, G.; Hartanto, H.B.; Kraus, M.; Weber, T.; Worek, M. Hard Photons in Hadroproduction of Top Quarks with Realistic Final States. *J. High Energy Phys.* **2018**, *2018*, 158. [CrossRef]
51. Bevilacqua, G.; Hartanto, H.B.; Kraus, M.; Weber, T.; Worek, M. Precise predictions for $t\bar{t}\gamma/t\bar{t}$ cross section ratios at the LHC. *J. High Energy Phys.* **2019**, *2019*, 188. [CrossRef]
52. Cacciari, M.; Salam, G.P. Pileup subtraction using jet areas. *Phys. Lett. B* **2008**, *659*, 119–126. 2007.09.077. [CrossRef]
53. CMS Collaboration. Extraction and validation of a new set of CMS pythia8 tunes from underlying-event measurements. *Eur. Phys. J. C* **2020**, *80*, 4. [CrossRef]
54. Bellm, J.; Gieseke, S.; Grellscheid, D.; Plätzer, S.; Rauch, M.; Reuschle, C.; Richardson, P.; Schichtel, P.; Seymour, M.H.; Sidmok, A.; et al. Herwig 7.0/Herwig++ 3.0 release note. *Eur. Phys. J. C* **2016**, *76*, 196. [CrossRef]
55. CMS Collaboration. Development and validation of HERWIG 7 tunes from CMS underlying-event measurements. *Eur. Phys. J.* **2021**, *81*, 312. [CrossRef] [PubMed]
56. Durieux, G.; Matsedonskyi, O. The top-quark window on compositeness at future lepton colliders. *J. High Energy Phys.* **2019**, *2019*, 72. [CrossRef]
57. Dawson, S.; Homiller, S.; Lane, S.D. Putting standard model EFT fits to work. *Phys. Rev. D* **2020**, *102*, 055012. [CrossRef]
58. Ethier, J.J.; Magni, G.; Maltoni, F.; Mantani, L.; Nocera, E.R.; Rojo, J.; Slade, E.; Vryonidou, E.; Zhang, C. Combined SMEFT interpretation of Higgs, diboson, and top quark data from the LHC. *J. High Energy Phys.* **2021**, *2021*, 89. [CrossRef]
59. Durieux, G.; Camacho, A.G.; Mantani, L.; Miralles, V.; López, M.M.; Llácer, M.M.; Poncelet, R.; Vryonidou, E.; Vos, M. Snowmass White Paper: Prospects for the measurement of top-quark couplings. *arXiv* **2022**, arXiv:2205.02140.
60. Cepeda, M.; Gori, S.; Ilten, P.; Kado, M.; Riva, F.; Khalek, R.A.; Aboubrahim, A.; Alimena, J.; Alioli, S.; Alves, A.; et al. Higgs Physics at the HL-LHC and HE-LHC. *arXiv* **2019**, arXiv:1902.00134.
61. Maltoni, F.; Mantani, L.; Mimasu, K. Top-quark electroweak interactions at high energy. *J. High Energy Phys.* **2019**, *2019*, 4. [CrossRef]
62. Durieux, G.; Perelló, M.; Marcel, V.; Zhang, C. Global and optimal probes for the top-quark effective field theory at future lepton colliders. *J. High Energy Phys.* **2018**, *2018*, 168. [CrossRef]
63. Amjad, M.S.; Bilokin, S.; Boronat, M.; Doublet, P.; Frisson, T.; García, I.; García, P.M.; Pöschl, R.; Richard, F.; Ros, E.; et al. A precise characterisation of the top quark electro-weak vertices at the ILC. *Eur. Phys. J. C* **2015**, *75*, 512. [CrossRef]
64. Amjad, M.S.; Boronat, M.; Frisson, T.; Garcìa, I.G.; Pöschl, R.; Ros, E.; Richard, F.; Rouëné, J.; Femenia, P.R.; Vos, M. A precise determination of top quark electro-weak couplings at the ILC operating at $\sqrt{s} = 500\,GeV$. *arXiv* **2013**, arXiv:1307.8102.
65. The CLIC Collaboration. Top-quark physics at the CLIC electron-positron linear collider. *J. High Energy Phys.* **2013**, *2013*, 3. [CrossRef]
66. The ILC International Development Team and the ILC community. The International Linear Collider: Report to Snowmass 2021. *arXiv* **2022**, arXiv:2203.07622.
67. Robson, A.; Roloff, P. Updated CLIC luminosity staging baseline and Higgs coupling prospects. *arXiv* **2018**, arXiv1812.01644.

68. Bernardi, G.; Brost, E.; Denisov, D.; Landsberg, G.; Aleksa, M.; d'Enterria, D.; Janot, P.; Mangano, M.L.; Selvaggi, M.; Zimmermann, F.; et al. The Future Circular Collider: A Summary for the US 2021 Snowmass Process. *arXiv*, **2022**, arXiv: 2203.06520.
69. Abada, A.; Abbrescia, M.; AbdusSalam, S.S.; Abdyukhanov, I.; Fernandez, J.A.; Abramov, A.; Aburaia, M.; Acar, A.O.; Adzic, P.R.; Agrawal, P.; et al. FCC-hh: The Hadron Collider. *Eur. Phys. J. Spec. Top.* **2019**, *228*, 755–1107. [CrossRef]
70. Cros, B.; Muggli, P. ALEGRO input for the 2020 update of the European Strategy. *arXiv* **2020**, arXiv:1901.08436.
71. Stratakis, D.; Mokhov, N.; Palmer, M.; Pastrone, N.; Raubenheimer, T.; Rogers, C.; Schulte, D.; Shiltsev, V.; Tang, J.; Yamamoto, A.; et al. A Muon Collider Facility for Physics Discovery. *arXiv* **2022**, arXiv:2203.08033.

universe

MDPI

Review

Measurements of the Cross-Section for the $t\bar{t}$ + Heavy-Flavor Production at the LHC

Jorgen D'Hondt [1] and Tae Jeong Kim [2,*]

[1] Inter-University Institute for High Energies (IIHE), Vrije Universiteit Brussel, Pleinlaan 2, B-1050 Brussel, Belgium; jorgen.dhondt@vub.be
[2] Department of Physics, Hanyang University, Seoul 04763, Republic of Korea
* Correspondence: taekim@hanyang.ac.kr

Abstract: At the LHC, the process of a Higgs boson decaying into bottom or charm quarks produced in association with a pair of top quarks, $t\bar{t}H$, allows for an empirical exploration of the heavy-flavor quark Yukawa couplings to the Higgs boson. Accordingly, the cross-sections for the $t\bar{t}$ + heavy-flavor production without the appearance of the Higgs boson have been measured at the LHC in various phase spaces using data samples collected in pp collisions at $\sqrt{s}$ = 7, 8 and 13 TeV with the ATLAS and CMS experiments. Flavor ratios of cross-sections of $t\bar{t}$ + heavy-flavors to $t\bar{t}$ + additional jets processes are also measured. In this paper, the measured cross-sections and ratios are reviewed and the prospects with more data are presented.

Keywords: top quark; heavy-flavor; Higgs boson

Citation: D'Hondt, J.; Kim, T.J. Measurements of the Cross-Section for the $t\bar{t}$ + Heavy-Flavor Production at the LHC. *Universe* **2023**, *9*, 242. https://doi.org/10.3390/universe9050242

Academic Editor: Daniel Cherdack

Received: 19 March 2023
Revised: 8 May 2023
Accepted: 14 May 2023
Published: 21 May 2023

1. Introduction

Decades of theoretical and experimental exploration of the most elementary particles and their properties yielded a detailed description of fundamental interactions, captured in a quantum field framework known as the Standard Model of particle physics. The success of this model in describing observations over many orders of magnitude in interaction energy cannot be overestimated. However, despite leading to a more profound understanding, the research field faces several problems and mysteries. Some are related to cosmological observations of dark matter in the universe and the ubiquity of matter over antimatter, some to the mathematical consistency of the model itself with respect to even the smallest variations in its parameters. Several puzzling features are related to the flavor structure of the Standard Model of particle physics, not least those present in the heavy-flavor sector. Through accurate measurements, we attempt to find cracks in the model where theoretical predictions may not match experimental observations. These discoveries may open new avenues to address the open problems and mysteries, either within the realm of quantum field theory or even by questioning the basic principles underlying this mathematical framework.

After the Higgs (H) boson discovery in 2012, the consistency check with the H boson in the standard model was one of the highest priorities at the Large Hadron Collider (LHC), especially in the heavy-flavor sector. From analyzing the proton collision data of the LHC, the couplings of a top quark and a bottom quark (third-generation quarks) in the standard model with the H boson were discovered in different processes [1,2]. However, the confirmation that both couplings are simultaneously consistent with the predictions is only possible at the LHC by measuring the unique process of the H boson production in association with a $t\bar{t}$ pair ($t\bar{t}H$), where the H boson decays to a pair of bottom (b) quarks. This decay channel of the Higgs boson gives the largest signature of the $t\bar{t}$ pair ($t\bar{t}H$). This process alone has yet to be discovered in the data, leading to a $t\bar{t}b\bar{b}$ final state. Understanding the $t\bar{t}b\bar{b}$ process in proton–proton collisions without the presence of an H boson is a prerequisite to the discovery. In addition, the charm (c) jets in the $t\bar{t}c\bar{c}$ process can also be misidentified as b jets, inducing a background in the analogy of the $t\bar{t}b\bar{b}$ process. Therefore, the measurements of cross-sections of the $t\bar{t}$ + heavy-flavor ($t\bar{t}$ + HF) process at the LHC are essential, yet challenging objectives.

Calculations of the inclusive production cross-section for top quark pairs with additional jets by matching matrix element generators to parton showers have been performed to next-to-leading-order (NLO) precision in quantum chromodynamics (QCD) [3–7]. Theoretical QCD calculations of the $t\bar{t}b\bar{b}$ process are available at NLO [8–15] but they suffer from large factorization and renormalization uncertainties due to the presence of two very different scales in this process. Therefore, precise measurements can also provide a good test of the NLO QCD theory itself. Full NLO QCD corrections to off-shell $t\bar{t}b\bar{b}$ production are available in Ref. [16,17]. Calculations of $t\bar{t}b\bar{b}$ with massive b quarks use parton density functions (PDFs) of the proton in the four flavor scheme (4FS), where b quarks are not part of the proton PDF. These matrix element level predictions of $t\bar{t}b\bar{b}$ with massive b quarks are matched to parton showers [18–20]. In addition, the associated production of $t\bar{t}b\bar{b}$ with one additional jet is available as well [21].

The cross-sections for the $t\bar{t}$ + HF production have been measured in various phase spaces using data samples collected in pp collisions by the ATLAS [22] and CMS [23] experiments at $\sqrt{s} = 7$, 8 and 13 TeV [24–33]. In order to obtain observable cross-section values, certain kinematic thresholds should be applied to the additional heavy-flavor jets. The interplay between the b jets from the top quark decay with the additional heavy-flavor jets is not trivial and, accordingly, the definition of the signal is challenging. The definitions are different in each measurement and between experiments. We will discuss the definitions in Section 2 in more detail. In order to achieve higher precision, flavor ratios of cross-sections of $t\bar{t}$ + HF to $t\bar{t}$ + additional jets processes are also measured. The cross-section ratio measurement was originally motivated as many kinematic distributions are expected to be similar for $t\bar{t}b\bar{b}$, $t\bar{t}c\bar{c}$ and $t\bar{t}jj$, leading to reduced systematic uncertainties in the ratio.

Most measurements focus on the $t\bar{t}b\bar{b}$ cross-section. The $t\bar{t}c\bar{c}$ process has been explored less due to the fact that the experimental signature of a c jet is sandwiched between that of b jets and light quark jets and gluons. With the recent development of charm jet taggers, the $t\bar{t}b\bar{b}$ and $t\bar{t}c\bar{c}$ processes can be more efficiently distinguished and the $t\bar{t}c\bar{c}$ cross-section has now been measured by CMS [32].

In this experimental review, we summarize the results for the inclusive and differential cross-section measurements of $t\bar{t}$ + HF production at the LHC submitted to journals or available to the public before May 2023.

2. Definition of the $t\bar{t}$ + Heavy-Flavor Signal

The measurements of the $t\bar{t}b\bar{b}$ and $t\bar{t}c\bar{c}$ cross-sections are performed for both regions of the visible and the full phase space. The resulting cross-sections at the particle level in the visible phase spaces have reduced theoretical and modeling uncertainties while the purpose of performing the measurement in the full phase space is to facilitate comparisons to theoretical calculations or measurements obtained in other decay modes. An example of the $t\bar{t}b\bar{b}$ and $t\bar{t}c\bar{c}$ processes in Feynman diagrams are shown in Figure 1. Final-state particles are defined in Section 2.1 and the processes in Section 2.2.

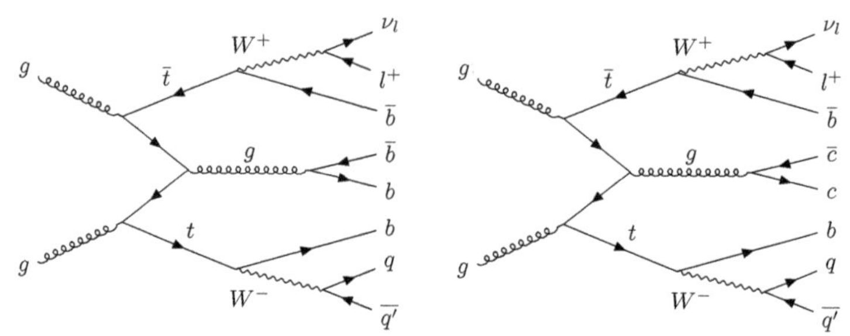

Figure 1. An example of the Feynman diagram of the tt̄bb̄ and tt̄cc̄ processes at the LHC in the lepton + jets channel.

2.1. Particle-Level Object Definition

In the definition of the visible phase space, all generated objects such as leptons and jets are required to be within the experimentally accessible kinematic region. In ATLAS, the objects are defined at the particle level which is based on the stable particles after the hadronization to reduce the dependence on the generation level information. Electrons and muons not emerging from hadron decays are considered. Furthermore, to reach the full particle-level definition, for all charged leptons, potential final-state photon radiation within a $\Delta R = 0.1$ cone around the lepton is added to the four-momenta of the lepton. In CMS, the electrons and muons are required to originate from a W boson at the generator level. The electrons or muons originating from the leptonic decays of τ leptons produced in $W \to \tau\nu$ decays are included. The procedure of adding final state photon radiation to the lepton is not performed in CMS except for the latest result in the lepton + jets channel [33], where the final-state photon radiation is added to the lepton at the particle level. The particle-level jets are defined by clustering stable particles, excluding neutrinos with the anti-kt algorithm with a distance parameter of 0.4 at a center of mass energy $\sqrt{s} = 13$ TeV and 0.5 at $\sqrt{s} = 7,\ 8$ TeV. To identify the heavy-flavor b and c particle-level jets, a so-called ghost matching is performed. The b- and c-hadrons are included in the jet clustering procedure after scaling their momenta to negligible values while preserving their directions. The b and c jets are then identified by the presence of the corresponding "ghost" hadrons among the jet constituents. The approach to defining the particle-level jets is the same in ATLAS and CMS. However, in terms of defining the tt̄ + HF quarks processes, there are subtle distinctions between different channels and experiments we will discuss in the following section.

2.2. Process Definition

In ATLAS, tt̄b(b̄) is defined by the presence of at least three (four) particle-level b jets. Events with only three b jets can come from the case wherein one of the b jets is out of acceptance or two b jets are merged together. For tt̄c, in the dilepton channel, the number of particle-level b jets should be less than 3 and at least one c jet while in the lepton + jets channel events should contain at least two c jets as the events with exactly one c jet would come from the $W \to c\bar{s}(\bar{c}s)$ decays. If the events with additional jets do not meet the criteria described above, the events are grouped into a tt̄l. In ATLAS, only measurements in the visible phase space are available and the origin of the heavy-flavor jets is not identified. Instead, the two b jets with the smallest ΔR separation or with the highest transverse momentum are selected. In CMS, for the visible phase space, the tt̄jj process is defined if the event contains at least four particle-level jets including two b jets, and the same number of leptons as required at the reconstructed level. The tt̄bb̄ process is defined by the presence of at least four b jets regardless of their origin in the dilepton, lepton + jets channel and hadronic channel, called "parton-independent". Additionally, in the hadronic channel, a parton-based definition of requiring two b jets originating from the top quark and two additional b jets is introduced. For the tt̄bl process, the event should contain only one additional b jet and at least one additional light-flavor jet or c jet. The $t\bar{t}ll$ process is the case where there are no additional b or c jets, but at least two additional light-flavor jets within the acceptance. In the $t\bar{t}c\bar{c}$ measurements, the $t\bar{t}c\bar{c}$ process is defined by the presence of at least two b jets and at least two c jets.

The cross-sections are measured in the visible phase space to reduce the systematic uncertainties that can be coming from the theory dependence on the acceptance. For the full-phase space measurements performed by CMS, the additional b jets are required not to be from the weak decay of the $t\bar{t}$ system at the generator level. There is no further requirement for the decay particles from the top quarks. Therefore, the measurements can be compared across different channels as well as with theory predictions. The cross-sections in the full phase space are obtained by taking into account the acceptance which can only be calculated based on simulations inducing an additional systematic uncertainty. The definitions of the signal phase space in ATLAS and CMS are summarized in Table 1. The measured cross-sections can not be compared directly between ATLAS and CMS due to the different phase space definitions. In the following sections, the results from the two experiments are reviewed.

Table 1. Signal definitions at particle level in each measurement. As a default, in the dilepton channel, at least two leptons are required (in ATLAS, 1 e and 1 μ are required) and in the lepton + jets channel, one lepton is required exclusively.

Phase Space	Process	ATLAS	CMS
Full	$t\bar{t}b\bar{b}$	-	$\geq$ 2b not from t [27–30]
	$t\bar{t}c\bar{c}$	-	$\geq$ 2c not from t [32]
Visible	$t\bar{t}b\bar{b}$ (di-lepton)	$\geq$ 3(4)b [24–26]	$\geq$ 4b [27–30]
	$t\bar{t}b\bar{b}$ (semi-lepton)	$\geq$ 5(6)j, $\geq$ 3(4)b [24–26]	$\geq$ 5(6)j, $\geq$ 3(4)b [30,33]
	$t\bar{t}b\bar{b}$ (semi-lepton)	-	$\geq$ 6(7)j, $\geq$ 3(4)b, $\geq$ 3l [33]
	$t\bar{t}b\bar{b}$ (fully hadron)	-	$\geq$ 8j, $\geq$ 4b [31]
	$t\bar{t}c\bar{c}$ (di-lepton)	-	$\geq$ 2b, $\geq$ 2c [32]

2.3. Monte Carlo Simulation

The signals of the $t\bar{t}$ + HF events were simulated using various Monte Carlo (MC) samples in ATLAS and CMS. Theoretical predictions are summarized in this section.

The nominal $t\bar{t}$ sample was generated using the POWHEG generator at next-to-leading-order (NLO) [34–36] at $\sqrt{s}$ = 13 TeV. The parton shower, fragmentation and the underlying events were simulated using PYTHIA 8.210 [37]. This sample is called POWHEG +PYTHIA 8 in the following. At $\sqrt{s}$ = 8 TeV, the events generated using the POWHEG generator were interfaced with PYTHIA 6 [38]. In CMS, the MADGRAPH [39] generator was also used as the nominal $t\bar{t}$ sample at $\sqrt{s}$ = 8 TeV. For the purpose of assessing the uncertainty due to the choice of the QCD MC model and to compare with unfolded data, alternative $t\bar{t}$ samples were generated. Two samples were generated using POWHEG +PYTHIA 8 with different renormalization and factorization scales. To estimate the effect of the choice of parton shower and hadronization algorithms, a $t\bar{t}$ sample was generated by interfacing POWHEG with HERWIG 7 [40,41] (referred to as POWHEG + 7 or as POWHEG + HERWIG ++ in this paper). The $t\bar{t}$ events were also generated with the SHERPA 2.2.1 generator [42], which models the zero and one additional parton process at NLO accuracy and up to four additional partons at LO accuracy. In addition to the samples above, a $t\bar{t}$ sample was also generated using the MADGRAPH _aMC@NLO [5], interfaced to PYTHIA 8. In CMS, the MADGRAPH _aMC@NLO generator is matched to HERWIG 6 and PYTHIA 6 as well at $\sqrt{s}$ = 8 TeV. All of the $t\bar{t}$ samples are normalized to a cross-section calculated at next-to-next-to-leading order (NNLO) [43,44].

A dedicated sample of $t\bar{t}b\bar{b}$ events was generated using SHERPA +OPENLOOPS [18]. The $t\bar{t}b\bar{b}$ matrix elements were calculated with massive b-quarks at NLO, using the COMIX [45] and OPENLOOPS [46] matrix element generators, and merged with the SHERPA parton shower, tuned by the authors [47]. This sample is referred to as SHERPA 2.2 $t\bar{t}b\bar{b}$ (4FS). A sample of $t\bar{t}b\bar{b}$ events was generated using the POWHEL [15], where the matrix elements were calculated at NLO with massless b-quarks and matched to the PYTHIA 8. This sample is referred to as POWHEL +PYTHIA 8 $t\bar{t}b\bar{b}$ (5FS). The POWHEL generator with massive b-quarks and matched to the PYTHIA 8 is referred to as POWHEL +PYTHIA 8 $t\bar{t}b\bar{b}$ (4FS). Another sample of $t\bar{t}b\bar{b}$ events using the POWHEG generator where $t\bar{t}b\bar{b}$ matrix elements were calculated at NLO with mass b-quarks. The events were matched to the PYTHIA 8. This sample is referred to as POWHEG +PYTHIA 8 $t\bar{t}b\bar{b}$ (4FS) to distinguish it from the nominal $t\bar{t}$ sample described above.

3. Portfolio of Cross-Section Measurements

In ATLAS, the $t\bar{t}b$ and $t\bar{t}b\bar{b}$ inclusive and differential cross-sections are measured using data corresponding to an integrated luminosity of 4.7 fb^{-1} of proton–proton collisions at a center-of-mass energy of 7 TeV [24] and 20.3 fb^{-1} at $\sqrt{s}$ = 8 TeV [25]. At $\sqrt{s}$ = 13 TeV, the cross-sections using data corresponding to an integrated luminosity of 36.1 fb^{-1} are measured in the $e\mu$ and in the lepton + jets channels [26].

In CMS, the $t\bar{t}b\bar{b}$ inclusive and differential cross-sections are measured using data collected at $\sqrt{s}$ = 8 TeV [28]. The inclusive cross-sections of the $t\bar{t}b\bar{b}$ production are also measured in the dilepton channel using early data corresponding to an integrated luminosity of 2.3 fb^{-1} at $\sqrt{s}$ = 13 TeV [29]. The inclusive analysis was updated in the dilepton channel and extended to the lepton + jets channel with data corresponding to an integrated luminosity of 35.9 fb^{-1} [30]. Measurements of the $t\bar{t}b\bar{b}$ process in the hadronic channel are performed using data corresponding to an integrated luminosity of 35.9 fb^{-1} at $\sqrt{s}$ = 13 TeV [31]. Measurements of the $t\bar{t}c\bar{c}$ production are also available at $\sqrt{s}$ = 13 TeV [32]. Recently, the measurements of the inclusive and differential cross-sections were updated in the lepton + jets channel with a full Run 2 data corresponding to an integrated luminosity of 138 fb^{-1} [33].

It is worth noting that only a small fraction of the available data has been used for all these $\sigma_{t\bar{t}b\bar{b}}$ and $\sigma_{t\bar{t}c\bar{c}}$ measurements except the lepton + jets channel.

3.1. Inclusive Cross-Section Measurement

In ATLAS, at $\sqrt{s}$ = 13 TeV, the cross-section measurements were performed in the $e\mu$ channel within the at least three b jet visible phase space and in lepton + jets channels within the at least four b jet visible phase space. To extract the $t\bar{t}$ + HF number of events, in both channels, a binned maximum likelihood fit is used on observables discriminating between signal and background. A combined template is created from the sum of all backgrounds. Three templates of $t\bar{t}b$, $t\bar{t}c$ and $t\bar{t}l$ events are created from all of $t\bar{t}$, $t\bar{t}$ in association with a vector boson ($t\bar{t}V$) and $t\bar{t}H$ simulations as those samples contain the signal process. In the $e\mu$ channel, $t\bar{t}c$ and $t\bar{t}l$ are merged together to fit to the distribution of the third highest b-tagging discriminant among the reconstructed jets in the event. The scale factors obtained from the fit are 1.33 $\pm$ 0.06 for the number of $t\bar{t}b$ events and 1.05 $\pm$ 0.04 for the number of the combined $t\bar{t}c$ + $t\bar{t}l$ events. In the lepton + jets channel, all three templates are used to fit to the 2D histograms of the third and fourth b-tagging discriminant. The best fit values are 1.11 $\pm$ 0.2 for the number of $t\bar{t}b$ events, 1.59 $\pm$ 0.06 for the number of $t\bar{t}c$ events and 0.962 $\pm$ 0.003 for the number of $t\bar{t}l$ events. The measured cross-section values for $t\bar{t}b$ for both channels are compatible with each other.

To facilitate the comparison with the theory $t\bar{t}b\bar{b}$ cross-section, the $t\bar{t}H$ and $t\bar{t}V$ processes are also subtracted from the measured cross-section. The measured inclusive cross-sections are shown in Figure 2. All of the inclusive cross-sections measured at $\sqrt{s}$ = 13 TeV in the visible phase space by the ATLAS experiment are summarized in Table 2. The cross-section measurement for the $\geq$three b jet phase space in the $e\mu$ channel has an uncertainty of 13%, which is the most precise measurement. The uncertainties are dominated by systematic uncertainties mainly from the $t\bar{t}$ modeling and b-tagging, as well as the jet energy scale.

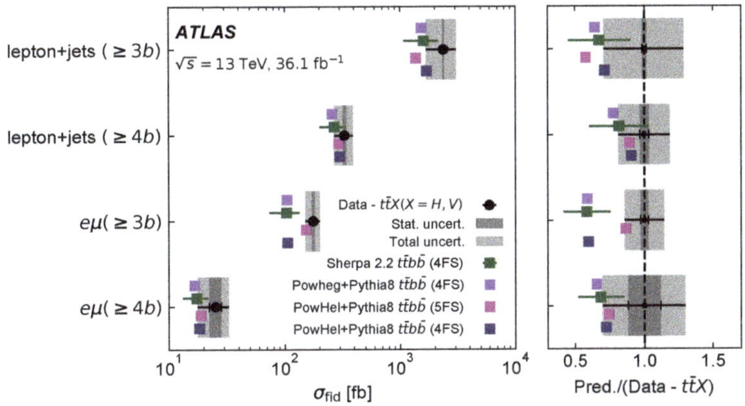

Figure 2. The visible phase space cross-sections measured by ATLAS compared with $t\bar{t}b\bar{b}$ predictions obtained using SHERPA 2.2, POWHEG +PYTHIA 8 and POWHEL +PYTHIA 8 $t\bar{t}b\bar{b}$. The $t\bar{t}H$ and $t\bar{t}V$ processes are subtracted from the measurement to facilitate the comparison with theory [26].

Table 2. Measured $t\bar{t}b\bar{b}$ cross-sections compared with predictions from SHERPA 2.2 in the visible phase space at $\sqrt{s}$ = 13 TeV by ATLAS. The $t\bar{t}H$ and $t\bar{t}V$ contributions are subtracted from the measurement. The statistical and dominating systematical uncertainties on the measurements are presented.

Channel	Measurements (pb)	Predictions (pb)	Phase Space
$t\bar{t}b\bar{b}$			
($e\mu$) [26]	$177 \pm 5 \pm 24$	103 ± 30	$\geq$3b
($e\mu$) [26]	$25 \pm 3 \pm 7$	17.3 ± 4.2	$\geq$4b
(lepton + jets) [26]	$2370 \pm 40 \pm 690$	1600 ± 530	$\geq$ 5j, $\geq$3b
(lepton + jets) [26]	$331 \pm 11 \pm 61$	270 ± 70	$\geq$ 6j, $\geq$4b

The ratio measurement of the cross-sections of $t\bar{t}b\bar{b}$ to $t\bar{t}jj$ production is also available using data collected at $\sqrt{s}$ = 8 TeV [25]. The ratio measurement is motivated to reduce the systematic uncertainties and the result is compared with predictions in Figure 3.

In CMS, the inclusive $t\bar{t}b\bar{b}$ cross-sections are measured in the different phase spaces of the dilepton, lepton + jets and hadronic channels using data collected at $\sqrt{s}$ = 13 TeV by CMS. In the dilepton channel, measurements at $\sqrt{s}$ = 8 TeV are also available. In the dilepton channel, the final state consists of two reconstructed leptons and at least four reconstructed b jets. With these two leptons, the dominating Z + jets background is estimated from data using control samples enriched in Z boson events. Among the at least four b jets, the first and the second jets in decreasing order of the b tagging discriminator tend to be the b jets from the top quark. Therefore, jets with the third and fourth largest b tagging discriminator are considered as the additional b jets. Using the two-dimensional distribution of these discriminators of two determined additional jets, the number of $t\bar{t}b\bar{b}$ events is extracted. Together with the ratio $\sigma_{t\bar{t}b\bar{b}}/\sigma_{t\bar{t}jj}$, the cross-sections $\sigma_{t\bar{t}b\bar{b}}$ and $\sigma_{t\bar{t}jj}$ are measured in the visible phase space. For the purpose of comparing the measurements with the theoretical prediction and with measurements in the other decay modes, the cross-sections in the full phase space are obtained by taking into account the acceptance, $\sigma_{full} = \sigma_{visible}/\mathcal{A}$, where $\mathcal{A}$ is the acceptance, defined as the number of events in the corresponding visible phase space divided by the number of events in the full phase space. The results for the full phase space are shown in Figure 4 (upper).

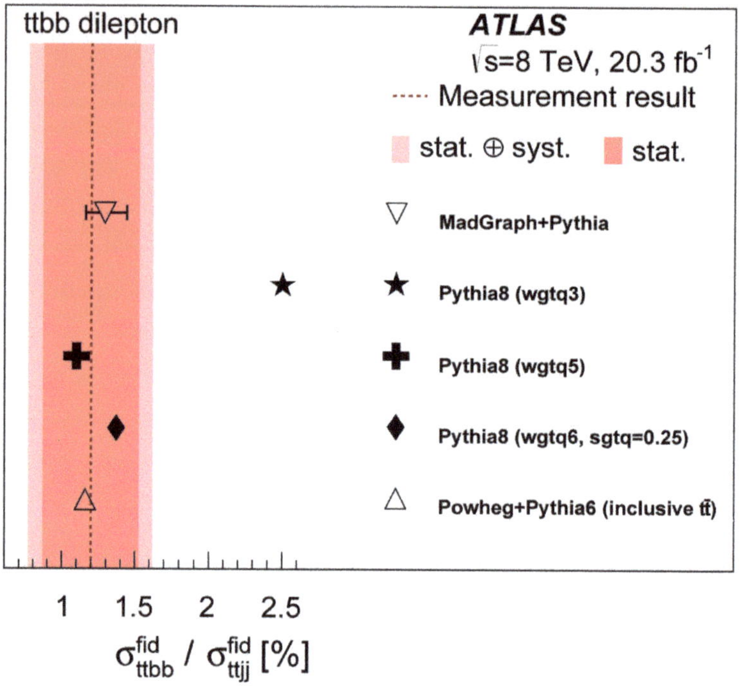

Figure 3. Measurement of the ratio between the $t\bar{t}b\bar{b}$ and $t\bar{t}j\bar{j}$ visible cross-sections at $\sqrt{s}$=8 TeV by ATLAS [25].

In the lepton + jets channel, the measurement was conducted with data corresponding to an integrated luminosity of 35.9 fb^{-1} at $\sqrt{s}$ = 13 TeV in CMS. In this channel, the identification of the origin of the jets is challenging because the final state with at least six jets including four b jets leads to ambiguities in the jet assignment. Moreover, the heavy-flavor jet can also originate from the W boson decay. In order to address this, the kinematic reconstruction method is used to identify the additional b jets. The algorithm assigns a χ^2 value according to the goodness of fit of each jet permutation to meet certain kinematic constraints. The solution selected is the one with the lowest χ^2 value. Once a jet topology is selected, the additional jets in the event are arranged in decreasing order of their b tagging discriminant value. Then, similar to the dilepton channel, only the information from two additional jets with the highest b tagging discriminant value is used to extract the $t\bar{t}b\bar{b}$ cross-section. The results for the ratio $\sigma_{t\bar{t}b\bar{b}}/\sigma_{t\bar{t}j\bar{j}}$, $\sigma_{t\bar{t}b\bar{b}}$ and $\sigma_{t\bar{t}j\bar{j}}$ are presented for both the visible phase space and the full phase space (see Figure 4). Recently, the measurement in the lepton + jets channel was updated with a full Run 2 data corresponding to an integrated luminosity of 138 fb^{-1} [33]. In this analysis, the cross-sections in four different visible phase spaces are measured extensively in four different phase spaces. The final states of each phase space are shown in Table 1. For the phase spaces with the requirement of three additional light jets, it is motivated for the study of additional QCD radiation in $t\bar{t}b$ or $t\bar{t}b\bar{b}$ events as these have been shown to be sensitive to the modeling of $t\bar{t}b\bar{b}$ production. The measured cross-sections in all phase spaces are larger than the predictions from the POWHEG +PYTHIA 8. All other predicted values in each phase space are available in Ref. [33].

In the hadronic channel, the multi-jet process is the main background. To remove the multi-jet events, the quark–gluon discriminant was used. The unsupervised learning algorithm was also further used to maximize the contribution of $t\bar{t}b\bar{b}$ events. The measured cross-sections follow two definitions of the $t\bar{t}b\bar{b}$ events in the fiducial phase space. One is based exclusively on stable generated particles after hadronization (parton-independent). This definition facilitates comparisons with predictions from event generators. The other uses parton-level information after radiation emission (parton-based). This definition is closer to the approach taken by searches for $t\bar{t}H$ production to define the contribution from the $t\bar{t}b\bar{b}$ process. To address the large combinatorial ambiguity in identifying the additional jets in the events, a boosted decision tree (BDT) was used. The cross-section is also reported for the total phase space by correcting the parton-based fiducial cross-section by the experimental acceptance. The results are presented in Figure 5.

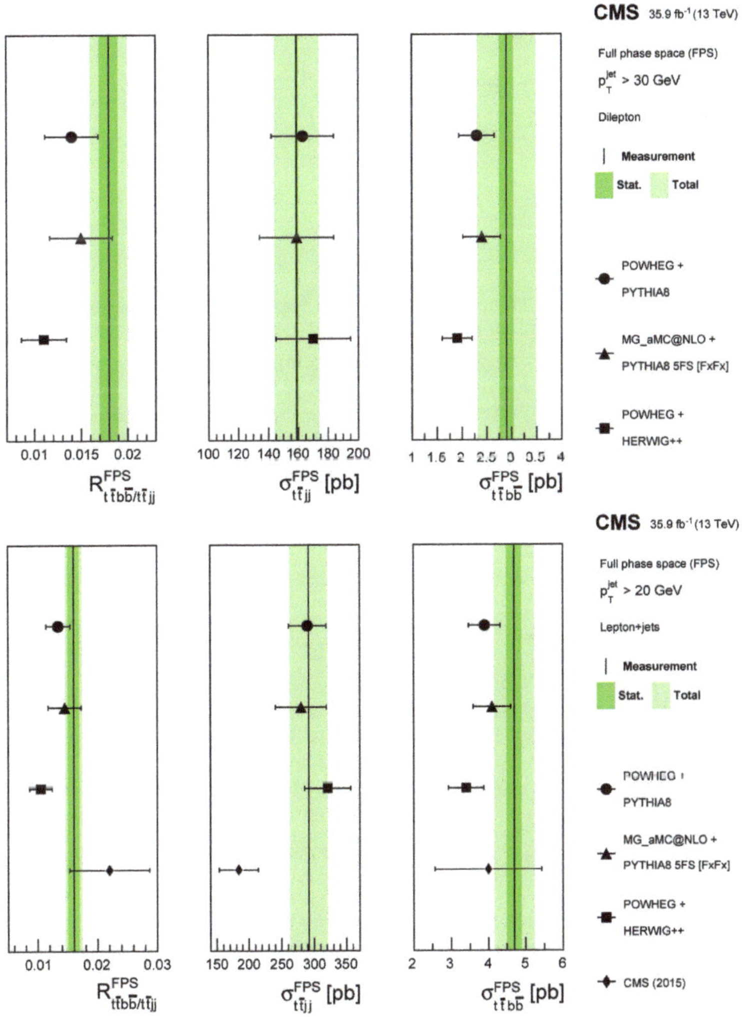

Figure 4. The measured $t\bar{t}b\bar{b}$ cross-sections in the full phase space in the dilepton channel (**upper**) and the lepton + jet channel (**lower**) from the CMS experiment. The dark (light) shaded bands show the statistical (total) uncertainties on the measured values [30].

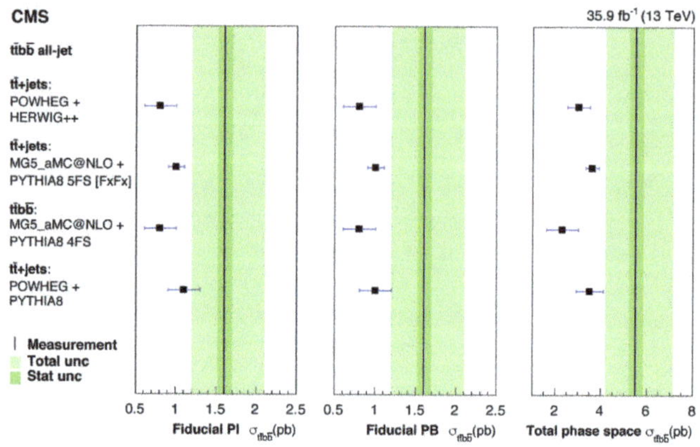

Figure 5. The measured $t\bar{t}b\bar{b}$ cross-sections in the hadronic channel in the parton-independent (**left**), parton-based (**center**) and full phase space (**right**) from the CMS experiment. The dark (light) shaded bands show the statistical (total) uncertainties on the measured values [31].

The cross-section of a top quark pair production with an additional pair of c jets has been measured for the first time by CMS. This measurement is challenging as the experimental signature of a b jet is very similar to that of a c jet. Two additional jets are selected using a deep neural network classifier. To separate the $t\bar{t}c\bar{c}$, $t\bar{t}b\bar{b}$ and $t\bar{t}ll$ events, a NN is trained using charm jet tagging information of the first and second additional jets, and kinematic variables such as the angular separation ΔR between two additional jets, as well as the NN score for the best jet permutation. This NN predicts the probabilities for five output classes of $t\bar{t}c\bar{c}$, $t\bar{t}cl$, $t\bar{t}b\bar{b}$, $t\bar{t}bl$ and $t\bar{t}ll$. Two discriminators are derived as follows.

$$\Delta_b^c = \frac{P(t\bar{t}c\bar{c})}{P(t\bar{t}c\bar{c}) + P(t\bar{t}b\bar{b})}, \qquad \Delta_L^c = \frac{P(t\bar{t}c\bar{c})}{P(t\bar{t}c\bar{c}) + P(t\bar{t}ll)}. \tag{1}$$

The $t\bar{t}c\bar{c}$, $t\bar{t}b\bar{b}$ and $t\bar{t}ll$ cross-sections are extracted from a fit to the two-dimensional distribution of these discriminators. The ratios R_b and R_c of, respectively, the measured $\sigma_{t\bar{t}b\bar{b}}$ and $\sigma_{t\bar{t}c\bar{c}}$ cross-sections with respect to the inclusive $t\bar{t}$ + two jets cross-section were also measured. The results are compared to theoretical predictions of either the POWHEG or MADGRAPH 5_aMC@NLO generators as shown in Figure 6.

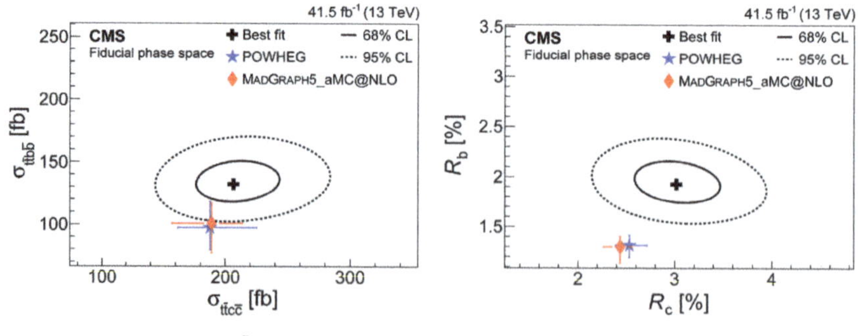

Figure 6. Results of the $t\bar{t}b\bar{b}$ versus $t\bar{t}c\bar{c}$ cross-section measured by CMS in the fiducial phase space, and their ratios to the inclusive $t\bar{t}$ + two jets cross-section [32].

All of the inclusive cross-sections measured in the visible phase space by the CMS experiment are summarized in Tables 3 and 4, and for the full phase space in Tables 5 and 6. Figure 7 also shows the comparison between the measured values in the full phase space and various theoretical predictions in CMS.

Table 3. Measured and predicted cross-sections in the visible phase space at $\sqrt{s}$ = 13 TeV by CMS. In the hadronic channel, the parton-based cross-section is shown. The predictions of POWHEG +PYTHIA 8 are shown.

Channel	Measurements (pb)	Predictions (pb)	b jet Requirement
$t\bar{t}b\bar{b}$			
dilepton [29]	$0.088 \pm 0.012 \pm 0.029$	0.070 ± 0.009	$p_T > 20$ GeV, $\|\eta\| < 2.5$
dilepton [30]	$0.040 \pm 0.002 \pm 0.005$	0.032 ± 0.004	$p_T > 30$ GeV, $\|\eta\| < 2.4$
lepton + jets			
$\geq$ 6j $\geq$ 4b [30]	$0.62 \pm 0.03 \pm 0.07$	0.52 ± 0.06	$p_T > 20$ GeV, $\|\eta\| < 2.5$
$\geq$ 5j, $\geq$ 3b [33]	$2.368 \pm 0.142 \pm 0.014$	1.791	$p_T > 25$ GeV, $\|\eta\| < 2.4$
$\geq$6j $\geq$ 3b $\geq$ 3l [33]	$1.036 \pm 0.090 \pm 0.012$	0.899	$p_T > 25$ GeV, $\|\eta\| < 2.4$
$\geq$ 6j $\geq$ 4b [33]	$0.289 \pm 0.036 \pm 0.006$	0.240	$p_T > 25$ GeV, $\|\eta\| < 2.4$
$\geq$ 7j $\geq$ 4b $\geq$ 3l [33]	$0.144 \pm 0.025 \pm 0.005$	0.129	$p_T > 25$ GeV, $\|\eta\| < 2.4$
hadronic [31]	$1.6 \pm 0.1^{+0.5}_{-0.4}$	1.0 ± 0.2	$p_T > 20$ GeV, $\|\eta\| < 2.5$
dilepton [32]	$0.132 \pm 0.010 \pm 0.015$	0.097 ± 0.021	$p_T > 20$ GeV, $\|\eta\| < 2.4$
$t\bar{t}c\bar{c}$			
dilepton [32]	$0.207 \pm 0.025 \pm 0.027$	0.187 ± 0.038	$p_T > 20$ GeV, $\|\eta\| < 2.4$

Table 4. Measured and predicted cross-section ratios in the visible phase space at $\sqrt{s}$ = 13 TeV by CMS. The predictions of POWHEG +PYTHIA 8 are shown.

Channel	Measurements (%)	Predictions (%)	b jet Requirement
$t\bar{t}b\bar{b}/t\bar{t}jj$			
dilepton [29]	$2.4 \pm 0.3 \pm 0.7$	1.4 ± 0.1	$p_T > 20$ GeV, $\|\eta\| < 2.5$
dilepton [30]	$1.7 \pm 0.1 \pm 0.1$	1.3 ± 0.2	$p_T > 30$ GeV, $\|\eta\| < 2.4$
lepton + jets [30]	$2.0 \pm 0.1 \pm 0.1$	1.7 ± 0.2	$p_T > 20$ GeV, $\|\eta\| < 2.5$
dilepton [32]	$1.93 \pm 0.15 \pm 0.18$	1.31 ± 0.12	$p_T > 20$ GeV, $\|\eta\| < 2.4$
$t\bar{t}c\bar{c}/t\bar{t}jj$			
dilepton [32]	$3.01 \pm 0.34 \pm 0.31$	2.53 ± 0.18	$p_T > 20$ GeV, $\|\eta\| < 2.4$

Table 5. Measured and predicted cross-sections in the full phase space at $\sqrt{s}$ = 13 TeV by CMS. The predictions of POWHEG +PYTHIA 8 prediction are shown.

Channel	Measurements (pb)	Predictions (pb)	b jet Requirement
$t\bar{t}b\bar{b}$			
dilepton [29]	$4.0 \pm 0.6 \pm 1.3$	3.2 ± 0.4	$p_T > 20$ GeV, $\|\eta\| < 2.5$
dilepton [30]	$2.9 \pm 0.1 \pm 0.5$	2.3 ± 0.4	$p_T > 30$ GeV, $\|\eta\| < 2.4$
lepton + jets [30]	$4.7 \pm 0.2 \pm 0.6$	3.9 ± 0.4	$p_T > 20$ GeV, $\|\eta\| < 2.5$
hadronic [31]	$5.5 \pm 0.3^{+1.6}_{-1.3}$	3.5 ± 0.6	$p_T > 20$ GeV, $\|\eta\| < 2.5$
dilepton [32]	$4.54 \pm 0.34 \pm 0.56$	3.34 ± 0.72	$p_T > 20$ GeV, $\|\eta\| < 2.4$
$t\bar{t}c\bar{c}$			
dilepton [32]	$10.1 \pm 1.2 \pm 1.4$	9.1 ± 1.8	$p_T > 20$ GeV, $\|\eta\| < 2.4$

Table 6. Measured and predicted cross-section ratios in the full phase space at $\sqrt{s}$ = 13 TeV by CMS. The predictions of POWHEG +PYTHIA 8 prediction are shown.

Channel	Measurements (%)	Predictions (%)	b jet Requirement
$t\bar{t}b\bar{b}/t\bar{t}jj$			
dilepton [29]	$2.2 \pm 0.3 \pm 0.6$	1.2 ± 0.1	$p_T > 20$ GeV, $\lvert\eta\rvert < 2.5$
dilepton [30]	$1.8 \pm 0.1 \pm 0.2$	1.4 ± 0.3	$p_T > 30$ GeV, $\lvert\eta\rvert < 2.4$
lepton + jets [30]	$1.6 \pm 0.1 \pm 0.1$	1.3 ± 0.2	$p_T > 20$ GeV, $\lvert\eta\rvert < 2.5$
dilepton [32]	$1.51 \pm 0.11 \pm 0.16$	1.03 ± 0.08	$p_T > 20$ GeV, $\lvert\eta\rvert < 2.4$
$t\bar{t}c\bar{c}/t\bar{t}jj$			
dilepton [32]	$3.36 \pm 0.38 \pm 0.34$	2.81 ± 0.20	$p_T > 20$ GeV, $\lvert\eta\rvert < 2.4$

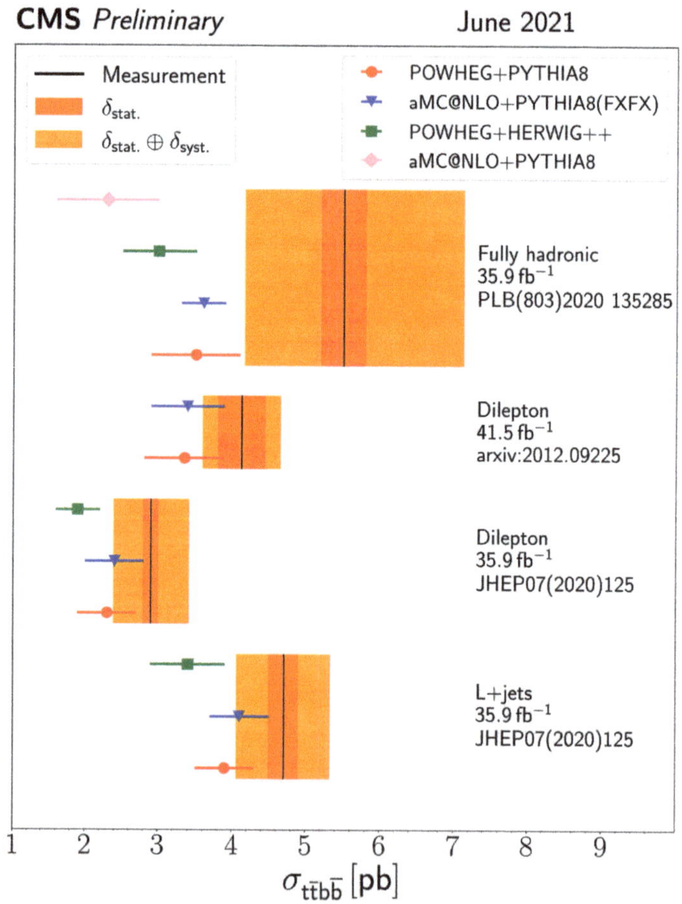

Figure 7. The $t\bar{t}b\bar{b}$ cross-sections in the various channels for the full phase space measured by CMS [48].

3.2. *Differential Cross-Section Measurements*

In addition to the inclusive cross-section measurements, the differential measurements of the $t\bar{t}$ + HF production cross-sections can also provide information on the perturbative QCD (pQCD) and enable the searches for potential new physics. The $t\bar{t}b\bar{b}$ differential cross-section measurements have been performed at $\sqrt{s}$ = 7, 8 and 13 TeV with the ATLAS experiment and at $\sqrt{s}$ = 8 and 13 TeV with the CMS experiment.

To measure the differential cross-sections, the measured distributions at the detector level need to be unfolded to the generator level where the detector effect is removed so that the resulting cross-section can be compared with theory predictions and results from other experiments. At the generator level, it is not trivial to define the additional b jets in the $t\bar{t}b\bar{b}$ process as we have b jets from the top quark decay. Moreover, the b jet could also emerge from the W boson decay. The additional b jets are expected to come from the gluon decay and can also come from the decay of the H boson or another boson.

In the b jet identification, there is a clear difference between the two experiments in ATLAS and CMS. In the ATLAS experiment, at the particle level, there is no attempt to identify the origin of the b jets relying on the simulation information. At this particle level, the two b jets with the highest p_T or the smallest ΔR are selected for the differential cross-section measurement. The highest p_T jets are considered as the b jets from the top quark while the b jets with the smallest ΔR are considered as the additional b jets not from the top quark decay to make use of the fact that the b jets from a gluon splitting tend to be collinear. While in CMS, the origin of the b jets is explicitly identified using the simulation information. For example, the b hadron is traced back through its ancestors in the simulation chain. In this way, only if the b jet is not from a top quark, the b jet is identified as one of the two additional b jets.

For the ATLAS measurements, the unfolded results are presented as normalized differential cross-sections in visible phase space as a function of the b jet multiplicity, global event properties and various kinematic variables. The measurements are conducted in the $e\mu$ channel with at least three reconstructed b jets and in the lepton + jets channel with at least four b jets. The sample with at least four b jets in the lepton + jets channel has high signal purity resulting in a measurement with smaller dependence on the simulation. The $e\mu$ channel benefits from an order of magnitude of a larger sample size containing at least three b jets.

Once the reconstructed level distributions of $t\bar{t}$ + HF events are extracted, then the measured distributions are unfolded to the particle level. The detector resolution effect and inefficiency are corrected by inverting the migration matrix which is optimized for a diagonal matrix. An iterative Bayesian unfolding technique [49] implemented in the ROOUNFOLD software package [50] is used in this process. Detector efficiencies and acceptance are then corrected using a bin-by-bin method. Figure 8 shows the normalized cross-section as a function of the b jet multiplicity compared with predictions from various generator set-ups. The first three panels show the ratios of various predictions to data. The last panel shows the ratio of predictions of normalized differential cross-sections from MADGRAPH 5_aMC@NLO+PYTHIA 8, including or not the contributions from the $t\bar{t}H$ and $t\bar{t}V$ processes. All predictions relying on the parton shower generation of jets for high multiplicities are lower compared to the measurements. This suggests that the b jet production by the parton shower is not optimal in these processes. The comparison of the predictions from various generators with the measurements are made after subtracting the simulation-estimated contributions of $t\bar{t}V$ and $t\bar{t}H$ production from the data. The impact of including these processes in the prediction increases with b jet multiplicity, resulting in a change of about 10% relative to the QCD $t\bar{t}$ prediction alone in the inclusive four b jet bin. The measurement in the $e\mu$ channel with at least three b jets tends to be more precise than in the lepton + jets channel with at least four b jets.

It is also of importance to verify the distributions of the p_T, the mass and the angular distance ΔR of the two b jets where the $b_1 b_2$ system is built from the two highest-p_T b jets and the two closest b jets in ΔR. The measured distributions of those three variables in the lepton + jets channel are shown in Figures 9–11. The differential cross-section as a function of the p_T of the $b_1 b_2$ system is measured with a precision of 10–15% over the full range in the $e\mu$ channel and with an uncertainty of 20–25% in the lepton + jets channel. In general, the differential distributions are well described by the different theoretical predictions, which vary significantly less compared to the size of the experimental uncertainty. All other distributions such as H_T or p_T of additional b jets are available in Ref. [26].

In CMS, the differential cross-sections are measured in the visible phase space as a function of various kinematic properties such as the p_T and η of the leading and subleading additional b jets, the angular distance ΔR between them and the invariant mass $m_{b\bar{b}}$ of the two additional b jets. In particular, the differential cross-sections as a function of the $m_{b\bar{b}}$ and ΔR are of interest as the two additional b jets from a gluon tend to be produced collinearly and those from the H boson have the resonance peak at 125 GeV.

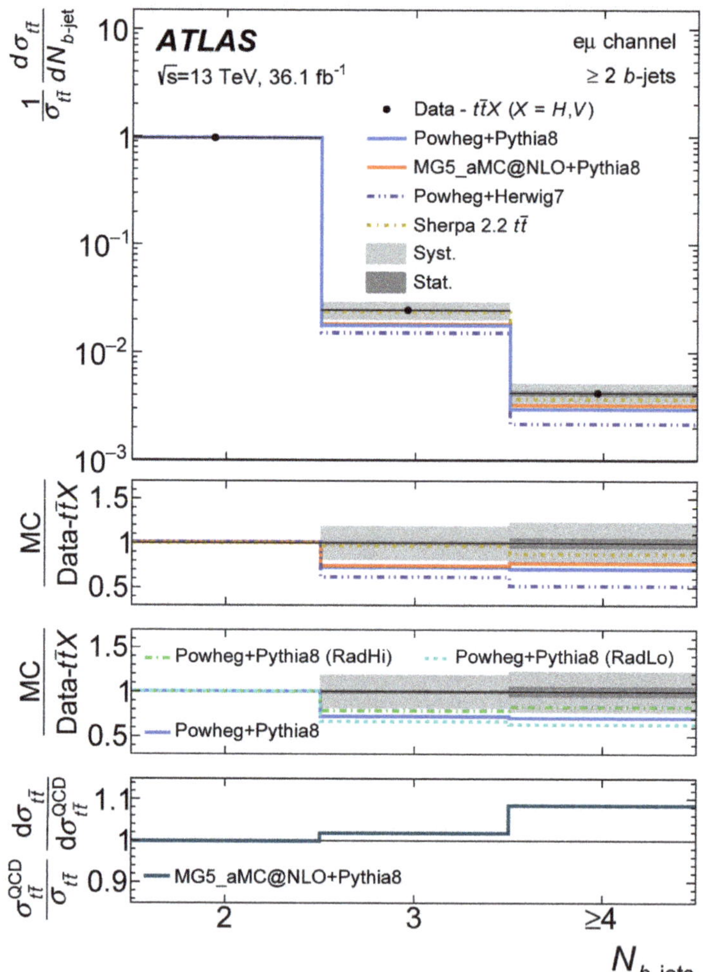

Figure 8. The relative differential cross-section as a function of the b jet multiplicity in events with at least two b jets in the $e\mu$ channel compared with various generators. The $t\bar{t}H$ and $t\bar{t}V$ contributions are subtracted from data. Uncertainty bands represent the statistical and total systematic uncertainties [26].

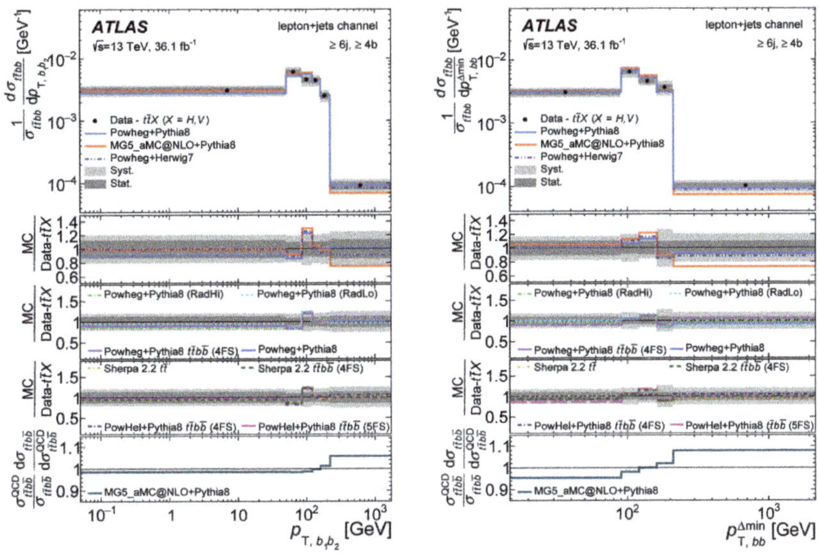

Figure 9. Relative differential cross-sections as a function of p_T of the two highest-p_T b jets (**left**) and the two closest b jets in ΔR (**right**) in the events with at least four b jets in the lepton + jets channel compared with various generators from the ATLAS measurements. The contributions from $t\bar{t}$H and $t\bar{t}$V are subtracted from data [26].

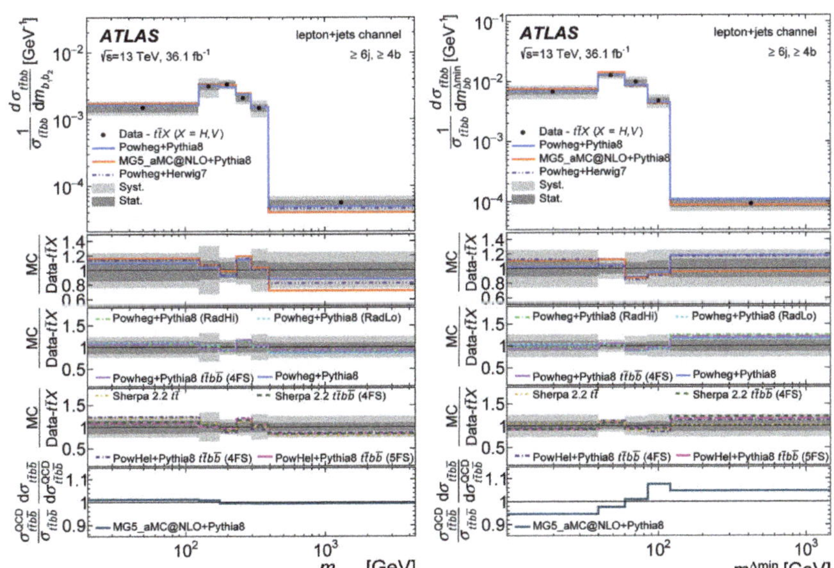

Figure 10. Relative differential cross-sections as a function of $m_{b_1b_2}$ of the two highest-p_T b jets (**left**) and the two closest b jets in ΔR (**right**) in the events with at least four b jets in the lepton + jets channel compared with various generators from the ATLAS measurements. The contributions from $t\bar{t}$H and $t\bar{t}$V are subtracted from data [26].

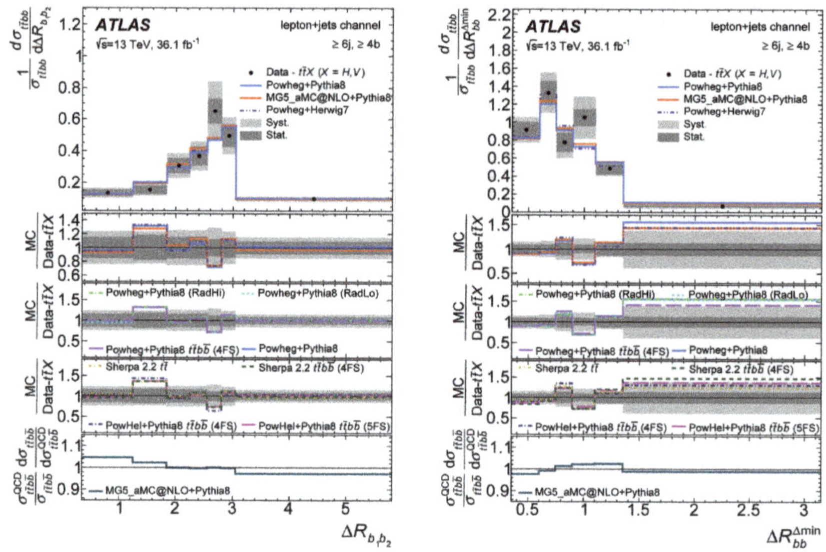

Figure 11. Relative differential cross-sections as a function of $\Delta R_{b_1,b_2}$ of the two highest$-p_T$ b jets (**left**) and the two closest b jets in ΔR (**right**) in the events with at least four b jets in the lepton + jets channel compared with various generators from the ATLAS measurements. The contributions from $t\bar{t}H$ and $t\bar{t}V$ are subtracted from data [26].

At the reconstruction level, it is very challenging to identify two additional b jets because there are four b jets from top quarks and a gluon splitting. To select the additional b jets, the multivariate approach of a BDT was used to maximize the correct assignment of additional b jets. The input variables to the BDT combine information from the two final-state leptons, the jets and E_T^{miss}. A total of twelve variables, e.g., the sum and difference of the invariant mass of the bl^+ and $\bar{b}l^-$ system, $m^{bl^+} \pm m^{\bar{b}l^-}$; the absolute difference in the azimuthal angle between them, $|\Delta\phi^{bl^+,\bar{b}l^-}|$; the p_T of the bl^+ and $\bar{b}l^-$ system, $p_T^{bl^+}$ and $p_T^{\bar{b}l^-}$ and the difference between the invariant mass of the two b jets and two leptons and the invariant mass of the $b\bar{b}$ pair, $m^{b\bar{b}l^+l^-} - m^{b\bar{b}}$, are used as input variables. The variables insensitive to the additional radiation are selected to avoid any dependence on the kinematics of the additional jets. The jets from the $t\bar{t}$ system are identified as the pair with the highest BDT discriminant. From the remaining jets, those b-tagged jets with the highest p_T are selected as being the leading additional ones. With this method, the correct assignment rate for the additional b jets in $t\bar{t}b\bar{b}$ events is around 40%.

A template fit to the b-tagged jet multiplicity distribution is performed to improve the data and simulation comparison. For the differential cross-section measurements, effects from detector efficiency and resolution are corrected by using the regularized inversion of the response matrix which is calculated from simulated $t\bar{t}$ events. The measured differential cross-sections as a function of the leading and subleading additional b jet p_T, the ΔR and invariant mass of two additional b jets are shown in Figure 12 for CMS. Measured cross-sections are compared with various theoretical predictions. The shape of the p_T distributions are well described by prediction. However, the measured values by CMS have larger uncertainties due to the use of a smaller data sample with respect to ATLAS.

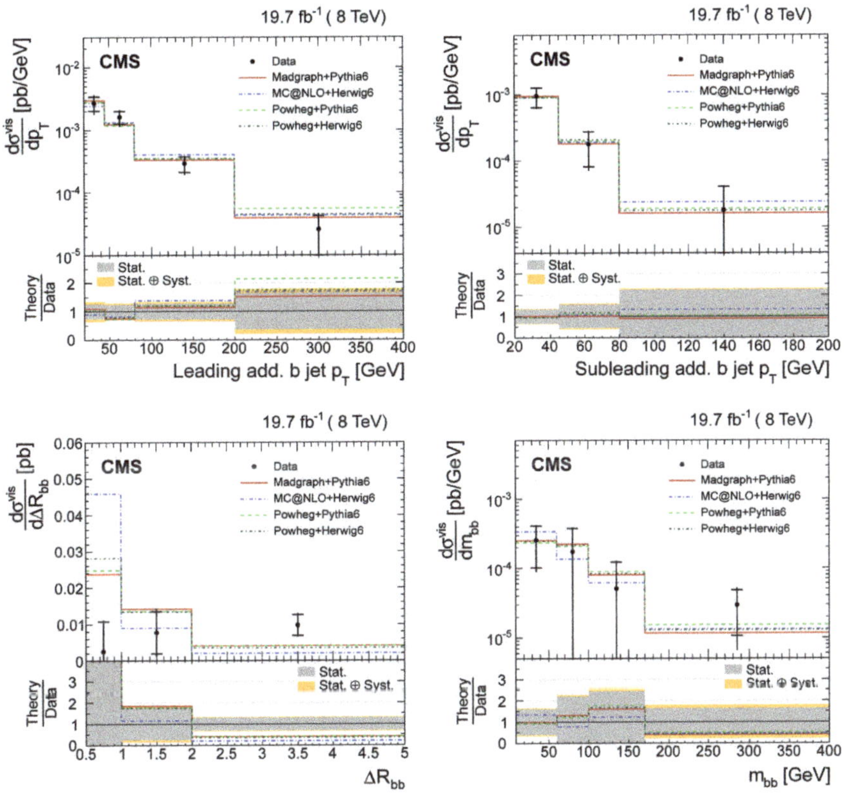

Figure 12. Absolute differential $t\bar{t}$ cross-sections as a function of the leading and subleading additional b jet p_T (**upper**) and angular separation and invariant mass of two additional b jets (**lower**) in the visible phase space at the CMS. Inner vertical bars indicate the statistical uncertainties [27].

In CMS, the differential cross-sections are measured with a full Run 2 data in the lepton + jets channel. In this analysis, two approaches are used to identify the additional b jets from the gluon splitting, while two b jets with the smallest angular separation are selected to reduce the systematic uncertainty on theory dependence, a multivariate algorithm based on a deep neural network (DNN) is also used to identify additional b jets not from top quarks by using the MC information.

To find the correct pair of b jets not from top quarks, only four b jets in the highest p_T order are used as candidate jets, which results in the six possible candidate jet combinations. The DNN makes use of two sets of input variables, targeting jet-specific input information and global event information separately. For jet-specific input information, the input variables consist of the p_T, η, a flag indicating whether it passes the tight b tagging working point, the angular separation (ΔR) with the charged lepton and the invariant mass with the charged lepton. These input variables are connected via five convolutional network layers (CNN) [51] followed by a long short-term memory (LSTM) cell [52]. For the global event information, the input variables consist of the scalar p_T sum of the four candidate b jets, the p_T, η, ϕ of the charged lepton, the $\Delta\phi$, $\Delta\eta$ and invariant mass of the dijet combinations, the Δ R of the dijet combinations and the charged lepton as well as the jet and b-tagged jet multiplicities. These input variables are connected to three dense network layers with 50 nodes each. Both of these sequences are concatenated at the end into one dense layer with 10 nodes, which is connected to an output layer consisting of six nodes, each representing one of the six possible candidate jet combinations. The pair of b-tagged jets with the highest

DNN output value per event is chosen as the correct assignment of the additional b jet pair and used further for the differential cross-section measurement.

The correct assignment of additional b jets in the DNN is about 49%, which represents a significant increase compared to choosing the two b jets closest in ΔR, which only yields about 41%. The measured differential cross-sections as a function of the leading and subleading additional b jet p_T, the ΔR and invariant mass of two additional b jets selected in the DNN are shown in Figure 13. The distributions are not well described by POWHEG + HERWIG 7 (referred to as POWHEG + H7 in Figure 13). More differential variables are available in Ref. [33].

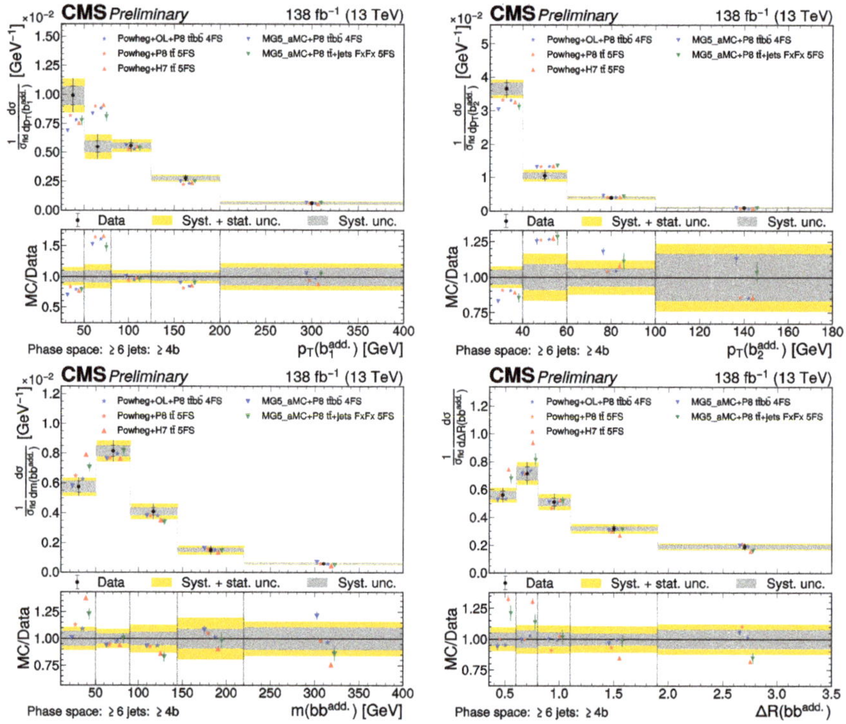

Figure 13. Normalized differential $t\bar{t}$ cross-sections as a function of the leading and subleading additional b jet p_T (**upper**) and angular separation and invariant mass of two additional b jets (**lower**) in the visible phase space at the CMS. Inner vertical bars indicate the statistical uncertainties [33].

4. Discussion

The results from the Run 1 and Run 2 data analyses of ATLAS and CMS of the measured inclusive cross-sections of $t\bar{t}$ + HF jets are higher than the theoretical predictions (see Tables 2–6).

It will be interesting to observe whether these differences become significant with additional data. There were also attempts to measure the differential cross-sections aiming to identify variables where the differences become larger. The measured differential cross-sections are in general consistent with theory predictions within its large statistical uncertainty. However, in the ΔR distribution, there is a discrepancy at the first bin. In particular, the HERWIG prediction tends to produce two additional b jets with smaller angles than the measured value as well as other predictions matched to PYTHIA.

In the realm of $t\bar{t}$ + HF, there is a large fraction of Run-2 data yet to be analyzed and we expect twice more data in Run-3 at the LHC in the coming years. We can envisage reducing not only the statistical uncertainty but also systematic uncertainties as more data may enable more data-driven techniques. More data will also make it possible to use a smaller bin width to enable hints about potential discrepancies in the inclusive measurements. More advanced heavy-flavor tagging may distinguish the c-flavor jet better from the b-flavor jets. Novel flavor tagging developments can indeed increase our understanding of pQCD and the potential to discover new physics. More synchronized definitions are also required to compare or combine results from ATLAS and CMS.

As we have a systematically higher measured value of cross-section of the $t\bar{t}$ + HF compared to prediction, more data from Run-3 and eventually from the High Luminosity-LHC may provide interesting opportunities to find cracks in our theoretical understanding. The discrepancy could be from the fact that signal samples are modeled only at NLO in QCD. We should also make use of the effective field theory (EFT) approach for possible new physics. To interpret experimental measurements in the context of physics beyond the standard model, the EFT approach is of interest as a model-independent approach [53]. Differential measurements may be crucial in this approach as the presence of the SMEFT operators can modify the kinematics in the standard model processes.

5. Conclusions

The inclusive and differential cross-sections for the $t\bar{t}$ + HF jet production have been measured extensively in ATLAS and CMS for the various phase spaces using data samples collected in pp collisions at $\sqrt{s} = 7$, 8 and 13 TeV. The ratio of the cross-sections of the $t\bar{t}$ + HF jets with respect to the cross-section of the $t\bar{t}$ + additional jets is also measured, aiming for reduced uncertainties as many kinematic distributions are expected to be similar between the $t\bar{t}$ + HF jets and the $t\bar{t}$ + additional jets. These measured cross-sections systematically tend to be higher than the predictions. The measurements are dominated by systematic uncertainties that could be reduced by deploying data-driven techniques to better control the impact of backgrounds and reconstruction-related systematic uncertainties. Having more data in the coming years with a better understanding of detectors and more sophisticated reconstruction techniques will bring us to the precision era, where possible new physics could finally be revealed.

Funding: This work was supported under the framework of the international cooperation program managed by the National Research Foundation of Korea (NRF-2022K2A9A1A06093535). J.D. is supported in part by the Strategic Research Program "High-Energy Physics" of the VUB and also by the FWO-Vlaanderen.

Conflicts of Interest: The authors declare no conflict of interest.

References

1. Sirunyan, A.M. et al. [CMS Collaboration]. Observation of $t\bar{t}$H Production. *Phys. Rev. Lett.* **2018**, *120*, 231801. [CrossRef] [PubMed]
2. Aaboud, M. et al. [ATLAS Collaboration]. Observation of Higgs boson production in association with a top quark pair at the LHC with the ATLAS detector. *Phys. Lett. B* **2015**, *784*, 173–191.
3. Höche, S.; Krauss, F.; Maierhöfer, P.; Pozzorini, S.; Schönherr, M.; Siegert, F. Next-to-leading order QCD predictions for top-quark pair production with up to two jets merged with a parton shower. *Phys. Lett. B* **2015**, *748*, 74–78. [CrossRef]
4. Höche, S.; Maierhöfer, P.; Moretti, N.; Pozzorini, S.; Siegert, F. Next-to-leading order QCD predictions for top-quark pair production with up to three jets. *Eur. Phys. J. C* **2017**, *77*, 145. [CrossRef]
5. Alwall, J.; Frederix, R.; Frixione, S.; Hirschi, V.; Maltoni, F.; Mattelaer, O.; Shao, H.S.; Stelzer, T.; Torrielli, P.; Zaro, M. The automated computation of tree-level and next-to-leading order differential cross sections, and their matching to parton shower simulations. *J. High Energy Phys.* **2014**, *2014*, 79. [CrossRef]
6. Frederix, R.; Frixione, S. Merging meets matching in MC@NLO. *J. High Energy Phys.* **2012**, *2012*, 61. [CrossRef]
7. Frixione, S.; Nason, P.; Ridolfi, G. A positive-weight next-to-leading-order Monte Carlo for heavy flavour hadroproduction. *J. High Energy Phys.* **2007**, *2017*, 126. [CrossRef]
8. Worek, M.; Bevilacqua, G. On the ratio of $t\bar{t}b\bar{b}$ and $t\bar{t}jj$ cross sections at the CERN Large Hadron Collider. *J. High Energy Phys.* **2014**, *2014*, 135.
9. Worek, M. Next-to-leading order QCD corrections to $t\bar{t}b\bar{b}$ production at the LHC. *arXiv* **2009**, arXiv:0910.4080. [CrossRef]
10. Worek, M. On the next-to-leading order QCD K-factor for $t\bar{t}b\bar{b}$ production at the TeVatron. *J. High Energy Phys.* **2012**, *2012*, 43. [CrossRef]

11. Bevilacqua, G.; Czakon, M.; Papadopoulos, C.G.; Pittau, R.; Worek, M. Assault on the NLO wishlist: Pp → $t\bar{t}b\bar{b}$. *J. High Energy Phys.* **2009**, *2019*, 109. [CrossRef]
12. Bredenstein, A.; Denner, A.; Dittmaier, S.; Pozzorini, S. NLO QCD corrections to ttbb production at the LHC: 1. Quark-antiquark annihilation. *J. High Energy Phys.* **2008**, *2008*, 108. [CrossRef]
13. Bredenstein, A.; Denner, A.; Dittmaier, S.; Pozzorini, S. NLO QCD corrections to pp → $t\bar{t}b\bar{b}$ + X at the LHC. *Phys. Rev. Lett.* **2009**, *103*, 012002. [CrossRef]
14. Bredenstein, A.; Denner, A.; Dittmaier, S.; Pozzorini, S. NLO QCD corrections to ttbb production at the LHC: 2. Full hadronic results. *J. High Energy Phys.* **2010**, *2010*, 21. [CrossRef]
15. Garzelli, M.V.; Kardos, A.; Trócsányi, Z. Hadroproduction of $t\bar{t}b\bar{b}$ final states at LHC: Predictions at NLO accuracy matched with Parton Shower. *J. High Energy Phys.* **2015**, *2015*, 83. [CrossRef]
16. Denner, A.; Lang, J.-N.; Pellen, M. Full NLO QCD corrections to off-shell ttbb production. *Phys. Rev. D* **2021**, *104*, 056018. [CrossRef]
17. Giuseppe, B.; Huan-Yu, B.; Hartanto, H.B.; Kraus, M.; Lupattelli, M.; Malgorzata, W. $t\bar{t}b\bar{b}$ at the LHC: On the size of corrections and b-jet definitions. *J. High Energy Phys.* **2021**, *2021*, 8. [CrossRef]
18. Cascioli, F.; Maierhöfer, P.; Moretti, N.; Pozzorini, S.; Siegert, F. NLO matching for $t\bar{t}b\bar{b}$ production with massive b quarks. *Phys. Lett. B* **2014**, *734*, 210–214. [CrossRef]
19. Bevilacqua, G.; Garzelli, M.V.; Kardos, A. $t\bar{t}b\bar{b}$ hadroproduction with massive bottom quarks with POWHEL. *arXiv* **2017**, arXiv:1709.06915.
20. Jezo, T.; Lindert, J.M.; Moretti, N.; Pozzorini, S. New NLOPS predictions for $t\bar{t}$ + b jet production at the LHC. *Eur. Phys. J. C* **2018**, *78*, 502. [CrossRef]
21. Buccioni, F.; Kallweit, S.; Pozzorini, S.; Zoller, M.F. NLO QCD predictions for $t\bar{t}b\bar{b}$ production in association with a light jet at the LHC. *J. High Energy Phys.* **2019**, *2019*, 15. [CrossRef]
22. Aad, G. et al. [The ATLAS Collaboration] ATLAS Collaboration. The ATLAS Experiment at the CERN LHC. *JINST* **2008**, *3*, S08003. Available online: https://iopscience.iop.org/article/10.1088/1748-0221/3/08/S08003 (accessed on 6 May 2023).
23. Chatrchyan, S. et al. [The CMS Collaboration] The CMS Experiment at the CERN LHC. *JINST* **2008**, *3*, S08004. Available online: https://iopscience.iop.org/article/10.1088/1748-0221/3/08/S08004/meta (accessed on 6 May 2023).
24. Aguilar-Saavedra, J.A. et al. [ATLAS Collaboration]. Study of heavy-flavor quarks produced in association with top-quark pairs at $\sqrt{s}$ = 7 TeV using the ATLAS detector. *Phys. Rev. D* **2014**, *89*, 072012.
25. Abbott, B. et al. [ATLAS Collaboration]. Measurements of fiducial cross-sections for $t\bar{t}$ production with one or two additional b-jets in pp collisions at $\sqrt{s}$ = 8 TeV using the ATLAS detector. *Eur. Phys. J. C* **2016**, *76*, 11. [CrossRef]
26. Aaboud, M. et al. [ATLAS Collaboration]. Measurements of inclusive and differential fiducial cross-sections of $t\bar{t}$ production with additional heavy-flavour jets in proton-proton collisions at $\sqrt{s}$ = 13 TeV with the ATLAS detector. *J. High Energy Phys.* **2019**, *2019*, 46. [CrossRef]
27. Hollar, J. et al. [CMS Collaboration]. Measurement of $t\bar{t}$ production with additional jet activity, including b quark jets, in the dilepton decay channel using pp collisions at $\sqrt{s}$ = 8 TeV. *Eur. Phys. J. C* **2016**, *76*, 379. [CrossRef]
28. Hollar, J. et al. [CMS Collaboration]. Measurement of the cross section ratio $\sigma_{t\bar{t}b\bar{b}}/\sigma_{t\bar{t}jj}$ in pp collisions at $\sqrt{s}$ = 8 TeV. *Phys. Lett. B* **2015**, *746*, 132–153. [CrossRef]
29. Belforte, S. et al. [CMS Collaboration]. Measurements of $t\bar{t}$ cross sections in association with *b* jets and inclusive jets and their ratio using dilepton final states in pp collisions at $\sqrt{s}$ = 13 TeV. *Phys. Lett. B* **2018**, *776*, 355–378. [CrossRef]
30. Sirunyan, A.M. et al. [CMS Collaboration]. Measurement of the cross section for $t\bar{t}$ production with additional jets and b jets in pp collisions at $\sqrt{s}$ = 13 TeV. *J. High Energy Phys.* **2020**, *2020*, 125. [CrossRef]
31. Bondu, O. et al. [CMS Collaboration]. Measurement of the $t\bar{t}b\bar{b}$ production cross section in the all-jet final state in pp collisions at $\sqrt{s}$ = 13 TeV. *Phys. Lett. B* **2020**, *803*, 135285. [CrossRef]
32. Sirunyan, A.M. et al. [CMS Collaboration]. First measurement of the cross section for top quark pair production with additional charm jets using dileptonic final states in pp collisions at $\sqrt{s}$ = 13 TeV. *Phys. Lett. B* **2021**, *820*, 136565. [CrossRef]
33. CMS Collaboration. CMS-PAS-TOP-22-009. Available online: http://cms-results.web.cern.ch/cms-results/public-results/preliminary-results/TOP-22-009/index.html (accessed on 6 May 2023)
34. Nason, P. A new method for combining NLO QCD with shower Monte Carlo algorithms. *J. High Energy Phys.* **2004**, *2004*, 40. [CrossRef]
35. Frixione, S.; Nason, P.; Oleari, C. Matching NLO QCD computations with parton shower simulations: The POWHEG method. *J. High Energy Phys.* **2007**, *2007*, 70. [CrossRef]
36. Alioli, S.; Nason, P.; Oleari, C.; Re, E. A general framework for implementing NLO calculations in shower Monte Carlo programs: The POWHEG BOX. *J. High Energy Phys.* **2010**, *2010*, 43. [CrossRef]
37. T. Sjöstrand, S. Mrenna and P. Z. Skands, An Introduction to PYTHIA 8.2. *Comput. Phys. Commun.* **2015**, *191*, 159. [CrossRef]
38. Sjöstrand, T.; Mrenna, S.; Skands, P.Z. Pythia 6.4 Physics and Manual. *J. High Energy Phys.* **2006**, *2006*, 26. [CrossRef]
39. Alwall, J.; Herquet, M.; Maltoni, F.; Mattelaer, O.; Stelzer, T. MadGraph v5: Going beyond. *J. High Energy Phys.* **2011**, *2011*, 128. [CrossRef]
40. Bähr, M.; Gieseke, S.; Gigg, M.A.; Grellscheid, D.; Hamilton, K.; Latunde-Dada, O.; Plätzer, S.; Richardson, P.; Seymour, M.H.; Sherstnev, A.; et al. Herwig++ Physics and Manual. *Eur. Phys. J. C* **2008**, *58*, 639–707. [CrossRef]

41. Bellm, J.; Gieseke, S.; Grellscheid, D.; Plätzer, S.; Rauch, M.; Reuschle, C.; Richardson, P.; Schichtel, P.; Seymour, M.H.; Siódmok, A.; et al. Herwig 7.0/Herwig++ 3.0 release note. *Eur. Phys. J. C* **2016**, *76*, 196. [CrossRef]
42. Gleisberg, T.; Höche, S.; Krauss, F.; Schönherr, M.; Schumann, S.; Siegert, F.; Winter, J. Event generation with SHERPA 1.1. *J. High Energy Phys.* **2009**, *2009*, 7. [CrossRef]
43. Cacciari, M.; Czakon, M.; Mangano, M.; Mitov, A.; Nason, P. Top-pair production at hadron colliders with next-to-next-to-leading logarithmic soft-gluon resummation. *Phys. Lett. B* **2012**, *710*, 612. [CrossRef]
44. Czakon, M.; Mitov, A. Top++: A program for the calculation of the top-pair cross-section at hadron collider. *Comput. Phys. Commun.* **2014**, *185*, 2930–2938. [CrossRef]
45. Gleisberg, T.; Höche, S. Comix, a new matrix element generator. *J. High Energy Phys.***2008**, *2008*, 39. [CrossRef]
46. Cascioli, F.; Maierhofer, P.; Pozzorini, S. Scattering Amplitudes with Open Loops. *Phys. Rev. Lett.* **2012**, *108*, 111601. [CrossRef]
47. Schumann, S.; Krauss, F. A parton shower algorithm based on Catani-Seymour dipole factorisation, *J. High Energy Phys.* **2008**, *2008*, 38. [CrossRef]
48. CMS Top Quark Physics Summary Figures. Available online: https://twiki.cern.ch/twiki/bin/view/CMSPublic/PhysicsResultsTOPSummaryFigures (accessed on 5 March 2023).
49. D'Agostini, G. A multidimensional unfolding method based on Bayes' theorem. *Nucl. Instrum. Meth. A* **1995**, *362*, 487–498. [CrossRef]
50. Adye, T. Unfolding algorithms and tests using RooUnfold. In Proceedings of the PHYSTAT 2011 Workshop on Statistical Issues Related to Discovery Claims in Search Experiments and Unfolding, CERN, Geneva, Switzerland, 17–20 January 2011; CERN-2011-006; CERN: Geneva, Switzerland, 2011; p. 313.
51. LeCun, Y.; Boser, B.; Denker, J.; Henderson, D.; Howard, R.; Hubbard, W.; Jackel, L. Handwritten digit recognition with a back-propagation network. In Proceedings of the 2nd International Conference on Neural Information Processing Systems, NIPS'89, (Cambridge, MA, USA), Denver, CO, USA, 27–30 November 1989; MIT Press: Cambridge, MA, USA, 1989; pp. 396–404.
52. Hochreiter, S.; Schmidhuber, J. Long short-term memory. *Neural Comput.* **1997**, *9*, 1735–1780. [CrossRef]
53. D'Hondt, J.; Mariotti, A.; Mimasu, K.; Moortgat, S.; Zhang, C. Learning to pinpoint effective operators at the LHC: A study of the $t\bar{t}b\bar{b}$ signature. *J. High Energy Phys.* **2018**, *2018*, 131. [CrossRef]

universe

MDPI

Review

Four-top quark physics at the LHC

Freya Blekman [1,*], Fréderic Déliot [2], Valentina Dutta [3] and Emanuele Usai [4]

1 Deutsches Elektronen-Synchrotron DESY, Notkestr. 85, 22607 Hamburg, Germany, and Universität Hamburg, Luruper Chaussee 149, 22761 Hamburg, Germany
2 Irfu, CEA Paris-Saclay, Université Paris-Saclay, CEDEX, 91191 Gif-sur-Yvette, France
3 Physics Department, Carnegie Mellon University, Wean Hall, 5000 Forbes Avenue, Pittsburgh, PA 15213-3890, USA
4 Department of Physics and Astronomy, The University of Alabama, 514 University Blvd, Tuscaloosa, AL 35487-0324, USA
* Correspondence: freya.blekman@desy.de

Abstract: The production of four top quarks presents a rare process in the Standard Model that provides unique opportunities and sensitivity to Standard Model observables including potential enhancement of many popular new physics extensions. This article summarises the latest experimental measurements of the four-top quark production cross section at the LHC. An overview is provided detailing interpretations of the experimental results regarding the top quark Yukawa coupling in addition to the limits on physics beyond the Standard Model. Further, prospects for future measurements and opportunities offered by this challenging final state are given herein.

Keywords: four top quarks; top quark; top quark Yukawa coupling; LHC; ATLAS; CMS; particle physics; SMEFT; collider physics; hep-ex; quantum chromodynamics

Citation: Blekman, F.; Déliot, F.; Dutta, V.; Usai, E. Four-top quark physics at the LHC. *Universe* **2022**, *8*, 638. https://doi.org/10.3390/universe8120638

Academic Editor: Jinmin Yang

Received: 22 July 2022
Accepted: 1 November 2022
Published: 30 November 2022

1. Introduction

The top quark was discovered in 1995 [1,2] and plays a pivotal role in particle physics at the energy frontier. In the LHC era, top quark pair production is under high scrutiny by the ATLAS and CMS collaborations [3–7]. Single-top quark production, originally observed at the Tevatron, is now also well established at the LHC [8–14]. Although studies surrounding the intrinsic physics properties represent an important area of research, top quarks play a key background role in many analyses that concern the search for new physics signatures.

The simultaneous production of four top quarks is an example of a rare multiparticle process in the Standard Model (SM), and also presents a promising avenue in the search for signals of new physics beyond the Standard Model (BSM). The production of four top quarks is interesting in its own right since experimental data are expected to challenge state-of-the-art perturbative QCD calculation techniques. A selection of representative diagrams is presented in Figure 1. Recent advanced calculations predict the $t\bar{t}t\bar{t}$ cross section at a centre-of-mass energy of $\sqrt{s} = 13$ TeV to be 12.0 ± 2.4 fb at next-to-leading order in QCD including NLO electroweak corrections, with the quoted uncertainty originating from renormalisation and factorisation scales [15–20]. When focusing on events with two $t\bar{t}$ pairs, warranted questions arise concerning the relevance of double parton scattering in the search for $t\bar{t}t\bar{t}$ events. With a simple PYTHIA model at leading order [21], the cross section for this process can be confirmed to be of the order of 3 ab. As the cross section for $t\bar{t} + t\bar{t}$ double parton scattering is over three-orders of magnitude smaller than the $t\bar{t}t\bar{t}$ cross section, this background is thus irrelevant in the search for four-top quark production.

The $t\bar{t}t\bar{t}$ state provides direct ways to constrain otherwise tricky to measure SM parameters such as the top quark Yukawa coupling and several SM Effective Field Theory parameters sensitive to the quartic couplings between top quarks. If the scale of new physics is beyond the capacity of direct observation, it can manifest as a deviation from the

SM, e.g., a modification of the $t\bar{t}t\bar{t}$ cross section created by virtual and direct (s-channel) contributions of undiscovered BSM particles. These measurements would provide crucial input to the understanding of the SM.

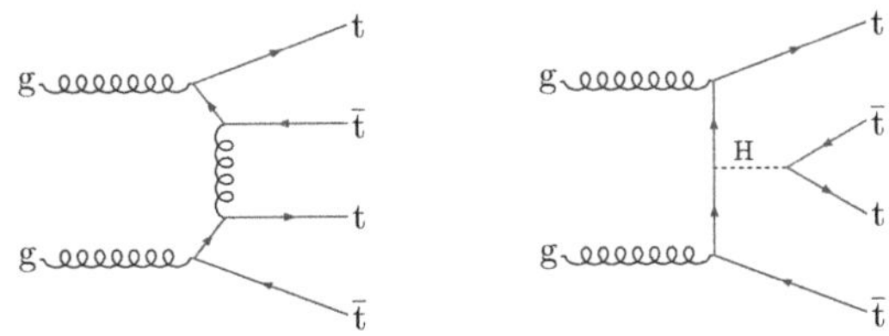

Figure 1. Selected Feynman graphs representing the main production modes of $t\bar{t}t\bar{t}$ production.

The current state-of-the art search approaches for $t\bar{t}t\bar{t}$ production show that the LHC Run 2 dataset is sufficient to establish evidence [22]. Due to the presence of four top quarks in the event, and consequently four W bosons and four b-quarks, the experimental challenges of the search for $t\bar{t}t\bar{t}$ production depend heavily on the W boson decay considered. Analyses considering multi-lepton and same-charge dilepton decays typically have the highest impact in the search significance and are characterised by a very low-branching fraction and acceptance, which is compensated by extremely low backgrounds from SM particles. On the other hand, the analyses of single-lepton and opposite-charge dilepton bear much higher branching fractions but considerable background from top quark pair production. Finally, the all-hadronic final state is largely driven by the reduction in overwhelming background from QCD multijet production.

This review paper summarises the current experimental status in the study of $t\bar{t}t\bar{t}$ production. It is also essential to look forward as $t\bar{t}t\bar{t}$, when firmly established, has substantial physics potential for future HL-LHC runs [23,24] and future hadron colliders [25]. Section 2 summarises the current status of $t\bar{t}t\bar{t}$ measurements. In Sections 3 and 4, interpretations of the SM measurements, BSM searches and opportunities for further analyses are discussed.

2. Current Status of Four-Top Quark Measurements

Searches for $t\bar{t}t\bar{t}$ production in proton–proton (pp) collisions at a center-of-mass energy of 13 TeV were conducted by the ATLAS and CMS collaborations in multiple final states. The most recent results, which supersede previous ones, are described below. The most sensitive searches from ATLAS and CMS target same-charge dilepton and multi-lepton final states [22,26] (Section 2.1), and were carried out with data collected during Run 2 of the LHC. The data samples used for these searches correspond to 139 fb^{-1} for ATLAS, and 137 fb^{-1} for CMS. Searches targeting single-lepton and opposite-charge dilepton final states [27,28] have also been conducted by both collaborations (Section 2.2). Finally, a search in the all-hadronic final state has also recently been conducted by CMS (Section 2.3).

2.1. *Searches for $t\bar{t}t\bar{t}$ Production in Same-Charge Dilepton and Multi-Lepton Final States*

These searches target final states with either two same-charge or at least three light-charged leptons (electrons or muons), corresponding to a combined branching fraction of $\approx$12% for $t\bar{t}t\bar{t}$ events. Since these final states have low levels of background from other SM processes, they are the most sensitive to $t\bar{t}t\bar{t}$ production.

2.1.1. Event Selection and Backgrounds

In addition to the lepton requirements, events selected for the searches are required to have jet activity consistent with the hadronization of b quarks from the top quark decays, or from hadronic decays of the W bosons that do not decay leptonically, as well as large overall event activity. A minimum requirement of at least two jets ($N_{\text{jet}} \geq 2$) is imposed for the CMS search, while a more stringent requirement of $N_{\text{jet}} \geq 6$ is imposed for the ATLAS search. The difference in N_{jet} requirements between the two experiments is driven by whether control regions are defined separately or included directly in the baseline selection. Both searches require at least two jets to be "tagged", or identified, as b-jets ($N_b \geq 2$). For the ATLAS search, a requirement of large event activity is imposed by requiring the scalar sum of the transverse momenta of jets and isolated leptons to exceed 500 GeV, while for the CMS search, a minimum requirement of 300 GeV is imposed on the the scalar sum of the transverse momenta of jets. The CMS search also requires the presence of missing transverse momentum ($p_T^{miss} > 50$ GeV). The latter is expected to arise from the presence of neutrinos from leptonic W boson decays, which would escape the detector without leaving a visible signature.

Backgrounds to these searches arise from processes in which $t\bar{t}$ is produced in association with bosons that decay leptonically, i.e., $t\bar{t}W$, $t\bar{t}Z$, and $t\bar{t}$H production, particularly when these processes are accompanied by the production of additional jets. These backgrounds are generally estimated using simulated events. In ATLAS, the $t\bar{t}W$ background receives a different treatment because theoretical studies [15,19,29–35] showed that electroweak corrections not included in the used simulation have a significant effect. Previous measurements [36] also showed that $t\bar{t}W$ production in association with jets could obtain a larger normalisation factor than predicted by the Monte Carlo (MC) simulation. For these reasons, normalisation of the $t\bar{t}W$ background in the ATLAS analysis is corrected using data in a dedicated control region (CR). In CMS, a dedicated CR is used to constrain normalisation of the $t\bar{t}Z$ background. The simulated samples are corrected to account for observed discrepancies in CMS data. In particular, the modeling of the multiplicity of additional jets from initial- or final-state radiation (ISR or FSR) in $t\bar{t}Z$ and $t\bar{t}W$ simulation is improved by reweighting the ISR/FSR jet multiplicity. Additionally, the modeling of the flavour of additional jets in $t\bar{t}W$, $t\bar{t}Z$, and $t\bar{t}H$ simulation is corrected based on the measured ratio of $t\bar{t}b\bar{b}$ and $t\bar{t}jj$ events, 1.7 ± 0.6 [37], where j represents a jet of any flavour.

Backgrounds may also originate from dilepton $t\bar{t}$ decays with one lepton that has an erroneously assigned charge, or from single-lepton $t\bar{t}$ decays with an additional "non-prompt" lepton. Here, a non-prompt lepton refers to a lepton produced in a hadron decay or from a photon conversion in a jet, or to a hadronic jet that is misidentified as a lepton. The background with charge-misidentified electrons is estimated by applying the electron charge-misidentification probabilities measured in simulation and corrected to account for discrepancies with data (or directly measured in data using $Z \rightarrow ee$ events) to opposite-charge dilepton events. The charge-misidentification probability for muons is an order of magnitude smaller, and therefore the background with charge-misidentified muons is considered to be negligible.

For the ATLAS measurement, the non-prompt lepton background is estimated using the so-called template method. This method relies on the simulation to model the kinematic distributions of background processes arising from non-prompt leptons and on CRs to determine their normalisations. These CRs are included in the fit together with the signal region (SR), and the normalisation factors are determined simultaneously with the $t\bar{t}t\bar{t}$ signal. For the CMS search, the non-prompt lepton background is estimated using the "tight-to-loose" ratio method [38]. The more stringent ("tight") lepton selection criteria used in the SRs are relaxed to define a "loose" selection enriched in non-prompt leptons. The efficiency of non-prompt leptons satisfying the "loose" criteria to also satisfy the "tight" criteria is measured in a control sample. The non-prompt lepton background contribution in the SRs is then estimated by applying weighting factors to events selected by requiring at least one lepton to pass the loose selection while failing the tight one.

2.1.2. Signal Extraction and Results

The ATLAS search separates signal from background events using a multivariate discriminant built in the signal region. The most important inputs to the boosted decision tree (BDT) are the best pseudo-continuous b-tagging discriminant scores [39] summed over all the jets in the event as well as the minimum distance between two leptons among all possible pairs. The $t\bar{t}t\bar{t}$ production cross section and the normalisation factors of the backgrounds are determined via a binned likelihood fit to the BDT score distribution in the SR and to discriminating variable distributions in background CRs. The systematic uncertainties in both the signal and background predictions are included as nuisance parameters. The measured $t\bar{t}t\bar{t}$ production cross section is $\sigma(t\bar{t}t\bar{t}) = 24 \pm 5(\text{stat})^{+5}_{-4}(\text{syst})$ fb = 24^{+7}_{-6} fb. The significance of the observed (expected) signal is found to be 4.3 (2.4) standard deviations. The normalisation factors of the different background sources determined from the fit are compatible with 1 except for $t\bar{t}W$. Apart from the theoretical uncertainty of the signal cross section, the largest systematic uncertainty impacting the signal extraction originates from the modelling of the $t\bar{t}W$ + jets process. Within the uncertainties of the background modelling, the impact of the uncertainty in $t\bar{t}t\bar{t}$ production is also significant. The distribution of the BDT score in the SR after performing the fit is shown in Figure 2 (left) where a good agreement between data and the fitted prediction is observed.

In the CMS search, a BDT classifier is trained to distinguish $t\bar{t}t\bar{t}$ from background events, using variables that include N_{jet}, N_{b}, N_{l}, $p_{\text{T}}^{\text{miss}}$, H_{T} (scalar sum of jet transverse momenta) and other kinematic properties of the jets and leptons in an event. Events are subdivided into 17 SRs based on the BDT discriminant output. Based on the results of a binned maximum-likelihood fit to the data combining all exclusive SRs and the $t\bar{t}Z$ CR, in which nuisance parameters representing systematic uncertainties are profiled, the cross-section measurement for $t\bar{t}t\bar{t}$ production is $\sigma(t\bar{t}t\bar{t}) = 12.6^{+5.8}_{-5.2}$ fb. The observed (expected) significance relative to the background-only hypothesis is 2.6 (2.7) standard deviations. Figure 2 (right) shows the distribution of events in the SRs and CR included in the fit for the BDT analysis, with the post-fit estimates for background and signal.

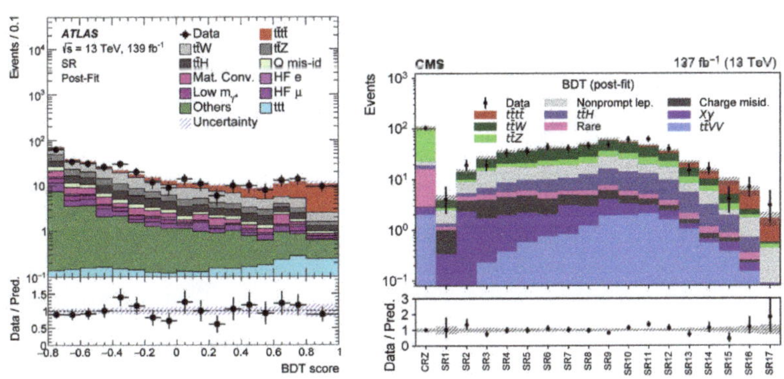

Figure 2. Comparison between data and prediction after the fit for the distribution of the BDT score in the signal region of the ATLAS multi-lepton analysis [22] (**left**), and for events in the $t\bar{t}Z$ CR and SRs of the CMS BDT-based multi-lepton analysis (**right**) [26].

2.2. *Searches for $t\bar{t}t\bar{t}$ Production in Single-Lepton and Opposite-Charge Dilepton Final States*

The target events of this search contain either exactly one (1L) or exactly two opposite-charge light-charged leptons (2LOS). In the latter case, the leptons can have different flavour. The total branching fraction of these final states is about 57% of the $t\bar{t}t\bar{t}$ events. Despite the much larger branching fraction with respect to the analysis in Section 2.1, this combination of final states nears lower sensitivity. This is caused by large cross-section SM processes with similar final states such as $t\bar{t}$ production with additional jets. In particular, $t\bar{t}$ + bb production is the major background and the proper modelling of this process and its separation from $t\bar{t}t\bar{t}$ are major challenges of this analysis.

2.2.1. Event Selection and Backgrounds

The final state is characterised by four b-quarks resulting from the decays of the four top quarks and by either six or four light jets arising from the hadronic decays of the W boson decays and also from the top quark decays. Thus, in ATLAS, events are required to have at least 10 jets (8 jets) in the 1L channel (2LOS channel), among which four are b-tagged. In CMS, events are required to have at least seven jets (muon channel) and eight jets (electron channel) for the 1L final state and at least four jets for the 2LOS final state. The difference in N_{jet} requirements between the two experiments is once again driven by whether control regions are defined separately or included in the simultaneous fit. The background in the high jet multiplicity regions was found to be mis-modelled by MC, thus ATLAS developed a strategy to reweight the $t\bar{t}$ MC generation using data to obtain a reliable $t\bar{t}$ +jets estimate. In addition, the rate of $t\bar{t}$ production in association with b-jets was observed to be underestimated in the MC simulation, so it is adjusted as well. The selected events are categorised according to the lepton and jet multiplicities with different b-tagging requirements. Corrections to the normalisation and kinematics of the $t\bar{t}$ + light, $t\bar{t} + \geq 1c$ and $t\bar{t} + \geq 1b$ jets are derived using data in regions with 2 b-tagged jets where there is low signal contamination and validated in regions with 3 b-tagged jets. The first reweighting adjusts the normalisation of $t\bar{t}$ production with heavy flavour jets. A sequential reweighting is then performed to mitigate the kinematic mismodelling in the distributions of number of jets, number of jets with large radius, scalar sum of all jet and lepton momenta and the average angular separation between two jets. In CMS, a dedicated correction for the modeling of $t\bar{t}$ in high jet multiplicity events is derived in a signal depleted region and applied to signal enriched regions. Additionally, the top quark transverse momentum spectrum of $t\bar{t}$ simulated events is corrected to match the observed spectrum in data.

2.2.2. Signal Extraction and Results

In the ATLAS analysis, the different $t\bar{t}$ +jets components after reweighting are further adjusted and constrained in a binned profile likelihood fit together with the extraction of the signal strength. A total of 21 control and signal regions are used in the fit (12 regions in the 1LOS channel and 9 regions in the 2LOS region). In the region most sensitive to $t\bar{t}t\bar{t}$ production, BDTs are used to discriminate signal from background events after applying reweighting. Several variables are inputs to the discriminant: global event variables, and kinematic properties of the reconstructed objects. Among the input variables, jets with large radius are used as proxies for hadronically decaying top quark with high momentum. The most powerful variable across all regions is the sum of the b-tagging score of the six jets with the highest scores. Several uncertainties are implemented as nuisance parameters in the fit and special care is taken for uncertainties in the $t\bar{t}$ background prediction since these uncertainties have the largest impact on the measurement sensitivity. Following the fit, the $t\bar{t}t\bar{t}$ cross section is measured to be: $\sigma(t\bar{t}t\bar{t}) = 26 \pm 8(\text{stat})^{+15}_{-13}(\text{syst})$ fb = 26^{+17}_{-15} fb which corresponds to an observed significance of 1.9 standard deviations relative to the background-only hypothesis (while 1.0 standard deviation is expected). The largest systematic uncertainty is revealed to originate from the modelling of $t\bar{t} + \geq 1b$ jets, mainly driven by the generator and flavour scheme uncertainty. The observed and expected event yields are shown in Figure 3.

This measurement is further combined with the result in the same-charge dilepton and multi-lepton channel (see Section 2.1) by performing a simultaneous profile likelihood fit across all regions of both analyses. Most of the relevant systematic uncertainties in these two analyses are uncorrelated. The combined $t\bar{t}t\bar{t}$ cross section is measured to be $\sigma(t\bar{t}t\bar{t}) = 24 \pm 4(\text{stat})^{+5}_{-4}(\text{syst})$ fb = 24^{+7}_{-6} fb. The observed (expected) significance of the result is 4.7σ (2.6σ) above the background-only hypothesis, presenting an improvement over the result in the same-charge dilepton and multi-lepton channel alone.

In CMS, events are categorised as a function of their jet multiplicity, b-tagged jet multiplicity, and top-tagged jet multiplicity (using a BDT algorithm to identify hadronically decaying top quarks). Different multiplicity ranges are considered for the single-lepton and the dilepton final states.

In order to reduce background from QCD multijet processes, in addition to the event categorisation, CMS requires that $H_{\text{T}} > 500$ GeV and $p_{\text{T}}^{\text{miss}} > 50$ GeV are imposed. The single-lepton analysis uses event-level BDTs to discriminate $t\bar{t}t\bar{t}$ events from the predominant $t\bar{t}$ background. The event-level BDT is trained using global event variables and employs top and bottom tagger outputs in addition to jet kinematics and angular relationships between leptons and jets. The dilepton final states analysis uses H_T, the sum of the transverse momentum of all jets in the event, except for the analysis performed on the 2016 data which follows the same strategy as the single-lepton analysis and uses a BDT. A binned likelihood fit to the event-level distributions is used to set limits and best fit to the $t\bar{t}t\bar{t}$ cross section and determine the significance of the signal over the no $t\bar{t}t\bar{t}$ hypothesis. The single-lepton state analysis has an observed significance of 1.2 standard deviations, while the expected significance from simulation, assuming the SM $t\bar{t}t\bar{t}$ cross section, is 1.4 standard deviations. The measured best fit to the signal cross section is 15^{+13}_{-11} fb. The opposite-sign dilepton state analysis has an observed significance of 1.8 standard deviations, while the expected significance from simulation, assuming the SM $t\bar{t}t\bar{t}$ cross section, is 0.6 standard deviations. The measured best fit to the signal cross section is 37^{+21}_{-20} fb. These are combined with all other final states which brings a significant improvement in the cross section limits and best fit measurement, described in Section 2.3.3.

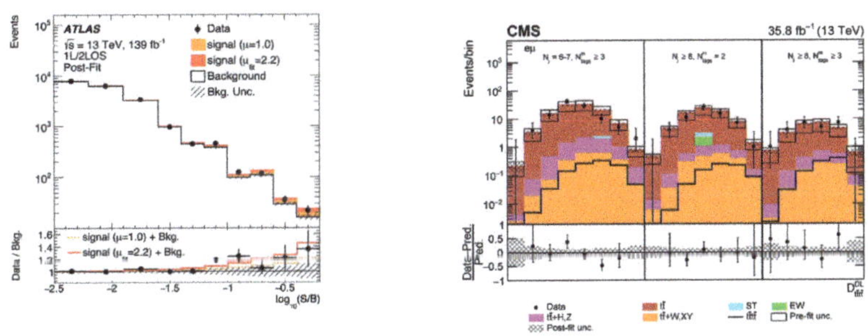

Figure 3. Observed and expected event yields as a function of $\log_{10}(S/B)$ where S and B are the post-fit signal and background yields in the single-lepton and opposite-charge dilepton ATLAS analysis [27] (**left**). Post-fit and observed distribution of the event-level BDT in three separate signal-enriched regions in the $e\mu$ final state, from the single-lepton and opposite-charge dilepton CMS analysis on the 2016 data (**right**) [28].

2.3. *Search for $t\bar{t}t\bar{t}$ Production in the All-Hadronic Final State*

For $t\bar{t}t\bar{t}$ measurements, the final state in which all four W bosons from the top quark decays subsequently decay hadronically represents a challenging but important exploration opportunity. Roughly 20% of $t\bar{t}t\bar{t}$ events are expected to decay into the all-hadronic final state. The main challenge of this final state lies in the experimental backgrounds: a very large background arises from purely QCD multijet events, while another significant source of background originates from $t\bar{t}$ events with fully hadronic top quark decays and additional jets. The QCD multijet background is especially challenging to model accurately with simulation, and thus data-driven methods are needed in order to obtain a robust background prediction in this final state. This final state for $t\bar{t}t\bar{t}$ measurements was recently studied for the first time by CMS [40].

2.3.1. Event Selection and Backgrounds

Events selected for this search are required to have no identified leptons, a minimum of nine jets with at least three of them being b-tagged, and $H_T > 700$ GeV. In order to distinguish potential signal events from the multijet background, the search makes use of dedicated techniques to identify the presence of hadronically decaying top quarks with either moderate or large Lorentz boosts. Moderately boosted hadronic top quark decays will typically produce three separate jets in the detector; these "resolved" top quark decays are identified with a custom BDT-based algorithm. In contrast, the decay products of significantly boosted top quarks can be reconstructed in a single large-radius jet, and are identified with the algorithm defined in Ref. [41]. In order to avoid double counting, resolved and boosted top quark candidates are required to be well separated in $\eta - \phi$ space. A minimum requirement of at least one resolved top candidate ($N_{\mathrm{RT}} \geq 1$) is imposed in selecting events for this search. Events are then categorized into 12 exclusive SR categories based on H_{T}, N_{RT}, and the number of boosted top quark candidates (N_{BT}).

The dominant backgrounds in the search originate from hadronic $t\bar{t}$ decays and QCD multijet events. Data-driven methods are employed to estimate the normalization and BDT shape for these backgrounds. An extension of the ABCD method [42] is applied to estimate the number of background events in the SR categories. A total of five CRs, defined using N_{jet} and N_{b}, are used for the estimation—two more than in the traditional ABCD method, in order to better account for correlations between the variables and higher-order effects. The background BDT shape in the SRs is predicted using a Deep Neural Network (DNN) [43]. The DNN is trained in the five CRs to learn shape transformations from $t\bar{t}$ simulation to the estimated QCD multijet plus $t\bar{t}$ shape in data (after the subtraction of other background contributions estimated from simulation). The H_{T} and BDT shapes are learned simultaneously in each N_{RT} and N_{BT} category, and the DNN is then used to predict the combined shape of the QCD multijet and $t\bar{t}$ background in the SR. Small additional background contributions originate from $t\bar{t}W$, $t\bar{t}Z$, and $t\bar{t}H$ events and from diboson production. These are estimated from simulation.

2.3.2. Signal Extraction and Results

An event-level BDT is used to extract signal from background in the SR via a simultaneous maximum-likelihood fit to the BDT shape in all SR categories. The BDT input variables include the multiplicity and kinematics of jets and b-tagged jets, the kinematics of top-tagged candidates, variables related to jet angular distributions, and event shape variables. The expected significance for this analysis is 0.4σ; however, a non-significant excess of 2.5σ above background is observed. This corresponds to a $t\bar{t}t\bar{t}$ production cross section of $\sigma(t\bar{t}t\bar{t}) = 70^{+30}_{-29}$ fb.

2.3.3. CMS Run 2 Combination

For LHC Run 2, the all-hadronic, single-lepton, opposite and same-charge dilepton and multi-lepton channel $t\bar{t}t\bar{t}$ measurements by the CMS Collaboration were explicitly designed to select orthogonal kinematic phase space. This also meant that control regions were chosen so as to not partially overlap with signal regions in other final states. This strategy allows all final states except those explicitly containing τ leptons to be combined by performing a simultaneous profile likelihood fit. The systematic uncertainties were correlated when appropriate. The combined $t\bar{t}t\bar{t}$ cross section is measured to be $\sigma(t\bar{t}t\bar{t}) = 17 \pm 5$ (stat+syst) fb, and the combination has an observed (expected) significance of 3.9σ (3.2σ) above the background-only hypothesis [40].

Despite the lower sensitivity of the all-hadronic, single-lepton and opposite-charge dilepton analyses compared to the same-charge dilepton and multi-lepton analysis, the combination of the different event signatures produces a significant improvement on the cross-section limits and best fit measurement. The systematic and statistical uncertainties are of similar magnitude, suggesting that in order to make further advances, improvements in analysis techniques are needed to suppress systematic uncertainties.

3. Interpretations

3.1. *Yukawa Coupling*

Four top-quark events can be produced with a virtual Higgs boson as mediator. So the $t\bar{t}t\bar{t}$ production rate is sensitive to the value of the coupling between the top quark and the Higgs boson (y_t) [44,45]. The advantages of the $t\bar{t}t\bar{t}$ process lie in the fact that it does not rely on any assumption on the Higgs width and that its cross section is proportional to the fourth-power of the top Yukawa coupling. It can also be used to probe the CP nature of

y_t. In addition to the $t\bar{t}H$ and tH processes, $t\bar{t}t\bar{t}$ production can then help to shed valuable light on the Higgs boson properties.

CMS has used its upper limit on the measured $t\bar{t}t\bar{t}$ production rate from the multi-lepton channel described in Section 2.1 to constrain y_t. As the $t\bar{t}H$ background cross section also depends on y_t, the fit performed to extract the $t\bar{t}t\bar{t}$ cross section is repeated with the $t\bar{t}H$ contribution scaled by $|y_t/y_{SM}|$ where y_{SM} is the SM value for the top quark Yukawa coupling. The resulting dependence on y_t from the measured signal and background is then compared to the theoretical prediction obtained at LO [44] scaled to the NLO value of $12^{+2.2}_{-2.5}$ fb. The obtained 95% CL limits with the central, upper and lower values of the theoretical prediction are found to be $|y_t/y_{SM}| < 1.7, 1.4$ and 2.0, respectively, [26]. Compared to the $|y_t/y_{SM}| < 1.6$ measured in differential $t\bar{t}$ production, these numbers are complementary and considered relatively model-independent compared to the values of $0.7 < |y_t/y_{SM}| < 1.1$ derived from direct measurements of $t\bar{t}H$ production [46–48]. Further analyses might investigate the use of $t\bar{t}t\bar{t}$ kinematics to better constrain y_t.

3.2. EFTs

Four top-quark production is sensitive to interactions between four heavy quarks (four-heavy-quark operators, QQQQ), to interactions between top quarks and light quarks (two-heavy-two-light four-quark operators, QQqq) and to operators that modify gluon–top quark interaction such as the chromomagnetic operator c_{tG} (see for instance [49]). Among these, the four-heavy-quark operators can only be constrained by $t\bar{t}t\bar{t}$ or $t\bar{t}b\bar{b}$ production, which renders EFT studies in four top-quark production especially interesting. The QQqq operators affect $t\bar{t}t\bar{t}$ and $t\bar{t}$ production and consequently would also modify the backgrounds in $t\bar{t}t\bar{t}$ analyses (mainly the fake background and $t\bar{t}$ production in the same-charge dilepton and multi-lepton channel, or the $t\bar{t}$ +jets and $t\bar{t}b\bar{b}$ background in the single-lepton and opposite-charge dilepton channel).

There are five QQQQ operators that preserve $SU(2)_L$: $\mathcal{O}^1_{QQ}$, $\mathcal{O}^8_{QQ}$, $\mathcal{O}^1_{Qt}$, $\mathcal{O}^8_{Qt}$ and $\mathcal{O}^1_{tt}$. However, if we consider only the $t\bar{t}t\bar{t}$ process and LO operators, then these operators are redundant. Only four operators are independent and we can write: $\mathcal{O}^8_{QQ} = \frac{1}{3}\mathcal{O}^1_{QQ}$ (see for instance [45,50]).

From an experimental perspective, few analyses have interpreted the search for the four top quark process in the context of EFTs. An ATLAS search for four top quark production in the single-lepton and opposite-charge dilepton final state using a partial 13 TeV data set [51] performs such an interpretation. The EFT signal is modeled through a four-top quark contact interaction operator [52]. The normalization of the non-resonant signal is regulated by the expression $|C_{4t}|/\Lambda^2$ where C_{4t} is the coupling constant and Λ is the energy scale of new physics. The analysis set limits at 95% CL on $|C_{4t}|/\Lambda^2 < 1.9$ TeV^{-2} (observed and expected).

The CMS search for four top quarks described in Section 2.1 reports an interpretation of the search for four top quark in terms of the Higgs oblique parameter $\hat{H}$. Within the context of an EFT, $\hat{H}$ is the Wilson coefficient of the only dimension-6 operator that modifies the Higgs boson propagator. This parameter modifies the off-shell behaviour of the Higgs boson. It can be proven that $t\bar{t}t\bar{t}$ is sensitive to $\hat{H}$ through the production modes containing the Higgs boson [53]. The CMS analysis uses simulations of the $t\bar{t}t\bar{t}$ process with modified $\hat{H}$ parameter. Additionally, the $t\bar{t}H$ cross section is scaled by a factor $(1 - \hat{H})^2$ to take into account the dependence on the oblique parameter. A 95% CL upper limit of $\hat{H} < 0.12$ is extracted from the analysis. The value is competitive with the constraint of $\hat{H} < 0.16$ [53] extracted using on-shell Higgs boson measurements [54].

The single-lepton plus opposite-charge dilepton analysis by CMS described in Section 2.2 studies the impact on $t\bar{t}t\bar{t}$ in EFT operators. Limits on the EFT operators are obtained neglecting any acceptance or BDT distribution shape deviations from the purely SM. At leading order, the four-top quark cross section in an EFT scenario can be parametrized as the SM cross section plus a combination of the coupling parameters of the four independent EFT operators that contribute to $t\bar{t}t\bar{t}$ C_k, where $k = \mathcal{O}^1_{tt}, \mathcal{O}^1_{QQ}, \mathcal{O}^1_{Qt}, \mathcal{O}^8_{Qt}$ with a set of four linear parametrization coefficients $\sigma_k^{(1)}$, and a set of nine bilinear coefficients $\sigma_{k,j}^{(2)}$. The linear and bilinear parametrization coefficient are extracted from MADGRAPH simulations [17]. The parametrized formula of the cross section is used to extract expected and observed 95% CL intervals on the coupling coefficient of the four EFT operators. Limits are provided for two different scenarios, the "independent" scenario where only one coefficient is non-null and the "marginalized" scenario where the other coefficients are constrained between a range where the perturbative expansion is stable $C_k/\Lambda^2 \in [-4\pi, 4\pi]$ TeV^{-2}. The observed limit intervals are reported in Table 1 for the independent and marginalized scenarios. The expected limits are compatible with the observed ones. Interestingly, the intervals obtained in the two scenarios are highly similar, showing that little correlation is present between the operators considered.

Table 1. Observed intervals at 95% CL for the coupling parameters of the four independent EFT operators contributing to $t\bar{t}t\bar{t}$ production. Intervals are reported for both the independent and marginalized scenarios [28].

Coupling Parameter	Marginalized C_k/Λ^2 (TeV^{-2})	Independent (TeV^{-2})
$C_{\mathcal{O}^1_{tt}}$	$[-2.2, 2.1]$	$[-2.1, 2.0]$
$C_{\mathcal{O}^1_{QQ}}$	$[-2.2, 2.0]$	$[-2.2, 2.0]$
$C_{\mathcal{O}^1_{Qt}}$	$[-3.7, 3.5]$	$[-3.5, 3.5]$
$C_{\mathcal{O}^8_{Qt}}$	$[-8.0, 6.8]$	$[-7.9, 6.6]$

Emerging Machine Learning techniques to probe EFT operators [55], such as those used by CMS in [56], provide a very promising path towards interpreting $t\bar{t}t\bar{t}$ searches in the context of EFT theories and take into account the effect of the operators on the event kinematics.

3.3. BSM Sensitivity

The rate of $t\bar{t}t\bar{t}$ production may be significantly enhanced in several BSM models. For example, new particles that couple to the top quark and have masses two-fold greater than the top quark mass, such as heavy scalars or pseudoscalars predicted in Type-II two-Higgs-doublet models (2HDM) [57–59] or simplified models of dark matter [60,61], can be produced on-shell in association with top quarks and subsequently decay into top quark pairs. This results in an increased $t\bar{t}t\bar{t}$ production cross section. Less-massive particles, such as a scalar (ϕ) or vector boson (Z') with couplings to the top quark [62], may also enhance the $t\bar{t}t\bar{t}$ cross section through off-shell contributions. Final states with four top quarks may also be produced through the decay of pair-produced gluinos in models of supersymmetry [63–72]; however, for sufficiently massive gluinos (>1 TeV), these are typically studied in searches requiring very large missing transverse momentum and boosted signatures [73–76].

The CMS search for $t\bar{t}t\bar{t}$ production in same-charge dilepton and multi-lepton final states [26] described in Section 2.1 has reported limits on the masses and couplings of a neutral ϕ or Z' with masses smaller than m_t that could contribute to $\sigma(t\bar{t}t\bar{t})$ through off-shell effects. Couplings larger than 1.2 are excluded for m_ϕ between 25 and 340 GeV. For $m_{Z'} = 25$ (300) GeV, couplings larger than 0.1 (0.9) are excluded. The search also probes models with new scalar or pseudoscalar (H/A) particles with masses greater than m_t decaying to $t\bar{t}$ and produced in association with a single-top quark or a top quark pair. Limits are placed in the plane of $\tan\beta$ vs. $m_{H/A}$ for Type-II 2HDM models in the alignment limit [77,78]. For $\tan\beta = 1$, H (A) masses up to 470 (550) GeV are excluded. Similar exclusions are placed on simplified models of DM with a Dirac fermion DM candidate (χ) in addition to H/A when setting the parameters g_{SM} and g_{DM}, representing the couplings of H/A to SM fermions and χ, respectively, to 1, and assuming $m_{H/A} < m_\chi$. Large portions of the parameter space of m_χ vs. $m_{H/A}$ are excluded when relaxing the $m_{H/A} < m_\chi$ assumption for specific choices of $g_{\mathrm{DM}} = 1$ or 0.5 with $g_{\mathrm{SM}} = 1$.

Four top quark production is also relevant in probing the existence of color-octet scalar states, commonly referred to as sgluons, which are predicted in some models of new physics such as non-minimal supersymmetric models featuring Dirac gauginos. In the supersymmetric case, a complex color-octet scalar is predicted that splits into two non-degenerate real components after SUSY breaking, a scalar and a pseudoscalar in the case that its couplings preserve CP [79]. The pseudoscalar, generally expected to be lighter, decays solely into quark pairs and predominantly into $t\bar{t}$, while the scalar, generally heavier, decays into both quarks and gluons. Sgluon production and decay could thus contribute to $t\bar{t}t\bar{t}$ production. CMS results probing $t\bar{t}t\bar{t}$ production in the same-charge dilepton and multi-lepton final states with 35.9 fb^{-1} of data collected at $\sqrt{s} = 13$ TeV [80] have been used to place constraints on sgluon pair production, conservatively excluding pseudoscalar sgluon masses up to 1.06 TeV at 95% CL [79]. The sensitivity to sgluon production can be improved in future measurements by adopting a dedicated search strategy exploiting the kinematic properties of the signal, such as features in the distribution of hadronic activity [79].

4. Future of Four-Top Quark Measurements

The high-luminosity LHC is expected to provide a fertile environment for $t\bar{t}t\bar{t}$ studies [23]. While the production cross section increases by a modest factor of 1.3 when the centre of mass energy of pp collisions is increased from 13 to 14 TeV (and by a factor of 1.19 from 13 TeV to 13.6 TeV), the signal-to-background ratio is expected to improve since this increase is smaller for most backgrounds. Four-top quark production also shows promise at the higher-energy future colliders currently under study, such as the HE-LHC ($\sqrt{s}$ = 27 TeV) and FCC-hh ($\sqrt{s}$ =100 TeV) [25]. Moreover, the high-collision energies also have the consequence that partons at lower Björken x values will be in the phase space for $t\bar{t}t\bar{t}$ production. This means that the theoretical uncertainties originating from sources such as parton density functions are expected to become substantially reduced, even after considering the lack of improvements beyond the current state of the art. At the HL-LHC, HE-LHC and FCC-hh, $t\bar{t}t\bar{t}$ measurements bear the potential for precision QCD tests and precise physics measurements including stringent SMEFT constraints on four-quark interactions [81].

With 3 ab^{-1} of integrated luminosity collected at the HL-LHC, analyses using leptonic final states will start relying on detailed prediction of the SM backgrounds that create same-charge leptons or multi-lepton backgrounds, such as $t\bar{t}V$ and multi-boson production. ATLAS projects that the $t\bar{t}t\bar{t}$ production cross section can be constrained to 11% total accuracy using events with two same-charge leptons or at least three leptons [24,82]. In the same final state and with the same luminosity, CMS expects the statistical uncertainty of a cut-and-count analysis to be of the order of 9% but warns that backgrounds estimated from simulation introduce substantial systematic uncertainties between 18% and 28% depending on the considered sources of theory uncertainty. At the HE-LHC, a similar analysis could be expected to constrain the $t\bar{t}t\bar{t}$ production cross section to within a 1–2% statistical uncertainty, and the systematic uncertainties also decrease due to the improved signal to background ratio [24,83]. A more recent ATLAS extrapolation [84] based on the Run 2 result described in Section 2.1 with different scenarios for the improvement of the systematic uncertainties, projects a $t\bar{t}t\bar{t}$ cross section uncertainty of 14% for the most optimistic case at the HL-LHC.

When the $t\bar{t}t\bar{t}$ production cross section is constrained to this accuracy, the measurements can again be employed to constrain the top-Higgs interaction. Using the same-charge and multi-lepton cross section values, the modification factor that quantifies the Higgs contribution to $\sigma_{t\bar{t}t\bar{t}}$, κ_t, can be estimated using the projected cross-section uncertainties. Assuming that the cross section is modified but acceptance and analysis efficiency do not change substantially, a direct bound on $\kappa_t \leq 1.41$ can be obtained at the HL-LHC and $\kappa_t \leq 1.15(1.12, 1.10)$ with a luminosity of 10 (20, 30) ab^{-1} at the HE-LHC, respectively. The measurement of κ_t provides a direct link to the top quark Yukawa coupling; however, it should be noted that these estimates are dependent on the order of theoretical calculations and were not determined using the complete NLO calculations [24,44]. A similar procedure can also be performed for modifications to $\sigma_{t\bar{t}t\bar{t}}$ from SMEFT contributions [24,83]. Depending on the operator, constraints for top-up quark operators can be very tight at the HL-LHC, down to $|\tilde{C}^{(1)}_{\mathrm{tu}}| < 2.5$, and for generic top-quark interactions down to $|\tilde{C}^{(1)}_{\mathrm{tq}}| < 2.2$. The four-top quark interaction coefficients can be constrained even more tightly down to approximately $|\tilde{C}_{\mathrm{tt}}| < 1.1$ at the HL-LHC or well below 1.0 for the HE-LHC, a substantial improvement compared to Table 1. The production of $t\bar{t}t\bar{t}$ can also be used to constrain the top quark dipole moment [85].

Many of the BSM theories that predict final states with $t\bar{t}t\bar{t}$ can be investigated at the HL-LHC. Most projections for these searches were performed in the high-purity leptonic final states and were limited by the current knowledge of $t\bar{t}V$ and $t\bar{t}t\bar{t}$ production, and could potentially be also explored in other final states for enhanced sensitivity. A study by ATLAS [86] investigates $t\bar{t}t\bar{t}$ at the HL-LHC in same-charge lepton and multi-lepton signatures to search for two additional scalars that can both decay to $t\bar{t}$ or enhance $t\bar{t}t\bar{t}$ production, and where $t\bar{t}t\bar{t}$ and $t\bar{t}V$ production would be the dominant background. These studies project that scalar dark matter mediators A and H from the previously mentioned two-Higgs doublet models can be observed with sensitivity for A masses between a few 100 GeV and 1 TeV for m_H = 600 GeV and $\sin\theta = 0.35$, or excluded over large range of $\sin\theta$ values for lower m_H. When extrapolating $t\bar{t}t\bar{t}$ production in a recast of a cut-and-count analysis by the CMS experiment in same-charge and multi-lepton final states, sgluons and similar coloured pseudoscalar octet particles could be excluded for masses under 1260 and 1470 GeV, respectively, for the HL-LHC and HE-LHC full datasets [24,79,87].

It is worth mentioning that a higher-energy hadron collider, such as the FCC-hh [25], would offer opportunities for an extremely diverse $t\bar{t}t\bar{t}$ measurement program. However, these studies are still very much in their infancy and are also beyond the scope of this review.

Opportunities

The production of three top quarks in the SM can occur in association with a light quark or a W boson [88,89]. One of the tree-level diagrams contributing to the $t\bar{t}tj$ process, where j is a light quark, is mediated by the triple gauge boson vertex. Despite a less busy final state with respect to $t\bar{t}t\bar{t}$, the production of three top quarks in the SM at the LHC at a centre-of-mass energy of 13 TeV is far rarer with $\sigma_{t\bar{t}tW} = 0.73$ fb and $\sigma_{t\bar{t}tj} = 0.47$ fb. While this process, unless enhanced by BSM physics, is very likely outside of the reach of LHC Run 2 and the upcoming Run 3 data-collecting periods, the potential exists for finding evidence of the process with the HL-LHC and HE-LHC full data sets [89]. Production of three top quark without extra jets or W bosons requires flavour changing neutral currents. Setting limits on the production of this process can help to constrain $uttt$ EFT operators according to [90,91].

Final states with one or more hadronically decaying τ leptons constitute ≈29% of $t\bar{t}t\bar{t}$ decays, and are not currently being exploited by the LHC searches. The exploration of these decay modes presents an interesting opportunity for future investigations, and could be relevant for interpretations in certain leptoquark models. Such models may be interesting in light of the anomalies observed in lepton flavour universality measurements.

5. Conclusions

In this paper, we reviewed the current status of searches for $t\bar{t}t\bar{t}$ production at hadron colliders. In particular, recent searches from the ATLAS and CMS Collaborations are summarised. The searches were performed using data collected over the time period spanning from 2016 to 2018 exploring several final states with one, two (same charge or opposite charge) and multiple leptons in the final state. Combinations of different final states, advanced machine learning techniques, and innovative background estimation techniques provide evidence of $t\bar{t}t\bar{t}$ production at 4.7 standard deviations in a measurement from ATLAS [22]. The most precise estimations of the cross section are $24 \pm 4(\mathrm{stat})^{+5}_{-4}(\mathrm{syst})$ fb

= 24^{+7}_{-6} fb and $\sigma(t\bar{t}t\bar{t}) = 17 \pm 5$ (stat + syst) fb by the ATLAS and CMS collaborations, respectively. These measurements are statistically consistent and can be compared with the current highest-order theoretical cross section of 12.0 $\pm$ 2.4 fb cross section at a centre-of-mass energy of $\sqrt{s} = 13$ TeV.

The $t\bar{t}t\bar{t}$ process can be exploited to measure relevant parameters of the SM and its effective field theory extension. The Higgs-mediated production diagram of $t\bar{t}t\bar{t}$ exposes the Yukawa coupling of the top quark, which is measured to be $|y_t/y_{SM}| < 1.7$ at 95% CL in a CMS analysis [26]. Effective field theory operators involving four heavy quarks or two heavy and two light quarks were constrained exploring the effect of such operators on the $t\bar{t}t\bar{t}$ cross section.

Additionally, the $t\bar{t}t\bar{t}$ process offers a direct portal to physics beyond the standard model. Models introducing additional light neutral scalar (ϕ) and vector (Z'), or heavy ($m > 2m_t$) scalar (H) and pseudoscalar (A) bosons in the context of 2HDM models have been constrained in a CMS analysis [26]. In the context of SUSY models, constraints can be placed on sgluon pair production.

The future of the LHC program and its high-luminosity and high-energy upgrades provide opportunities for precise SM and EFT measurements and searches for new physics with the $t\bar{t}t\bar{t}$ process. Projections of current results for the HL- and HE-LHC programs predict the ability to measure the $t\bar{t}t\bar{t}$ cross section with a precision of 11% using the full data set of the HL-LHC and constrain $|y_t/y_{SM}| < 1.1$ at the HE-LHC. The exploration of $t\bar{t}t\bar{t}$ final states with τ leptons or no leptons (all-hadronic) will further enrich the four top quark physics program. Finally, the large integrated luminosity accumulated by the LHC project and its extension will allow researchers to explore even rarer related processes, for which $t\bar{t}t\bar{t}$ is a background process such as $t\bar{t}tV$ and $t\bar{t}tq$ production.

Overall, the $t\bar{t}t\bar{t}$ process offers a wide breadth of opportunities for building a strong physics program including both precise measurements of important SM parameters and its EFT extensions, and the direct probing of different types of BSM theories. Although the $t\bar{t}t\bar{t}$ research program has commenced only recently, highly promising results have already been obtained with the current data collected by the LHC.

Funding: This research received no external funding.

Acknowledgments: We kindly thank the editors of this dedicated issue of Universe for the invitation to contribute this overview article. F.B. acknowledges support from DESY (Hamburg, Germany) a member of the Helmholtz Association HGF, and support by the Deutsche Forschungsgemeinschaft (DFG, German Research Foundation) under Germany's Excellence Strategy—EXC 2121 "Quantum Universe"—390833306. V.D. acknowledges support from the US Department of Energy Grant Number DE-SC0011702 to University of California Santa Barbara, Santa Barbara, CA, USA. E.U. acknowledges support from the US Department of Energy Grant Number DE-SC0010010 to Brown University, Providence, RI, USA. F.D. acknowledges the support of CEA-DRF/IRFU, France.

Conflicts of Interest: The authors declare no conflict of interest.

References

1. CDF Collaboration. Observation of top quark production in $\bar{p}p$ collisions. *Phys. Rev. Lett.* **1995**, *74*, 2626. [CrossRef]
2. D0 Collaboration. Observation of the top quark. *Phys. Rev. Lett.* **1995**, *74*, 2632. [CrossRef] [PubMed]
3. ATLAS, CMS Collaboration. Combination of inclusive top-quark pair production cross-section measurements using ATLAS and CMS data at $\sqrt{s}$= 7 and 8 TeV. *arXiv* **2022**, arXiv:2205.13830.
4. ATLAS Collaboration. Measurement of the $t\bar{t}$ production cross-section in the lepton+jets channel at $\sqrt{s} = 13$ TeV with the ATLAS experiment. *Phys. Lett. B* **2020**, *810*, 135797. [CrossRef]
5. ATLAS Collaboration. Measurement of the $t\bar{t}$ production cross-section and lepton differential distributions in $e\mu$ dilepton events from pp collisions at $\sqrt{s} = 13$ TeV with the ATLAS detector. *Eur. Phys. J. C* **2020**, *80*, 528. [CrossRef]
6. CMS Collaboration. Measurement of the $t\bar{t}$ production cross section using events with one lepton and at least one jet in pp collisions at $\sqrt{s}$ = 13 TeV. *J. High Energy Phys.* **2017**, *9*, 51. [CrossRef]
7. CMS Collaboration. Measurement of the $t\bar{t}$ production cross section using events in the $e\mu$ final state in pp collisions at $\sqrt{s} = 13$ TeV. *Eur. Phys. J. C* **2017**, *77*, 172. [CrossRef]
8. D0 Collaboration. Observation of Single Top Quark Production. *Phys. Rev. Lett.* **2009**, *103*, 92001. [CrossRef]

9. CDF Collaboration. First Observation of Electroweak Single Top Quark Production. *Phys. Rev. Lett.* **2009**, *103*, 92002. [CrossRef]
10. ATLAS, CMS Collaboration. Combinations of single-top-quark production cross-section measurements and $|f_{LV}V_{tb}|$ determinations at $\sqrt{s}$ = 7 and 8 TeV with the ATLAS and CMS experiments. *J. High Energy Phys.* **2019**, *5*, 88. [CrossRef]
11. ATLAS Collaboration. Measurement of the inclusive cross-sections of single top-quark and top-antiquark t-channel production in pp collisions at $\sqrt{s}$ = 13 TeV with the ATLAS detector. *J. High Energy Phys.* **2017**, *4*, 86. [CrossRef]
12. ATLAS Collaboration. Measurement of the cross-section for producing a W boson in association with a single top quark in pp collisions at $\sqrt{s} = 13$ TeV with ATLAS. *J. High Energy Phys.* **2018**, *1*, 63. [CrossRef]
13. CMS Collaboration. Measurement of the single top quark and antiquark production cross sections in the t channel and their ratio in proton-proton collisions at $\sqrt{s} = 13$ TeV. *Phys. Lett. B* **2020**, *800*, 135042. [CrossRef]
14. CMS Collaboration. Measurement of the production cross section for single top quarks in association with W bosons in proton-proton collisions at $\sqrt{s} = 13$ TeV. *J. High Energy Phys.* **2018**, *10*, 117. [CrossRef]
15. Frederix, R.; Pagani, D.; Zaro, M. Large NLO corrections in $t\bar{t}W^{\pm}$ and $t\bar{t}t\bar{t}$ hadroproduction from supposedly subleading EW contributions. *J. High Energy Phys.* **2018**, *2*, 31. [CrossRef]
16. Bevilacqua, G.; Worek, M. Constraining BSM Physics at the LHC: Four top final states with NLO accuracy in perturbative QCD. *J. High Energy Phys.* **2012**, *7*, 111. [CrossRef]
17. Alwall, J.; Frederix, R.; Frixione, S., Hirschi, V.; Maltoni, F.; Mattelaer, O.; Shao, H.S.; Stelzer, T.; Torrielli, P.; Zaro, M. The automated computation of tree-level and next-to-leading order differential cross sections, and their matching to parton shower simulations. *J. High Energy Phys.* **2014**, *7*, 79. [CrossRef]
18. Maltoni, F.; Pagani, D.; Tsinikos, I. Associated production of a top-quark pair with vector bosons at NLO in QCD: Impact on t$\bar{\text{t}}$H searches at the LHC. *J. High Energy Phys.* **2016**, *2*, 113. [CrossRef]
19. Frederix, R.; Frixione, S.; Hirschi, V.; Pagani, D.; Shao, H.S.; Zaro, M. The automation of next-to-leading order electroweak calculations. *J. High Energy Phys.* **2018**, *7*, 185. [CrossRef]
20. Ježo, T.; Kraus, M. Hadroproduction of four top quarks in the powheg box. *Phys. Rev. D* **2022**, *105*, 114024. [CrossRef]
21. Sjostrand, T.; Mrenna, S.; Skands, P.Z. A Brief Introduction to PYTHIA 8.1. *Comput. Phys. Commun.* **2008**, *178*, 852. [CrossRef]
22. ATLAS Collaboration. Evidence for $t\bar{t}t\bar{t}$ production in the multilepton final state in proton–proton collisions at $\sqrt{s} = 13$ TeV with the ATLAS detector. *Eur. Phys. J. C* **2020**, *80*, 1085. [CrossRef]
23. Apollinari, G.; Béjar Alonso, I.; Brüning, O.; Lamont, M.; Rossi, L. High-Luminosity Large Hadron Collider (HL-LHC): Preliminary Design Report. *CERN Yellow Rep. Monogr.* **2015**, *5*, 285. [CrossRef]
24. Azzi, P. Report from Working Group 1: Standard Model Physics at the HL-LHC and HE-LHC. *CERN Yellow Rep. Monogr.* **2019**, *7*, 1–220. [CrossRef]
25. FCC Collaboration. FCC-hh: The Hadron Collider: Future Circular Collider Conceptual Design Report Volume 3. *Eur. Phys. J. ST* **2019**, *228*, 755. [CrossRef]
26. CMS Collaboration. Search for production of four top quarks in final states with same-sign or multiple leptons in proton-proton collisions at $\sqrt{s} = 13$ TeV. *Eur. Phys. J. C* **2020**, *80*, 75. [CrossRef]
27. ATLAS Collaboration. Measurement of the t$\bar{\text{t}}$t$\bar{\text{t}}$ production cross section in pp collisions at $\sqrt{s}$ = 13 TeV with the ATLAS detector. *J. High Energy Phys.* **2021**, *11*, 118. [CrossRef]
28. CMS Collaboration. Search for the production of four top quarks in the single-lepton and opposite-sign dilepton final states in proton-proton collisions at $\sqrt{s} = 13$ TeV. *J. High Energy Phys.* **2019**, *11*, 82. [CrossRef]
29. Frederix, R.; Tsinikos, I. Subleading EW corrections and spin-correlation effects in $t\bar{t}W$ multi-lepton signatures. *Eur. Phys. J. C* **2020**, *80*, 803. [CrossRef]
30. Broggio, A.; Ferroglia, A.; Frederix, R.; Pagani, D.; Pecjak, B.D.; Tsinikos, I. Top-quark pair hadroproduction in association with a heavy boson at NLO+NNLL including EW corrections. *J. High Energy Phys.* **2019**, *8*, 39. [CrossRef]
31. Kulesza, A.; Motyka, L.; Schwartländer, D.; Stebel, T.; Theeuwes, V. Associated top quark pair production with a heavy boson: Differential cross sections at NLO + NNLL accuracy. *Eur. Phys. J. C* **2020**, *80*, 428. [CrossRef]
32. von Buddenbrock, S.; Ruiz, R.; Mellado, B. Anatomy of inclusive $t\bar{t}W$ production at hadron colliders. *Phys. Lett. B* **2020**, *811*, 135964. [CrossRef]
33. Cordero, F.F.; Kraus, M.; Reina, L. Top-quark pair production in association with a $W^{\pm}$ gauge boson in the POWHEG-BOX. *Phys. Rev. D* **2021**, *103*, 94014. [CrossRef]
34. Bevilacqua, G.; Bi, H.Y.; Cordero, F.F.; Hartanto, H.B.; Kraus, M.; Nasufi, J.; Reina, L.; Worek, M. Modeling uncertainties of $t\bar{t}W^{\pm}$ multilepton signatures. *Phys. Rev. D* **2022**, *105*, 14018. [CrossRef]
35. Denner, A.; Pellicciolli, G. Combined NLO EW and QCD corrections to off-shell t$\bar{\text{t}}$W production at the LHC. *Eur. Phys. J. C* **2021**, *81*, 354. [CrossRef]
36. ATLAS Collaboration, *Analysis of $t\bar{t}H$ and $t\bar{t}W$ Production in Multilepton Final States with the ATLAS Detector*, ATLAS Conference Note ATLAS-CONF-2019-045, 2019.
37. CMS Collaboration. Measurements of $t\bar{t}$ cross sections in association with b jets and inclusive jets and their ratio using dilepton final states in pp collisions at $\sqrt{s}$ = 13 TeV. *Phys. Lett. B* **2018**, *776*, 355–378. [CrossRef]
38. CMS Collaboration. Search for new physics in same-sign dilepton events in proton–proton collisions at $\sqrt{s} = 13$ TeV. *Eur. Phys. J. C* **2016**, *76*, 439. [CrossRef]

39. ATLAS Collaboration. ATLAS b-jet identification performance and efficiency measurement with $t\bar{t}$ events in pp collisions at $\sqrt{s} = 13$ TeV. *Eur. Phys. J. C* **2019**, *79*, 970. [CrossRef]
40. CMS Collaboration. Evidence for the simultaneous production of four top quarks in proton-proton collisions at $\sqrt{s}$ = 13 TeV, CMS Physics Analysis Summary CMS-PAS-TOP-21-005, 2022
41. CMS Collaboration. Machine learning-based identification of highly Lorentz-boosted hadronically decaying particles at the CMS experiment, CMS Physics Analysis Summary CMS-PAS-JME-18-002, 2019.
42. Choi, S.; Oh, H. Improved extrapolation methods of data-driven background estimations in high energy physics. *Eur. Phys. J. C* **2021**, *81*, 643. [CrossRef]
43. Choi, S.; Lim, J.; Oh, H. Data-driven Estimation of Background Distribution through Neural Autoregressive Flows. *arXiv* **2020**, arXiv:2008.03636.
44. Cao, Q.-H.; Chen, S.-L.; Liu, Y. Probing Higgs Width and Top Quark Yukawa Coupling from $t\bar{t}H$ and $t\bar{t}t\bar{t}$ Productions. *Phys. Rev. D* **2017**, *95*, 53004. [CrossRef]
45. Cao, Q.-H.; Chen, S.-L.; Liu, Y.; Zhang, R.; Zhang, Y. Limiting top quark-Higgs boson interaction and Higgs-boson width from multitop productions. *Phys. Rev. D* **2019**, *99*, 113003. [CrossRef]
46. CMS Collaboration. Measurement of the top quark Yukawa coupling from $t\bar{t}$ kinematic distributions in the dilepton final state in proton-proton collisions at $\sqrt{s} = 13$ TeV. *Phys. Rev. D* **2020**, *102*, 92013. [CrossRef]
47. CMS Collaboration. Measurement of the top quark Yukawa coupling from $t\bar{t}$ kinematic distributions in the lepton+jets final state in proton-proton collisions at $\sqrt{s} = 13$ TeV. *Phys. Rev. D* **2019**, *100*, 72007. [CrossRef]
48. CMS Collaboration. Measurement of the Higgs boson production rate in association with top quarks in final states with electrons, muons, and hadronically decaying tau leptons at $\sqrt{s} = 13$ TeV. *Eur. Phys. J. C* **2021**, *81*, 378. [CrossRef]
49. Hartland, N.P.; Maltoni, F.; Nocera, E.R.; Rojo, J.; Slade, E.; Vryonidou, E.; Zhang, C. A Monte Carlo global analysis of the Standard Model Effective Field Theory: The top quark sector. *J. High Energy Phys.* **2019**, *4*, 100. [CrossRef]
50. D'Hondt, J.; Mariotti, A.; Mimasu, K.; Moortgat, S.; Zhang, C. Learning to pinpoint effective operators at the LHC: A study of the $t\bar{t}b\bar{b}$ signature. *J. High Energy Phys.* **2018**, *11*, 131. [CrossRef]
51. ATLAS Collaboration. Search for four-top-quark production in the single-lepton and opposite-sign dilepton final states in pp collisions at $\sqrt{s} = 13$ TeV with the ATLAS detector. *Phys. Rev. D* **2019**, *99*, 52009. [CrossRef]
52. Degrande, C.; Gerard, J.-M.; Grojean, C.; Maltoni, F.; Servant, G. Non-resonant New Physics in Top Pair Production at Hadron Colliders. *J. High Energy Phys.* **2011**, *3*, 125. [CrossRef]
53. Englert, C.; Giudice, G.F.; Greljo, A.; Mccullough, M. The $\hat{H}$-Parameter: An Oblique Higgs View. *J. High Energy Phys.* **2019**, *9*, 41. [CrossRef]
54. ATLAS Collaboration. Combined measurements of Higgs boson production and decay using up to 80 fb^{-1} of proton-proton collision data at $\sqrt{s} = 13$ TeV collected with the ATLAS experiment. *Phys. Rev. D* **2020**, *101*, 12002. 101.012002. [CrossRef]
55. Brehmer, J.; Cranmer, K.; Louppe, G.; Pavez, J. A Guide to Constraining Effective Field Theories with Machine Learning. *Phys. Rev. D* **2018**, *98*, 52004. [CrossRef]
56. CMS Collaboration. Probing effective field theory operators in the associated production of top quarks with a Z boson in multilepton final states at $\sqrt{s}$ = 13 TeV. *J. High Energy Phys.* **2021**, *12*, 83. [CrossRef]
57. Dicus, D.; Stange, A.; Willenbrock, S. Higgs decay to top quarks at hadron colliders. *Phys. Lett. B* **1994**, *333*, 126–131. [CrossRef]
58. Craig, N.; D'Eramo, F.; Draper, P.; Thomas, S.; Zhang, H. The Hunt for the Rest of the Higgs Bosons. *J. High Energy Phys.* **2015**, *6*, 137. [CrossRef]
59. Craig, N.; Hajer, J.; Li, Y.-Y.; Liu, T.; Zhang, H. Heavy Higgs bosons at low tan β: From the LHC to 100 TeV. *J. High Energy Phys.* **2017**, *1*, 18. [CrossRef]
60. Boveia, A.; Shepherd, W.; D'Eramo, F.; Worm, S.; du Pree, T.; Salek, D.; Wang, L.T.; De Simone, A.; Zurek, K.; Khoze, V.V.; et al. Recommendations on presenting LHC searches for missing transverse energy signals using simplified s-channel models of dark matter. *Phys. Dark Univ.* **2020**, *27*, 100365. [CrossRef]
61. Albert, A. et al. [LHC Dark Matter Working Group]. Recommendations of the LHC Dark Matter Working Group: Comparing LHC searches for dark matter mediators in visible and invisible decay channels and calculations of the thermal relic density. *Phys. Dark Univ.* **2019**, *26*, 100377. [CrossRef]
62. Alvarez, E.; Faroughy, D.A.; Kamenik, J.F.; Morales, R.; Szynkman, A. Four Tops for LHC. *Nucl. Phys. B* **2017**, *915*, 19. [CrossRef]
63. Ramond, P. Dual theory for free fermions. *Phys. Rev. D* **1971**, *3*, 2415. [CrossRef]
64. Gol'fand, Y.A.; Likhtman, E.P. Extension of the algebra of Poincaré group generators and violation of P invariance. *JETP Lett.* **1971**, *13*, 323.
65. Neveu, A.; Schwarz, J.H. Factorizable dual model of pions. *Nucl. Phys. B* **1971**, *31*, 86. [CrossRef]
66. Volkov, D.V.; Akulov, V.P. Possible universal neutrino interaction. *JETP Lett.* **1972**, *16*, 438.
67. Wess, J.; Zumino, B. A Lagrangian model invariant under supergauge transformations. *Phys. Lett. B* **1974**, *49*, 52–54. [CrossRef]
68. Wess, J.; Zumino, B. Supergauge transformations in four dimensions. *Nucl. Phys. B* **1974**, *70*, 39. [CrossRef]
69. Fayet, P. Supergauge invariant extension of the Higgs mechanism and a model for the electron and its neutrino. *Nucl. Phys. B* **1975**, *90*, 104. [CrossRef]
70. Nilles, H.P. Supersymmetry, supergravity and particle physics. *Phys. Rep.* **1984**, *110*, 1–162. [CrossRef]
71. Martin, S.P. A Supersymmetry primer. *Adv. Ser. Direct. High Energy Phys.* **1998**, *18*, 1–153. [CrossRef]

72. Farrar, G.R.; Fayet, P. Phenomenology of the Production, Decay, and Detection of New Hadronic States Associated with Supersymmetry. *Phys. Lett. B* **1978**, *76*, 575–579. [CrossRef]
73. ATLAS Collaboration. Search for new phenomena in final states with large jet multiplicities and missing transverse momentum using $\sqrt{s}$ = 13 TeV proton-proton collisions recorded by ATLAS in Run 2 of the LHC. *J. High Energy Phys.* **2020**, *10*, 62. [CrossRef]
74. CMS Collaboration. Searches for physics beyond the standard model with the M_{T2} variable in hadronic final states with and without disappearing tracks in proton-proton collisions at $\sqrt{s} = 13$ TeV. *Eur. Phys. J. C* **2020**, *80*, 3. [CrossRef] [PubMed]
75. CMS Collaboration. Search for supersymmetry in proton-proton collisions at 13 TeV in final states with jets and missing transverse momentum. *J. High Energy Phys.* **2019**, *10*, 244. [CrossRef]
76. CMS Collaboration. Search for supersymmetry in pp collisions at $\sqrt{s} = 13$ TeV with 137 fb^{-1} in final states with a single lepton using the sum of masses of large-radius jets. *Phys. Rev. D* **2020**, *101*, 52010. [CrossRef]
77. Bauer, M.; Haisch, U.; Kahlhoefer, F. Simplified dark matter models with two Higgs doublets: I. Pseudoscalar mediators. *J. High Energy Phys.* **2017**, *5*, 138. [CrossRef]
78. LHC Dark Matter Working Group. LHC Dark Matter Working Group: Next-generation spin-0 dark matter models. *Phys. Dark Univ.* **2020**, *27*, 100351. [CrossRef]
79. Darmé, L.; Fuks, B.; Goodsell, M. Cornering sgluons with four-top-quark events. *Phys. Lett. B* **2018**, *784*, 223–228. [CrossRef]
80. CMS Collaboration. Search for standard model production of four top quarks with same-sign and multilepton final states in proton–proton collisions at $\sqrt{s} = 13$ TeV. *Eur. Phys. J. C* **2018**, *78*, 140. [CrossRef]
81. ATLAS Collaboration. Physics at a 100 TeV pp Collider: Standard Model Processes. In *HL-LHC Prospects for the Measurement of the Standard Model Four-Top-Quark Production Cross-Section*; ATLAS Conference Note ATL-PHYS-PUB-2018-047, 2018.
82. ATLAS Collaboration. *HL-LHC Prospects for the Measurement of the Standard Model Four-Top-Quark Production Cross-Section*; ATLAS Conference Note ATL-PHYS-PUB-2018-047, 2018.
83. CMS Collaboration. Projections of sensitivities for tttt production at HL-LHC and HE-LHC, CMS Physics Analysis Summary CMS-PAS-FTR-18-031, 2018.
84. ATLAS Collaboration. Extrapolation of ATLAS sensitivity to the measurement of the Standard Model four top quark cross section at the HL-LHC, ATLAS Conference Note ATL-PHYS-PUB-2022-004, 2022.
85. Malekhosseini, M.; Ghominejad, M.; Khanpour, H.; Najafabadi, M.M. Constraining top quark flavor violation and dipole moments through three and four-top quark productions at the LHC. *Phys. Rev. D* **2018** *98*, 95001. [CrossRef]
86. ATLAS Collaboration. *ATLAS Sensitivity to Two-Higgs-Doublet Models with an Additional Pseudoscalar Exploiting Four Top Quark Signatures with 3ab*$^{-1}$ *of* $\sqrt{s} = 14$ *TeV Proton-Proton Collisions*; ATL-PHYS-PUB-2018-047, 2018.
87. Calvet, S.; Fuks, B.; Gris, P.; Valery, L. Searching for sgluons in multitop events at a center-of-mass energy of 8 TeV. *J. High Energy Phys.* **2013**, *4*, 43. [CrossRef]
88. Barger, V.; Keung, W.-Y.; Yencho, B. Triple-Top Signal of New Physics at the LHC. *Phys. Lett. B* **2010**, *687*, 70–74. [CrossRef]
89. Boos, E.; Dudko, L. Triple top quark production in standard model. *Int. J. Mod. Phys. A* **2022**, *37*, 2250023 [CrossRef]
90. Cao, Q.-H.; Chen, S.-L.; Liu, Y.; Wang, X.-P. What can We Learn from Triple Top-Quark Production? *Phys. Rev. D* **2019**, *100*, 55035. [CrossRef]
91. Kohda, M.; Modak, T.; Hou, W.-S. Searching for new scalar bosons via triple-top signature in $cg \rightarrow tS^0 \rightarrow tt\bar{t}$. *Phys. Lett. B* **2018**, *776*, 379–384. [CrossRef]

universe

MDPI

Review

Diffractive and Photon-Induced Production of Top Quark

Michael Pitt [1,2]

[1] Department of Physics, Ben-Gurion University of the Negev, Beer-Sheva 84105, Israel; michael.pitt@cern.ch
[2] Department of Physics, The University of Kansas, Lawrence, KS 66045, USA

Abstract: The top quark plays a central role in particle physics, as many experiments at the Large Hadron Collider scrutinize its properties within the Standard Model. Although most of the measurements of the top quarks today concentrate on production modes initiated by quarks or gluons, this review will highlight the lesser-explored modes initiated by pomerons or photons. It aims to provide an in-depth look into both the phenomenological studies and the existing experimental measurements, emphasizing the necessity of exploring the diffractive and photon-induced production of top quarks to enhance the accuracy of top-quark measurements.

Keywords: top quark; diffraction; photon exchange; EFT; CMS

1. Introduction

Since its discovery in 1995 at Fermilab's Tevatron collider, the top quark has continued to receive significant attention, as it is a promising pathway for investigating the SM by conducting high-precision measurements of its properties and potentially unveiling new laws of physics by discerning any deviation from SM predictions.

The present measurements of top quarks produced from the collisions of protons at the LHC predominantly focus on production modes initiated by quarks or gluons. However, the production of top quarks initiated by color-neutral particles remains largely unexplored and is the subject of this review. Production modes involving color-neutral particles are typically categorized according to the type of particle involved. For example, top quark pairs can also be produced through photon (γ) or pomeron ($I\!P$) exchange, and the different production categories are depicted in Figure 1.

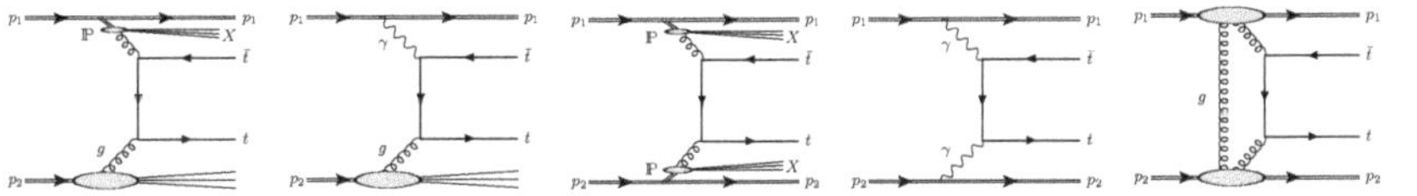

Figure 1. Categories of top quark pair production involving color-neutral particles. From left to right: single-diffractive production, photoproduction, double pomeron exchange, central exclusive production via photon exchange, and central exclusive production via pomeron exchange.

The photon is an elementary particle responsible for electromagnetic interactions. On the other hand, the pomeron is an object that can be viewed as a color-singlet multi-gluon state, introduced in the early 1960s to describe high-energy hadron-scattering amplitudes (an in-depth review can be found in reference [1]). Over time, with deeper insights into diffractive processes, it became evident that the pomeron has an internal structure that includes quarks as well. Consequently, any process initiated by quarks or gluons can be generated by exchanging pomerons.

The discussion of the diffractive production of top quarks predates its actual discovery [2] yet, up to the present day, it has lacked experimental verification. The production of top quarks through color-neutral particles can contribute to the total inclusive cross-section on a level of up to a few percent. The events resulting from a photon or pomeron exchange

Citation: Pitt, M. Diffractive and photon-induced production of top quark. *Universe* **2023**, *9*, 483. https://doi.org/10.3390/universe9110483

Academic Editor: Nikolaos Kidonakis

Received: 11 October 2023
Revised: 31 October 2023
Accepted: 2 November 2023
Published: 16 November 2023

exhibit a distinct radiation pattern, often characterized by an absence of hadronic activity in certain regions of pseudorapidity, known as large rapidity gaps. Additionally, these events may be identified by the presence of an intact proton. The following article will review the theory and experimental data available on these rare production modes.

2. Monte Carlo Event Generators

Diffractive and exclusive processes have been incorporated into various Monte Carlo (MC) event generators. Different MC generators may implement these processes differently based on their underlying physics models. However, they all adhere to the factorization principle [3], where the production cross-section of a system X in hard proton–proton scattering can be viewed as follows:

$$\sigma(\mathrm{pp} \to \mathrm{X}) = \sum_{\mathrm{i,j}} \int \mathrm{dx_i} f_{\mathrm{i}}(\mathrm{x_i}, \mu) \int \mathrm{dx_j} f_{\mathrm{j}}(\mathrm{x_j}, \mu) \cdot \sigma_{\mathrm{hard}}(\mathrm{ij} \to \mathrm{X}), \tag{1}$$

where i, j label the partons that initiate hard scattering, $\sigma_{\mathrm{hard}}(\mathrm{ij} \to \mathrm{X})$ is the parton level cross-section computed perturbatively in terms of powers of $\alpha_S(\mu)$, μ is the energy scale of the process, and $f(\mathrm{x}, \mu)$ are the parton distribution functions (PDFs) of the colliding particles. For color-neutral interaction, these parton densities are often replaced by the following:

$$\mathrm{dx} f(\mathrm{x}, \mu) = \int \mathrm{d}t \int \mathrm{d}\beta \mathbb{F}(\xi, t) f_d(\beta, \mu), \tag{2}$$

where $\mathbb{F}(\xi, t)$ is the flux of color-neutral mediators emitted by a proton as a function of longitudinal momentum (ξ) and the momentum transfer (t). Here, $\beta = \mathrm{x}/\xi$ and $f_d(\beta, \mu)$ is the parton density function of the color-neutral object (technically related to the diffractive PDF—dPDF). In the case of direct photon/pomeron exchange processes, whether they are exclusive or semi-exclusive, the value of $f_d(\beta, \mu)$ is set to 1, but the flux will incorporate additional form-factors, denoted by $S^2(\mathrm{b})$, which are derived in terms of impact parameter and their computation varies across different event generators.

The pomeron flux $\mathbb{F}_{\mathbb{P}/\mathrm{p}}(\xi, k_T)$ is the probability that a pomeron with a given value of ξ and t couples to the proton. Several different parametrizations have been proposed over the years [4,5]. The most recent empirical expression was suggested by the H1 collaboration [6]:

$$\mathbb{F}_{\mathbb{P}/\mathrm{p}}(\xi, t) = \xi^{1-2\alpha(t)} A e^{Bt} \tag{3}$$

where $\alpha(t) = 1 + \varepsilon + \alpha' t$ is the pomeron trajectory, and ε, α', A, and B are obtained from fitting the data. Recent fits include sub-leading contributions (commonly denoted as "Reggeon" exchange), which became pronounced at high ξ, but the H1 data weakly constrain them.

The pomeron structure function is the probability of extracting a parton from a pomeron. In the simplest model, the most general form of the dPDF is [6]

$$\beta f_{i/\mathbb{P}}(\beta) = A_i \beta^{B_i} (1 - \beta)^{C_i} \tag{4}$$

where A_i, B_i, and C_i are fit parameters. In the latest fit to the diffractive data, the dPDFs are modeled by incorporating a light flavor quark distribution (assuming zero intrinsic densities for c- and b-quarks within the pomeron) and the gluon distribution. Figure 2 illustrates a few existing pomeron structure functions fitted by the H1 collaboration.

The following MC event generators can simulate processes involving the diffractive and photon-induced production of top quark.

The SuperChic event generator [7,8] is used for studying central exclusive and semi-exclusive production processes in proton–proton, proton–ion, and ion–ion collisions. It implements the central exclusive production via Pomeron exchange using the improved perturbative QCD estimates provided by the "Durham model" [9]. In the latest version, SuperChic v4, the photon-induced production of top quark pairs has also been made available [10]. For photon-induced processes, the event generator applies form-factors based on the "structure–function" approach [11].

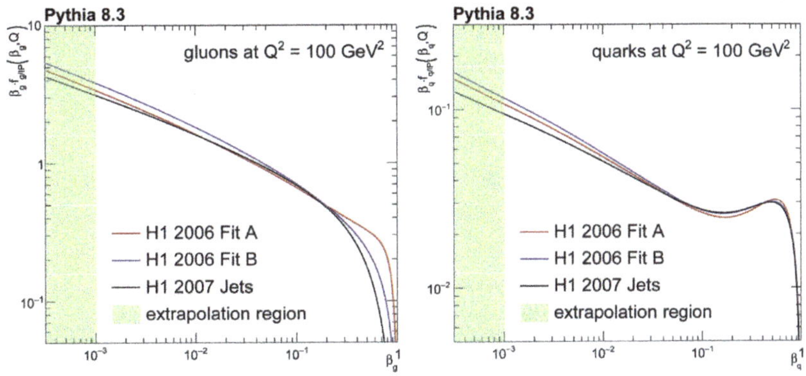

Figure 2. Diffractive parton densities ($\beta \cdot f(\beta, \mu = Q)$) in the pomeron as a function of the fraction of the momentum carried by the gluon (**left**) or quark (**right**), extracted from a HERA fit to combined structure–function data from H1 [6].

The Madgraph5_aMC@NLO event generator [12] is designed to generate matrix elements for both SM and beyond SM processes, including NLO QCD and EW corrections for parton-initiated processes [13,14]. While it can generate processes initiated by photons, it does not cover those initiated by pomerons. Elastic photon fluxes are obtained using the equivalent photon approximation (EPA) [15], and inelastic photon-initiated collisions can be generated from γ distribution functions inside the proton, such as, e.g., the LUXqed one [16]. A novel event generator, gamma-UPC [17], has been recently introduced, which derives photon fluxes from electric dipole and charge form-factors for protons and ions and includes realistic hadronic survival probabilities for them. This model has been successfully interfaced with Madgraph5_aMC@NLO v3. In addition, within this framework, the $\gamma\gamma \rightarrow t\bar{t}$ process can be computed at NLO perturbative QCD accuracy.

Forward Physics Monte Carlo (FPMC) [18] is a specialized Monte Carlo event generator developed for simulating exclusive and diffractive production processes. It possesses the capability to simulate all elementary $2 \rightarrow 2$ and $2 \rightarrow 1$ processes available in HERWIG6.5 [19], particularly the SM top quark pair production, and t-channel single top production. In FPMC, the original HERWIG code, which simulates two-photon exchange in electron–positron collisions, was modified. Pomeron fluxes were introduced with a particular proton structure in diffractive events (based on H1 fits) and are used in hadron collisions in this case. Recently, the anomalous exclusive production of $\gamma\gamma \rightarrow t\bar{t}$ was implemented in FPMC as well [20].

Pythia 8 [21], one of the most used event generators, is widely used for simulating events involving various interactions and particles, including hard diffraction [22], resolved and direct photons [23], and photoproduction [24]. The survival factors for Pomeron-induced processes are implemented as a part of the multiparton interaction (MPI) framework.

Table 1 summarizes different MC generators and outlines the processes that can be simulated by each.

Table 1. A list of various MC generators and available top quark production modes of processes with top quarks at the final state. In the table, X represent proton or pomeron remnants in a dissociative process.

Generator	$\gamma\gamma \to t\bar{t}$	$I\!P I\!P \to t\bar{t}$	$I\!P I\!P \to t\bar{t}X$	$\gamma p \to t\bar{t}X$	$I\!P p \to t\bar{t}X$
Superchic v4	✓	✓	—	—	—
MadGraph/gamma-UPC	✓	—	—	✓	—
FPMC	✓	✓	✓	✓	✓
Pythia8	✓	—	—	✓	✓
	$I\!P I\!P \to tqX$	$I\!P I\!P \to tWX$	$\gamma p \to tqX$	$\gamma p \to tWX$	$I\!P p \to tqX$
MadGraph/gamma-UPC	—	—	✓	✓	—
FPMC	✓	—	✓	—	✓
Pythia8	—	—	✓	—	✓

3. Tagging Diffractive and Photon-Induced Processes

Diffractive and photon-initiated processes often exhibit low hadronic activity, and some events are characterized by an intact proton emerging from the primary interaction. Forward Proton Detectors (FPD) are utilized at the LHC to identify these events by detecting forward protons. These detectors, such as the ATLAS Forward Proton detector (AFP) [25] and the CMS-TOTEM Precision Proton Spectrometer (CT-PPS) [26], are positioned approximately 200 m from the proton–proton interaction point. In diffractive or photon-initiated interactions, protons lose a fraction of their nominal momentum and are deflected differently by the LHC magnets, diverting them from the main bunch of protons. The FPD are near-beam detectors usually housed in Roman Pots vessels [27] which could approach the proton beam up to a few mm, aiming to measure the slight displacement of protons that participated in the interaction. Figure 3 illustrates a schematic layout of the beamline between the interaction point and the FPDs installed in LHC sector 56, corresponding to the negative z direction in the CMS.

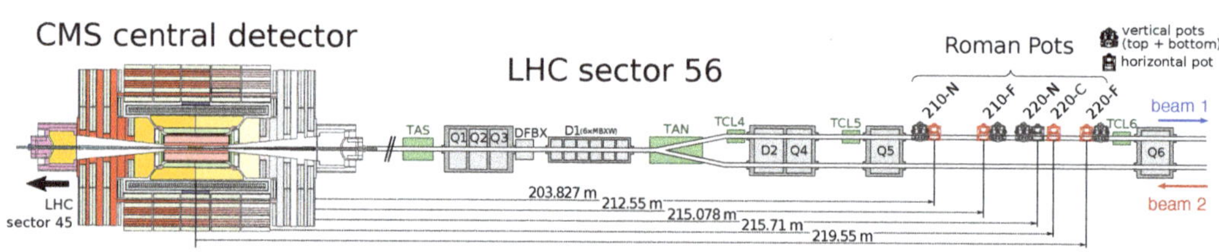

Figure 3. Schematic layout of the CMS Proton Precision Spectrometer (PPS) located in LHC sector 56. The accelerator magnets are displayed in gray, while the collimator system parts are highlighted in green. The detector units, marked in red, are those used by PPS during Run 2. The figure is taken from Ref. [28].

The FPDs are equipped with tracking and timing capabilities that determine the proton scattering angle and momentum loss based on the hit position of protons in the detector planes. The hit position of a proton is affected by its momentum loss, denoted by $\xi = \Delta p_z/p$, two scattered angles at the interaction point $\left(\theta_x^*, \theta_y^*\right)$, and coordinates of the proton–proton collision vertex situated on the plane perpendicular to the beam trajectory, denoted by (x^*, y^*):

$$\begin{aligned} \delta x &= D_x(\xi) + v_x(\xi) \cdot x^* + L_x(\xi) \cdot \theta_x^* \\ \delta y &= D_y(\xi) + v_y(\xi) \cdot y^* + L_y(\xi) \cdot \theta_y^* \end{aligned} \tag{5}$$

where D_x and D_y are the horizontal and vertical dispersion, respectively. The terms v_x and v_y stand for the horizontal and vertical magnifications, respectively, while $L_{x,y}$ are the effective lengths. These parameters are functions of the proton's momentum loss at different positions from the interaction point and are determined by simulating the proton's trajectory in the LHC magnetic field, as described in reference [29].

A minimum of five independent spatial measurements of the scattered proton are required to accurately determine proton kinematics, necessitating at least three tracking stations. However, even with just two tracking stations, it is possible to reconstruct proton kinematics by approximating Equation (5), such as considering $x^* = 0$, as demonstrated in reference [29].

The timing detectors measure the arrival times of the protons to the FPD. The time difference between the arrival of two protons from the same interaction vertex, in double-tagged events, is tied to the z-position of their production vertex. The correlation between the vertex position deduced from the proton arrival times and the vertex position determined from final state particles produced in association with the protons serves as an effective tool to separate diffractive interactions from non-diffractive ones [30]. This is particularly useful in rejecting protons originating from additional proton–proton interactions occurring during the same bunch crossing in high-intensity runs (high pileup runs) at the LHC. In addition to double-tagged events, the proton timing detectors can differentiate between pileup protons in single-tagged events, provided that the arrival time of the final state particles is also measurable [31].

4. Diffractive and Photo-Production of Top Quark Pairs

4.1. Single-Diffraction and Photoproduction of Top Quark Pairs

The diffractive production of top quarks can constitute up to a few percent of the total inclusive production cross-section. This aspect is crucial when aiming for precision measurements involving top quarks or when searching for new physics phenomena. For a top quark mass of $m_{\text{top}} = 172.5$ GeV, the inclusive production cross-section of top quark pairs, calculated at next-to-next leading order (NNLO) in QCD with resummation at next-to-next-to-leading logarithmic (NNLL) soft gluon terms [32], and more recently with third-order soft gluon corrections and the additional inclusion of electroweak corrections [33], ranges from around 180 to around 1000 pb, for proton–proton collision energies of $\sqrt{s} = 7$ TeV and $\sqrt{s} = 14$ TeV, respectively. The dominant diffractive production modes of top quark pairs are via single-diffractive dissociation and photoproduction, with their respective leading order (LO) cross-sections at $\sqrt{s} = 13$ TeV calculated to be 5 pb and 1.45 pb, respectively [34].

The study outlined in reference [34] investigated the single diffractive and photoproduction processes. The analysis assumed an FPD acceptance to an intact proton with momentum loss between 3% and 10%. This span is defined by a set of constraints; the lower limit is determined by the minimum distance of the detectors from the beam, while the upper limit is constrained by the beam collimators that shield the magnets from the intense radiation. The study revealed that, for photon-induced and single-diffractive processes, the acceptance rates are 30% and 20%, respectively, driven by the different photon and pomeron density fluxes. The intact proton kinematics for the pomeron- and photon-initiated production of top quark pairs were computed for the proton–proton beam conditions used during LHC Run 2 (2015–2018) and are depicted in Figure 4.

Based on the $t\bar{t}$ selection criteria and reconstruction efficiencies obtained in the measurement of differential cross sections of top quark pair production, in association with jets in pp collisions at $\sqrt{s} = 13$ by the ATLAS collaboration [35], event yields were calculated to be 150 ± 20 for single diffractive and 94 ± 3 photo-production of top quark pairs, given an integrated luminosity of 1 fb^{-1}. At the LHC, the predominant background for diffractive and photon-induced events stems from the multiple proton–proton collisions occurring in a single bunch crossing. This results in a combinatorial background, where a non-diffractive collision may be accompanied by additional diffractively scattered protons from other soft proton–proton interactions. To observe top quarks produced in association with an intact proton, a dataset with a low average number of interactions per bunch crossing is desired, ideally below 0.01. However, no data samples meeting these conditions have been collected so far.

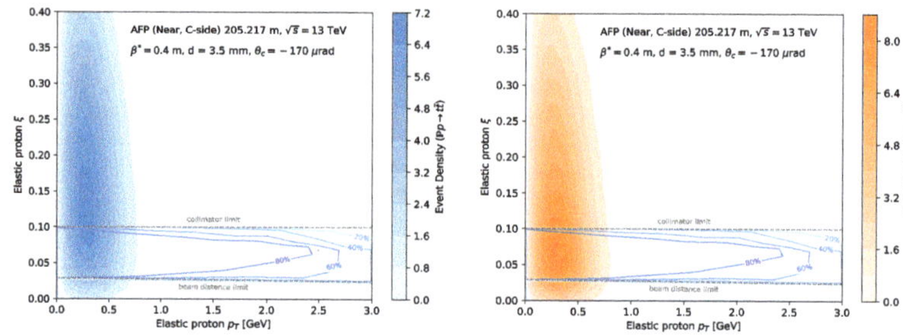

Figure 4. The event density for the production of top quark pairs in single-diffractive (**left**) and photoproduction (**right**) modes as a function of the proton's p_T and relative momentum loss (ξ). For context, acceptance contours for ATLAS AFP detectors using LHC Run 2 optics are superimposed for comparison. Figures are taken from reference [34].

4.2. Single-Diffraction Production of Single Top Quark

The study of the production of single top quarks at the LHC holds significant importance, as it probes the heavy quark content of the proton (proton PDF). This stems from the fact that the process involves interactions of heavy quarks originating from dissociated protons. The diffractive production of single top quarks offers further insights into the heavy quark content of the pomeron. Figure 5 illustrates the diagrams representing these interactions.

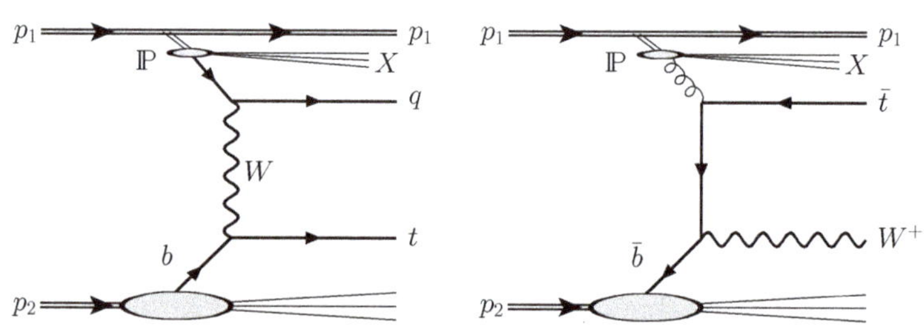

Figure 5. Representative diagrams of the single diffractive production of t-channel single top (**left**), and the W-associated production of single top (**right**).

Besides its use as a valuable tool for probing PDFs, the single diffractive production of single-top quarks can also shed light on the intrinsic pomeron content of heavy flavor quarks. The origin of the heavy flavor quark, whether from the proton or the pomeron, can be discerned by measuring the kinematics of the light jet in the t-channel single-top quark production. The difference between diffractive and non-diffractive Parton Distribution Functions (PDFs) leads to a disparity in the hardness of partons originating from non-diffractive protons. As a result, the central system is typically boosted in the direction of the non-diffractive proton when partons from the latter are more energetic. Therefore, by tagging the light jet and the intact proton, one can determine whether it is a proton or a pomeron with intrinsic heavy flavor content. The pseudorapidity distribution of the light jet for generated single-diffractive single-top-quark events, where the bottom quark is initiated by a diffractive or non-diffractive proton, is shown in Figure 6.

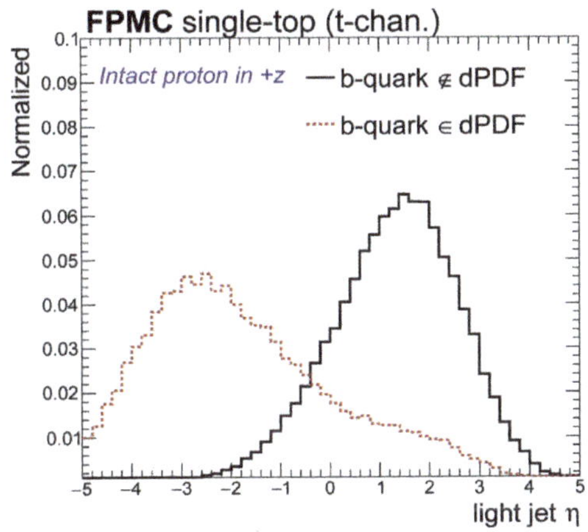

Figure 6. Light jet pseudorapidity in hard single diffractive $pp \to tq$ production process, distinguishing between different intrinsic quark flavors of the pomeron, either b-quarks (dashed red line) or gluons (solid black line).

Given the low cross-section of the single diffractive production of single-top events and the high density of gluons within the pomeron, such processes have not received significant attention in the literature until the present day.

4.3. Double Pomeron Exchange Processes

The diffractive events associated with two intact protons can arise when two color-singlets are exchanged. These events typically have a lower combinatorial background, but the signal cross-section drops substantially. As discussed in references [36,37], the production cross-section for $\gamma\gamma \to t\bar{t} + X$, $\gamma I\!P \to t\bar{t} + X$, and $I\!P I\!P \to t\bar{t} + X$ are 0.34 fb, 52 fb, and 28.4 fb, respectively. The authors of reference [36] estimated the sensitivity to diffractive production of top quarks in so-called semi-leptonic $t\bar{t}$ decays, $t\bar{t} \to jjb\ell\nu_\ell\bar{b}$, where one top quark decays hadronically into two light quarks and a b-quark, and the other into a b-quark with a W boson, which then decays leptonically into a lepton and neutrino. The size of this combinatorial background depends exponentially on the number of pileup collisions. Ab analysis followed the semi-leptonic event selection: four jets with p_T >25 GeV and $|\eta| < 2.5$, one lepton (electron or muon) with p_T >25 GeV and $|\eta| < 2.5$, at least two jets b-tagged jets, FPD acceptance between 1.5% and 15%, and a limited number of tracks associated with a primary interaction vertex (N_{TRK}). Due to the nature of the hard color-singlet exchange, the signature of diffractive interactions is in low hadronic activity, and N_{TRK} is typically used as a discriminating variable. The distribution of the number of tracks for different pileup events per interaction is shown in Figure 7.

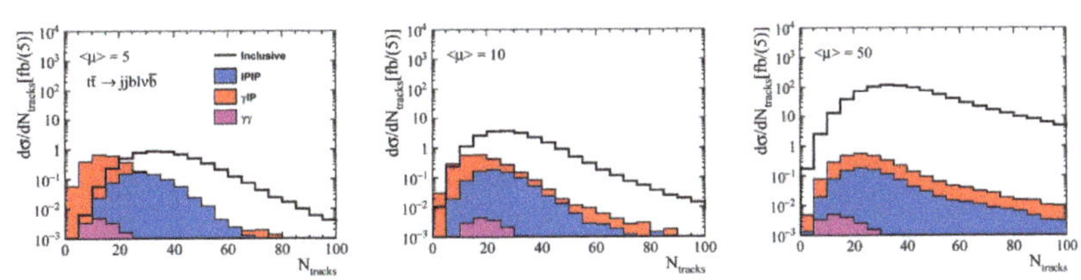

Figure 7. Distribution of the number of tracks for three different pileup profiles after event preselection, as detailed in the referenced publication. The plot is taken from reference [36].

The measurement of the diffractive production of top quarks requires a significant number of signal events and minimal background contamination. The authors of reference [36] have shown that the statistical significance for observing a diffractive signature was $\sigma = 11$ for a pileup rate of $\mu = 5$ and an integrated luminosity of 10 fb^{-1}. However, this significance drops to $\sigma = 6$ with a pileup rate of $\mu = 10$ and an integrated luminosity of 30 fb^{-1}.

In a subsequent study, the authors considered a pileup scenario of 200 interactions per bunch crossing, which is relevant for the High Luminosity LHC (HL-LHC) conditions [37]. To mitigate the effects of pileup, the authors integrated proton time of flight (ToF) measurements, effectively rejecting the protons stemming from pileup interaction. With a presumed timing resolution of 10 ps, the analysis managed to attain a statistical significance of $\sigma = 3$ even in this high pileup scenario, using the full HL-LHC integrated luminosity of 4000 fb^{-1}.

These studies emphasize the challenges posed by the pileup in measuring the diffractive production of top quarks and the importance of developing techniques to mitigate these effects in order to achieve a high sensitivity to the diffractive production of top quarks.

4.4. Central Exclusive Production of Top Quarks

The central exclusive production of top quark pairs has the lowest production cross-section among different diffractive and photon-induced production processes. Typical cross-sections span between 0.1 and 0.4 fb [17,34,36,38–41] for elastic photon exchange and of the order of 0.01–0.001 fb for Pomeron exchange [34,41]. One of the prominent advantages of proton tagging in central exclusive production processes is that the beam kinematics are exclusively transferred to the top quarks and protons. Given the absence of other intermediary particles, there is a direct correlation between the fractional momentum loss of the scattered proton and the top quarks, which can be expressed as follows:

$$\xi(t,\bar{t}) = \frac{1}{\sqrt{s}} \sum_{t,\bar{t}} [E_t \pm p_Z], \tag{6}$$

where s denotes the center-of-mass energy of the collision, and the two solutions for $\pm p_Z$ pertain to the protons moving in the positive or negative z direction. Equation (6) can be reformulated in terms of di-top mass ($m_{t\bar{t}}$) and rapidity ($y_{t\bar{t}}$), yielding the following:

$$m_{t\bar{t}} = \sqrt{s\xi_1\xi_2}, \qquad y_{t\bar{t}} = \frac{1}{2}\log(\xi_1\xi_2), \tag{7}$$

where ξ_1 and ξ_2 are the momentum losses of the two measured protons.

The study presented in reference [41] calculated both the inclusive and exclusive production cross-sections of top quark pairs via photon fusion, arguing for a considerable reduction in signal efficiency when applying a veto on charged particles or outgoing jets. They emphasized that the inclusive production of top quarks, either via gluon fusion or quark anti-quark annihilation, has an immense contribution at high pileup rates, which limits the potential to observe the Standard Model contribution of the $\gamma\gamma \to t\bar{t}$ production process at the LHC.

In the search for the exclusive production of top quark pairs, the CMS experiment analyzed data from 2017, corresponding to an integrated luminosity of 29.4 fb^{-1}, collected at a center-of-mass energy of 13 TeV [42]. The analysis focused on two channels: a dileptonic channel, where both top quarks decayed leptonically ($t \to b\ell\nu$), and a semileptonic channel, where one top quark decayed leptonically and the other hadronically. For the semileptonic channel, events were triggered by the presence of an electron or a muon with transverse momentum (p_T) above 30 GeV or 27 GeV, respectively. In addition, while the dileptonic channel used events triggered by two lepton triggers, the semileptonic channel used events triggered by a lepton and jet trigger, with the lepton p_T threshold being reduced to 30 GeV and at least one jet satisfying the momentum cut of $p_T > 35$ GeV.

In the dileptonic channel, the final selection required the presence of at least two oppositely charged leptons, where at least one of them is required to have $p_T > 30$ GeV and $|\eta| < 2.1$, and the dilepton system, where theform is required to have an invariant mass $m_{\ell\ell} > 20$ GeV. For the events with two reconstructed leptons of the same flavor, $m_{\ell\ell}$ must be outside a 30 GeV mass window around the Z boson mass peak. In the semileptonic channel, the final selection required the presence of exactly one lepton (electron or muon), at least two jets passing the b-tagging selection criteria, and at least two jets failing the b-tagging selection criteria. The b-tagging selection criteria were based on the *DeepCSV* algorithm [43].

The analysis utilizes a boosted decision tree (BDT) algorithm to discriminate exclusive from inclusive production. As input variables, the kinematics of leptons and jets are used, as well as the kinematic variables obtained from proton reconstruction and the ones obtained by reconstructing the top quark pairs. The resulting BDT distributions for each channel are shown in Figure 8.

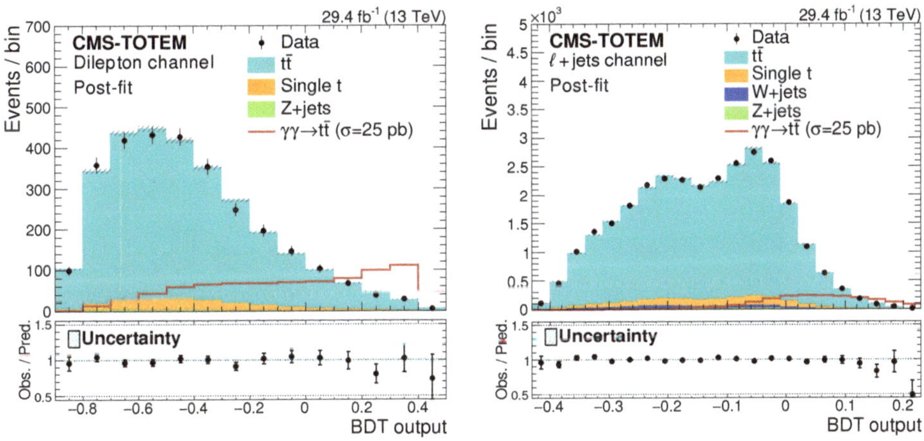

Figure 8. Distribution of the BDT score in the signal region for simulated events after the fit and for data for dileptonic (**left**) and semi-leptonic (**right**) channels. The figure is taken from reference [42].

The dominant background in the search for the exclusive production of top quark pairs is the combinatorial background, which arises from non-diffractive $t\bar{t}$ events produced in conjunction with two uncorrelated protons from pileup interactions. These events were modeled by mixing the reconstructed protons measured from the data with non-diffractive $t\bar{t}$ processes simulated in MC, normalized to the pileup proton tagging rate extracted from the data, creating a "mixed" MC samples with pileup proton information.

The probability of detecting at least one proton per arm in data ranges from 40 to 70%, depending on the LHC magnets settings and instantaneous luminosity, results in very high background rates from non-diffractive events, as the central exclusive production of top quark pairs comprises an order of 10^{-5} percent of the inclusive cross-section. Due to the high combinatorial background, with an average pileup interaction rate of $\mu \sim 35$, and a significantly small signal cross-section, an upper bound on the production cross-section was set at 0.59 pb at a 95% confidence level. This corresponds to about 3000 times the SM cross-section. The resulting observed and expected limits are shown in Figure 9. With the anticipated improvements in FPD timing capabilities and the larger amount of data expected to be collected at the HL-LHC, the potential for observing SM-exclusive top quark production will be increased [44–46].

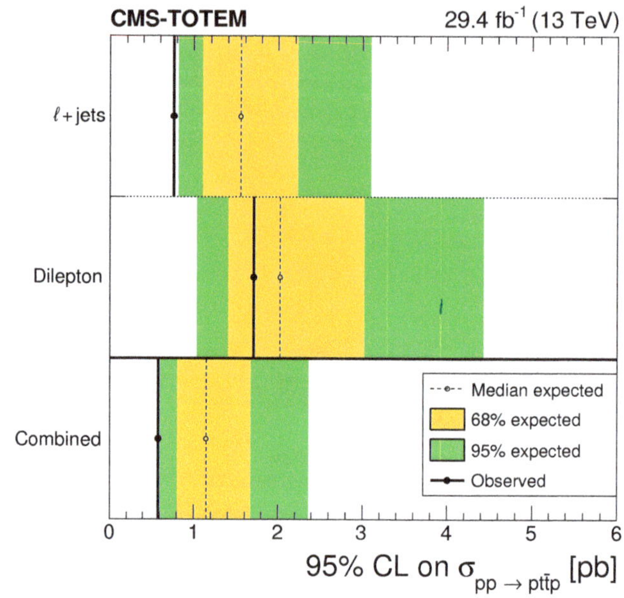

Figure 9. Expected 95% CL upper limit for the signal cross-section, for the two event selections (semi-leptonic and di-leptonic), and their combination. The green and yellow bands show the $\pm\sigma$ and $\pm 2\sigma$ intervals, respectively. The figure is taken from reference [42].

5. New Physics Searches with Photon-Induced Production of Top Quarks

Photon-induced production of top quark offers a unique opportunity to study flavor-changing neutral currents (FCNC) within the context of the SM. In the SM, production channels such as $u\gamma \to t$ or $c\gamma \to t$ are considerably suppressed. Therefore, the observation of top quark production without any accompanying quarks or bosons might indicate the presence of photon-mediated FCNC interactions.

Searches for single top quark production through FCNC have been conducted in reference [47], focusing on gluon-initiated processes. However, the photon-induced production of direct single top quarks through FCNC was proposed and investigated in reference [48]. Given the pronounced resemblance to the single top production process in the SM, a discriminator based on a neural network classifier was developed. Yet, the sensitivity was found to be comparable to the gluon-initiated processes. Nevertheless, as highlighted in reference [34], proton tagging in photon-initiated processes could improve the prospects for detecting FCNC interactions in photon-induced top quark production.

The central exclusive production of top quarks has a cross-section below 1 fb, necessitating a large amount of data and effective discrimination against pileup protons. Nonetheless, photon–photon fusion processes at the LHC offer opportunities to probe the SM and search for various beyond-SM physics models with anomalous $t\gamma$ couplings that could enhance production cross-section. These processes could have a distinct final state characterized by the exclusive topology, which includes the absence of proton beam remnants. Utilizing the kinematic correlation between reconstructed protons and top quarks could result in a search with low backgrounds.

The anomalous $\gamma t\bar{t}$ couplings were explored in reference [39] in terms of its impact on the electromagnetic dipole moments of the top quarks in pp collisions at the center of mass energies anticipated for HL-LHC (14 TeV) and HE-LHC (33 TeV) with integrated luminosities of 100, 300, and 3000 fb^{-1}. The analysis utilizes proton tagging, assuming FPD acceptance ranges from 1.5% to either 15% or 50%. The finding revealed a heightened sensitivity to the electric dipole moment of the top quark, which is the source of CP violation, compared to the magnetic dipole moment.

In reference [20], anomalous $\gamma\gamma t\bar{t}$ couplings were examined through the lens of dimension-8 operators within the framework of SM-effective field theory and also considering a new broad neutral resonance produced from the fusion of two photons that decays into a pair of top quarks, $\gamma\gamma \to \phi \to t\bar{t}$. The analysis assumed an FPD acceptance to fractional proton momentum loss between 1.5 and 20%, and a timing resolution of 20 and 60 ps, compatible with the design scenario presented in reference [26]. Proton time of flight (ToF) measurements in two tagged proton events can be used to reconstruct vertex z coordinate (v_z) and vertex production time (v_t), which can be computed from the following:

$$v_t = \frac{c}{2}(t_+ - t_-), \qquad v_t = \frac{1}{2}(t_+ + t_-) - \frac{z_{\mathrm{PPS}}}{c}, \tag{8}$$

where $c = 0.299792$ mm/ps denotes the speed of light, t_+ (t_+) indicates the ToF of the proton with positive (negative) p_z, and z_{PPS} is the distance from the interaction point to the FPD timing plane.

The study underscored a significant background suppression—about two orders of magnitude—achieved by incorporating proton ToF measurements with timing detectors having a nominal resolution of 20 ps. This highlights the pivotal role of FPD timing capabilities in refining the exclusive production of top quarks to search for physics beyond the SM. The sensitivity in terms of the dimension-8 operator coefficients was of the order of 10 TeV^{-4}. Furthermore, a scenario of a broad neutral scalar with a mass m and two typical couplings to photons was considered:

$$f_{\gamma\gamma} = \frac{m}{4\pi}, \qquad \Gamma_{\gamma\gamma} = 4\pi m \qquad \text{(Maximally broad width)} \tag{9}$$

$$f_{\gamma\gamma} = \frac{m}{\sqrt{4\pi}}, \qquad \Gamma_{\gamma\gamma} = m \qquad \text{(Moderately broad width)} \tag{10}$$

The neutral scalar can be considered broad for both scenarios since $\Gamma > m$. A Feynman diagram for the exclusive production of new particle ϕ with the decay to a pair of top quarks is shown in Figure 10.

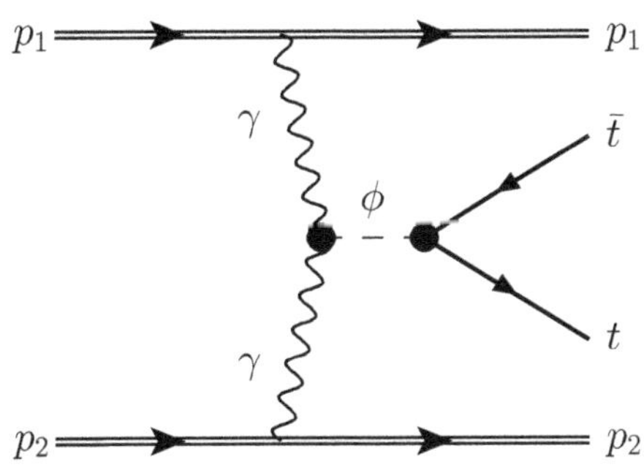

Figure 10. Feynman diagram for the exclusive production of ϕ via photon fusion; it decays to a pair of top quarks.

In this scenario as well, the timing detectors played a crucial role in amplifying the search sensitivity to resonance mass of $m_\phi > 1.5$ TeV. The projected sensitivity to new scalar ϕ in the exclusive $t\bar{t}$ analysis is shown in Figure 11.

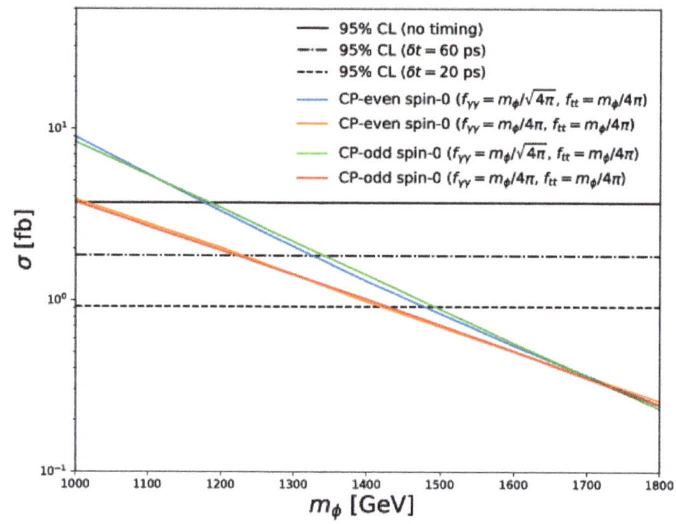

Figure 11. Projected sensitivity to the $pp \to p\, t\bar{t}\, p$ cross-section at 95% CL as a function of the mass of the neutral scalar. The figure is taken from reference [20].

6. Conclusions

The production of top quarks at the LHC is a primary area of interest in contemporary particle physics. Most analyses primarily focus on production modes initiated by quarks or gluons. However, there is a relatively uncharted territory concerning production modes initiated by color-neutral objects like pomerons or photons. A deep understanding of these processes is vital for precision-driven studies of the top quark's properties. Furthermore, several photon-induced production modes involving top quarks at the final state could broaden the phase space explored in searches for physics beyond the SM, offering substantial discovery potential at the LHC and beyond.

Funding: This research received no external funding.

Institutional Review Board Statement: Not applicable.

Informed Consent Statement: Not applicable.

Data Availability Statement: Not applicable.

Conflicts of Interest: The author declares no conflict of interest.

References

1. Donnachie, S.; Dosch, H.G.; Nachtmann, O.; Landshoff, P. *Pomeron Physics and QCD*; Cambridge University Press: Cambridge, UK, 2004; Volume 19.
2. Collins, P.D.B.; Spiller, T.P. A model for diffractive top-quark production. *J. Phys. Nucl. Phys.* **1984**, *10*, 1667. [CrossRef]
3. Collins, J.C.; Soper, D.E.; Sterman, G.F. Factorization of Hard Processes in QCD. *Adv. Ser. Direct. High Energy Phys.* **1989**, *5*, 1–91. [CrossRef]
4. Ingelman, G.; Schlein, P.E. Jet Structure in High Mass Diffractive Scattering. *Phys. Lett. B* **1985**, *152*, 256–260. [CrossRef]
5. Donnachie, A.; Landshoff, P.V. Hard diffraction: Production of high pT jets, W or Z, and Drell-Yan pairs. *Nucl. Phys. B* **1988**, *303*, 634–652. [CrossRef]
6. Aktas, A.; Andreev, V.; Anthonis, T.; Antunovic, B.; Aplin, S.; Asmone, A.; Astvatsatourov, A.; Babaev, A.; Backovic, S.; Baghdasaryan, A.; et al. Measurement and QCD analysis of the diffractive deep-inelastic scattering cross-section at HERA. *Eur. Phys. J. C* **2006**, *48*, 715–748. [CrossRef]
7. Harland-Lang, L.A.; Khoze, V.A.; Ryskin, M.G. Exclusive physics at the LHC with SuperChic 2. *Eur. Phys. J. C* **2016**, *76*, 9. [CrossRef]

8. Harland-Lang, L.A.; Khoze, V.A.; Ryskin, M.G. Exclusive LHC physics with heavy ions: SuperChic 3. *Eur. Phys. J. C* **2019**, *79*, 39. [CrossRef]
9. Khoze, V.A.; Martin, A.D.; Ryskin, M.G. Can the Higgs be seen in rapidity gap events at the Tevatron or the LHC? *Eur. Phys. J. C* **2000**, *14*, 525–534. [CrossRef]
10. Harland-Lang, L.A.; Tasevsky, M.; Khoze, V.A.; Ryskin, M.G. A new approach to modelling elastic and inelastic photon-initiated production at the LHC: SuperChic 4. *Eur. Phys. J. C* **2020**, *80*, 925. [CrossRef]
11. Han, T.; Valencia, G.; Willenbrock, S. Structure function approach to vector boson scattering in p p collisions. *Phys. Rev. Lett.* **1992**, *69*, 3274–3277. [CrossRef]
12. Alwall, J.; Frederix, R.; Frixione, S.; Hirschi, V.; Maltoni, F.; Mattelaer, O.; Shao, H.S.; Stelzer, T.; Torrielli, P.; Zaro, M. The automated computation of tree-level and next-to-leading order differential cross sections, and their matching to parton shower simulations. *J. High Energy Phys.* **2014**, *7*, 79. [CrossRef]
13. Hirschi, V.; Mattelaer, O. Automated event generation for loop-induced processes. *J. High Energy Phys.* **2015**, *10*, 146. [CrossRef]
14. Frederix, R.; Frixione, S.; Hirschi, V.; Pagani, D.; Shao, H.S.; Zaro, M. The automation of next-to-leading order electroweak calculations. *J. High Energy Phys.* **2018**, *7*, 185; Erratum in *J. High Energy Phys.* **2021**, *11*, 85. [CrossRef]
15. Budnev, V.; Ginzburg, I.; Meledin, G.; Serbo, V. The two-photon particle production mechanism. Physical problems. Applications. Equivalent photon approximation. *Phys. Rep.* **1975**, *15*, 181–282. [CrossRef]
16. Manohar, A.V.; Nason, P.; Salam, G.P.; Zanderighi, G. The Photon Content of the Proton. *J. High Energy Phys.* **2017**, *12*, 46. [CrossRef]
17. Shao, H.S.; d'Enterria, D. Gamma-UPC: automated generation of exclusive photon-photon processes in ultraperipheral proton and nuclear collisions with varying form factors. *J. High Energy Phys.* **2022**, *9*, 248. [CrossRef]
18. Boonekamp, M.; Dechambre, A.; Juranek, V.; Kepka, O.; Rangel, M.; Royon, C.; Staszewski, R. FPMC: A Generator for forward physics. *arXiv* **2011**, arXiv:1102.2531.
19. Corcella, G.; Knowles, I.G.; Marchesini, G.; Moretti, S.; Odagiri, K.; Richardson, P.; Seymour, M.H.; Webber, B.R. HERWIG 6.5 release note. *arXiv* **2002**, arXiv:hep-ph/0210213v2.
20. Baldenegro, C.; Bellora, A.; Fichet, S.; von Gersdorff, G.; Pitt, M.; Royon, C. Searching for anomalous top quark interactions with proton tagging and timing detectors at the LHC. *J. High Energy Phys.* **2022**, *8*, 21. [CrossRef]
21. Bierlich, C.; Chakraborty, S.; Desai, N.; Gellersen, L.; Helenius, I.; Ilten, P.; Lönnblad, L.; Mrenna, S.; Prestel, S.; Preuss, C.T.; et al. A comprehensive guide to the physics and usage of PYTHIA 8.3. *Scipost Phys. Codebases* **2022**, *8*, 1–287. [CrossRef]
22. Rasmussen, C.O. Hard Diffraction in Pythia 8. *EPJ Web Conf.* **2016**, *120*, 02002. [CrossRef]
23. Helenius, I. Photon-photon and photon-hadron processes in Pythia 8. *CERN Proc.* **2018**, *1*, 119. [CrossRef]
24. Helenius, I.; Rasmussen, C.O. Hard diffraction in photoproduction with Pythia 8. *Eur. Phys. J. C* **2019**, *79*, 413. [CrossRef]
25. The ATLAS Collaboration. *Technical Design Report for the ATLAS Forward Proton Detector*; CERN: Geneva, Switzerland, 2015.
26. The CMS and TOTEM Collaborations. *CMS-TOTEM Precision Proton Spectrometer*; CERN: Geneva, Switzerland, 2014.
27. Amaldi, U.; Biancastelli, R.; Bosio, C.; Matthiae, G.; Allaby, J.V.; Bartel, W.; Block, M.M.; Cocconi, G.; Diddens, A.N.; Dobinson, R.W.; et al. Measurements of the proton proton total cross-sections by means of Coulomb scattering at the Cern intersecting storage rings. *Phys. Lett. B* **1973**, *43*, 231–236. [CrossRef]
28. The CMS Collaboration. Development of the CMS detector for the CERN LHC Run 3. *arXiv* **2023**, arXiv:2309.05466.
29. CMS and TOTEM Collaborations. Proton reconstruction with the CMS-TOTEM Precision Proton Spectrometer. *arXiv* **2022**, arXiv:2210.05854.
30. Černý, K.; Sýkora, T.; Taševský, M.; Žlebčík, R. Performance studies of Time-of-Flight detectors at LHC. *J. Instrum.* **2021**, *16*, P01030. [CrossRef]
31. Pasechnik, R.; Taševský, M. Multi-dimensional hadron structure through the lens of gluon Wigner distribution. *arXiv* **2023**, arXiv:2310.10793.
32. Czakon, M.; Mitov, A. Top++: A Program for the Calculation of the Top-Pair Cross-Section at Hadron Colliders. *Comput. Phys. Commun.* **2014**, *185*, 2930. [CrossRef]
33. Kidonakis, N.; Guzzi, M.; Tonero, A. Top-quark cross sections and distributions at approximate N3LO. *Phys. Rev. D* **2023**, *108*, 054012. [CrossRef]
34. Howarth, J. Elastic Potential: A proposal to discover elastic production of top quarks at the Large Hadron Collider. *arXiv* **2020**, arXiv:2008.04249.
35. The ATLAS Collaboration. Measurements of differential cross sections of top quark pair production in association with jets in pp collisions at $\sqrt{s} = 13$ TeV using the ATLAS detector. *J. High Energy Phys.* **2018**, *10*, 159. [CrossRef]
36. Gonçalves, V.P.; Martins, D.E.; Rangel, M.S.; Tasevsky, M. Top quark pair production in the exclusive processes at the LHC. *Phys. Rev. D* **2020**, *102*, 074014. [CrossRef]
37. Martins, D.E.; Tasevsky, M.; Goncalves, V.P. Challenging exclusive top quark pair production at low and high luminosity LHC. *Phys. Rev. D* **2022**, *105*, 114002. [CrossRef]
38. de Favereau de Jeneret, J.; Lemaitre, V.; Liu, Y.; Ovyn, S.; Pierzchala, T.; Piotrzkowski, K.; Rouby, X.; Schul, N.; Vander Donckt, M. High energy photon interactions at the LHC. *arXiv* **2009**, arXiv:0908.2020.
39. Fayazbakhsh, S.; Monfared, S.T.; Mohammadi Najafabadi, M. Top quark anomalous electromagnetic couplings in photon-photon scattering at the LHC. *Phys. Rev. D* **2015**, *92*, 014006. [CrossRef]

40. d'Enterria, D.; Lansberg, J.P. Study of Higgs boson production and its b anti-b decay in gamma-gamma processes in proton-nucleus collisions at the LHC. *Phys. Rev. D* **2010**, *81*, 014004. [CrossRef]
41. Łuszczak, M.; Forthomme, L.; Schäfer, W.; Szczurek, A. Production of $t\bar{t}$ pairs via $\gamma\gamma$ fusion with photon transverse momenta and proton dissociation. *J. High Energy Phys.* **2019**, *2*, 100. [CrossRef]
42. The CMS and TOTEM Collaborations. Search for central exclusive production of top quark pairs in proton-proton collisions at $\sqrt{s} = 13$ TeV with tagged protons. *arXiv* **2023**, arXiv:2310.11231.
43. The CMS Collaboration. Identification of heavy-flavour jets with the CMS detector in pp collisions at 13 TeV. *J. Instrum.* **2018**, *13*, P05011. [CrossRef]
44. The CMS collaboration. The CMS Precision Proton Spectrometer at the HL-LHC—Expression of Interest. *arXiv* **2021**, arXiv:2103.02752.
45. Pitt, M. Physics at the HL-LHC with Proton Tagging. *Acta Phys. Pol. Proc. Suppl.* **2023**, *16*, A7–A12. [CrossRef]
46. Deile, M.; Taševský, M. *High Luminosity Forward Physics*; World Scientific: Singapore, 2023; Chapter 18, pp. 271–279. [CrossRef]
47. Aad, G.; Abbott, B.; Abdallah, J.; Abdinov, O.; Aben, R.; Abolins, M.; AbouZeid, O.S.; Abramowicz, H.; Abreu, H.; Abreu, R.; et al. Search for single top-quark production via flavour-changing neutral currents at 8 TeV with the ATLAS detector. *Eur. Phys. J. C* **2016**, *76*, 55; Erratum in *Eur. Phys. J. C* **2022**, *82*, 70. [CrossRef] [PubMed]
48. Goldouzian, R.; Clerbaux, B. Photon initiated single top quark production via flavor-changing neutral currents at the LHC. *Phys. Rev. D* **2017**, *95*, 054014. [CrossRef]

universe

MDPI

Review

Top Quark Asymmetries

Thorsten Chwalek [1,*,†] and Frédéric Déliot [2,*,†]

1 Institut für Experimentelle Teilchenphysik, Karlsruher Institut für Technologie, 76131 Karlsruhe, Germany
2 Irfu, CEA Paris-Saclay, Université Paris-Saclay, CEDEX, 91191 Gif-sur-Yvette, France
* Correspondence: thorsten.chwalek@cern.ch (T.C.); frederic.deliot@cern.ch (F.D.)
† These authors contributed equally to this work.

Abstract: The production of top quark pairs ($t\bar{t}$) via the quark-antiquark initial state is not symmetric under the exchange of top quark and antiquark. Calculations of this next-to-leading order effect predict asymmetries of about one to a few percent, depending on the centre-of-mass energy and the selected phase space. Experimentally, this charge asymmetry of $t\bar{t}$ production manifests itself in differences in angular distributions between top quarks and antiquarks. Sensitive observables are the rapidities of the produced top quarks and antiquarks as well as their energies. In dileptonic $t\bar{t}$ events, the asymmetry of the $t\bar{t}$ system is reflected in a similar asymmetry in the system of the produced lepton pair, with the crucial advantage of a simpler reconstruction procedure. In this article we review the measurements of this effect in different final states and using different observables by the ATLAS and CMS Collaborations in LHC collisions at three different centre-of-mass energies.

Keywords: top quark; asymmetry; LHC; ATLAS; CMS

Citation: Chwalek, T.; Déliot, F. Top Quark Asymmetries. *Universe* **2022**, *8*, 622. https://doi.org/10.3390/universe8120622

Academic Editor: Jinmin Yang

Received: 21 September 2022
Accepted: 16 November 2022
Published: 25 November 2022

1. Introduction

Symmetry is a well known principle in Nature and has also a significant impact on our every-day-life, where symmetric states are considered "right" or "beautiful", be it in arts, architecture, or other fields of our life. Moreover, symmetry is a powerful concept of the mathematical description of the world we live in and many of the great achievements in humankind's endeavour to understand and formally describe the underlying principles of physics have been made thanks to symmetries. A prominent example is the standard model of particle physics (SM) being based on gauge symmetries. "Asymmetry" on the other hand is—according to the Merriam-Webster dictionary—defined as "the lack or absence of symmetry" [1], while this—on the first glance—sounds like a shortcoming of something that fails to be symmetric, a deeper look reveals that asymmetry as a principle of Nature is as important as symmetry. Just think of the asymmetry between matter and antimatter in our Universe that makes the Universe and life possible in the first place. Asymmetries in elementary particle physics often shape the way towards a deeper understanding of the fundamental objects and their interactions. One well-known example for such an asymmetry in particle physics is the asymmetry between top quarks and top antiquarks in the production of $t\bar{t}$ pairs. When produced via the charge symmetric fusion of two gluons, top quark (t) and antiquark ($\bar{t}$) behave symmetrically, while they loose this symmetry when produced from quark-antiquark-annihilation, as the top quark is tied to the incoming quark and the top antiquark is connected to the incoming antiquark, hence defining preferred directions for the production of top quarks and top antiquarks.

The fact that the production of top quark-antiquark pairs ($t\bar{t}$) from quark-antiquark initial states is not symmetric under exchange of top quark and antiquark when looking at cross sections as functions of angular variables was described and predicted already more than thirty years ago [2,3], followed by first measurements at the Fermilab Tevatron by the CDF [4] and D0 [5] Collaborations in 2008. Initially, the CDF results [6] have been a bit large in value with respect to the prediction and caused some stir in the physics community at that time. Triggered by the observed discrepancy between SM predictions and experimental results, several attempts have been made to explain the experimental findings with BSM physics, see for example Refs. [7–10]. However, in the following, updated theory calculations including further corrections and superseding new experimental results from the Tevatron experiments came closer [11], calming down the excitement that has been raised by the first experimental results.

In the meanwhile, data taking at the LHC started with the first proton-proton collisions in 2009 and the top quark and its properties came also in focus of the LHC experiments. Differences in angular distributions of top quarks and antiquarks could also be measured at the LHC, although under less favorable conditions, reducing the measurable size of the effect significantly. The challenging environment is partially compensated by the vast amount of $t\bar{t}$ candidate events produced in LHC collisions with the effect of a reduced statistical component of the measurement's uncertainty.

In this article, we summarize the most recently published results by the ATLAS and CMS Collaborations on different manifestations of this $t\bar{t}$ production asymmetry, derived at three different centre-of-mass energies, and evaluate what we have learned from them concerning the top quark and potential indications for physics from beyond the standard model in the top quark sector.

2. Phenomenology and Theory Overview

When produced from the annihilation of a quark and an antiquark, the top quark and top antiquark interact with the colour fields of the incoming partons. This interaction leads to a correlation of the directions of motion of the outgoing top quark and of the incoming quark and of the directions of motion of the outgoing top antiquark and of the incoming antiquark, yielding differences in angular distributions of top quark and antiquark. This charge asymmetry, i.e., an asymmetry between top quark and antiquark, can also be calculated quantitatively. It occurs at next-to-leading-order (NLO) perturbation theory as an effect of the interference of initial-state (ISR) and final-state (FSR) radiation diagrams and of the interference of the Born and box diagrams for quark-antiquark initial states. While the ISR-FSR interference contributes negatively to the asymmetry, the contribution from the Born-box interference is positive. As the gluon-gluon initial state is charge symmetric, no asymmetry is present in these events. The $t\bar{t}$ processes (without extra additional jets) with a quark-gluon initial state feature a very small asymmetry, which - compared to the asymmetry from quark-antiquark initial states- can be neglected in most of the cases.

On parton level, the $t\bar{t}$ charge asymmetry can be defined as as a forward-backward asymmetry (A_{FB}):

$$A_{\mathrm{FB}} = \frac{\sigma(x > 0) - \sigma(x < 0)}{\sigma(x > 0) + \sigma(x < 0)}, \tag{1}$$

where x can be any observable (for example (pseudo-)rapidity) that defines two hemispheres with $x > 0$ defining the hemisphere of "forward" direction and $x < 0$ the hemisphere of "backward" direction and the σs are the corresponding $t\bar{t}$ production cross sections for production in either of the two hemispheres.

How this parton level asymmetry actually manifests itself on particle level and how big the measurable effect is, depends on the colliding particles and their energy.

At the Tevatron collider at Fermilab protons and antiprotons were collided at a centre-of-mass energy of 1.96 TeV. Given the parton distribution functions (PDFs) of proton and antiproton at this energy, a valid assumption is, that the direction of the proton beam is also the direction of the incoming initial-state valence quark, while the direction of the antiproton beam defines the direction of the initial-state valence antiquark: "the charge asymmetry will be reflected not only in the partonic rest frame but also in the center-of-mass system of the proton and antiproton" [3]. The asymmetry is thus visible in the ratios of produced top quarks to top antiquarks in the two hemispheres. The experimental observable that has been used for the Tevatron analyses is $\Delta\eta$, the difference of the pseudorapidities of top quark and top antiquark in each event.

A $t\bar{t}$ forward-backward asymmetry can also be defined at the LHC, although not as straight-forward as for the Tevatron. While at the Tevatron, as discussed above, the "forward" direction is simply given by the proton beam direction, such a "global" definition of a preferred spatial direction for all events is not possible, because of the charge symmetric proton-proton collisions at the LHC. Here, it is a priori not clear whether the incoming (anti)quark comes from a proton from the clockwise injected beam or from the anti-clockwise injected beam. The parton level definition of "forward" and "backward" hemispheres is therefore only valid for single events. Consequently also on particle level the forward and backward directions can only be defined for single events. Therefore one needs to exploit a reconstructable observable that is sensitive to the forward direction as defined on parton level. The longitudinal momentum of the produced $t\bar{t}$ system is suited for that purpose as the incoming quark parton has on average a higher momentum fraction of the proton compared to the antiquark parton and the $t\bar{t}$ system is thus on average boosted along the direction of motion of the incoming quark.

However, also without making the effort to define forward and backward directions in LHC collisions, one can still identify a difference in angular distributions between top quark and top antiquark. The interacting quarks in proton-proton collisions can be valence quarks or sea quarks, while the interacting antiquarks are always sea quarks. Thus, the initial state quarks carry on average a higher momentum fraction of the proton compared to initial state antiquarks. This difference in average momentum together with the already mentioned correlation of the momentum of the incoming (anti-)quark with that of the produced top (anti-)quark, results in higher momenta on average of the produced top quarks, while the top antiquarks are produced more centrally. Experimentally this effect can be observed through the difference in the widths of the rapidity (y) distributions of top quarks compared to the one of top antiquarks.

Finally, in $t\bar{t}$ events with an additional high-p_{T} jet, where the quark gluon initial state ($qg \rightarrow t\bar{t}j$) plays the dominating role, exists an asymmetry in the scattering angles between top quarks and top antiquarks with respect to the direction of motion of the additional jet. Energy and momentum conservation connects the asymmetry of the top quark and antiquark scattering angles in the $t\bar{t}j$ rest frame to an observable energy difference of top quark and antiquark, ΔE. Hence this effect is named energy asymmetry [12]. Exploiting the boost of the outgoing quark-jet in the direction of the incoming valence quark one can define ΔE differentially and thus maximise the statistical sensitivity to the energy asymmetry.

In the SM, the positive contributions from the Born-box interference outweigh the negative contributions from the ISR-FSR interference and the predicted asymmetry values from theoretical calculations lie in the range of very few percent (see for example Refs. [13–15]). The actual values depend on the colliding hadrons, on the centre-of-mass energy, on the examined phase-space, on the exploited observable, and on the terms and corrections included in the calculation. Although the asymmetry occurs at NLO precision in QCD calculations, it has been shown that electroweak (EW) contributions as well as next-to-next-to-leading order (NNLO) QCD contributions are of significant size [13,15]. However, the analyses described in this review make in general use of the state-of-the-art NLO-QCD Monte Carlo generators for the simulation of the $t\bar{t}$ signal. Predictions from these generators are therefore in general smaller than predictions from theory calculations including higher order QCD effects and EW contributions. For that reason, in the analyses described in this review wherever possible the experimental results are compared to predictions from calculations rather than to predictions from simulation. Physics contributions from beyond the SM (BSM) could however significantly enhance the asymmetry between top quark and antiquark while sustaining other boundary conditions like the total $t\bar{t}$ production cross section or other kinematic distributions of the produced top(anti)quarks. As the top quark plays a special role among all quarks of the SM due to its large mass and correspondingly its Yukawa coupling being close to unity, it is believed to be particularly sensitive to new physics contributions. Depending on the mass scale of hypothetical new particles, different approaches are used in the theory calculations to predict the impact on the $t\bar{t}$ charge asymmetry. Assuming the new physics being very heavy and out of reach of the LHC collisions, an effective field theory (EFT) ansatz is used to identify those operators [16,17] that could have a significant impact on the $t\bar{t}$ charge asymmetry while keeping the predictions for various kinematic distributions and cross sections in agreement with observations. Potential light new particles could be exchanged via s, t, or u channel. Prominent examples of such additional new particles are a colour-octet vector G, a neutral Z' boson, a charged W' boson, a colour-triplet scalar ω, or a colour-sextet scalar Ω [18].

3. Measurements of the $t\bar{t}$ Charge Asymmetry at the LHC

The ATLAS and CMS Collaborations have measured the $t\bar{t}$ charge asymmetry at three different centre-of-mass energies at the LHC, at 7, 8, and 13 TeV. $t\bar{t}$ candidate events with either two charged leptons (dilepton channel) or one charged lepton (lepton+jets channel) are analyzed. In the analyses of both collaborations, "lepton" refers to electrons and muons, while tau leptons are only considered indirectly through their decay products (electrons or muons) when decaying leptonically. The difference of the absolute values of the rapidity of the top quarks and antiquarks serves as the observable sensitive to the $t\bar{t}$ charge asymmetry,

$$A_C^{t\bar{t}} = \frac{N(\Delta|y| > 0) - N(\Delta|y| < 0)}{N(\Delta|y| > 0) + N(\Delta|y| < 0)}, \quad (2)$$

where $\Delta|y| = |y_t| - |y_{\bar{t}}|$.

In order to construct this observable, the $t\bar{t}$ system needs to be fully reconstructed. For this purpose different approaches are used in the different analyses. The common property of all these reconstruction methods is that the reconstructed fourvectors of top quarks and antiquarks as well as the resulting kinematic observable are always diluted with respect to their true values and they need to be corrected for. The impacts from a $\Delta|y|$ dependent detector efficiency and resolution effects of the kinematic reconstruction are rectified in the different analyses by deploying unfolding methods.

For events in the dilepton channel, it is also possible to measure directly $\Delta|\eta|$ between the positively and the negatively charged lepton from the top quark decays: $\Delta|\eta| = |\eta_{\ell^+}| - |\eta_{\ell^-}|$. As the leptons from the top quark decay inherit the direction of motion of the mother top quark, a similar asymmetry is expected also for this lepton based $\Delta|\eta|$ observable,

$$A_C^{\ell\ell} = \frac{N(\Delta|\eta| > 0) - N(\Delta|\eta| < 0)}{N(\Delta|\eta| > 0) + N(\Delta|\eta| < 0)}. \quad (3)$$

However, the direction of motion of the leptons is not fully identical to that of the mother top (anti)quarks, as a result the predicted asymmetry is smaller by about 40% compared to the asymmetry of top quark and antiquark. This disadvantage is compensated for by the advantage that no reconstruction of the $t\bar{t}$ system is required and thus no diluting effects from the reconstruction resolution have to be considered when analysing the asymmetry of the leptons. Only the detector acceptance needs to be corrected for in the unfolding, making this procedure simpler and reducing the impact on the result from the chosen unfolding method.

It is also interesting to measure $A_C^{t\bar{t}}$ differentially as a function of kinematic variables of the $t\bar{t}$ system that are sensitive to the $t\bar{t}$ charge asymmetry. Suited for this purpose are the transverse momentum $p_T^{t\bar{t}}$, the rapidity $y_{t\bar{t}}$, and the invariant mass $m_{t\bar{t}}$ of the $t\bar{t}$ system. Furthermore, the boost in z direction of the $t\bar{t}$ system, $\beta_{z,t\bar{t}}$, is used in differential measurements. The transverse momentum of the $t\bar{t}$ system gives a handle on the ratio of the positive contribution to the total asymmetry from the interference between Born and box diagrams and the negative contribution from the interference between ISR and FSR. Events with additional hard radiation feature on average also a higher transverse momentum of the $t\bar{t}$ system, thus for $t\bar{t}$ events at high transverse momentum the negative contribution from the ISR-FSR interference dominates [13]. As $t\bar{t}$ pairs produced via $q\bar{q}$ annihilation are most often found at large rapidities, while the charge symmetric production via gluon fusion is dominant in the central region, one can expect an enhancement of the charge asymmetry with increasing $|y_{t\bar{t}}|$ [13]. This is because of the on average higher momentum of the valence quarks with respect to the momentum of the sea antiquarks, resulting in a boost of the $t\bar{t}$ system in the z direction. For the very same reason, also the boost of the $t\bar{t}$ system in z direction is a powerful observable to enhance the size of the asymmetry. Finally, the invariant mass of the $t\bar{t}$ system is also sensitive to production mechanism with an enhanced fraction of $q\bar{q}$ produced $t\bar{t}$ pairs at high values of $m_{t\bar{t}}$ and thus also enhanced values for the asymmetry. In addition, the effect of new contributions from BMS physics is expected to be stronger for events with high $m_{t\bar{t}}$ values or with large values of the z coordinate of the velocity of the $t\bar{t}$ system [19].

In order to correct for efficiency and resolution effects in these differential measurements, sophisticated multidimensional unfolding procedures need to be deployed as not only the $\Delta|y|$ distribution but also the values of the kinematic variables of the $t\bar{t}$ system need to be corrected at the same time.

The following sections summarize the $t\bar{t}$ charge asymmetry measurements at 7, 8, and 13 TeV in the dilepton and lepton+jets channels, carried out by the ATLAS and CMS Collaborations.

3.1. Measurements at 7 TeV Centre-of-Mass Energy

The ATLAS and CMS Collaborations have both explored the full dataset of collisions at 7 TeV centre-of-mass energy to measure the $t\bar{t}$ charge asymmetry in the dilepton and lepton+jets channels. The dataset collected by the ATLAS experiment corresponds to an integrated luminosity of 4.6 fb^{-1}, while the CMS dataset corresponds to 5 fb^{-1}. The $t\bar{t}$ charge asymmetry $A_C^{t\bar{t}}$ is measured in both channels while the lepton asymmetry $A_C^{\ell\ell}$ is an exclusive observable of the dilepton channel. Theory calculations at NLO QCD including mixed QCD-QED and QCD-weak interaction corrections [14] predict for the SM at a centre-of-mass energy of 7 TeV values of

$$A_C^{t\bar{t}} = 0.0123 \pm 0.0005(\text{scale}), \tag{4}$$
$$A_C^{\ell\ell} = 0.0070 \pm 0.0003(\text{scale}). \tag{5}$$

The quoted scale uncertainties consider variation of the renormalization and factorization scales by factors of 0.5 and 2 with respect to the reference value, set to the top quark mass. Another calculation, differing in few technical aspects like the central value of the renormalization and factorization scales being set to the partonic centre-of-mass energy and the usage of a leading order parton distribution function instead of a next-to-leading order one, yields a similar value for the $t\bar{t}$ charge asymmetry, $A_C^{t\bar{t}} = 0.0115 \pm 0.0006(\text{scale})$ [13].

These predictions are made for inclusive $t\bar{t}$ production without any constraints on the phase space of the process. Hence, the asymmetries measured experimentally in a certain phase space region, as defined by the event selection requirements of the analyses, need to be extrapolated to the full phase space of $t\bar{t}$ production to be directly comparable to the predictions. For this purpose, corrections are applied to the reconstructed distributions of the sensitive observables.

In this chapter, the measurements of $A_C^{t\bar{t}}$ and $A_C^{\ell\ell}$ at $\sqrt{s} = 7$ TeV by the ATLAS and CMS Collaborations are summarized.

3.1.1. Measurements of the $t\bar{t}$ and Lepton Charge Asymmetry in the Dilepton Channel by ATLAS

The ATLAS Collaboration measured the lepton and $t\bar{t}$ charge asymmetry in the 7 TeV dataset in the dilepton channel [20]. Electrons, muons, and jets, reconstructed using the anti-k_T algorithm with a distance parameter of $R = 0.4$, are considered in the analysis. To suppress backgrounds from fake and non-prompt leptons, the electrons and muons are required to be isolated from any other activity in the detector around their tracks. Exactly two, oppositely charged, isolated leptons are required and according to the flavour of the leptons, the selected dataset is divided into three channels: ee, $e\mu$, $\mu\mu$. To suppress background events from Drell-Yan and Z boson production, in the ee and $\mu\mu$ channels, the invariant mass of the lepton pair needs to be larger than 15 GeV and needs to fall outside a window of 20 GeV width centered on the Z boson mass. In addition, the missing transverse energy $E_{\mathrm{T}}^{\mathrm{miss}}$ of candidate events needs to be larger than 60 GeV in these channels.

8125 candidate events remain in the selected dataset, with about six times more signal than background expected. The main background contribution comes from the electroweak production of single top quarks, followed by diboson production. The two backgrounds from Z boson production and fake or non prompt leptons are efficiently reduced by the requirements mentioned above and comprise the smallest contribution to the background considered in this analysis.

The sensitive observable of the measurement of the lepton charge asymmetry, the difference in pseudorapidity of the two charged leptons $\Delta|\eta|$, is directly accessible from the fourvectors of the selected leptons. For the difference of the rapidity of the two top quarks $\Delta|y|$, the sensitive observable of the measurement of the $t\bar{t}$ charge asymmetry, a kinematic reconstruction of the $t\bar{t}$ system needs to be applied, starting from the objects measured in the detector and imposing energy conservation at each decay vertex. The resulting kinematic equations are underconstrained due to the two neutrinos present in the signal process. For that reason further assumptions are made on the reconstructed top quark and W boson masses and a scan over all possible hypotheses for the two final state neutrinos is performed. Every hypothesis gets a weight that reflects the degree of agreement between the $E_{\mathrm{T}}^{\mathrm{miss}}$ calculated from the reconstructed neutrinos and the $E_{\mathrm{T}}^{\mathrm{miss}}$ as observed in the event. In addition, all possible associations of jets and leptons are considered where the jet energies are allowed to float within their resolutions. For each event the hypothesis with the best neutrino weight is used in the further analysis chain. Studies on simulated $t\bar{t}$ events show that with this method, solutions for 80% of all events can be found, while for 20% of the events the method does not produce a valid solution. Consequently, the events without solution of the kinematic reconstruction are not considered for the analysis.

Figure 1 shows the reconstructed values of $\Delta|\eta|$ between the two leptons (left) and $\Delta|y|$ between top quark and antiquark (right) over their generated true values. This response matrix describes the resolution of the reconstruction and how likely migrations between different bins of true and reconstructed values of the sensitive observable are. The response matrix for the lepton asymmetry is dominated by the diagonal elements with each being larger than 90%, thus showing a very good resolution of the reconstruction. Hence, the main correction needed in this analysis is the correction of the acceptance effects.

Acceptance effects and the small migration effects are corrected for by bin-by-bin correction factors, which are applied to the background subtracted numbers of selected data events in each bin. The correction factors are derived from simulation and are given by the ratio of predicted signal events from the true distribution in a certain bin divided by the predicted signal events from the reconstructed distribution in the same bin.

As can be seen from Figure 1 (right), the resolution of the observable relying on the $t\bar{t}$ system reconstruction is worse with respect to that of the lepton based observable. Hence, instead of bin-wise correction factors, the fully Bayesian unfolding (FBU, [21]) is applied to correct for detector resolution effects. For that purpose, the response matrix as shown in Figure 1 (right), describing the detector resolution effects, is derived from the nominal simulated $t\bar{t}$ sample from generated and reconstructed values of the sensitive observable. By applying this matrix to the true distribution and adding a background model, the total expected yield of events in the selected phase space can be expressed and compared via a maximum likelihood estimator to the number of observed events per bin. Within the FBU, the maximum likelihood estimator is regularized with an additional curvature based regularization function that dampens statistical fluctuations. On top of the FBU additional acceptance correction factors are applied on the unfolded distribution. The total corrections to the bins of the $\Delta|y|$ distribution depend on the actual bin and channel and vary between 10 and 100.

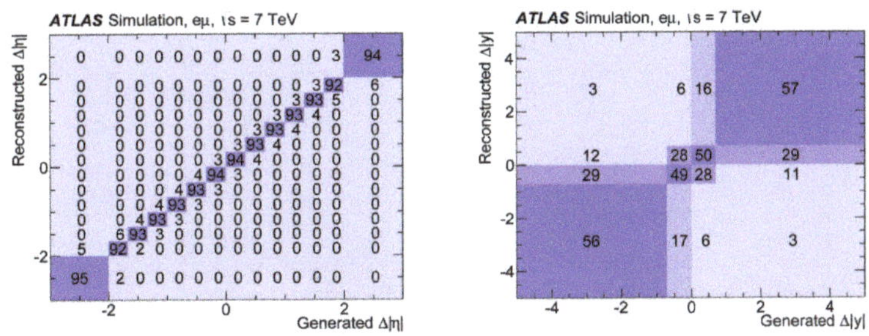

Figure 1. Response matrices for the lepton $\Delta|\eta|$ observable (**left**) and the $t\bar{t}$ $\Delta|y|$ observable (**right**) in the $e\mu$ channel. Each column of the matrices is normalized to unity and values are reported as percentage [20].

For both measurements, the linearity of the applied methods is checked using samples of simulated $t\bar{t}$ events, reweighted such that they exhibit different true asymmetries and comparing the results with the true asymmetry. The impacts of various sources of systematic uncertainties vary for the different channels with the overall tendency that the lepton based measurement is dominated by the uncertainties in the lepton reconstruction and the uncertainty in the modeling of the non prompt and fake leptons in the case of the *ee* channel, while for the $t\bar{t}$ charge asymmetry measurement several sources of uncertainty are of similar importance, including uncertainties in the reconstruction of leptons, jets, and missing transverse energy, as well uncertainties in the modeling of the non prompt and fake leptons background component.

Figure 2 shows the normalized $\Delta|\eta|$ and $\Delta|y|$ distributions in the $e\mu$ channel after the described corrections have been applied.

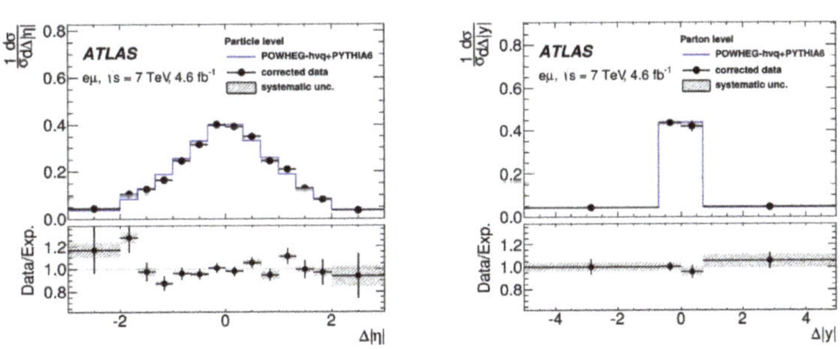

Figure 2. Normalized differential cross-sections for lepton $\Delta|\eta|$ (**left**) and $t\bar{t}$ $\Delta|y|$ (**right**) in the $e\mu$ channel after correcting for detector effects. The distributions predicted by POWHEG-hvq + PYTHIA6 are compared to the data in the top panels. The bottom panels show the ratio of the corrected data to the predictions. The error bars correspond to the statistical uncertainties and the hatched area to the systematic uncertainties [20].

The asymmetries are measured individually in the three channels and are later on combined using the best linear unbiased estimator (BLUE) method [22,23]. The individual results and their combination are listed in Table 1. For both measurements the result from the $e\mu$ channel contributes with the largest weight to the combination, while the result from the *ee* channel plays in both cases only a minor role. The combined results are comparable with the predictions for the SM, but also with zero asymmetry, as can be seen in Figure 3.

Table 1. Results for the lepton-based asymmetry $A_C^{\ell\ell}$ and the $t\bar{t}$ asymmetry $A_C^{t\bar{t}}$ after correcting for detector, resolution, and acceptance effects. The values in the *ee*, *eμ*, and *μμ* channels as well as the combined value are presented with their statistical and systematic uncertainties [20].

Channel	$A_C^{\ell\ell}$	$A_C^{t\bar{t}}$
ee	$0.101 \pm 0.052 \pm 0.021$	$0.025 \pm 0.069 \pm 0.027$
$e\mu$	$0.009 \pm 0.019 \pm 0.009$	$0.007 \pm 0.032 \pm 0.018$
$\mu\mu$	$0.047 \pm 0.030 \pm 0.012$	$0.043 \pm 0.045 \pm 0.013$
Combined	$0.024 \pm 0.015 \pm 0.009$	$0.021 \pm 0.025 \pm 0.017$

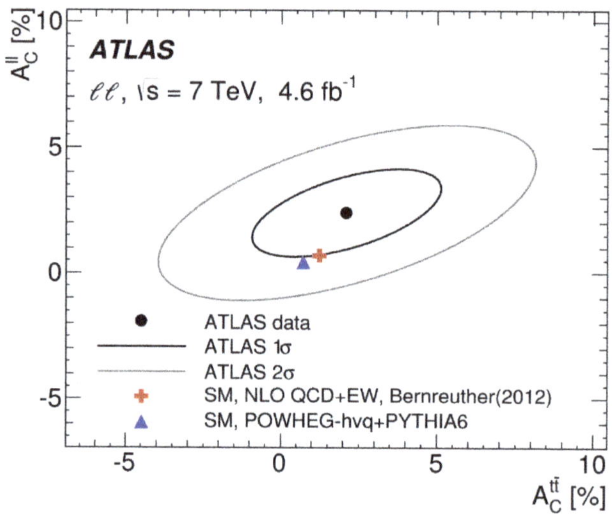

Figure 3. Comparison of the inclusive $A_C^{\ell\ell}$ and $A_C^{t\bar{t}}$ measurement results to the theory predictions (SM NLO QCD+EW prediction [14]) and the prediction of the POWHEG-hvq + PYTHIA generator. Ellipses corresponding to 1σ and 2σ combined statistical and systematic uncertainties of the measurement, including the correlation between $A_C^{\ell\ell}$ and $A_C^{t\bar{t}}$, are also shown [20].

3.1.2. Measurements of the $t\bar{t}$ and Lepton Charge Asymmetry in the Dilepton Channel by CMS

Lepton and $t\bar{t}$ charge asymmetry were also measured by the CMS Collaboration in the 7 TeV dataset [24]. In addition to the inclusive results for both asymmetries, the lepton charge asymmetry was measured differentially as function of the invariant mass, absolute value of the rapidity, and transverse momentum of the $t\bar{t}$ system. Similar requirements are made to the recorded events as in the ATLAS analysis discussed above to select a dataset enriched in $t\bar{t}$ dilepton candidate events and to suppress the dominant background contributions. Exactly two isolated charged leptons are required and at least one of the two or more jets, reconstructed using the k_T algorithm with a distance parameter of $R = 0.5$, needs to be identified as stemming from the hadronization of a b quark, using the Combined Secondary Vertex Tagger (CSV) [25]. CMS employs the particle flow technique [26] to reconstruct jets and the missing transverse momentum.

The selected dataset consists of 9824 candidate events with an estimated signal to background ratio of about 13, thus the dataset in the CMS analysis is slightly larger compared to that of the ATLAS analysis and features a two times higher purity in terms of $t\bar{t}$ dilepton events. The main background contributions come from the production of single top quarks and $t\bar{t}$ events without dileptonic decays, followed by diboson production. In this analysis, no separate measurements are performed in the three lepton flavour channels

ee, $e\mu$, and $\mu\mu$, instead the events of the three channels are added up and the resulting distributions are then fitted.

In order to resolve the ambiguity of the missing transverse momentum and the two neutrinos present in the events and the ambiguity in combining the b tagged jets with the charged leptons for the reconstruction of the $t\bar{t}$ system, the Analytical Matrix Weighting Technique (AMWT) [27] is employed. This technique finds the most probable solution for a top quark mass of 172.5 GeV out of the up to eight possible solutions for the reconstruction of the $t\bar{t}$ system. The momenta of jets and the missing transverse momentum are allowed to float within their respective uncertainties to reduce the fraction of events with no analytic solution. The about 14% of events for which the AMWT fails to find a solution are only considered in the inclusive lepton charge asymmetry measurement and omitted for all measurements that involve reconstructed quantities of the $t\bar{t}$ system.

As in the ATLAS analysis, the reconstructed distributions of $\Delta|\eta|$ and $\Delta|y|$ need to be corrected for acceptance and migration effects. The smearing matrix with generated and reconstructed $\Delta|\eta|$ ($\Delta|y|$) values, derived from simulated $t\bar{t}$ events and a matrix with acceptance times efficiency on the diagonal elements are used to correct the reconstructed distributions via a regularized unfolding algorithm based on singular-value decomposition (SVD) [28]. In the CMS analysis, for the correction of both observables, the same unfolding technique is employed.

For the differential measurements of the lepton charge asymmetry as functions of $m_{t\bar{t}}$, $|y_{t\bar{t}}|$, and $p_T^{t\bar{t}}$, two-dimensional distributions of reconstructed values need to be unfolded, using the same unfolding technique as for the inclusive measurements. In order to not amplify statistical uncertainties, the number of bins in $\Delta|\eta|$ and $\Delta|y|$ is reduced to two, while three bins are used for the $m_{t\bar{t}}$, $|y_{t\bar{t}}|$, and $p_T^{t\bar{t}}$ distributions.

The performance of the applied unfolding method is checked using samples of simulated $t\bar{t}$ events, reweighted such that they feature asymmetries between -0.2 and 0.2, and comparing the resulting asymmetry measurement with the true asymmetry of the respective sample. The impact of various experimental and theoretical sources of systematic uncertainties are evaluated. For the measurement of the lepton asymmetry, the by far dominating contribution to the overall systematic uncertainty is the uncertainty in the factorization and renormalization scales of the used simulation. In the case of the $t\bar{t}$ charge asymmetry factorization and renormalization scale uncertainties are also the dominant contribution together with the uncertainty in the jet energy scale.

Figure 4 shows the distributions of $\Delta|\eta|$ and $\Delta|y|$ after background subtraction and unfolding, compared to the distributions predicted by MC@NLO. The resulting asymmetry values of the inclusive measurements, listed in Table 2, are comparable with the predictions by the SM but are within their uncertainties also comparable with zero asymmetry.

The dependence of the measured lepton asymmetry on three kinematic variables of the $t\bar{t}$ system are shown in Figure 5. Within their relatively large uncertainties also the differential results are comparable with the predictions of the simulation.

Table 2. Results for the lepton-based asymmetry $A_C^{\ell\ell}$ and the $t\bar{t}$ asymmetry $A_C^{t\bar{t}}$ after correcting for detector, resolution, and acceptance effects. The values are presented with their statistical and systematic uncertainties [24].

Channel	$A_C^{\ell\ell}$	$A_C^{t\bar{t}}$
$\ell\ell$	$0.009 \pm 0.010 \pm 0.006$	$-0.010 \pm 0.0017 \pm 0.008$

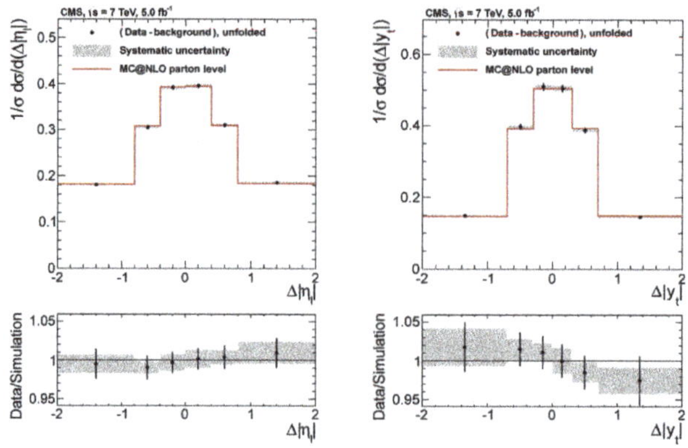

Figure 4. Background-subtracted and unfolded differential measurements of $\Delta|\eta|$ (**left**) and $\Delta|y|$ (**right**), both normalised to unit area (points), and the parton-level predictions from MC@NLO (histograms). Bottom: the ratio between the data and the MC@NLO prediction for lepton $\Delta|\eta|$ (**left**) and $t\bar{t}$ $\Delta|y|$ (**right**). The error bars represent the statistical uncertainties in the data, while the systematic uncertainties are represented by the hatched band. The first and last bins include underflow and overflow events, respectively [24].

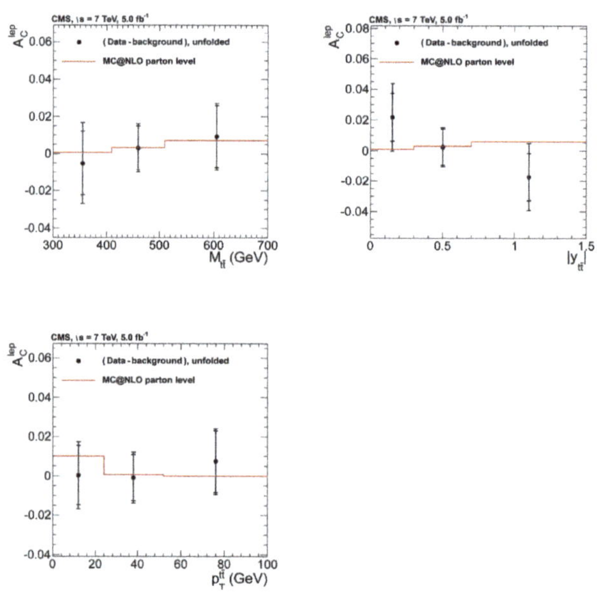

Figure 5. Dependence of the unfolded $A_C^{\ell\ell}$ values (points) on $m_{t\bar{t}}$ (**top left**), $|y_{t\bar{t}}|$ (**top right**), and $p_T^{t\bar{t}}$ (**bottom**), and the parton-level predictions from MC@NLO (histograms). The inner and outer error bars represent the statistical and total uncertainties, respectively. The last bin of each plot includes overflow events [24].

3.1.3. Measurement of the $t\bar{t}$ Charge Asymmetry in the Lepton+Jets Channel by ATLAS

The ATLAS Collaboration measured the $t\bar{t}$ charge asymmetry in the 7 TeV dataset also for events with a single lepton signature (lepton+jets channel) [29]. Exactly one isolated high-p_T electron or muon, missing transverse momentum, and at least four jets are required. On top of this "pretag" selection, at least one of the selected jets needs to be identified as stemming from the hadronization of a b quark for the event to enter the "tag" selection. For the purpose of identifying jets from b quarks, a combination of three different b-tagging algorithms is used [30,31].

From the 111,817 events passing the criteria of the "pretag" selection, 59,497 events feature in addition at least one identified b jet. The estimated signal to background ratio in the "tag" selection dataset is about 4. The dominant background contribution comes from the production of W bosons in association with jets, followed by single top quark production and the production of QCD multijets.

As the production cross section for positively charged W bosons is larger than that for negatively charged W bosons, the asymmetry in the numbers of selected events with positively and negatively charged leptons can be used to estimate the background contribution from W+jets production directly from the data. This estimation is done in the pretag selected dataset and then extrapolated to the tag selection by applying tagging efficiencies derived from simulated W+jets events. As the flavour composition of the jets accompanying the W bosons is not well predicted in the simulation, the flavour fractions are derived in data and applied to the simulation. The amount of multijet production from QCD processes is determined using the matrix method, which, based on tight and more loosely defined lepton samples, estimates the efficiency for leptons from multijet background to pass the tight selection criteria of the event selection applied in this analysis.

The tt system is reconstructed by applying a kinematic fit based on likelihood determination for the different reconstruction options for each event [32]. While for the inclusive measurement all $t\bar{t}$ candidate events are used, the differential measurements are only carried out for events with a likelihood value above a certain threshold to reject events that are not reconstructed well.

As for the analysis in the dilepton channel, again the FBU method is used to correct the reconstructed distributions for acceptance and resolution effects. As the number of selected events is larger compared to that of the dilepton analysis, four bins instead of only two are used for the $\Delta|y|$ distribution, in the inclusive measurement as well as in the differential measurement.

Several sources of systematic uncertainty are considered as nuisance parameters in the likelihood function used for the unfolding and the systematic uncertainty of the measurement is then determined by means of a marginalization procedure applied to this likelihood. The dominating contributions come from the uncertainty in the energy scale and resolution of leptons and jets, and from the missing transverse momentum and pileup modeling.

The measured asymmetry values are summarized in Table 3. In addition to the inclusive result for the full phase space, also the asymmetries for events with $m_{t\bar{t}} > 600$ GeV and for events with $\beta_{z,t\bar{t}} > 0.6$, i.e., for events from phase space regions where an enhanced asymmetry is predicted, are measured separately and compared to the SM predictions. The quoted uncertainty for the measured values represents the total uncertainty including statistical and systematic components. Figure 6 shows the results of the differential measurements. All measured values are within their uncertainties comparable to SM predictions as well as to zero asymmetry.

3.1.4. Measurement of the $t\bar{t}$ Charge Asymmetry in the Lepton+Jets Channel by CMS

The charge asymmetry was also measured inclusively and differentially in the lepton+jets channel by the CMS Collaboration [33]. After the typical event selection requirements of exactly one isolated electron or muon, at least four jets, of which at least one has to be tagged as b jet, and substantial missing transverse momentum, 57,687 events remain for the measurement. About 20% of the selected events come from background processes, with the biggest contribution coming from the production of W bosons in association with jets, followed by the electroweak production of single top quarks and multijet production.

Table 3. Measured inclusive charge asymmetry values for the electron and muon channels combined after unfolding for the full phase space, for events with $\beta_{z,t\bar{t}} > 0.6$, and for events with $m_{t\bar{t}} > 600$ GeV [29], along with the respective predictions [14]. The uncertainties of the measurements include statistical and systematic components.

Phase Space	Measured $A_C^{t\bar{t}}$	SM Prediction for $A_C^{t\bar{t}}$
Full	0.006 ± 0.010	0.0123 ± 0.0005
$m_{t\bar{t}} > 600$ GeV	0.018 ± 0.022	$0.0175^{+0.0005}_{-0.0004}$
$\beta_{z,t\bar{t}} > 0.6$	0.011 ± 0.018	$0.020^{+0.006}_{-0.007}$

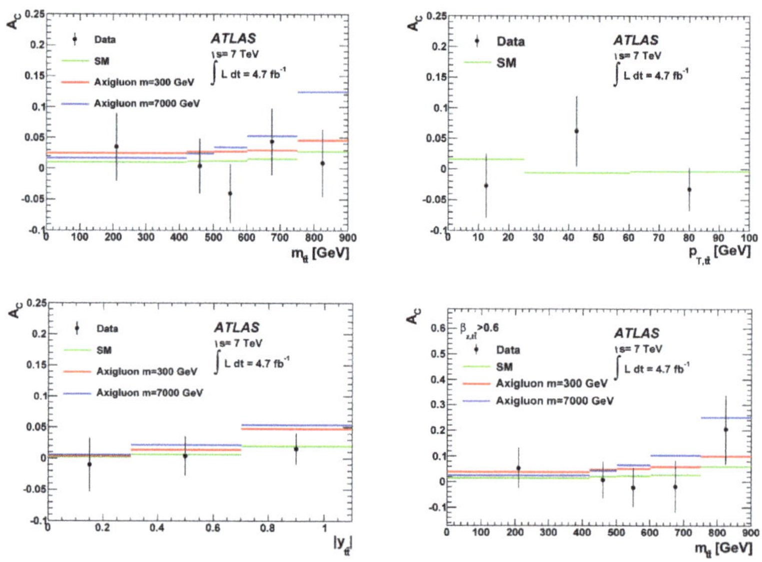

Figure 6. Distributions of $A_C^{t\bar{t}}$ as a function of $m_{t\bar{t}}$ (**top left**), $p_T^{t\bar{t}}$ (**top right**), and $|y_{t\bar{t}}|$ (**bottom left**) after unfolding, for the electron and muon channels combined. The $A_C^{t\bar{t}}$ distribution as a function of $m_{t\bar{t}}$ for $\beta_{z,t\bar{t}} > 0.6$, is also shown (**bottom right**). The $A_C^{t\bar{t}}$ values after the unfolding (points) are compared with the SM predictions (green lines) and the predictions for a colour–octet axigluon with a mass of 300 GeV (red lines) and 7000 GeV (blue lines), respectively. The thickness of the lines represents the factorization and renormalization scale uncertainties in the corresponding theoretical predictions. The values plotted are the average $A_C^{t\bar{t}}$ in each bin. The error bars include both the statistical and the systematic uncertainties [29].

The leptonically decaying top quark is reconstructed from the fourvector of the electron or muon and from the transverse momentum vector. All assignments of selected jets to the final state quarks in a $t\bar{t}$ event are considered and the hypothesis with the highest probability to correctly describe the $t\bar{t}$ system is chosen for each event. It is worth mentioning that all events are considered in the analysis and no restriction is made on the quality of the event reconstruction. Disturbing effects from event selection and detector and reconstruction method resolution are corrected for by employing a regularized generalized matrix inversion unfolding. For the differential measurements not only the $\Delta|y|$ distribution is unfolded but also the kinematic variables of the $t\bar{t}$ system. As the binning schemes in $\Delta|y|$ differ depending on the bin of the kinematic $t\bar{t}$ variable, a non trivial definition of "neighboring bin" had to be found to correctly consider the influence of adjacent bins in the regularization procedure.

The dominating source of systematic uncertainty comes from the dependency of the asymmetry on the three kinematic $t\bar{t}$ variables and its impact on the used unfolding method and from the uncertainty in the selection and identification of the leptons. Depending on the actual bin, also uncertainties in the jet energy scale contribute significantly to the differential measurements.

Figure 7 (top left) shows the unfolded $\Delta|y|$ spectrum. From this distribution, the inclusive $t\bar{t}$ charge asymmetry is measured to be $A_C^{t\bar{t}} = 0.004 \pm 0.010(\text{stat}) \pm 0.011(\text{syst})$. The $A_C^{t\bar{t}}$ values as functions of $m_{t\bar{t}}$, $|y_{t\bar{t}}|$, and $p_T^{t\bar{t}}$ are shown in Figure 7 (top right and bottom). All results are consistent with predictions for the SM, but are also consistent with zero asymmetry.

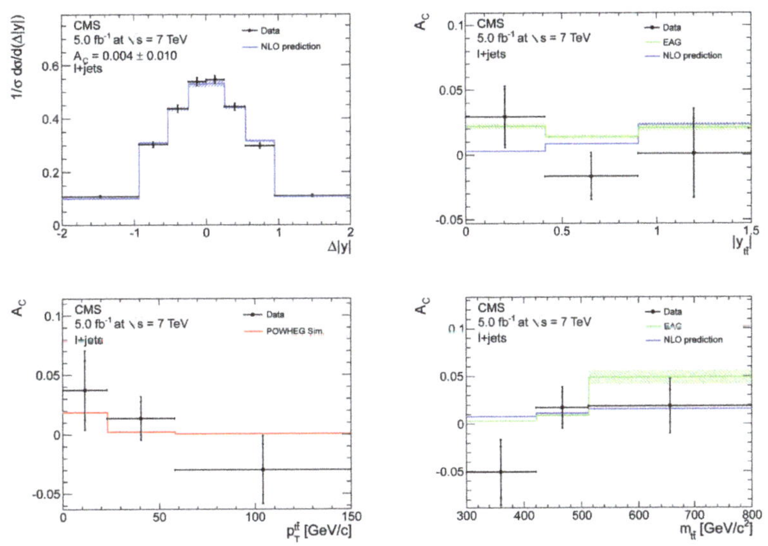

Figure 7. Unfolded inclusive $\Delta|y|$ distribution (**top left**), corrected asymmetry as a function of $|y_{t\bar{t}}|$ (**top right**), $p_T^{t\bar{t}}$ (**bottom left**), and $m_{t\bar{t}}$ (**bottom right**) [33]. The measured values are compared to NLO calculations for the SM — based on the calculations of Ref. [13] — and to the predictions of a model featuring an effective axial-vector coupling of the gluon (EAG) [34]. The error bars on the differential asymmetry values indicate the statistical and total uncertainties, determined by adding statistical and systematic uncertainties in quadrature. The shaded areas indicate the theoretical uncertainties on the NLO calculations.

3.1.5. Combination of the Inclusive Measurements of the $t\bar{t}$ Charge Asymmetry in the Lepton+Jets Channel by ATLAS and CMS

The results of the inclusive $A_C^{t\bar{t}}$ measurements in the lepton+jets channel at 7 TeV by the ATLAS and CMS Collaborations have been combined within the LHC working group for top quark physics (LHC*top*WG) [35]. The BLUE technique with the method implemented in Ref. [36] is used to find the combination of the two results with the smallest total uncertainty. Most systematic uncertainties are considered uncorrelated in the combination, except for the uncertainty in the modeling of the $t\bar{t}$ signal and the W+jets background (both 50% correlated between ATLAS and CMS) and the uncertainty in the used PDFs (100% correlated between ATLAS and CMS).

The result of the combination of the two measurements in the lepton+jets channel at 7 TeV is $A_C^{t\bar{t}} = 0.005 \pm 0.007(\text{stat}) \pm 0.006(\text{syst})$, with the ATLAS result contributing with a weight of 0.65 and the CMS result with a weight of 0.35. The χ^2 with one degree of freedom of the combination is 0.012, corresponding to a p-value of 0.91. The improvement in the total uncertainty of the combined results with respect to the individual results is 18% for the ATLAS measurement and 40% for the CMS analysis.

Figure 8 summarizes all ATLAS and CMS measurements of $A_C^{t\bar{t}}$ and $A_C^{\ell\ell}$ at 7 TeV centre-of-mass energy and their compatibility with theory calculations for the standard model.

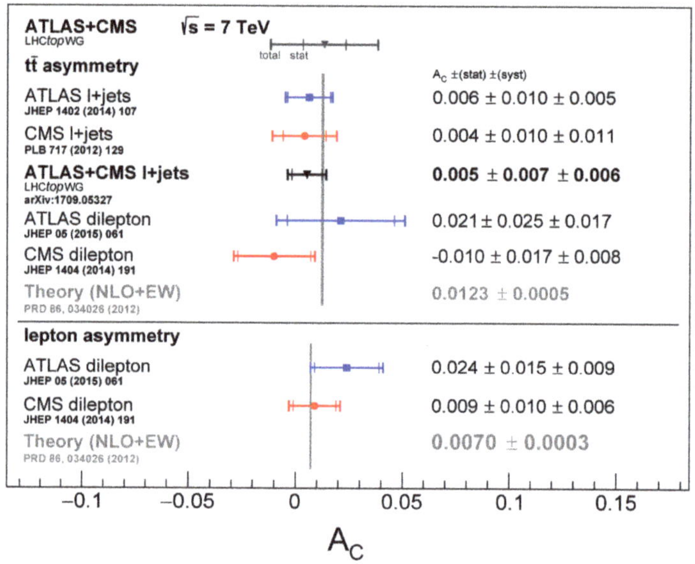

Figure 8. Summary of inclusive results at 7 TeV centre-of-mass energy [37].

3.2. Measurements at 8 TeV Centre-of-Mass Energy

The ATLAS and CMS Collaborations performed various top quark asymmetry measurements using the LHC dataset at $\sqrt{s} = 8$ TeV of around 20 fb^{-1} in the dilepton and lepton+jets channels. As for the 7 TeV measurements, the $t\bar{t}$ charge asymmetry $A_C^{t\bar{t}}$ is measured in both channels while the lepton asymmetry $A_C^{\ell\ell}$ is measured exclusively in the dilepton channel. At 8 TeV centre-of-mass energy, the SM prediction computed at NLO QCD including mixed QCD-QED and QCD-weak interaction corrections [14] is slightly smaller than at 7 TeV:

$$A_C^{t\bar{t}} = 0.0111 \pm 0.0004(\text{scale}), \tag{6}$$

$$A_C^{\ell\ell} = 0.0064 \pm 0.0003(\text{scale}). \tag{7}$$

The quoted scale uncertainties consider variation of the renormalization and factorization scales by factors of 0.5 and 2 with respect to the reference value, set to the top quark mass, as well as PDF uncertainties. As for 7 TeV, another calculation at NLO, using a different PDF set and setting the renormalizaton and factorization scales to a different value is available [13], yielding a similar result for the $t\bar{t}$ charge asymmetry, $A_C^{t\bar{t}} = 0.0102 \pm 0.0005(\text{scale})$. These two predictions are derived by evaluating the asymmetry in powers of the strong and electroweak couplings at NLO only in the numerator and evaluating the denominator using the LO matrix element. In a third calculation, at NNLO [15], both, the numerator and the denominator are calculated at full QCD NNLO precision without any expansion in powers of strong and electroweak couplings. Furthermore, dynamical factorization and renormalization scales and an NNLO PDF are used to derive the $t\bar{t}$ charge asymmetry prediction of $A_C^{t\bar{t}} = 0.0095^{+0.0005}_{-0.0007}(\text{scale})$.

In this chapter, we summarize the ATLAS and CMS results in the dilepton and lepton+jets channel at 8 TeV.

3.2.1. Measurements of the $t\bar{t}$ and Lepton Charge Asymmetry in the Dilepton Channel by ATLAS

As at 7 TeV, the ATLAS Collaboration measurement in the dilepton channel at 8 TeV studied both the lepton and the $t\bar{t}$ charge asymmetry [38]. Three different final states are considered in this analysis: events with two electrons (ee), with one electron and one muon ($e\mu$), and with two muons ($\mu\mu$). Exactly two of these leptons, isolated and of opposite electric charge, should be present in the final state. At least two reconstructed jets are also required. In the same-flavour channels, the invariant mass of the two leptons must lie outside the Z boson mass window. In these channels, it is also required that $E_{\mathrm{T}}^{\mathrm{miss}} > 30$ GeV and that at least one of the jets is b-tagged to further suppress Drell-Yan and Z boson production as well as diboson background. In the $e\mu$ channel, the background suppression is achieved by requiring that the scalar sum of the p_{T} of the two leading lets and leptons to be larger than 130 GeV. After this selection, the selected number of events in the ee, $\mu\mu$, and $e\mu$ channels, respectively, are: 12,785; 14,453; and 42,363. The background from Z+jets, single top, and diboson production are evaluated using simulated events. Because of possible mismodeling of the $E_{\mathrm{T}}^{\mathrm{miss}}$ distribution in Z bosons production in association with heavy-flavour jets, the normalisation of the inclusive and heavy-flavour component of the Drell-Yan background in the same-flavour channels is computed using data from control regions and applied afterwards to correct the simulated events. For the background coming from fake or non-prompt leptons the shape of distributions of kinematic observables is taken from simulation where at least one of the leptons is required not to be matched to the generated ones. Scale factors to adjust the normalisation are derived from data in a specific control region requiring two leptons with the same electric charge.

The lepton charge asymmetry $A_{\mathrm{C}}^{\ell\ell}$ is obtained directly from the pseudorapidity of the leptons while $A_{\mathrm{C}}^{t\bar{t}}$ requires the reconstruction of the top quarks. This kinematic reconstruction is performed by solving the system of equations that relates the particle momenta at the decay vertices. The system is underconstrained because of the presence of two neutrinos escaping detection. The system is hence solved numerically using the kinematic (KIN) method [39,40]. The experimental uncertainties of the measured jets and $E_{\mathrm{T}}^{\mathrm{miss}}$ are taken into account by sampling the phase space according to their resolution. For each of the sampling points up to four solutions can be found. The KIN method chooses the solution that leads to the lowest reconstructed mass of the $t\bar{t}$ system. There is also an ambiguity in the assignment of the lepton with the b-tagged jet. The assignment is chosen that has the most reconstructed trials. The performance of the KIN method is quantified by evaluating the efficiency of reconstructing $t\bar{t}$ events and the probability of reconstructing the correct sign of $\Delta|y|$. These probabilities are found to be 90% and 76%.

The ATLAS measurements are performed inclusively and differentially as a function of the invariant mass of the $t\bar{t}$ system, of the transverse momentum of the $t\bar{t}$ system, and the boost in z direction of the $t\bar{t}$ system. The measurements are performed in the full phase space after correcting for reconstruction and acceptance effects to parton level, as well as in a fiducial region after correction to particle level. The fiducial region is defined to closely match the region accessed by the ATLAS detector. Such fiducial results allow to reduce the dependency to MC generators and avoid large extrapolation to the full phase space. The fiducial volume definition closely follows the requirements applied on the reconstructed objects but using particle level objects. Using these particle objects, the top quark reconstruction is performed using the pseudotop algorithm described in Ref. [41]. As for the 7 TeV results, the correction for detector resolution and acceptance effects is performed using the FBU technique. The migration matrix is obtained from the nominal $t\bar{t}$ simulated sample. The combination between the ee, $e\mu$, and $\mu\mu$ channels is performed by mapping the reconstructed distributions of the three channels to the same corrected distribution. The systematic uncertainties are treated as nuisance parameters in the maximum likelihood estimation. For each differential measurement, the choice of binning for $\Delta|\eta|$ or $\Delta|y|$ is optimized by minimizing the expected statistical uncertainty. For the optimized binning choices, more than 50% of the events are found to lie within the diagonal bins of the migration matrix for $\Delta|y|$, and more than 97% for $\Delta|\eta|$.

Various sources of systematic uncertainties affect the measurements. The statistical uncertainty gives the largest contribution followed by the uncertainties on kinematic reconstruction and signal modeling uncertainties. Figure 9 summarizes the results in the full phase space, which appear to be compatible with the SM predictions. Figure 10 shows the unfolded distributions of $\Delta|\eta|$ and $\Delta|y|$ for the inclusive measurements in the fiducial phase space compared to NLO MC generator predictions [42].

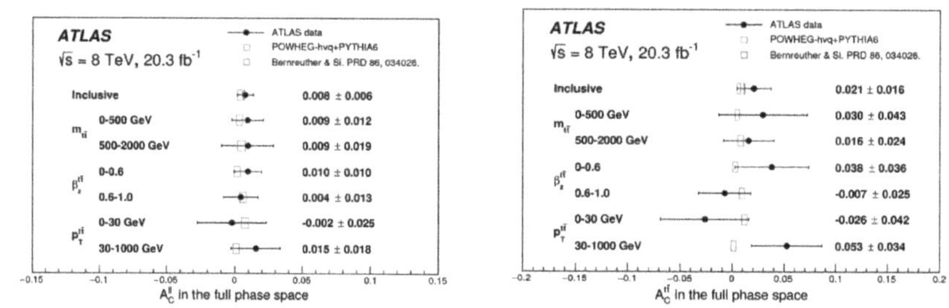

Figure 9. Summary of the differential results for the lepton asymmetry (**left**) and the $t\bar{t}$ asymmetry (**right**) in the full phase space [38]. The prediction in blue are obtained using the POWHEG-hvq generator at NLO [42]. The inclusive result is compared to the NLO+EW prediction [14].

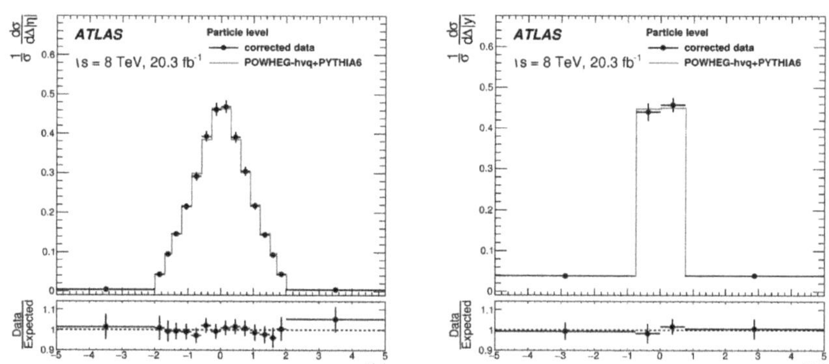

Figure 10. Unfolded data distribution for $\Delta|\eta|$ (**left**) and $\Delta|y|$ (**right**) [38] compared to the predictions from the POWHEG-hvq generator at NLO [42].

The inclusive results for $A_C^{\ell\ell}$ and $A_C^{t\bar{t}}$ in the full phase space are compared in Figure 11 to the SM predictions and to two BSM predictions compatible with the Tevatron results. One model contains a light octet with mass below the $t\bar{t}$ production threshold, while the other one introduces a heavy octet with mass beyond the LHC reach [43]. In this figure the correlation between $A_C^{\ell\ell}$ and $A_C^{t\bar{t}}$ is taken into account and is found to be 48%. The results are compatible with the SM but do not exclude the two BSM models considered.

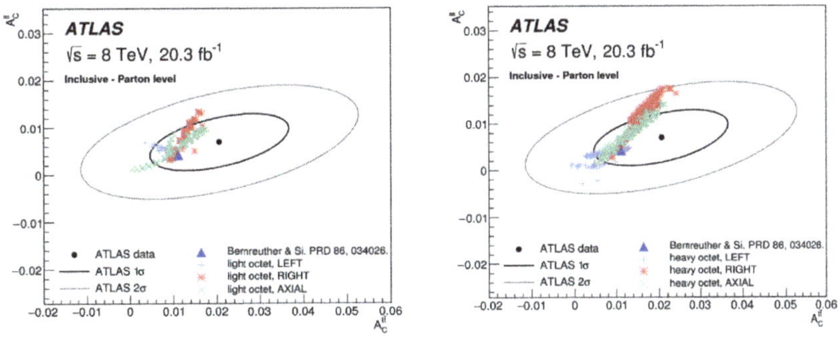

Figure 11. Comparison of the $A_C^{\ell\ell}$ and $A_C^{t\bar{t}}$ inclusive results in the full phase space with the SM NLO QCD+EW predictions [14] and two BSM benchmark scenarios [43]. Ellipses correspond to the 1 and 2 σ total measurement uncertainties.

3.2.2. Measurements of the $t\bar{t}$ and Lepton Charge Asymmetry in the Dilepton Channel by CMS

The CMS Collaboration also performed the inclusive and differential measurements of both the leptonic and $t\bar{t}$ asymmetries at 8 TeV [44]. The analysis requires events with exactly two isolated, oppositely charged leptons with $p_T > 20$ GeV and with an invariant mass of the dilepton system above 20 GeV and outside the Z boson mass window. At least two jets with $p_T > 30$ GeV are required with one of them identified as coming from a b quark. High missing transverse momentum ($p_T^{miss} > 40$ GeV) is required to suppress the Drell-Yan background in the channel with same-flavour leptons. In total, 43,898 events are selected with these requirements with an estimated background contribution of about 9%. In the $t\bar{t}$ system reconstruction method of this analysis, the solutions for the neutrino momenta are found analytically assuming a top quark mass of 172.5 GeV. Each of the events can have up to eight possible solutions. To choose the most probable one, the matrix weighting technique [27] is used as in the 7 TeV measurement. The signs of $\Delta|\eta|$ and $\Delta|y|$ are correctly reconstructed in 99.5% and 74.9% of the selected $t\bar{t}$ events, respectively. In about 16% of all events, no solution for the reconstruction of the $t\bar{t}$ system exists and these events are consequently only used for the inclusive measurement of the lepton charge asymmetry.

Samples of simulated events are used to evaluate the background contamination in the selected sample. Several data control regions are used to correct the normalisation of the Z+jets background and the background coming from fake or non-prompt leptons. After subtraction of the background, the distributions are corrected for detector acceptance, event selection efficiency, and finite detector resolution at parton level using an unfolding technique. In this process the binning of each of the distributions is adjusted according to the observable resolution. The unfolding is performed using the TUNFOLD package [45] using regularization based on the curvature of the simulated signal distributions. The optimized regularisation strength is found to be relatively weak.

The differential measurements are performed as a function of the invariant mass of the $t\bar{t}$ system, as well as the absolute rapidity and transverse momentum of the $t\bar{t}$ system in the laboratory frame. Three bins are used for each of these differential distributions.

Several sources of systematic uncertainties are considered in the measurements. The ones that impact the most the inclusive $A_C^{t\bar{t}}$ and $A_C^{\ell\ell}$ results are the uncertainty in the unfolding procedure coming from the limited number of simulated $t\bar{t}$ events, as well as the uncertainty from the $t\bar{t}$ modeling coming from variation of the renormalization and factorization scales, while the differential measurements are largely limited by the statistical component of the uncertainty.

The unfolded $\Delta|\eta|$ and $\Delta|y|$ distributions at parton level are shown in Figure 12, compared to the SM predictions at NLO+EW level [14] as well as to predictions from the NLO

MC@NLO generator [46]. The resulting inclusive values for $A_C^{t\bar{t}}$ and $A_C^{\ell\ell}$ are given in Table 4. The results are compatible with the SM expectations as well as with zero asymmetry.

Table 4. Results for the lepton-based asymmetry $A_C^{\ell\ell}$ and the $t\bar{t}$ asymmetry $A_C^{t\bar{t}}$ after correcting for detector, resolution, and acceptance effects. The values are presented with their statistical and systematic uncertainties [44].

Channel	$A_C^{\ell\ell}$	$A_C^{t\bar{t}}$
$\ell\ell$	$0.003 \pm 0.006 \pm 0.003$	$0.011 \pm 0.011 \pm 0.007$

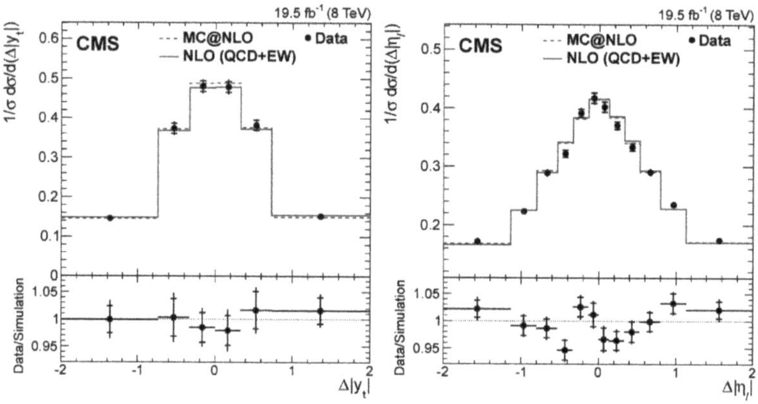

Figure 12. Unfolded data distribution for $\Delta|y|$ (**left**) and $\Delta|\eta|$ (**right**) [44] compared to the predictions from the MC@NLO at NLO [46] and the NLO+EW SM predictions [14].

The differential results as a function of $m_{t\bar{t}}$, $|y_{t\bar{t}}|$, and $p_T^{t\bar{t}}$ are presented in Figure 13 and show also reasonable agreement with the predictions.

3.2.3. Measurement of the $t\bar{t}$ Charge Asymmetry in the Lepton+Jets Channel by ATLAS

The ATLAS Collaboration has performed two types of measurements in the lepton+jets channel at 8 TeV. The first one measured inclusively and differentially the $t\bar{t}$ asymmetry $A_C^{t\bar{t}}$ in the full phase space [47]. The second measurement focused on the $t\bar{t}$ asymmetry in events where the top-quark pair is produced with a large invariant mass. i.e., highly boosted top-quark pair production [48]. The boosted channel is expected to have a higher sensitivity to the expected SM asymmetry due to a higher fraction of quark-antiquark initiated processes.

The standard analysis (so called resolved channel [47]) selects exactly one electron or muon candidate and at least four jets with $p_T > 25$ GeV. The selected events are separated according to the number of b-tagged jets they contain (zero, one or at least two). In order to suppress the background from multijet and Z+jets backgrounds in events with zero or one b-tagged jet, further requirements are applied on E_T^{miss} and on the reconstructed transverse mass of the W boson from the hadronically decaying top quark. After the event selection, 216,465 and 193,418 events remain in the two signal regions with either one or at least two b tagged jets, with approximated background contributions of about 34% and 11%, respectively. The main background comes from W+jets events. The shape of the distributions of kinematic observables for this background is estimated using simulated events while its normalisation is adjusted using data exploiting the difference in production cross sections between W^+ and W^-. This adjustment is used in situ and embedded in the unfolding procedure that corrects for detector acceptance and efficiency. The multijet background is also estimated from data using the matrix method [47]. The other minor backgrounds coming from single-top quark, Z+jets or diboson production are evaluated using simulation.

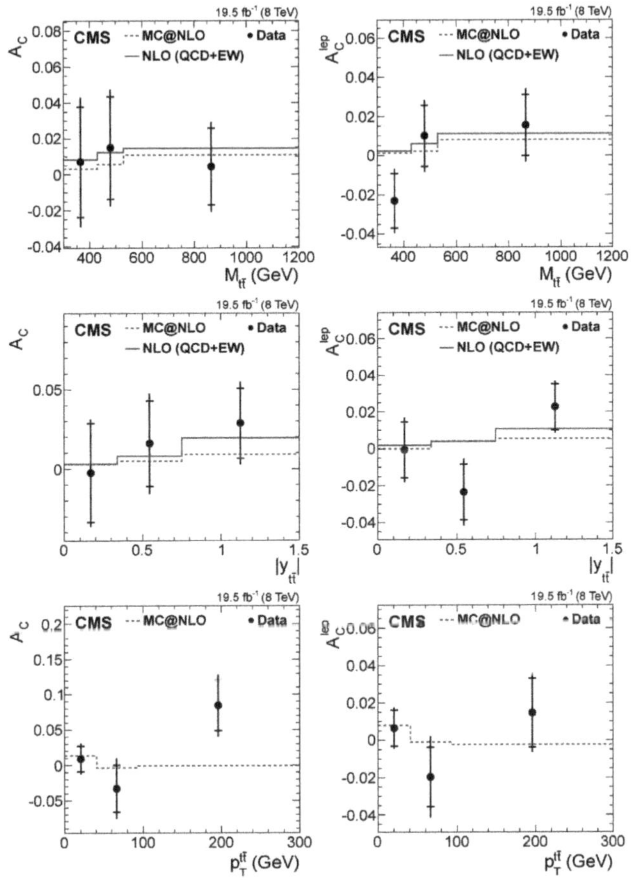

Figure 13. $t\bar{t}$ and lepton charge asymmetries as functions of $m_{t\bar{t}}$, $|y_{t\bar{t}}|$, and $p_T^{t\bar{t}}$ [44] compared to the predictions from MC@NLO [46] and the NLO+EW calculation for the SM [14].

As in the other channels, to measure the $t\bar{t}$ asymmetry inclusively and differentially, the full $t\bar{t}$ system needs to be reconstructed. This reconstruction is achieved applying a kinematic fit [49] using at most five jets. The average correct-sign fraction is estimated to be around 72% for events with exactly one b-tagged jets and 75% for events with at least two b-tagged jets. The lepton charge is used to determine the flavour (quark or antiquark) of the semilaptonically decaying top quark candidate.

The measurements are corrected for acceptance and detector resolution effects to parton level by unfolding using the FBU technique. The asymmetry is determined from maximizing the FBU extended likelihood where the events are separated based on the lepton charge and the b-tagged jet multiplicity (zero, one, at least two). The $\Delta|y|$ distribution is split in four bins, also for each differential measurement. The calibration of the W+jets prediction is achieved by adjusting in the fit the flavour components of the jets associated with the W boson. The b-tagged jet multiplicity provides information about the heavy and light-flavour composition of the W+jets background while the lepton charge asymmetry is used to determine the normalisation of each component. Various sources of systematic uncertainties are included as nuisance parameters. The dominant source is found to come from the uncertainty in jet energy scale and resolution.

The resulting inclusive $t\bar{t}$ asymmetry is measured to be $A_C^{t\bar{t}} = 0.009 \pm 0.005(\text{stat} + \text{syst})$, compatible with the SM prediction. The measurement is limited by the statistical uncertainty. The differential measurements of the charge asymmetry at parton level as a function of the invariant $t\bar{t}$ mass, the boost of the $t\bar{t}$ system, and the transverse momentum of this system are shown in Figure 14 compared with the SM prediction and with two benchmark BSM scenarios. The results are compatible with the SM and are not able to distinguish between SM and these two BSM scenarios within the quoted uncertainties.

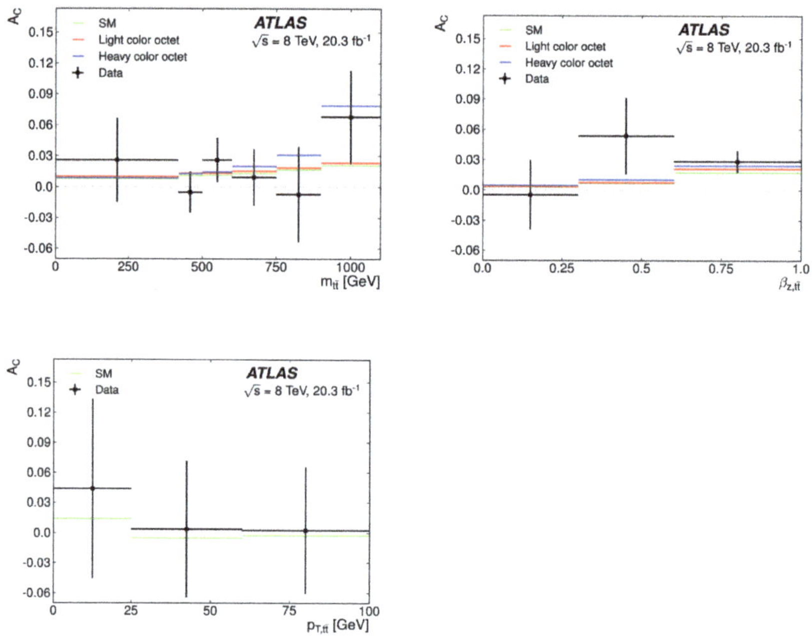

Figure 14. Measured $A_C^{t\bar{t}}$ as a function of $m_{t\bar{t}}$, $\beta_{z,t\bar{t}}$, and $p_T^{t\bar{t}}$ [47] compared with SM predictions [14] and two benchmark BSM scenarios [43].

The ATLAS measurement using highly boosted top quark pairs [48] focuses on lepton+jets events where the hadronic top-quark decay is reconstructed as a single large-radius jet (R = 1.0) and tagged as coming from an actual top quark using jet substructure variables. The leptonic top-quark decay is reconstructed from a standard small-radius jet (R = 0.4), a charged lepton (electron or muon), and missing transverse momentum from the escaping neutrino. The longitudinal component of the neutrino momentum is calculated using a constraint on the W boson mass and solving a quadratic equation. The selected large-radius jet should have $p_T > 300$ GeV and must be separated from both the charged lepton and the small-radius jet. A substructure analysis of the large-radius jet is used to tag the boosted top quark. At least one of the small-radius jets associated with the leptonically decaying top quark must be b-tagged. The selected events yield a good efficiency and resolution for the $t\bar{t}$ invariant mass of around 6% for $m_{t\bar{t}} \sim 1$ TeV. The background is estimated using simulation except for the normalisation of the W+jets background and heavy flavour fractions that are adjusted from data and the multijet background that is fully estimated from data using the matrix method.

For electron and muon channel combined, 7741 events are selected, with about 13% background contribution. This number of observed events is found to be approximately 10% less than the number predicted by the simulation, a result of the known mismodeling of the top quark p_T spectrum in the simulation. Since the asymmetry is computed as a ratio, it is not sensitive to the absolute cross section. The difference in shape has been tested to have a negligible impact. The $\Delta|y|$ distribution as a function of $m_{t\bar{t}}$ is corrected for acceptance and detector effects to parton level using FBU, as in the resolved analysis, in the phase space: $m_{t\bar{t}} > 750$ GeV and $-2 < \Delta|y| < 2$.

The $A_C^{t\bar{t}}$ values from the unfolded distribution in four $m_{t\bar{t}}$ bins are shown in Figure 15. The measurement generally agrees with the SM prediction. The largest observed difference reaches 1.6 σ in the third bin. In the measured phase space, $m_{t\bar{t}} > 750$ GeV and $|\Delta|y|| < 2$, the inclusive asymmetry is $A_C^{t\bar{t}} = 0.042 \pm 0.032$, where the dominant uncertainty is coming from modeling uncertainty and the statistical component. The result agrees well with the SM prediction.

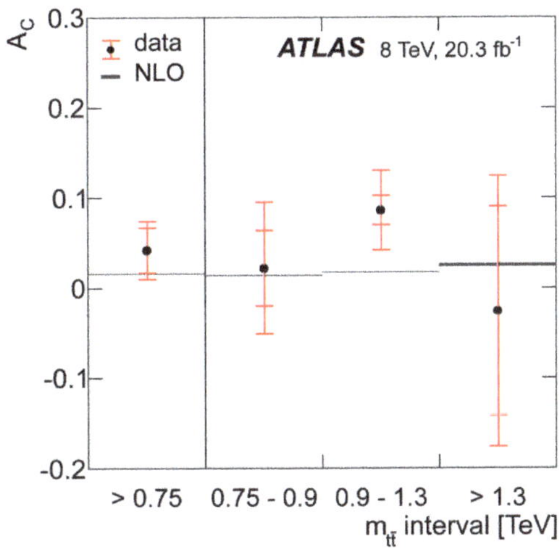

Figure 15. $A_C^{t\bar{t}}$ values from the unfolded distributions as a function of $m_{t\bar{t}}$ [48] compared to NLO calculation [13].

3.2.4. Measurement of the $t\bar{t}$ Charge Asymmetry in the Lepton+Jets Channel by CMS

The CMS Collaboration has performed two types of measurements in the lepton+jets channel at 8 TeV using different techniques to access the asymmetry. The first analysis measured inclusively and differentially the $t\bar{t}$ asymmetry $A_C^{t\bar{t}}$ at parton level after applying an unfolding technique [50]. For the second measurement, the same dataset was analyzed to get the inclusive $t\bar{t}$ asymmetry value but using a template technique based on a parametrization of the SM [51]. This second analysis also differs in selecting more events.

The analysis using an unfolding technique [50] measured $A_C^{t\bar{t}}$ inclusively and as functions of $m_{t\bar{t}}$, $|y_{t\bar{t}}|$, and $p_T^{t\bar{t}}$. The measurements are performed at parton level both in a fiducial phase space that emulates the restriction of the detector phase space and also extrapolated to the full phase space.

The event selection requires one electron or one muon and four or more jets with at least one of them being b-tagged. In total, 362,244 events are selected with about 20% contribution from background processes. The fiducial region is defined using particle objects with a selection mimicking the selection applied at reconstruction level. It contains around 10% of the events of the full phase space and roughly 50% of the events in the fiducial region pass the reconstruction level selection.

The distributions of kinematic observables for the background processes are modelled using simulated events while their rates are estimated using a simultaneous fit of the transverse mass of the W boson from the hadronic top quark decay and of the invariant mass of the combination of three jets that have the largest vectorial p_T sum. The multijet background is modelled using data in a control region with non isolated leptons. The top quark reconstruction is performed using the same likelihood method as in the 7 TeV analysis.

After background subtraction, the distributions corrected for acceptance and detector effects are determined through unfolding using a generalized matrix inversion method. Regularization is applied to limit the statistical uncertainties due to the unfolding procedure. It is implemented by minimizing the statistical correlations between bins in the unfolding spectrum. The correctness of the unfolding procedure has been verified using pseudo-experiments. The unfolded $\Delta|y|$ distributions in the fiducial and full phase spaces are shown in Figure 16 compared with SM predictions [13,14]. The measured inclusive $t\bar{t}$ asymmetry in the fiducial and full phase space, respectively, are summarized in Table 5. In both cases the experimental results are a bit low with respect to the predictions for the SM [13,14,52], especially in the fiducial region. However, the deviation is below two standard deviations.

Table 5. Results of the inclusive $t\bar{t}$ charge asymmetry measurements in the fiducial and full phase space [50], compared to the respective SM predictions [13,14,52].

Phase Space	Measured $A_C^{t\bar{t}}$	Calculated $A_C^{t\bar{t}}$
fiducial	$-0.0035 \pm 0.0072(\text{stat}) \pm 0.0031(\text{syst})$	0.0101 ± 0.0010
full	$0.0010 \pm 0.0068(\text{stat}) \pm 0.037(\text{syst})$	0.0111 ± 0.0004
		0.0102 ± 0.0005

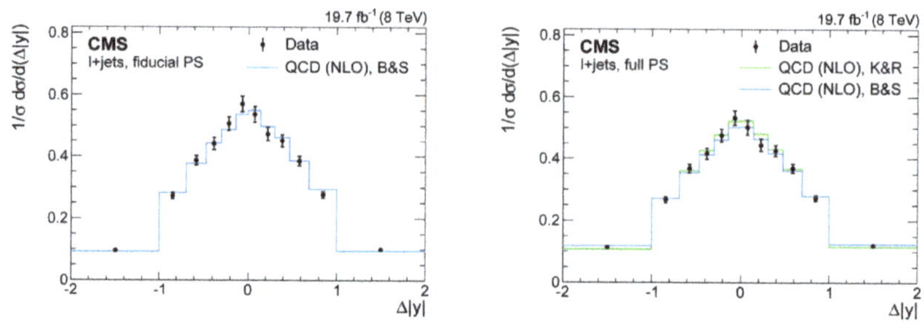

Figure 16. Unfolded inclusive $\Delta|y|$ distributions in the fiducial and full phase spaces [50] compared with NLO predictions from the SM [13,14].

The differential measurements as a function of $m_{t\bar{t}}$, $|y_{t\bar{t}}|$, and $p_T^{t\bar{t}}$ in the full phase space are displayed in Figure 17. The result of the asymmetry as a function of $m_{t\bar{t}}$ is shown with two different binnings. The distributions agree with the SM prediction and are also compared to a model containing an effective axial-vector coupling of the gluon [53,54].

The other CMS measurement in the lepton+jets channel is based on a template technique [51] using the bounded observable $\mathrm{Y}_{t\bar{t}}$ defined as $\mathrm{Y}_{t\bar{t}} = \tanh \Delta|y|$. The $\mathrm{Y}_{t\bar{t}}$ distribution can be translated as a linear combination of the symmetric and asymmetric components of the probability distribution $\rho(\mathrm{Y}_{t\bar{t}})$ (see Figure 18). The asymmetry $A_C^{t\bar{t}}$ corresponds then to the level of asymmetric component that best fits the observed data distribution.

The selected events contain one isolated electron or muon and at least four jets among which at least one is b-tagged. In total, 667,096 events are selected, of which about 35% are estimated to come from background processes. The top quarks are reconstructed using the most likely assignment of the jets to the $t\bar{t}$ decay partons. For each assignment the four-momenta of the jets are corrected according to the partons that are assigned to using flavour-dependent scale factors derived from $t\bar{t}$ simulation. The chosen assignment is determined using a likelihood criterion taking into account the b-tagging information and the constraints from the reconstruction hadronically decaying W boson and top quark. The energy resolution of the jets for the chosen assignment is further improved using a kinematic fit under the W boson and top quark mass constraints.

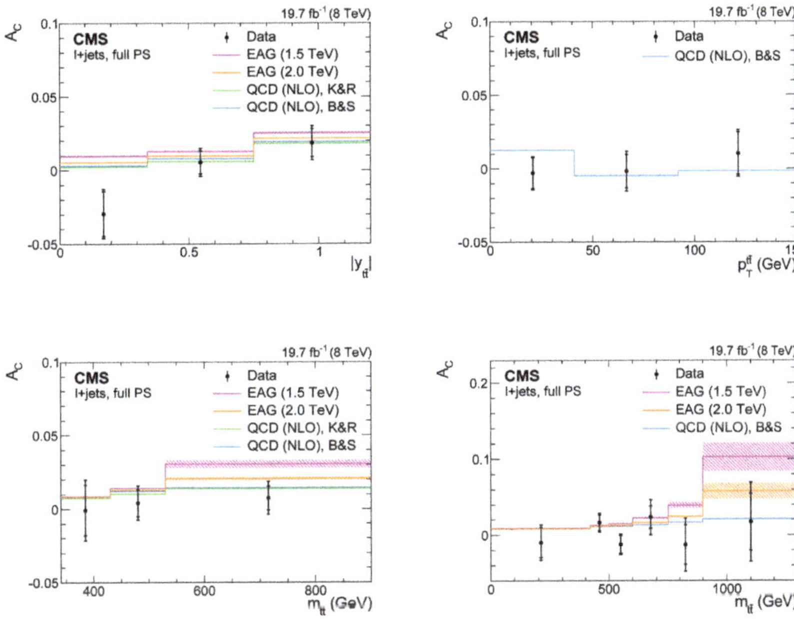

Figure 17. Measured $A_C^{t\bar{t}}$ in the full phase space as a function of $m_{t\bar{t}}$, $|y_{t\bar{t}}|$, and $p_T^{t\bar{t}}$ [50] compared with NLO predictions for the SM [13,14] and a model containing an effective axial-vector coupling of the gluon (EAG) [53,54].

The amount of $t\bar{t}$, W+jets, and multijet events after selection is determined using a likelihood discriminant built from the hadronically decaying W boson transverse mass and from the probability that at least one of the possible jet-parton assignments is the correct one. The other minor backgrounds are estimated using simulated events.

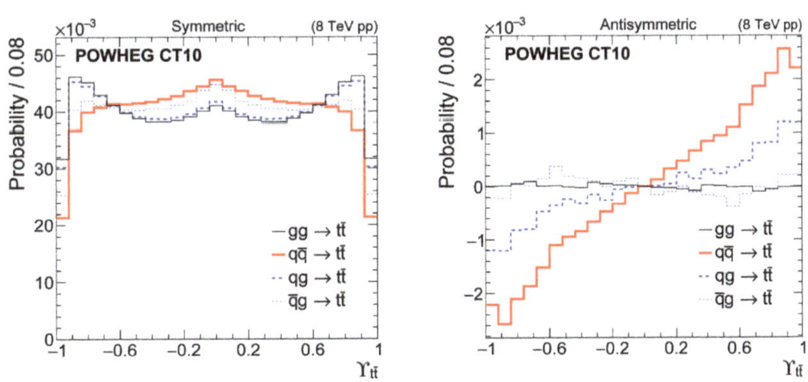

Figure 18. Symmetric and antisymmetric components of the binned probability distribution in the observable $\Upsilon_{t\bar{t}}$ [51] constructed using the POWHEG generator [55].

After the determination of the sample composition, another likelihood fit on the $\Upsilon_{t\bar{t}}$ distribution is performed to measure $A_C^{t\bar{t}}$. The performance of the method is checked on simulated samples showing negligible bias. The antisymmetric component of the $\Upsilon_{t\bar{t}}$ distribution is shown in Figure 19 for data and for the fit model in the e+jets and μ+jets channels. The measured $t\bar{t}$ asymmetry is found to be $A_C^{t\bar{t}} = 0.0033 \pm 0.0026(\text{stat}) + 0.0033(\text{syst})$. This result is compatible with the SM and with the result from the other CMS measurement, though more precise. The template method incorporates more information from the model than the analysis using unfolding leading to a reduced statistical uncertainty but at the expense of greater model dependence. The dominant source of systematic uncertainty is coming from the statistical uncertainty in the templates.

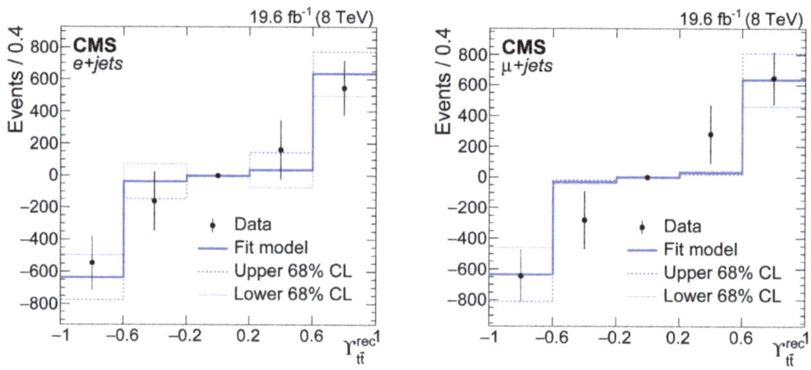

Figure 19. Antisymmetric component of the $Y_{t\bar{t}}$ distribution in the e+jets and μ+jets channels [51].

3.2.5. Combination of the Measurements of the $t\bar{t}$ Charge Asymmetry in the Lepton+Jets Channel by ATLAS and CMS

The results of the inclusive $A_C^{t\bar{t}}$ measurements in the lepton+jets channel at 8 TeV by the ATLAS and CMS Collaborations have been combined [35] using the same machinery as for the combination of the 7 TeV results. From the two CMS analyses in the lepton+jets channel, the one with the the smaller total uncertainty, the analysis based on a template method, is used for the combination with the result from the ATLAS Collaboration. With respect to the combination of the 7 TeV results, the treatment of the correlations of the systematic uncertainties between the two analyses is more fine grained. Especially the uncertainty in the jet energy scale is split into various sources that are grouped into four categories depending on the assumed correlation between the experiments, while again, most uncertainty sources are found to be uncorrelated, some of the signal and background modelling uncertainties are found to be correlated. The uncertainty in the chosen MC event generator, the uncertainty in the simulation of ISR and FSR, as well as the chosen PDF model are treated as fully correlated in the combination. The same applies to the uncertainty in the background contributions from single top quark production and the production of Z bosons in association with jets. As already mentioned, the individual sources of the jet energy uncertainty are grouped together into four categories. One category includes all jet energy related uncertainties that are considered uncorrelated, one category includes all uncertainties that are found to be partially correlated, using 50% as correlation parameter in the combination, and two categories contain the mostly and fully correlated uncertainties, which both enter with a 100% correlation assumption the combination.

The result of the combination of the two measurements in the lepton+jets channel at 8 TeV is $A_C^{t\bar{t}} = 0.0055 \pm 0.0023(\text{stat}) \pm 0.0025(\text{syst})$, with the ATLAS result contributing with a weight of 0.39 and the CMS result with a weight of 0.61. The χ^2 with one degree of freedom of the combination is 0.88, corresponding to a p-value of 0.35. The improvement in the total uncertainty of the combined results with respect to the individual results is 32% for the ATLAS measurement and 17% for the CMS analysis.

Figure 20 shows the allowed regions in the two dimensional plane spanned by the LHC charge asymmetry at 8 TeV and the Tevatron forward backward asymmetry for the SM and several BSM theories, together with the LHC combined $A_C^{t\bar{t}}$ value and A_{FB} results from the

Tevatron. The combined $A_C^{t\bar{t}}$ value puts strong constraints on the parameter space of several BSM scenarios, including models with a charged W' boson with right-handed couplings, heavy colour-octet vector gluons G_μ with axial couplings, colour-singlet Higgs boson models like isodoublets ϕ, colour-triplet scalars ω^4, and colour-sextet scalars Ω^4 with right-handed flavour-violating tu couplings (for more details on the BMS models, see Refs. [10,56]).

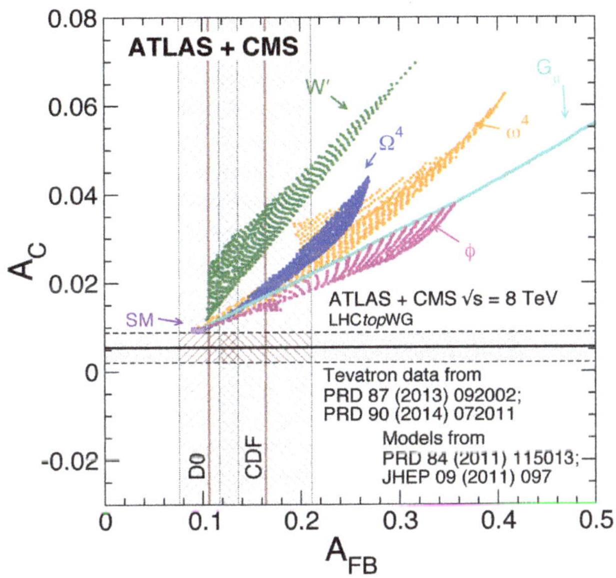

Figure 20. $A_C^{t\bar{t}}$-A_{FB} plane with allowed parameter regions for various BSM scenarios. The LHC combination is shown together with the latest Tevatron results from CDF and D0 [35].

Figure 21 summarizes the inclusive A_C measurements by the ATLAS and CMS Collaborations at 8 TeV in the dilepton and lepton+jets channels and the combination of the latter and compares the experimental results with theory predictions. Within the quoted uncertainties, the results are well comparable with the predictions.

In a second combination, the differential measurements of $A_C^{t\bar{t}}$ as a function of the invariant mass of the $t\bar{t}$ system of the ATLAS [47] and CMS [50] Collaborations are combined [35]. In both analyses, six bins are used for the $m_{t\bar{t}}$ distribution, with the same value ranges for the different bins. Although the same tools as for the combination of the inclusive measurements are used, the way how the correlations between the analyses are taken into account differs significantly. First of all, for the combination of the differential measurements the second CMS analysis is used, the one based on an unfolding method. As a result, two correlation assumptions need to be adjusted: the correlation of the parton shower and hadronization uncertainty is assumed to be 100% (was treated not correlated in the combination of inclusive results) and the uncertainty in the single top and Z+jets background modeling is considered uncorrelated (was assumed 100% correlated in the combination of the inclusive results).

However, the conceptual difference in the treatment of the uncertainty correlation is the additional consideration of bin-to-bin correlations. The correlation assumptions described above are strictly speaking only valid to describe the correlations between the same bins in two analyses (for instance between bin 2 of the ATLAS analysis and bin 2 of the CMS analysis). Two further categories of correlations need to be taken into account to describe the correlation model of differential measurements correctly. Within an analysis, the different bins are correlated with each other, which makes up the first category of bin-to-bin correlations. The same is true across experiments, making up the second category of bin-to-bin correlations. The above mentioned correlations between the same bins in two analyses are a special case of this second category. In contrast to these diagonal elements in a matrix of ATLAS versus CMS correlations, for the off-diagonal elements, the correlation assumption needs to be modified by the bin-to-bin uncertainties within the two analyses. This non-trivial treatment of correlations is detailed in Ref. [35].

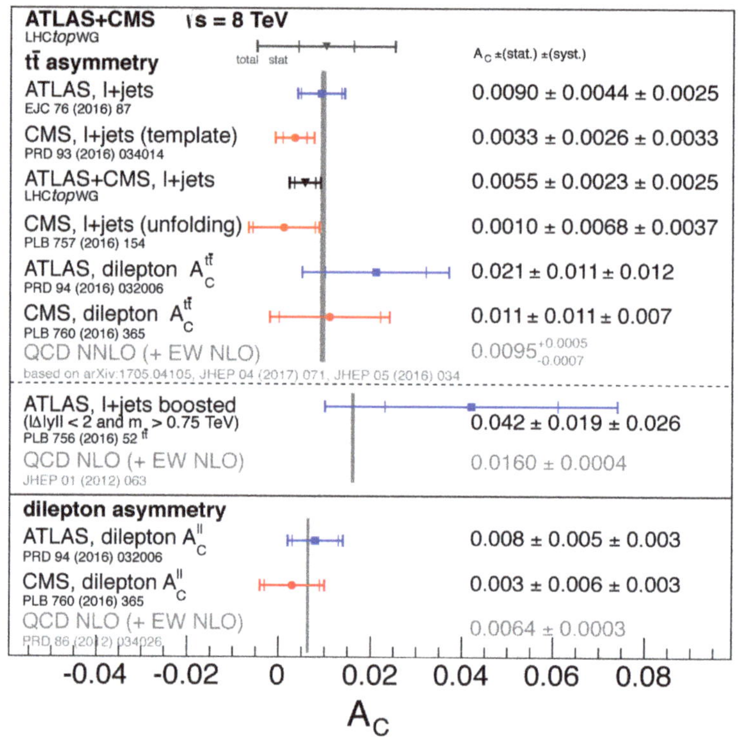

Figure 21. Summary of inclusive results at 8 TeV centre-of-mass energy [37].

The individual $t\bar{t}$ charge asymmetry values as a function of the invariant mass of the $t\bar{t}$ system along with the combined values can be found in Figure 22 (left). Depending on the bin, the ATLAS result contributes with weights between 0.22 and 0.59 to the combined value, while the weights of the CMS results vary between 0.41 and 0.78. The overall tendency is that the ATLAS result dominates the low mass region, while the CMS result dominates the high mass region. The total χ^2 of the combination with six degrees of freedom is 4.01 and corresponds to a p-value of 0.69. The improvement in precision ranges, again depending on the bin, between 20 and 52% for the ATLAS analysis and between 9 and 31% for the CMS analysis. In Figure 22, (right), the resulting combined charge asymmetry as function of $m_{t\bar{t}}$ is compared to two theory predictions for the SM, calculated at NLO [14] and NNLO [57–59], respectively, and to two versions of a colour-octet model [43]. The latter is chosen as its prediction for A_{FB} is in agreement with the Tevatron results and its deviations with respect to the SM prediction for $A_C^{t\bar{t}}$ are—for the heavy version—on the order of the measurement uncertainty. Within this uncertainty the combined values are comparable with the SM prediction as well as with both BSM scenarios.

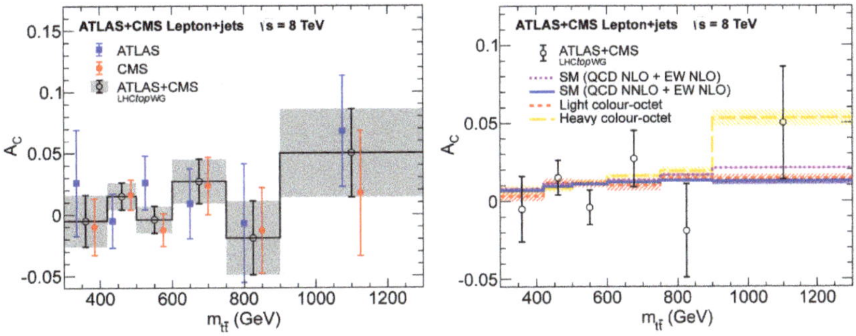

Figure 22. $A_C^{t\bar{t}}$ results as a function of the invariant mass of the $t\bar{t}$ system by ATLAS, CMS, and the LHC combined value (**left**). The LHC combined $A_C^{t\bar{t}}$ as a function of the invariant mass of the $t\bar{t}$ system compared to two calculations for the SM and to calculations for a colour octet model for two different mass scales [35].

3.3. Measurements at 13 TeV Centre-of-Mass Energy

The fraction of $q\bar{q}$ initiated top quark pair production decreases with increasing centre-of-mass energy, and so does the $t\bar{t}$ charge asymmetry. To compensate for the decreasing size of the effect to be measured, one needs to define regions of phase space in which the $q\bar{q} \to t\bar{t}$ process is enhanced with respect to the overall phase space for top quark pair production. As discussed earlier, events with high values of the reconstructed invariant mass of the $t\bar{t}$ system or with a large boost of the $t\bar{t}$ system in z direction feature enhanced fractions of $q\bar{q}$ initiated $t\bar{t}$ production and thus larger values of the predicted $t\bar{t}$ charge asymmetry. The additional advantage is that these phase space regions are also sensitive to contributions from new physics. Consequently, the ATLAS and CMS Collaborations measured the $t\bar{t}$ charge asymmetry at 13 TeV in samples of events with boosted $t\bar{t}$ candidates, i.e., events with high values of the reconstructed invariant mass of the $t\bar{t}$ system or events with large boosts in z direction. Depending on the actual definition of the phase space used for the measurement, the predictions for $A_C^{t\bar{t}}$ span one order of magnitude and vary between around 0.001 and 0.015.

The following two sections summarize the $A_C^{t\bar{t}}$ measurements by the ATLAS and CMS Collaborations at $\sqrt{s} = 13$ TeV.

3.3.1. Measurement of the $t\bar{t}$ Charge Asymmetry in Lepton+Jets and Dilepton Events by ATLAS

The ATLAS Collaboration measured the $t\bar{t}$ and the lepton charge asymmetry using the full Run 2 dataset of 139 fb^{-1} at 13 TeV [60]. The $A_C^{t\bar{t}}$ measurement is performed in the lepton+jets channel both in the resolved and boosted topologies. It is also measured using dilepton events in the resolved topology. The measurements are performed inclusively and differentially as a function of $m_{t\bar{t}}$, $\beta_{z,t\bar{t}}$, and $p_T^{t\bar{t}}$. The $A_C^{\ell\ell}$ measurement is performed only in the resolved channel inclusively and differentially as a function of the following kinematic quantities of the dilepton pair: $m_{\ell\ell}$, $\beta_{z,\ell\ell}$, and $p_T^{\ell\ell}$.

A common event selection is applied in the lepton+jets channel for both the resolved and boosted topologies requiring exactly one isolated lepton and at least one small-radius jet that has to be b-tagged. Further requirements on E_T^{miss} and on the reconstructed transverse mass of the W boson are applied to reduce the fake lepton background. The in total 4,126,511 selected events are classified based on their topologies (resolved or boosted) and on their number of b-tagged jets. The resolved topology is requiring at least four small-radius jets with $p_T > 25$ GeV. The assignment of the jets to the corresponding partons from the decaying top quarks is assessed using a BDT that aims to discriminant signal from the combinatorial background, separately for events with one or two b-tagged jets. This BDT combines kinematic variables and b-tagging information with weights from the kinematic likelihood fit [49] used for the lepton+jets measurement at 8 TeV. The best permutation in each event is required to have a BDT score higher than 0.3. This represents 50 % of correctly assigned jets for $t\bar{t}$ signal events that pass the selection. The boosted topology requires, together with the b-tagged small-radius jet, at least one large-radius jet with $p_{\mathrm{T}} > 350$ GeV. This large jet should be tagged as coming from the decay of a boosted top quark using an operating point with 80% efficiency. Additional requirements are applied on the separation of the large-radius jet with the lepton and with the small-radius jet since the two top quarks are expected to be produced back to back. The fourvector of the hadronically decaying top quark is taken to be the one of the large-radius jet. The semileptonically decaying top quark fourvector is reconstructed from the lepton kinematic, the small-radius jet and calculating the neutrino fourvector from a W boson mass constrain. The invariant mass of the $t\bar{t}$ system is requested to be larger than 500 GeV, which separates the resolved and boosted channel. The lepton+jets channel features an estimated background contribution of about 14%. Dilepton events are required to have at least two small-radius jets with $p_{\mathrm{T}} > 25$ GeV, among which at least one is b-tagged, and two opposite-charge leptons. For events with same-flavour leptons, the invariant mass of the dilepton system is required to be outside the Z boson mass window and an $E_{\mathrm{T}}^{\mathrm{miss}}$ cut is applied. The top quark kinematics is reconstructed using the neutrino weighting algorithm [61]. The 837,177 selected events are classified in four regions according to their lepton flavour ($e\mu$ and $ee + \mu\mu$) and b-tagged jet multiplicity (1 exclusive b-tagged jet, 2 inclusive b-tagged jets). The background contamination is estimated to be about 6%.

All signal and background processes are modelled using samples of simulated events except the non-prompt lepton and fake-lepton backgrounds, which are estimated using data-driven methods. The matrix method is used in the lepton+jets channel. In the dilepton channel, the fake background in the simulation is scaled by the data over simulation ratio computed in a control region with same sign leptons, separately for the $\mu\mu$ channel and for the ee and $e\mu$ channels, where an additional correction due to charge misidentification is also applied. Scaling factors are also applied to the Z+jets background in the dilepton channel, to allow for significant theoretical uncertainties in the modelling of Z boson production with heavy-flavour jets.

The differential $\Delta|y|$ distributions are corrected for acceptance and detector effects using the FBU technique where systematic uncertainties that affect the measurements are treated as nuisance parameters. In this process the $t\bar{t}$ signal normalisation is kept as a free parameter common to all bins and the lepton+jets and dilepton regions are combined. The number of bins is chosen as a compromise between the statistical uncertainty on the measured asymmetries and the bias in the measurements. Several assumptions about the correlations of the systematic uncertainties are made when combining the different regions. The experimental uncertainties are treated as fully correlated across all regions. For the signal and background modelling uncertainties several decorrelation schemes are studied. The modelling uncertainties are assumed to be correlated among the regions unless they are constrained by more than 30% in the FBU marginalisation. For a small number of them where the decorrelation scheme is increasing the total uncertainty by 5% to 20%, the more conservative scheme is chosen. The leading sources of systematic uncertainties come from the signal and background modelling uncertainties.

The combined inclusive $A_C^{t\bar{t}}$ asymmetry from single lepton and dilepton events is measured to be 0.0068 ± 0.0015 in agreement with the SM calculation at NNLO accuracy in the strong coupling with NLO electroweak corrections of $0.0064^{+0.0005}_{-0.0006}$ [15]. The quoted SM uncertainties include renormalization and factorization scale variations and PDF uncertainties, while the measurement uncertainty includes both the statistical and systematic uncertainties of 0.0010 each. The SM computation is performed by expanding the numerator and denominator at a given order in perturbation theory. The measurement differs from zero by 4.7 standard deviations, providing strong evidence for $t\bar{t}$ charge asymmetry at the LHC. Figure 23 (left) shows the $A_C^{t\bar{t}}$ differential measurement as a function of $m_{t\bar{t}}$ which is also consistent with the NNLO expectation.

The $A_C^{\ell\ell}$ asymmetry is measured in the dilepton channel only and gives $0.0054 \pm 0.0012(\text{stat}) \pm 0.0023(\text{syst})$, while the SM calculation at NLO in QCD, including NLO EW corrections predicts $0.0040^{+0.0002}_{-0.0001}$ [14], where the uncertainties come from renormalization and factorization scale variations and PDF uncertainties. The $A_C^{\ell\ell}$ differential measurement as a function of $m_{\ell\ell}$ is shown in Figure 23 (right).

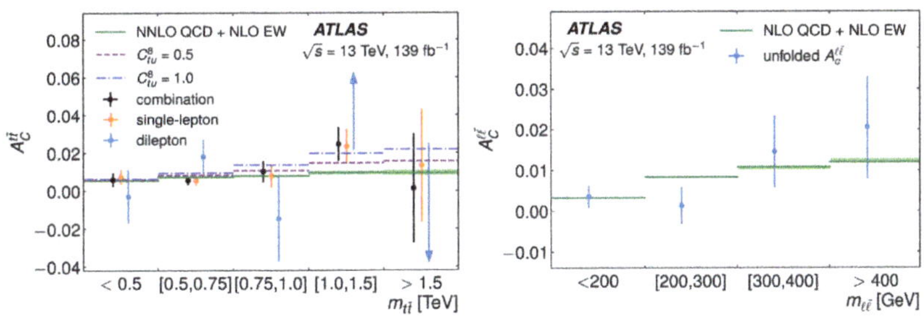

Figure 23. $A_C^{t\bar{t}}$ asymmetry as a function of $m_{t\bar{t}}$ (**left**) and $A_C^{\ell\ell}$ asymmetry as a function of $m_{\ell\ell}$ (**right**) from the unfolded distributions [60]. For $A_C^{t\bar{t}}$, the impact of a relevant Effective Field Theory coefficient for two different values is also shown. For the $A_C^{t\bar{t}}$ dilepton measurement, the central points for the two last bins is outside the range of the figure.

The combined $A_C^{t\bar{t}}$ results are interpreted within an SM effective field theory (SMEFT) using new operators connecting four quarks with different chiralities. Such operators could be the sign of new particles or interactions extending the SM [16,62,63]. For the interpretation of these results, dimension-six operators are considered. If Λ represents the characteristic new-physics scale, contributions to a given observable of Λ^{-2} come from the interference of the dimension-six operators with the SM amplitudes. The squares of dimension-six operators lead to contributions of order Λ^{-4}. Higher order contributions are neglected. 15 dimension-six operators are considered: eight $q\bar{q}t\bar{t}$ operators with left-handed left-handed (LL) or right-handed right-handed (RR) structure, six other $q\bar{q}t\bar{t}$ operators with LR structure and one tensor that modifies the top-gluon interaction. The relation between operators and the charge asymmetry is determined at NLO accuracy and parameterised with analytic function. Fits are performed either retaining only the Λ^{-2} contributions or including both Λ^{-2} and Λ^{-4} contributions. The limits on individual coefficients are derived using a χ^2 minimisation combining all results of the differential measurement as a function of $m_{t\bar{t}}$. They are shown on Figure 24. The limits that include both Λ^{-2} and Λ^{-4} contributions are generally about a factor two stronger than the bounds derived from the inclusive measurement because of the enhanced sensitivity at large $m_{t\bar{t}}$. The sensitivity of this analysis is also compared to the one of the energy asymmetry described in Section 5 performed using $t\bar{t}$+jet events. Due to the extra jet in these type of events, the QCD structure of the energy asymmetry is not the same as for $A_C^{t\bar{t}}$. So it appears that these two asymmetry observables probe different directions in chiral and colour space.

3.3.2. Measurement of the $t\bar{t}$ Charge Asymmetry in Lepton+Jets Events with Large Values of the Invariant Mass of the $t\bar{t}$ System by CMS

The CMS Collaboration analyzed the full Run-2 dataset corresponding to 138 fb^{-1} and measured the $t\bar{t}$ charge asymmetry in the lepton+jets channel for events with high values of the invariant mass of the $t\bar{t}$ system [64]. For events with boosted top quarks and top antiquarks, the decay products of the top (anti)quark are collimated and depending on the decay mode and the transverse momentum partially or fully merged. At the leptonic decay leg, the charged lepton from the leptonically decaying W boson is close to the b jet from the top quark decay, thus no isolation requirement is applied on the leptons in the event selection. For the hadronic decay leg, depending on the transverse momentum of the top (anti)quark, three possible topologies exist. For very large p_T values, the decay products end up reconstructed in a single, large-radius jet ("boosted" topology), while for low p_T values the three final state quarks give rise to three individual jets, the so-called "resolved" topology. For p_T values in between, the two quarks from the hadronically decaying W boson are reconstructed as one single jet, while the b jet from the top quark decay is resolved from the W boson jet, the so-called "semi-resolved" category. Using large-radius jet reconstruction and jet substructure techniques, the above mentioned topologies are experimentally examined, and the large-radius jets are either identified as the collimated decay products of a hadronically decaying top (anti)quark (t tag) or as collimated decay products of a hadronically decaying W boson (W tag). Events with one t tag and no W tag fall in the "boosted" category, events with one W tag and no t tag belong to the "semi-resolved" category, and events with neither t nor W tag are considered as "resolved" events. Events with more than one t or W tag are discarded.

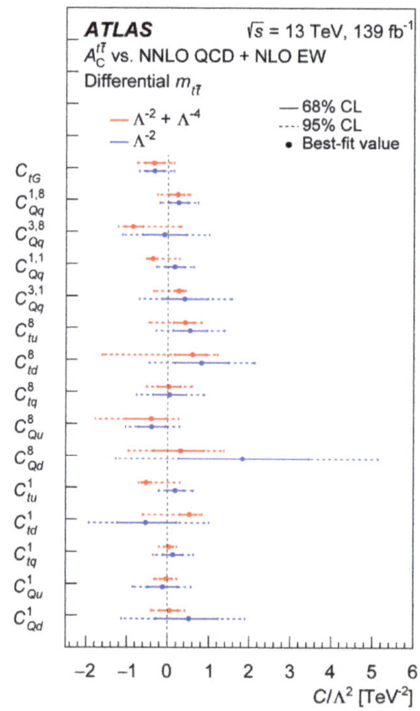

Figure 24. 68% and 95% individual limits on Wilson coefficients in the SMEFT framework from the differential $A_C^{t\bar{t}}$ asymmetry as a function of $m_{t\bar{t}}$ [60]. Only one coefficient is varied at a time while the other are kept to zero.

Depending on the topology of the $t\bar{t}$ event candidate, the reconstruction of the fourvectors of top quark and antiquark differs. For events with a boosted topology, the t tagged large-radius jet is assigned to the hadronically decaying top quark, and the small-radius jets with $\Delta R > 0.8$ from the hadronically decaying top quark are considered for the leptonically decaying top quark. In events with a "semi-resolved" topology, the large-radius jet with the W tag is assigned to the hadronically decaying W boson and all small-radius jets with $\Delta R > 0.8$ from the hadronically W boson are considered for either the b jet from the hadronically decaying top quark or the b jet from the leptronically decaying top quark. For events in the "resolved" category all small-radius jets are considered for assigning them to the final state quarks from the two decaying top quarks. For each event the hypothesis with the best χ^2, a measure for how close the masses of the two reconstructed top quarks are to each other and to the value determined simulation, is chosen and a cut on this variable is applied to suppress background events.

A binned maximum likelihood fit is performed simultaneously to the $\Delta|y|$ distributions in twelve channels (two lepton flavours, three data taking periods, two $m_{t\bar{t}}$ regions) to extract the $t\bar{t}$ charge asymmetry from the selected data set. The migration between true and generated values is taken into account when constructing the likelihood. The fit is performed for the fiducial volume as well as for the full phase space. For the latter one, additional corrections for acceptance and event selection efficiency are applied to the numbers of events in the twelve bins. For events with an invariant mass of the reconstructed $t\bar{t}$ system of $m_{t\bar{t}} > 750$ GeV, the measured value for the $t\bar{t}$ charge asymmetry in the full phase space of $t\bar{t}$ production is $A_C^{t\bar{t}} = 0.69^{+0.65}_{-0.69}$, where the uncertainty represents the total uncertainty. The corresponding theoretical prediction at NNLO in QCD perturbation theory with NLO electroweak corrections, calculated using the methods documented in Ref. [15] is $A_C^{t\bar{t}} = 0.94^{+0.05}_{-0.07}$. The experimental results for different $m_{t\bar{t}}$ ranges in the fiducial region as well as in the full phase space are summarized in Figure 25. The results are within their uncertainties comparable with the SM predictions but also with zero asymmetry. Despite of the limited statistical significance of the results, they nevertheless demonstrate the potential to measure top quark properties in boosted event topologies.

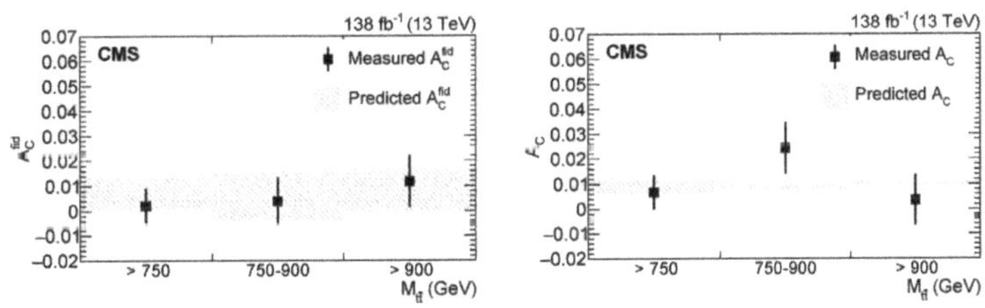

Figure 25. Measured $A_C^{t\bar{t}}$ in the fiducial phase space (**left**) and in the full phase space (**right**) presented in different mass regions after combining the μ+jets and e+jets channels [64]. The vertical bars represent the total uncertainties, with the inner tick mark indicating the statistical uncertainty in the observed data. The measured values are compared to the theoretical prediction, including NNLO QCD and NLO EW corrections, calculated with the methods described in Ref. [15]. The theoretical prediction in the fiducial region is obtained by fitting Asimov data that passed the signal candidate selection.

4. Measurement of the $t\bar{t}$ Forward-Backward Asymmetry at the LHC

To date, the LHC measurements of the charge asymmetry A_C have been discussed, where top quark and antiquark yield different widths of the rapidity distributions as a result of the underlying physics that are different for top quarks and antiquarks. Historically, the $t\bar{t}$ production asymmetry was described [3] and measured [4,5] as a forward-backward asymmetry. As discussed earlier, also at LHC forward and backward directions can be defined, and thus also a forward-backward asymmetry can be predicted and observed. Because of the charge symmetric proton-proton collision, the definition of forward and backward hemispheres is, however, only possible for individual events and needs to be indirectly accessed by the longitudinal momentum of the $t\bar{t}$ system.

In this chapter, we describe an A_{FB} measurement by the CMS Collaboration, using data from LHC collisions at $\sqrt{s} = 13$ TeV corresponding to an integrated luminosity of 35.9 fb^{-1} [65] . Candidate $t\bar{t}$ events with a lepton+jets signature are selected. The idea is to measure the forward-backward asymmetry in $q\bar{q} \to t\bar{t}$ events, by exploiting observables that have the power to discriminate the $q\bar{q}$ initial state for $t\bar{t}$ production from gg and gq initial states as well as from non-$t\bar{t}$ backgrounds, and are sensitive to the forward and backward direction of the produced top quark and antiquark.

The observable used to define whether the top quark is produced in forward or backward direction is the cosine of the angle between the top quark and the direction of motion of the incoming quark in the centre-of-mass frame of the $t\bar{t}$ system, c^*. The differential cross section for the $q\bar{q} \to t\bar{t}$ process as a function of c^* can be written as

$$\frac{d\sigma}{dc^*} \approx f_{\text{sym}}(c^*) + \left[\int_{-1}^{1} f_{\text{sym}}(x)dx\right] c^* A_{\text{FB}}^{(1)}(m_{\text{t}\bar{\text{t}}}), \tag{8}$$

a linear combination of symmetric and antisymmetric functions of the production angle c^* with $A_{\text{FB}}^{(1)}$ being a parameter. From this definition of the differential cross section and Equation (1) follows that the forward-backward asymmetry $A_{\text{FB}} \approx A_{\text{FB}}^{(1)}$, the linearized forward-backward asymmetry. A comparison between fitted $A_{\text{FB}}^{(1)}$ values and A_{FB} values determined from the numbers of events with top quarks produced in forward direction and events with top quarks produced in backward direction in generated signal events shows that the above approximation is valid.

The proton-proton collisions at LHC confront the analysers with two main challenges: Firstly, the production of $t\bar{t}$ pairs is dominated by the charge symmetric gg process, followed by the qg subprocess, which features only a tiny asymmetry and secondly, the forward and backward directions cannot be defined globally but need to be defined event by event. To overcome this challenge, the analysis exploits observables that help to discriminate the $q\bar{q} \to t\bar{t}$ process not only from non-$t\bar{t}$ backgrounds but also from $t\bar{t}$ production from gg and gq initial states, and that are sensitive to the direction of the incoming quark parton. The observables of choice are the invariant mass of the $t\bar{t}$ system, $m_{t\bar{t}}$, the scaled longitudinal momentum of the $t\bar{t}$ system in the laboratory frame, $x_{\text{F}} = 2p_{\text{L}}/\sqrt{s}$, and c^*.

Distributions of these three observables are shown in Figure 26, for simulated $t\bar{t}$ events, separately for $q\bar{q}$, gg, and qg initial states. The events have been generated using the `POWHEG` Monte Carlo generator [66]. In all three distributions a differentiation between the $q\bar{q}$ initial state and the other two is clearly visible, while the distributions from gg and qg initial states are very similar. For that reason, the latter two are treated as one single, combined process in the analysis. The parton distribution function of the proton leads to a—on average—higher momentum fractions of the incoming quark with respect to the incoming antiquark and thus the longitudinal momentum of the $t\bar{t}$ system in the laboratory frame is correlated with the direction of the incoming quark in $q\bar{q}$ initial states. Figure 26, lower right, shows the performance of choosing the longitudinal direction of the $t\bar{t}$ system in the laboratory frame as the direction of the initial state quark, based on the dilution factor D, defined as the difference between the number of correctly assigned quark directions and incorrectly assigned quark directions, normalized to the sum of both numbers, as a function of $|x_{\text{F}}|$. As can be seen, especially for the region with higher $|x_{\text{F}}|$, which is dominated by $q\bar{q}$ initial states, the efficiency for the choice of the quark direction based in the longitudinal direction of motion of the $t\bar{t}$ system is very high.

In order to construct the reconstructed versions of the observables discussed above, the candidate $t\bar{t}$ system in each event needs to be reconstructed from the objects detected by the detector. As contributions from potential new, massive particles are expected for large values of $m_{t\bar{t}}$, and as the fraction of $q\bar{q}$ initial states increases with increasing momentum of the $t\bar{t}$ system, it makes sense to group the events in categories according to the Lorentz boost of the event and thus the mass and momentum of the candidate $t\bar{t}$ pair. The analyzers define two "boosted" categories with high Lorentz boost of the event, where the decay products of the hadronically decaying top quark or top antiquark are all merged into one single fat jet that is identified as originating from a top quark, "top tagged", (type-1), or where these decay products are fully or partially merged into a fat jet that is not top tagged (type-2). The third category, containing most of the $t\bar{t}$ events, is defined as events, where all decay products are distinguishable, thus also called "resolved" category.

The reconstruction of the kinematic quantities of the $t\bar{t}$ pair is done via a maximum likelihood fit, in which the momenta of the decay products are allowed to vary within their resolution, and all combinations of jets are considered. Owing to the different event topologies in the three categories introduced above, the constraints for the assignment of the selected jets to the final state quarks differ. While for type-2 and type-3 events all jet-quark assignments are considered to reconstruct the leptonic and hadronic top quark candidates, for type-1 events, the top tagged jet is chosen to represent the hadronically decaying top quark. In addition, the selected lepton and the missing transverse momentum enter the reconstruction of the leptonic part of the $t\bar{t}$ candidate in all three cases. From the reconstructed $t\bar{t}$ candidates, the observables relevant for the analysis can then be constructed. The reconstructed observables are named m_r, x_r, and c_r^*. As discussed above, the direction of the incoming quark (needed for the calculation of c_r^*), is chosen according to the longitudinal momentum direction of the reconstructed $t\bar{t}$ system.

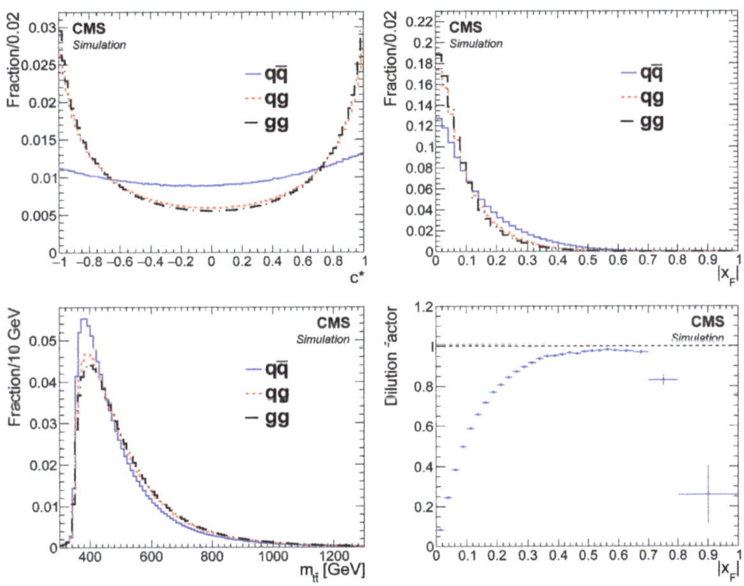

Figure 26. Normalized distributions at generator level of c^* (**upper left**), $|x_F|$ (**upper right**), and $m_{t\bar{t}}$ (**lower left**) for three different $t\bar{t}$ production modes. The dillution factor for the assumption that the quark direction in qg events is given by the longitudinal direction of the $t\bar{t}$ pair as a function of $|(x_F|$ (**upper right**) [65].

For the extraction of A_{FB}, a three-dimensional (in the three dimensions m_r, x_r, c_r^*) distribution function, being the sum of individual distribution functions for the signal and several background processes, each scaled with the corresponding scaling parameter, is used. $t\bar{t}$ dilepton and full hadronic, single top quark, and Z+jets background contributions are considered using a single background template, W+jets events are considered using another background template. The third background template used in the fit, is constructed from side-band data and represents the QCD mutlijet production. In order to construct the signal template for this fit, a fourth observable, the lepton charge Q, is exploited. Assuming a charge-parity symmetry of the detector with respect to the acceptance for events with negatively charged leptons and events with positively charged leptons, the charge of the lepton can be used to construct symmetric and antisymmetric distributions for the signal $q\bar{q} \to t\bar{t}$ process. The parameter $A_{FB}^{(1)}$ enters the fit function as linear parameter with the antisymmetric part of the signal template. The data are fitted in twelve different channels in total—3 categories (type-1, type-2, type-3), 2 flavours (electron, muon), 2 lepton charges—simultaneously. The systematic uncertainties are taken into account as nuisance parameters of the fit. Using 1000 sets of pseudo data for each value of $A_{FB}^{(1)}$, generated from the template models with the corresponding $A_{FB}^{(1)}$ input value, a Neyman construction is derived and the final result can be read off this construction. Figure 27 shows the Neyman construction, with the actual fit result drawn in as horizontal line and the corresponding true $A_{FB}^{(1)}$ value with uncertainties, projected on the x axis.

The resulting forward-backward asymmetry of $A_{\rm FB}^{(1)} = 0.048^{+0.095}_{-0.087}({\rm stat})^{+0.020}_{-0.029}({\rm syst})$ is well comparable with the expectations from the standard model.

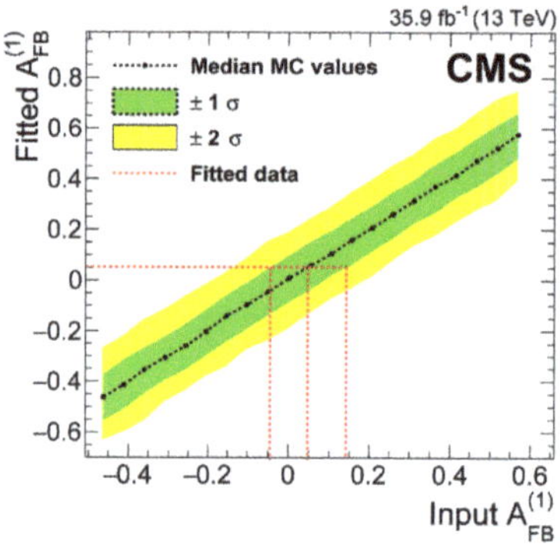

Figure 27. Neyman construction for the $A_{\rm FB}^{(1)}$ parameter of interest in 1000 pseudo-experiments generated with systematic uncertainty nuisance parameters allowed to vary. The horizontal dotted line indicates the value of the parameter determined from the fit and the vertical dotted line indicates where this value intersects with the central value and uncertainty contour from the pseudo-experiment.

5. Measurement of the $t\bar{t}$ Energy Asymmetry at the LHC

Another way to study the $t\bar{t}$ charge asymmetry is to use an observable linked to the energy difference between the top quarks and antiquarks: $\Delta E = E_t - E_{\bar{t}}$. The energy asymmetry [12] mainly occurs through the $qg \to t\bar{t}q$ process which is more abundant than the $gg \to t\bar{t}$ process at the LHC. It is therefor expected to be larger than asymmetries based on rapidity. The presence of an additional jet allows us to investigate QCD effects at leading order (LO) while the asymmetry in $q\bar{q} \to t\bar{t}$ is only appearing at NLO. The energy difference ΔE is connected to the angle of the final-state particles $(t, \bar{t}, j)$ by energy and momentum conservation and could be interpreted as an asymmetry of the top quark and antiquark scattering angles with respect to the jet direction. In the process $pp \to t\bar{t}j$, the energy asymmetry can be defined as a function of the jet angle θ_j as [12]:

$$A_E(\theta_j) = \frac{\sigma_{t\bar{t}j}(\theta_j|\Delta E > 0) - \sigma_{t\bar{t}j}(\theta_j|\Delta E < 0)}{\sigma_{t\bar{t}j}(\theta_j|\Delta E > 0) + \sigma_{t\bar{t}j}(\theta_j|\Delta E < 0)}, \tag{9}$$

where $\sigma_{t\bar{t}j}(\theta_j)$ is the differential $t\bar{t}j$ cross section as a function of θ_j. Both ΔE and θ_j are defined in the $t\bar{t}j$ rest frame, which corresponds to the partonic centre-of-mass frame in tree-level processes. The energy asymmetry can be formulated without involving the direction of the incoming quark. The outgoing quark-jet is boosted in the direction of the incoming valence quark. This boost translates into the rapidity of the $t\bar{t}j$ system in the laboratory frame: $y_{t\bar{t}j}$. Hence it is possible to optimise the definition of the energy asymmetry in Equation (9) by associating the forward $y_{t\bar{t}j} > 0$ and backward $y_{t\bar{t}j} < 0$ final state with the forward θ_j and backward $\pi - \theta_j$ scattered jet [67,68]:

$$\sigma^{\rm opt}(\theta_j) = \sigma(\theta_j|y_{t\bar{t}j} > 0) + \sigma(\pi - \theta_j|y_{t\bar{t}j} < 0), \qquad \theta_j \in [0, \pi]. \tag{10}$$

Equation (9) can then be redefined as:

$$A_E(\theta_j) = \frac{\sigma^{\rm opt}(\theta_j|\Delta E > 0) - \sigma^{\rm opt}(\theta_j|\Delta E < 0)}{\sigma^{\rm opt}(\theta_j|\Delta E > 0) + \sigma^{\rm opt}(\theta_j|\Delta E < 0)} \tag{11}$$

to maximise the statistical sensitivity to the energy asymmetry.

The ATLAS Collaboration has performed a measurement of this energy asymmetry using 139 fb^{-1} of proton-proton collision data of the LHC at $\sqrt{s} = 13$ TeV [69]. The energy asymmetry is measured in $t\bar{t}j$ production in a fiducial phase space defined at particle level. The analysis is performed in the semileptonic $t\bar{t}$ decay channel where one of the top quarks is leading to a W boson that decays leptonically (leptonic top quark). The hadronic decay products of the other top quark are required to be collimated in one large-radius jet, which characterizes a top quark with large momentum. This phase space is also referred to as the boosted regime. In this boosted regime, it is easier to identify the additional jet not coming from the hadronically decaying top quark.

Events are selected requiring an isolated electron or a muon with $p_{\mathrm{T}} > 27$ GeV with no other high momentum leptons. The hadronic top quark candidate is selected as a large-radius jet with $p_{\mathrm{T}} > 350$ GeV separated from the lepton. This large-radius jet is required to be tagged as a top quark candidate. Such top-tagging is performed with a DNN that relies on jet substructure variable inputs [70] and has a efficiency of 80% for a large-radius jet matched to a truth top quark. The jet from the leptonic top quark candidate is defined as a small-radius jet close to the lepton but isolated from the hadronic top quark candidate. The additional jet produced with the $t\bar{t}$ event is requested to be a small-radius jet with $p_{\mathrm{T}} > 100$ GeV separated from the large-radius jet and from the lepton and different from the leptonic top quark candidate. This p_{T} cut was chosen as a trade-off between statistical precision and increase of the energy asymmetry for high-p_{T} values. One of the small-radius jets is requested to be tagged as coming from a b quark. It should be either the jet of the leptonic top quark candidate or it should be within the hadronic top quark candidate. The missing transverse momentum is used as an estimation of the transverse momentum of the neutrino momentum coming from the W boson decay from the leptonic top quark. The leptonic top quark four-momentum is defined as the sum of the four-vectors of the lepton, the neutrino, and the jet assigned to the letponic top quark candidate. The charge of the leptonic top quark is assessed using the lepton charge, while the opposite charge is assigned to the hadronic top quark candidate.

The $t\bar{t}j$ process has been generated using the `POWHEG-BOX` generator [66] while MADGRAPH5_aMC@NLO [71] was used for the interpretation of the results in terms of SMEFT. The first main source of background to $t\bar{t}j$ production in the semileptonic channel comes from events with a prompt electron or muon from a W or Z boson decay (tW, s-channel single top production, W/Z+jets, diboson production, or $t\bar{t}W/Z/H$). This background is estimated from simulation. The production of W bosons in association with jets is the main contribution corresponding to 5% of the events in the signal region, followed by t events corresponding to roughly 3%. The second main source, contributing to 2% in the signal region, is coming from events with fake or non-prompt leptons. This background is evaluated using a data-driven method, the matrix method [72].

The asymmetry is extracted by computing the ratio defined in Equation (11) using the number of events $N^{\mathrm{opt}}(\theta_j)$ in three bins of θ_j: $[0, \pi/3, 3/5\pi, \pi]$, each divided into four ΔE bins (two bins with $\Delta E < 0$ and two bins with $\Delta E > 0$). The number of events observed at the detector level is corrected for detector effects to particle level in a fiducial phase space using the FBU technique. Experimental and theoretical systematic uncertainties are included through nuisance parameters that are marginalized. The unfolding has been found to be robust against the inclusion of standard model Effective Field Theory operators within their current limits (see below). The data distributions at detector level input to the unfolding are found to be in agreement with the SM expectations.

The uncertainty on the measurement is dominated by the statistical component. The largest systematic uncertainties are the uncertainty in jet energy resolution, in $t\bar{t}$ modelling, and in the fake background estimation. The results are shown in Figure 28. The measurement is found to be in good agreement with the SM expectation with a p-value of 0.80. In the first bin the measured asymmetry differs from zero by 2.1 standard deviations.

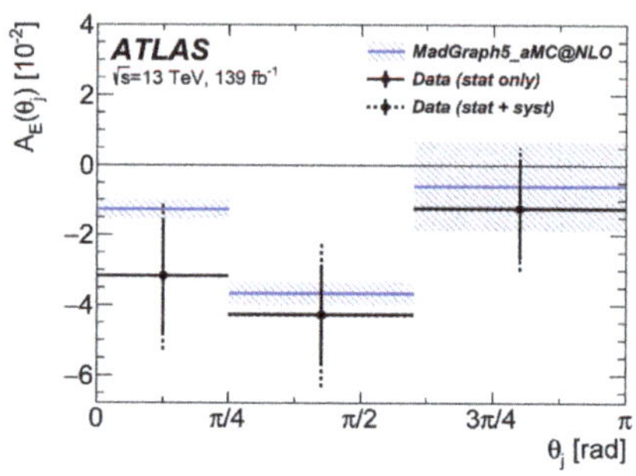

Figure 28. Measured energy asymmetry (black points) in three bins of the jet angle θ_j, compared to the prediction from simulation (blue lines) [69].

The sensitivity of the energy asymmetry to new physics at a high energy scale is investigated in the SMEFT framework. The operators considered in this interpretation are of dimension six, while higher-dimensional operators are neglected. The analysis also assumes CP invariance so that all coefficient are considered real. In the Warsaw basis [73], 15 dimension-six operators can affect $t\bar{t}j$ production at tree level [17]. The energy asymmetry is particularly sensitive to the chirality and colour charges of the involved operators. To illustrate this sensitivity, a selection of six four-quark operators involving different chiral structures (LL, RR, LR) and colour structure (singlet or octet) are chosen. To assess the sensitivity of the energy asymmetry to the Wilson coefficients of these operators, a χ^2 is built between the measured asymmetry and the SMEFT predictions in the three measured jet-angle bins taking into account the correlations among the measurements and among the predictions. The obtained bounds on individual Wilson coefficients are summarized in Figure 29. Overall the energy asymmetry is sensitive to the operator range $[-2, 2]$ at 95% C.L. Bounds for several pairs of operators are also derived. These show complementary constraints between the asymmetry built using rapidity and the energy asymmetry probing different directions in chiral and colour space. The energy asymmetry is then able to resolve nearly blind directions left by other top-quark observables and so will provide useful additional information in global EFT fits.

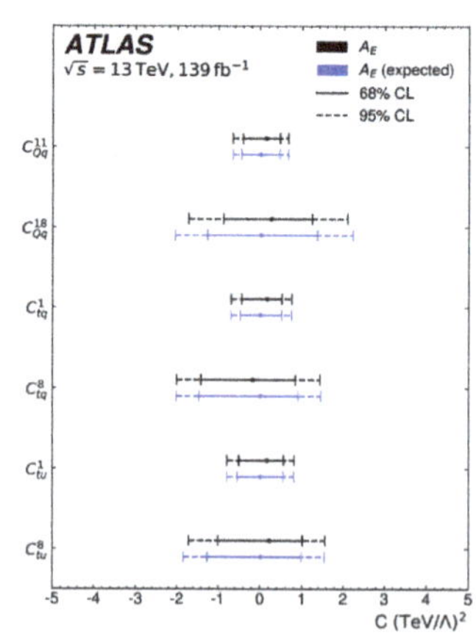

Figure 29. Bounds at 68% CL (solid) and 95% CL (dashed) on individual four-quark Wilson coefficient from the ATLAS energy asymmetry, from a combined fit to the measured (black) and expected (blue) energy asymmetry in three jet-angle bins [69].

6. Measurement of the $t\bar{t}W$ Charge Asymmetry at the LHC

As discussed above, $t\bar{t}$ production at the LHC is dominated by the gluon fusion process which is charge symmetric. On the other hand, the production of a $t\bar{t}$ pair in association with a W boson referred to as $t\bar{t}W$ is initiated at LO by a $q\bar{q}'$ initial state, while the quark-gluon channels open up at NLO. The gluon fusion initial state does not contribute until NNLO. The $t\bar{t}W$ production can then serve as an interesting tool to measure the $t\bar{t}$ charge asymmetry. Indeed the charge asymmetry in $t\bar{t}W$ events is expected to be larger than in $t\bar{t}$ production [74,75]. In addition the W boson in such process can be radiated from the $q\bar{q}'$ initial state and then serves as a polariser of $q\bar{q}'$ and thus of $t\bar{t}$. This polarisation further enhances the asymmetry between the decay products of the top quarks and top antiquarks, leading to an enhanced leptonic asymmetry when looking at leptonically decaying top and antitop quarks. The SM NLO QCD prediction of the leptonic asymmetry in the full phase space at $\sqrt{s} = 13$ TeV [74,75] is $-0.1316^{-0.0081}_{+0.0112}$ where the quoted uncertainties are coming from scale variations.

The drawback of the $t\bar{t}W$ process is however its much smaller cross section (around 0.6 pb) compared to $t\bar{t}$ production (around 830 pb). Besides being a rare process, making predictions for the $t\bar{t}W$ cross section is also challenging, as large corrections arise from higher powers of both the strong and weak couplings [76]. Hence $t\bar{t}W$ measurements represent sensitive tests of QCD predictions connected with the electroweak sector.

The ATLAS Collaboration has performed a search for the leptonic charge asymmetry in $t\bar{t}W$ production using 139 fb^{-1} of proton-proton collision data of the LHC at $\sqrt{s} = 13$ TeV [77]. The measurement is performed in the trilepton channel both at detector level and at particle level after unfolding. Events are selected with exactly three isolated light leptons (electrons or muons) with p_{T} larger than 30, 20, and 15 GeV. At least two jets and one b jet is required. The selected events are then classified into four signal regions according to their jet and b-jet multiplicity as well as their amount of $E_{\mathrm{T}}^{\mathrm{miss}}$. Four control regions are also defined in order to fit the dominant backgrounds simultaneously with the signal. Because two of the leptons are expected to come from the decays of the top quarks and antiquarks, the sum of the three lepton charges is required to be ± 1. The invariant mass of the opposite sign same flavour lepton pair is required to be higher than 30 GeV and outside the Z boson peak region (except for the control region for the $t\bar{t}Z$ background where such mass should be around 90 GeV). Two control regions are used to evaluate the amount of non-prompt electrons or muons arising from heavy-flavour hadron decays. These regions are defined by requiring that the third lepton fails the isolation criteria. A last control region is targeting the estimation of γ conversions by requiring that at least one of the leptons is an electron failing the material conversion rejection criteria.

In order to compute $\Delta|\eta| = |\eta_{\ell^+}| - |\eta_{\ell^-}|$ and then $A_{\mathrm{C}}^{\ell\ell}$, the two opposite sign leptons that are coming from the $t\bar{t}$ decay need to be separated from the one coming from the W decay. This is addressed using a BDT that takes five variables as input. They are defined as the masses and ΔR of the systems formed by the leptons and the closest or second closest b jets. The fraction of events in which the lepton with the highest BDT score originates from a top quark or top antiquark decay is estimated to be 71%, using simulation. The second lepton needed to compute $\Delta|\eta|$ is taken to be the lepton with the opposite charge from the lepton selected by the BDT.

A profile-likelihood fit is used to extract the signal as well as the normalisation for the most relevant background processes: ie. $t\bar{t}Z$, non-prompt electrons and muons as well as electrons from γ-conversions. The template shapes for these backgrounds are taken from simulated events. Each of the four signal regions are separated into $\Delta|\eta| > 0$ and $\Delta|\eta| \leq 0$. Separate normalisation factors in the $\Delta|\eta| > 0$ and $\Delta|\eta| \leq 0$ regions are allowed to float freely in the fit to avoid any bias from an assumption of SM asymmetries for these processes. The post-fit predictions and data yields for $\Delta|\eta| > 0$ and $\Delta|\eta| \leq 0$ in the four signal region are shown in Figure 30. Tests using MC have been performed to validate that the extracted asymmetry value is not biased by the absolute normalisation of the $t\bar{t}W$ process.

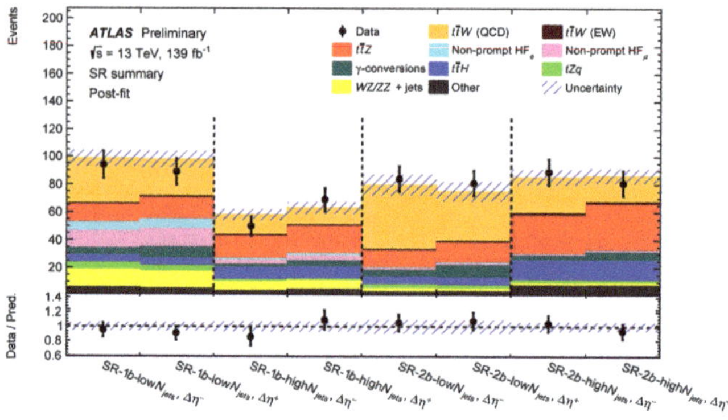

Figure 30. Comparison between data and the post-fit predictions for $\Delta|\eta| \leq 0$ and $\Delta|\eta| > 0$ in the four signal regions used to compute the $t\bar{t}W$ charge asymmetry in ATLAS [77].

The leptonic charge asymmetry in $t\bar{t}W$ event is measured to be: $A_C^{\ell\ell}(t\bar{t}W) = -0.123 \pm 0.136(\text{stat}) \pm 0.051(\text{syst})$ in agreement with the predictions from Sherpa simulation [78] in this phase space: $-0.084^{+0.005}_{-0.003}(\text{scale}) \pm 0.006(\text{MCstat})$. The measurement is statistically limited. The leading systematic uncertainty is coming from the comparison of the fit performed with separated control regions for $\Delta|\eta| > 0$ and $\Delta|\eta| \leq 0$ (as in the default case) with an alternative fit performed where each background is normalised with a single normalisation factor (i.e., not separated between $\Delta|\eta| > 0$ and $\Delta|\eta| \leq 0$). Other important sources of systematic uncertainties are coming from the $t\bar{t}W$ and $t\bar{t}Z$ modelling uncertainties.

To obtain the charge asymmetry at particle level in a specific fiducial phase space close to the reconstructed one, an unfolding procedure is performed to correct for detector and acceptance effects. For this result, a simpler method is adopted to associate the leptons to the top quarks. A lepton is chosen to come from a top quark if the mass of the system formed by this lepton and the closest b jet is the closest to the most probable mass value according to the $t\bar{t}W$ simulation. This procedure has an efficiency of approximately 65% to identify the correct leptons. The unfolding procedure is based on a profile-likelihood approach as in Ref. [79]. In this approach, each bin in the particle-level distribution is folded through the response matrix, resulting in the bins at detector level. The normalisations for the main backgrounds and the analysis regions are split into $\Delta|\eta| > 0$ and $\Delta|\eta| \leq 0$ in the same way as for the detector-level results. An injection test is performed to verify that charge asymmetries potentially deviating from the SM prediction can be recovered in the unfolding procedure. The charge asymmetry value unfolded to particle level in the fiducial phase space is found to be: $A_C^{\ell\ell}(t\bar{t}W) = -0.112 \pm 0.170(\text{stat}) \pm 0.055(\text{syst})$ compared to the SM prediction from Sherpa [78]: $-0.063^{+0.007}_{-0.004}(\text{scale}) \pm 0.004(\text{MCstat})$. The leading sources of systematic uncertainties are the same as for the detector level result. As both results are limited by the statistical uncertainty, the potential of $t\bar{t}W$ events to measure the charge asymmetry is promising in the years to come.

7. Conclusions

The ATLAS and CMS Collaborations have performed several measurements of $t\bar{t}$ asymmetries using LHC collision events with either a single lepton or exactly two leptons, collected at 7, 8 , and 13 TeV centre-of-mass energy. The fact that $t\bar{t}$ production via quark antiquark annihilation is not invariant under exchange of top quark and top antiquark manifests itself in different asymmetries. Both collaborations measured the lepton and $t\bar{t}$ charge asymmetries. The CMS Collaboration analyzed in addition the forward-backward asymmetry of $t\bar{t}$ events, while the ATLAS Collaboration investigated the $t\bar{t}$ energy asymmetry and the charge asymmetry in $t\bar{t}$ events where an additional W boson is produced. The measured asymmetries are typically corrected for efficiency and acceptance effects and thus extrapolated to either a fiducial or the full phase space. In addition both collaborations dedicated analyses to investigate the $t\bar{t}$ charge asymmetry in events with a boosted topology. While inclusive asymmetry measurements were performed at 7 TeV, the increase of the dataset available at 8 and 13 TeVallowed to perform differential measurements in phase spaces where asymmetries are predicted to be enhanced. Hence the expected asymmetry decrease with increased centre-of-mass energy could be compensated. The corrections of the reconstructed distributions required the application of unfolding methods where especially for the differential measurements and their combination significant development work was needed in order to make the existing algorithms applicable. The results of the presented analyses have shown that the $t\bar{t}$ charge asymmetry—although smaller in size compared to the effect at the Tevatron—is also measureable at the LHC, even differentially as function of suited kinematic variables of the $t\bar{t}$ system. The LHC measurements yield no significant hint of contributions from physics beyond the standard model.

Although the precision for some of the analyses is—especially given the small size of the effect—quite impressive, it is apparent that the analyses are dominated by statistical uncertainties in the phase spaces where the asymmetries are expected to be enhanced. Their sensitivity thus is limited by the size of the selected dataset. Given that the data taking periods at 7, 8, and 13 TeV are over, a reduction of the statistical component of the measurements will only be possible by analyzing the data of the recently started Run 3 and the future high luminosity LHC phase. The to be expected larger samples of selected $t\bar{t}$ candidate events will enable measurements to focus on the interesting regions in phase space with enhanced $t\bar{t}$ charge asymmetry, e.g., mainly events with boosted topologies. Furthermore, the vast amount of available $t\bar{t}$ signal events will facilitate multi-differential measurements of the $t\bar{t}$ charge asymmetry and will thus help to increase our knowledge of this property and the nature of the top quark itself. Higher collision energies on the contrary will not bring further advantage as the size of the effect decreases with increasing centre-of-mass energy.

Looking beyond the future LHC extensions, the next particle collider will most probably be an electron-positron machine. Assuming the centre-of-mass energy of the collider exceeds the $t\bar{t}$ production threshold, it will be able to study pairwise top-quark production via electron-positron annihilation for the first time. The interaction of the produced top quarks and antiquarks with the electric fields of the incoming electron and positron will also result in preferred spatial directions for the production of top (anti)quarks, thus giving rise to an asymmetry between top quark and antiquark. The thorough analysis of this asymmetry in top quark pair production through electron-positron annihilation will complement the information collected by analyzing the mostly QCD-induced charge asymmetry in $t\bar{t}$ production from quark-antiquark annihilation, as discussed in this review.

Thus, the future of $t\bar{t}$ production asymmetries remains interesting. Incoming measurements with increased precision will help to improve the level of details under which the standard model of particle physics is scrutinized and the search for new physics beyond what is currently known.

Funding: This research received no external funding.

Institutional Review Board Statement: Not applicable.

Informed Consent Statement: Not applicable.

Data Availability Statement: Not applicable.

Conflicts of Interest: The authors declare no conflict of interest.

References

1. Merriam-Webster.com Dictionary. Asymmetry. Available online: https://www.merriam-webster.com/dictionary/asymmetry (accessed on 18 May 2022).
2. Halzen, F.; Hoyer, P.; Kim, C. Forward-backward asymmetry of hadroproduced heavy quarks in QCD. *Phys. Lett. B* **1987**, *195*, 74. [CrossRef]
3. Kühn, J.H.; Rodrigo, G. Charge asymmetry in hadroproduction of heavy quarks. *Phys. Rev. Lett.* **1998**, *81*, 49–52. [CrossRef]

4. The CDF Collaboration. Forward-Backward Asymmetry in Top-Quark Production in $p\bar{p}$ Collisions at $\sqrt{s} = 1.96$ TeV. *Phys. Rev. Lett.* **2008**, *101*, 202001. [CrossRef] [PubMed]
5. The D0 Collaboration. Measurement of the Forward-Backward Charge Asymmetry in Top-Quark Pair Production. *Phys. Rev. Lett.* **2008**, *100*, 142002. [CrossRef]
6. The CDF Collaboration. Evidence for a mass dependent forward-backward asymmetry in top quark pair production. *Phys. Rev. D* **2011**, *83*, 112003. [CrossRef]
7. Aguilar-Saavedra, J.A.; Perez-Victoria, M. Probing the Tevatron t tbar asymmetry at LHC. *JHEP* **2011**, *5*, 34. [CrossRef]
8. Brivio, I.; Bruggisser, S.; Maltoni, F.; Moutafis, R.; Plehn, T.; Vryonidou, E.; Westhoff, S.; Zhang, C. O new physics, where art thou? A global search in the top sector. *JHEP* **2020**, *2*, 131 [CrossRef]
9. Zhang, C.; Willenbrock, S. Effective-Field-Theory Approach to Top-Quark Production and Decay. *Phys. Rev. D* **2011**, *83*, 034006. [CrossRef]
10. Aguilar-Saavedra, J.A.; Pérez-Victoria, M. Asymmetries in $t\bar{t}$ production: LHC versus Tevatron. *Phys. Rev. D* **2011**, *84*, 115013. [CrossRef]
11. The CDF and D0 Collaborations. Combined Forward-Backward Asymmetry Measurements in Top-Antitop Quark Production at the Tevatron. *Phys. Rev. Lett.* **2018**, *120*, 042001. [CrossRef]
12. Berge, S.; Westhoff, S. Top-quark charge asymmetry goes forward: two new observables for hadron colliders. *JHEP* **2013**, *7*, 179. [CrossRef]
13. Kühn, J.H.; Rodrigo, G. Charge asymmetries of top quarks at hadron colliders revisited. *JHEP* **2012**, *1*, 63. https://doi.org/10.1007/JHEP01(2012)063. [CrossRef]
14. Bernreuther, W.; Si, Z.G. Top quark and leptonic charge asymmetries for the Tevatron and LHC. *Phys. Rev. D* **2012**, *86*, 034026. [CrossRef]
15. Czakon, M.; Heymes, D.; Mitov, A.; Pagani, D.; Tsinikos, I.; Zaro, M. Top-quark charge asymmetry at the LHC and Tevatron through NNLO QCD and NLO EW. *Phys. Rev. D* **2018**, *98*, 014003. [CrossRef]
16. Brivio, I.; Trott, M. The Standard Model as an Effective Field Theory. *Phys. Rept.* **2019**, *793*, 1–98. [CrossRef]
17. Barducci, D.; Brivio, I.; Cirigliano, V.; Dekens, W.; de Vries, J.; Englert, C.; Fabbrichiesi, M.; Grojean, C.; Haisch, U.; Jiang, Y.; et al. Interpreting top-quark LHC measurements in the standard-model effective field theory. *arXiv* **2018**, arXiv:1802.07237.
18. Aguilar-Saavedra, J.A.; Amidei, D.; Juste, A.; Perez-Victoria, M. Asymmetries in top quark pair production at hadron colliders. *Rev. Mod. Phys.* **2015**, *87*, 421–455. [CrossRef]
19. Aguilar-Saavedra, J.; Juste, A.; Rubbo, F. Boosting the tt charge asymmetry. *Phys. Lett.* **2012**, *707*, 92–98. [CrossRef]
20. The ATLAS Collaboration. Measurement of the charge asymmetry in dileptonic decays of top quark pairs in pp collisions at $\sqrt{s} = 7$ TeV using the ATLAS detector. *JHEP* **2015**, *5*, 61. [CrossRef]
21. Choudalakis, G. Fully Bayesian Unfolding. *arXiv* **2012**, arXiv:1201.4612.
22. Lyons, L.; Gibaut, D.; Clifford, P. How to combine correlated estimates of a single physical quantity. *Nucl. Instrum. Meth. A* **1988**, *270*, 110–117. [CrossRef]
23. Valassi, A. Combining correlated measurements of several different physical quantities. *Nucl. Instrum. Meth. A* **2003**, *500*, 391–405. [CrossRef]
24. The CMS Collaboration. Measurements of the $t\bar{t}$ charge asymmetry using the dilepton decay channel in pp collisions at $\sqrt{s} = 7$ TeV. *JHEP* **2014**, *4*, 191. [CrossRef]
25. The CMS Collaboration. Identification of b-quark jets with the CMS experiment. *J. Instrum.* **2013**, *8*, P04013–P04013. [CrossRef]
26. The CMS Collaboration. *Commissioning of the Particle-Flow Reconstruction in Minimum-Bias and Jet Events from pp Collisions at 7 TeV*; Technical Report; CERN: Geneva, Switzerland, 2010.
27. The CMS Collaboration. Measurement of the $t\bar{t}$ production cross section and the top quark mass in the dilepton channel in pp collisions at $\sqrt{s} = 7$ TeV. *JHEP* **2011**, *7*, 49. [CrossRef]
28. Höcker, A.; Kartvelishvili, V. SVD approach to data unfolding. *Nucl. Instruments Methods Phys. Res. Sect. A Accel. Spectrometers Detect. Assoc. Equip.* **1996**, *372*, 469–481. [CrossRef]
29. The ATLAS Collaboration. Measurement of the top quark pair production charge asymmetry in proton-proton collisions at $\sqrt{s} = 7$ TeV using the ATLAS detector. *JHEP* **2014**, *2*, 107. [CrossRef]
30. The ATLAS Collaboration. *Commissioning of the ATLAS High-Performance b-Tagging Algorithms in the 7 TeV Collision Data*; Technical Report; CERN: Geneva, Switzerland, 2011. All Figures Including Auxiliary Figures Are Available. Available online: https://atlas.web.cern.ch/Atlas/GROUPS/PHYSICS/CONFNOTES/ATLAS-CONF-2011-102 (accessed on 20 September 2022).
31. The ATLAS Collaboration. *Calibrating the b-Tag Efficiency and Mistag Rate in* 35 pb^{-1} *of Data with the ATLAS Detector*; Technical Report; CERN: Geneva, Switzerland, 2011. All Figures Including Auxiliary Figures Are Available. Available online: https://atlas.web.cern.ch/Atlas/GROUPS/PHYSICS/CONFNOTES/ATLAS-CONF-2011-089 (accessed on 20 September 2022).
32. The ATLAS Collaboration. Measurement of the charge asymmetry in top quark pair production in pp collisions at $\sqrt{s} = 7$ TeV using the ATLAS detector. *Eur. Phys. J. C* **2012**, *72*, 2039. [CrossRef] [PubMed]
33. The CMS Collaboration. Inclusive and Differential Measurements of the $t\bar{t}$ Charge Asymmetry in Proton-Proton Collisions at $\sqrt{s} = 7$ TeV. *Phys. Lett. B* **2012**, *717*, 129–150. [CrossRef]
34. Brooijmans, G.; Gripaios, B.; Moortgat, F.; Santiago, J.; Skands, P.; Albornoz Vasquez, D.; Allanach, C.; Alloul, A.; Arbey, A.; Azatov, A.; et al. Les Houches 2011: Physics at TeV Colliders New Physics Working Group Report. In Proceedings of the 7th Les Houches Workshop on Physics at TeV Colliders, Les Houches, France, 30 May–17 June 2011; pp. 221–463.
35. The ATLAS and CMS Collaborations. Combination of inclusive and differential $t\bar{t}$ charge asymmetry measurements using ATLAS and CMS data at $\sqrt{s} = 7$ and 8 TeV. *JHEP* **2018**, *4*, 033. [CrossRef]
36. Nisius, R. On the combination of correlated estimates of a physics observable. *Eur. Phys. J. C* **2014**, *74*, 3004. [CrossRef]

37. Workgroup on Top physics at the LHC (LHCTopWG). LHCTopWG Summary Plots/Top charge asymmetry. Available online: https://twiki.cern.ch/twiki/bin/view/LHCPhysics/LHCTopWGSummaryPlots#Top_Charge_Asymmetry (accessed on 7 June 2022).
38. The ATLAS Collaboration. Measurements of the charge asymmetry in top-quark pair production in the dilepton final state at $\sqrt{s} = 8$ TeV with the ATLAS detector. *Phys. Rev. D* **2016**, *94*, 032006, [CrossRef]
39. The CDF Collaboration. Measurement of the top quark mass using template methods on dilepton events in proton antiproton collisions at $\sqrt{s} = 1.96$-TeV. *Phys. Rev. D* **2006**, *73*, 112006. [CrossRef]
40. The CDF Collaboration. W boson polarization measurement in the $t\bar{t}$ dilepton channel using the CDF II Detector. *Phys. Lett. B* **2013**, *722*, 48–54. [CrossRef]
41. The ATLAS Collaboration. Differential top-antitop cross-section measurements as a function of observables constructed from final-state particles using pp collisions at $\sqrt{s} = 7$ TeV in the ATLAS detector. *JHEP* **2015**, *6*, 100. [CrossRef]
42. Nason, P. A New method for combining NLO QCD with shower Monte Carlo algorithms. *JHEP* **2004**, *11*, 40. [CrossRef]
43. Aguilar-Saavedra, J.A. Portrait of a colour octet. *JHEP* **2014**, *8*, 172. [CrossRef]
44. The CMS Collaboration. Measurements of $t\bar{t}$ charge asymmetry using dilepton final states in pp collisions at $\sqrt{s} = 8$ TeV. *Phys. Lett. B* **2016**, *760*, 365–386. [CrossRef]
45. Schmitt, S. TUnfold: An algorithm for correcting migration effects in high energy physics. *JINST* **2012**, *7*, T10003. [CrossRef]
46. Frixione, S.; Nason, P.; Webber, B.R. Matching NLO QCD and parton showers in heavy flavor production. *JHEP* **2003**, *8*, 007. [CrossRef]
47. The ATLAS Collaboration. Measurement of the charge asymmetry in top-quark pair production in the lepton-plus-jets final state in pp collision data at $\sqrt{s} = 8$ TeV with the ATLAS detector. *Eur. Phys. J. C* **2016**, *76*, 87. [CrossRef] [PubMed]
48. The ATLAS Collaboration. Measurement of the charge asymmetry in highly boosted top-quark pair production in $\sqrt{s} = 8$ TeV pp collision data collected by the ATLAS experiment. *Phys. Lett. B* **2016**, *756*, 52–71. [CrossRef]
49. Erdmann, J.; Guindon, S.; Kroeninger, K.; Lemmer, B.; Nackenhorst, O.; Quadt, A.; Stolte, P. A likelihood-based reconstruction algorithm for top-quark pairs and the KLFitter framework. *Nucl. Instrum. Meth. A* **2014**, *748*, 18–25. [CrossRef]
50. The CMS Collaboration. Inclusive and differential measurements of the $t\bar{t}$ charge asymmetry in pp collisions at $\sqrt{s} = 8$ TeV. *Phys. Lett. B* **2016**, *757*, 154–179. [CrossRef]
51. The CMS Collaboration. Measurement of the charge asymmetry in top quark pair production in pp collisions at $\sqrt{s} = 8$ TeV using a template method. *Phys. Rev. D* **2016**, *93*, 034014. [CrossRef]
52. Bernreuther, W.; Si, Z.G. Distributions and correlations for top quark pair production and decay at the Tevatron and LHC. *Nuclear Phys. B* **2010**, *837*, 90–121. [CrossRef]
53. Gabrielli, E.; Raidal, M.; Racioppi, A. Implications of the effective axial-vector coupling of gluon on top-quark charge asymmetry at the LHC. *Phys. Rev. D* **2012**, *85*, 074021. [CrossRef]
54. Gabrielli, E.; Raidal, M. Effective axial-vector coupling of gluon as an explanation to the top quark asymmetry. *Phys. Rev. D* **2011**, *84*, 054017. [CrossRef]
55. Frixione, S.; Nason, P.; Ridolfi, G. A Positive-weight next-to-leading-order Monte Carlo for heavy flavour hadroproduction. *JHEP* **2007**, *9*, 126. [CrossRef]
56. Aguilar-Saavedra, J.A.; Perez-Victoria, M. Simple models for the top asymmetry: Constraints and predictions. *JHEP* **2011**, *9*, 97. [CrossRef]
57. Czakon, M.; Heymes, D.; Mitov, A.; Pagani, D.; Tsinikos, I.; Zaro, M. Top-pair production at the LHC through NNLO QCD and NLO EW. *JHEP* **2017**, *10*, 186. [CrossRef]
58. Czakon, M.; Heymes, D.; Mitov, A. Dynamical scales for multi-TeV top-pair production at the LHC. *JHEP* **2017**, *4*, 71. [CrossRef]
59. Czakon, M.; Fiedler, P.; Heymes, D.; Mitov, A. NNLO QCD predictions for fully-differential top-quark pair production at the Tevatron. *JHEP* **2016**, *5*, 034. [CrossRef]
60. The ATLAS Collaboration. Evidence for the charge asymmetry in $pp \to t\bar{t}$ production at $\sqrt{s} = 13$ TeV with the ATLAS detector. *arXiv* **2022**, arXiv:2208.12095.
61. The D0 Collaboration. Precise measurement of the top quark mass in dilepton decays using optimized neutrino weighting. *Phys. Lett. B* **2016**, *752*, 18–26. [CrossRef]
62. Buchmuller, W.; Wyler, D. Effective Lagrangian Analysis of New Interactions and Flavor Conservation. *Nucl. Phys. B* **1986**, *268*, 621–653. [CrossRef]
63. Leung, C.N.; Love, S.T.; Rao, S. Low-Energy Manifestations of a New Interaction Scale: Operator Analysis. *Z. Phys. C* **1986**, *31*, 433. [CrossRef]
64. The CMS Collaboration. Measurement of the $t\bar{t}$ charge asymmetry in events with highly Lorentz-boosted top quarks in pp collisions at $\sqrt{s} = 13$ TeV. *arXiv* **2022**, arXiv:2208.02751.
65. The CMS Collaboration. Measurement of the top quark forward-backward production asymmetry and the anomalous chromoelectric and chromomagnetic moments in pp collisions at $\sqrt{s} = 13$ TeV. *JHEP* **2020**, *6*, 146. [CrossRef]
66. Frixione, S.; Nason, P.; Oleari, C. Matching NLO QCD computations with Parton Shower simulations: the POWHEG method. *JHEP* **2007**, *11*, 70. [CrossRef]
67. Alte, S.; Berge, S.; Spiesberger, H. Top quark charge asymmetry: searching for light axigluons in $t\bar{t}$ + jet production at the LHC. *JHEP* **2014**, *9*, 84. [CrossRef]
68. Berge, S.; Westhoff, S. Observing the Top Energy Asymmetry at the LHC. *Phys. Rev. D* **2017**, *95*, 014035. [CrossRef]
69. The ATLAS Collaboration. Measurement of the energy asymmetry in $t\bar{t}j$ production at 13 TeV with the ATLAS experiment and interpretation in the SMEFT framework. *Eur. Phys. J. C* **2022**, *82*, 374. [CrossRef]

70. The ATLAS Collaboration. Performance of top-quark and W-boson tagging with ATLAS in Run 2 of the LHC. *Eur. Phys. J. C* **2019**, *79*, 375. [CrossRef]
71. Alwall, J.; Frederix, R.; Frixione, S.; Hirschi, V.; Maltoni, F.; Mattelaer, O.; Shao, H.S.; Stelzer, T.; Torrielli, P.; Zaro, M. The automated computation of tree-level and next-to-leading order differential cross sections, and their matching to parton shower simulations. *JHEP* **2014**, *7*, 79. [CrossRef]
72. The ATLAS Collaboration. *Estimation of Non-Prompt and Fake Lepton Backgrounds in Final States with Top Quarks Produced in Proton-Proton Collisions at $\sqrt{s}$=8 TeV with the ATLAS Detector*; Technical Report; CERN: Geneva, Switzerland, 2014; All Figures Including Auxiliary Figures Are Available; Available online: https://atlas.web.cern.ch/Atlas/GROUPS/PHYSICS/CONFNOTES/ATLAS-CONF-2014-058 (accessed on 20 September 2022).
73. Grzadkowski, B.; Iskrzynski, M.; Misiak, M.; Rosiek, J. Dimension-Six Terms in the Standard Model Lagrangian. *JHEP* **2010**, *10*, 85. [CrossRef]
74. Maltoni, F.; Mangano, M.L.; Tsinikos, I.; Zaro, M. Top-quark charge asymmetry and polarization in $t\bar{t}W^{\pm}$ production at the LHC. *Phys. Lett. B* **2014**, *736*, 252–260. [CrossRef]
75. Bevilacqua, G.; Bi, H.Y.; Hartanto, H.B.; Kraus, M.; Nasufi, J.; Worek, M. NLO QCD corrections to off-shell $t\bar{t}W^{\pm}$ production at the LHC: Correlations and asymmetries. *Eur. Phys. J. C* **2021**, *81*, 675. [CrossRef]
76. Broggio, A.; Ferroglia, A.; Frederix, R.; Pagani, D.; Pecjak, B.D.; Tsinikos, I. Top-quark pair hadroproduction in association with a heavy boson at NLO+NNLL including EW corrections. *JHEP* **2019**, *8*, 39. [CrossRef]
77. The ATLAS Collaboration. *Search for Leptonic Charge Asymmetry in $t\bar{t}W$ Production in Final States with Three Leptons at $\sqrt{s}$ = 13 TeV*; Technical Report; CERN: Geneva, Switzerland, 2022; All Figures Including Auxiliary Figures Are Available; Available online: https://atlas.web.cern.ch/Atlas/GROUPS/PHYSICS/CONFNOTES/ATLAS-CONF-2022-062/ (accessed on 20 September 2022).
78. Bothmann, E.; et al. Event Generation with Sherpa 2.2. *SciPost Phys.* **2019**, *7*, 34. [CrossRef]
79. The ATLAS Collaboration. *Measurement of the Charge Asymmetry in Top Quark Pair Production in Association with a Photon with the ATLAS Experiment*; Technical Report; CERN: Geneva, Switzerland, 2022; All Figures Including Auxiliary Figures Are Available; Available online: https://atlas.web.cern.ch/Atlas/GROUPS/PHYSICS/CONFNOTES/ATLAS-CONF-2022-049/ (accessed on 20 September 2022).

universe MDPI

Review

Flavour-Changing Neutral Scalar Interactions of the Top Quark

Nuno Filipe Castro [1] and Kirill Skovpen [2,*]

[1] Laboratório de Instrumentação e Física Experimental de Partículas (LIP), Departamento de Física, Escola de Ciências, Universidade do Minho, 4710-057 Braga, Portugal
[2] Department of Physics and Astronomy, Faculty of Sciences, Ghent University, Sint-Pietersnieuwstraat 33, 9000 Gent, Belgium
* Correspondence: kirill.skovpen@cern.ch

Abstract: A study of the top-quark interactions via flavour-changing neutral current (FCNC) processes provides an intriguing connection between the heaviest elementary particle of the standard model (SM) of particle physics and the new scalar bosons that are predicted in several notable SM extensions. The production cross sections of the processes with top-scalar FCNC interactions can be significantly enhanced to the observable level at the CERN Large Hadron Collider. The present review summarises the latest experimental results on the study of the top-quark interactions with the Higgs boson via an FCNC and describes several promising directions to look for new scalar particles.

Keywords: top quark; FCNC; Higgs boson; scalar; LHC; new physics

1. Introduction

Conservation laws and flavour-symmetry structures represent the core element of any theoretical model that provides a description of interactions involving elementary particles. An experimental study of fundamental interactions is an excellent probe of higher-order symmetries, potentially leading to the construction of a more complete model of nature, resolving the remaining unanswered questions of the remarkably successful standard model (SM). Flavour-violating processes in the quark sector in the electroweak interactions are allowed through the charged weak currents. Such flavour-changing transitions proceed via an exchange of a W boson between the two fermionic states. The weak eigenstates are treated as left-handed doublets, allowing transitions between the up- and down-type quarks, while the mass eigenstates are represented by a superposition of weak eigenstates connected via a unitary matrix. The rotation from one type of state to another is then expressed as the Cabibbo–Kobayashi–Maskawa (CKM) matrix which governs the flavour-mixing processes through the flavour-changing charged weak transitions [1]. The processes, where a fermion changes its flavour via an exchange of a neutral boson, are therefore absent at the tree level in the SM due to the unitarity of the rotational matrices and are called the flavour-changing neutral currents (FCNC) [2].

The effect of flavour mixing in the quark sector was first introduced using a three-quark model that only included the u, d, and s quarks [3]. The experimental studies of the $K_L \to \mu^+\mu^-$ decays and neutral kaon-mixing processes, however, indicated important difficulties in satisfying the theoretical predictions for the FCNC transitions [4]. The problem was theoretically solved in the 1970s by introducing the fourth type of quark, the charm (c) quark, in order to restore the quark–lepton symmetry of the weak interaction. It was shown that an additional contribution associated with an exchange of a c quark at the one-loop level almost completely cancels the respective contributions connected to the lighter quarks. This effect leads to a significant suppression of FCNC transitions at higher orders—the Glashow–Iliopoupos–Maiani (GIM) mechanism. The discovery of the c quark, just a few years later, confirmed these theoretical speculations [5,6]. The four-quark model was later extended to include five quark flavours, after the discovery of the bottom (b)

Citation: Castro, N.F.; Skovpen, K. Flavour-Changing Neutral Scalar Interactions of the Top Quark. *Universe* **2022**, *8*, 609. https://doi.org/10.3390/universe8110609

Academic Editors: Efe Yazgan and Suyong Choi

Received: 18 October 2022
Accepted: 16 November 2022
Published: 21 November 2022

quark [7]. It took a bit longer for the top (t) quark to be experimentally observed in 1995, completing the SM to contain six quark flavours [8,9]. In a full representation of the quark sector, the tree-level transitions between different quark flavours are only allowed through the weak flavour-changing charged interaction, while the tree-level FCNC transitions are completely missing in the SM and are only possible as loop corrections.

The FCNC effects are predicted in the leptonic sector as lepton flavour-violating transitions. However, the branching fraction of such processes is expected at the level of $\simeq 10^{-54}$, in the case of the $\mu \to e\gamma$ decay, due to an extreme suppression from the neutrino mass difference to the power of four, and is experimentally inaccessible [10–12]. The FCNC transitions in the decays of the hadronic states with s, c, or b quarks are observed experimentally [13–18]. The studies of these processes are, however, affected by the large uncertainties in the theoretical calculations of the branching ratios of the hadron decays, mainly driven by the non-perturbative long-distance strong interaction contributions.

The lifetime of the top quark ($\tau_t \simeq 5 \times 10^{-25}$ s), which is shorter than the typical formation time of the bound states ($\tau_{had} = 1/\Lambda_{QCD} \simeq 10^{-24}$ s), makes the processes with the top-quark production an excellent probe to search for FCNC effects. The absence of hadronic activity leading to the formation of bound states involving top quarks makes the study of the FCNC processes less affected by radiative QCD corrections. The FCNC effects can be probed in the top-quark production processes, as well as in the decays of the top quarks. The amplitude of an FCNC transition is proportional to the squared mass of the quark involved in the loop diagram. A remarkable suppression of the top-quark FCNC decays is explained by the fact that the only possible one-loop contributions are associated with the lighter quarks, leading to the branching fractions of $\mathcal{B}(t \to cX) \simeq 10^{-15} - 10^{-12}$ [19], where X represents either a gluon (g), photon (γ), Z, or a Higgs boson (h). Theoretical predictions for the top-quark FCNC effects are available with the next-to-leading order (NLO) precision [20,21], as well as the approximate next-to-next-to-next-to-leading order calculations for some of these processes [22,23].

The study of the flavour structure of the SM is one of the strongest probes of the beyond-the-SM (BSM) theories. A strong suppression of the top-quark FCNC transitions is a perfect condition to search for various possible deviations from the SM predictions. Several experimental studies of the properties of the FCNC decays of b hadrons have sparked a series of intriguing anomalies in the measured probabilities of the rare $b \to s\ell^+\ell^-$ FCNC transitions, as well as in the measurements of the ratios $\mathcal{B}(B^+ \to K^+\mu^+\mu^-)/\mathcal{B}(B^+ \to K^+e^+e^-)$ [24], $\mathcal{B}(B^0 \to K^{*0}\mu^+\mu^-)/\mathcal{B}(B^0 \to K^{*0}e^+e^-)$ [25], as well as the branching fractions [26–28]. A common analysis of these results reveals a potential tension with respect to the SM [29–33]. The experimental searches for FCNC effects in the top-quark sector therefore represent an important channel to probe the anomalous interactions of the third-generation quarks.

2. Experimental Studies of the Top-Quark FCNC Processes

The top-quark FCNC effects can be probed directly in the production of a single top-quark, as well as in the top-quark decays, as shown in Figure 1. Studies of the top-quark FCNC decays are typically associated with similar sensitivities to the top-quark FCNC couplings with an up and a charm quark. Experimental sensitivities to these couplings mainly differ in terms of the performance of various reconstruction methods used for the identification of hadronic jets originating from quarks of a different flavour. At hadron colliders, the single top-quark FCNC production process is mostly sensitive to the top-quark FCNC coupling with an up quark (or an up antiquark) due to an enhanced sensitivity due to the proton distribution function of the colliding protons (or antiprotons). The importance of these two production channels depends on a specific type of the top-FCNC coupling that is probed in an experiment.

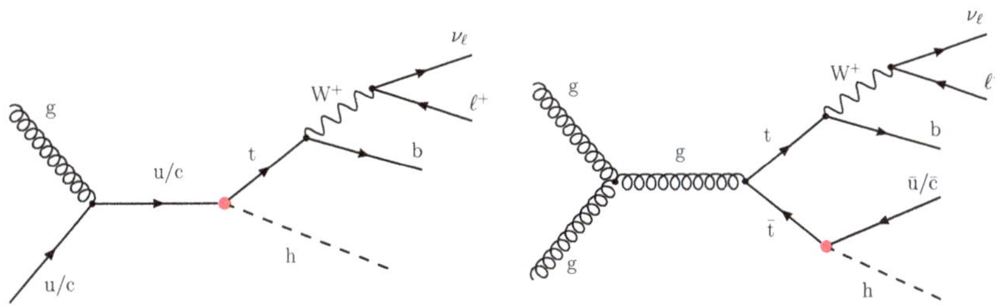

Figure 1. Representative leading-order Feynman diagrams for (**left**) single top quark and (**right**) top-quark pair production processes, involving top-Higgs FCNC couplings. The case of leptonic decays of the *W* boson is shown.

Before the LHC, the top-FCNC couplings were studied in electron–positron collisions at LEP2 [34–37], in deep inelastic scattering processes at HERA [38–42], and in proton–antiproton collisions at Tevatron [43–46]. The electron–positron colliders allow for a study of the top-γ and top-Z couplings in the processes with the production of a single top quark, $\mathrm{e^+e^-} \to \mathrm{t\bar{c}(\bar{u})}$. The study of the deep inelastic scattering of electrons on protons has an enhanced sensitivity to the same type of couplings in the processes of $\mathrm{ep} \to \mathrm{et} + \mathrm{X}$, as well as to the top-gluon FCNC couplings in the $\mathrm{ep} \to \mathrm{etq(g)} + \mathrm{X}$ processes. The obtained experimental constraints were recently improved by almost one order of magnitude after the analysis of the LHC proton–proton collision data [47–56].

The top-Higgs FCNC transitions receive the largest suppression in the SM with respect to the other top-quark FCNC processes because of the large mass of the Higgs boson. These transitions are among the rarest processes predicted in the SM in the quark sector, and therefore, the study of these processes is associated with a generally enhanced sensitivity to potential new physics effects. The discovery of the Higgs boson at the LHC paved a way to a comprehensive study of the top-Higgs FCNC processes at the ATLAS and CMS experiments, which resulted in the first experimental constraints on these anomalous couplings [57–65]. The direct searches for the top-Higgs FCNC effects are performed in top-quark decays, as well as in the associated production of single top quarks with a Higgs boson. Many of the performed studies were targeting the top-quark FCNC decays in $\mathrm{t\bar{t}}$ events. In recent studies of the 13 TeV data, the analysis of the single top-quark-associated production with a Higgs boson was also included [62–65].

2.1. $\mathrm{h} \to \gamma\gamma$

Search channels that are relevant to the top-Higgs FCNC couplings are usually defined based on the Higgs boson decay channels. The Higgs boson decays to pairs of photons provide a clean experimental signature to look for the top-Higgs FCNC effects. In addition to the two photons, these final states consist of up to one isolated lepton with additional hadronic jets. The analysis strategy is primarily based on the reconstruction of the Higgs boson diphotonic invariant mass. The contributions from various background processes are fitted in the mass sidebands in the data, followed by its extrapolation to the signal region. In these fits, the background contributions that are associated with the SM Higgs boson production must be accounted for, representing one of the dominant resonant backgrounds in the search region. The uncertainty associated with the choice of the fit function, the statistical uncertainty in the data, as well as the background contributions from the SM processes involving the Higgs boson represent the main uncertainties in the study of these final states.

The searches for top-Higgs FCNC processes in the $\mathrm{h} \to \gamma\gamma$ channel were carried out by ATLAS [59] and CMS [63] in the single-lepton and hadronic final states, including a pair of photons (Figure 2). The integrated luminosity of the recorded 13 TeV data corresponds to 36 and 137 $\mathrm{fb^{-1}}$, respectively. The identification of isolated photon objects and the common vertex of the photon pair is the core part of the analysis. The photon and the common vertex identification algorithms are based on the multivariate analysis (MVA) approaches. The obtained mass resolution allows the observation of a resonance structure in the diphoton invariant mass spectra in simulated signal events corresponding to the Higgs boson decay. The contributing nonresonant background processes include the diphoton production with jets, as well as the top-quark pair and the vector boson production processes with additional photons. The SM production of the Higgs boson represents the dominant resonant background. The nonresonant backgrounds are estimated directly from the data by performing a fit to the reconstructed diphoton invariant mass spectrum. The fitted function represents the sum of a double-sided Crystal Ball function that corresponds to the signal prediction, the resonant background from the SM Higgs production, and a parameterised function describing the nonresonant background obtained in a data control region. The main uncertainties include the b tagging and jet energy corrections, as well as the photon identification systematic uncertainties. The uncertainty in the limited number of events in the data also represents an important limiting factor in the final sensitivity in these searches. An additional contribution to the total systematic uncertainty is associated with theoretical uncertainties in the prediction of the resonant background processes with the SM Higgs boson production. The unbinned likelihood fit to the data using the described signal and background diphoton mass spectra is performed, and the constraints are set on the top-quark FCNC decay branching fractions. The observed (expected) limits obtained by ATLAS are $\mathcal{B}(\mathrm{t} \to \mathrm{hc}) < 2.2 \times 10^{-3}$ (1.6×10^{-3}) and $\mathcal{B}(\mathrm{t} \to \mathrm{hu}) < 2.4 \times 10^{-3}$ (1.7×10^{-3}). The observed (expected) constraints obtained in the CMS analysis are $\mathcal{B}(\mathrm{t} \to \mathrm{hc}) < 7.3 \times 10^{-4}$ (5.1×10^{-4}) and $\mathcal{B}(\mathrm{t} \to \mathrm{hu}) < 1.9 \times 10^{-4}$ (3.1×10^{-4}). An enhanced sensitivity obtained in the CMS analysis is explained by a larger data sample used in the study, as well as due to the inclusion of the top-Higgs FCNC process with the associated production of a single top quark and a Higgs boson. The latter has led to an improved sensitivity to $\mathcal{B}(\mathrm{t} \to \mathrm{hu})$.

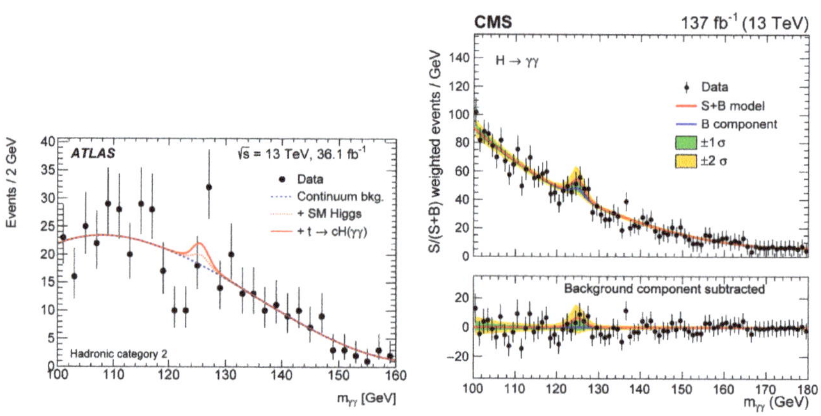

Figure 2. Distributions with the invariant diphoton mass showing the results of the fit to data in the top-Higgs FCNC study of the $\mathrm{h} \to \gamma\gamma$ channel at (**left**) ATLAS [59] and (**right**) CMS [63]. The ATLAS results are presented for hadronic final states, while the CMS results include a combination of all considered channels with events weighted by the associated significance of each event category.

2.2. $\mathrm{h} \to \mathrm{WW/ZZ}/\tau\tau$

Multilepton final states arise from the Higgs boson decays to a pair of W or Z bosons, as well as to τ leptons. The event categories in these studies are associated with the final states with two same-sign or three selected leptons. The same-sign lepton channel has dominant background contributions originating from processes with nonprompt and misidentified leptons, while the three-lepton channel is mainly affected by the presence of diboson events as well as nonprompt leptons. These backgrounds are estimated from the data. The search channels involving one hadronic τ lepton identified in the Higgs boson decay receive the dominant background contributions from the processes with misidentified τ lepton decays, as well as from events with the SM production of top quarks. In the case when the decays of both τ leptons result in hadronic final states, a significant background contribution is also associated with the Z boson decays to the pairs of τ leptons.

The searches for top-Higgs FCNC couplings in the multilepton channels were performed at ATLAS [58] and CMS [61] using 36 fb^{-1} of 13 TeV and 20 fb^{-1} of 8 TeV data, respectively. The events are split into the final states with two same-sign (2lSS) and three (3l) leptons. The dominant backgrounds are associated with the nonprompt and misidentified leptons, as well as with the leptons originating from photon conversions. The prompt-lepton backgrounds correspond to events with an associated production of top-quark pairs and a W, a Z, or a Higgs boson, with additional contributions arising from the processes with diboson production. The baseline selection criteria require the presence of two or three leptons and at least two jets, with one or two b-tagged jets. The prompt lepton identification plays an important role in these studies in suppressing the dominant nonprompt lepton backgrounds. The rejection of nonprompt leptons is usually achieved through an application of an MVA approach, which exploits a number of kinematic variables that provide a separation power between the two lepton categories, such as lepton isolation and properties of the reconstructed jet in proximity of the selected lepton. The statistical and systematic uncertainties associated with the prediction of the backgrounds with nonprompt leptons are among the dominant uncertainties in these searches. Two separate boosted decision tree (BDT) discriminants involving various reconstructed kinematic variables are trained in the 2lSS and 3l channels to further suppress various backgrounds. The BDT distributions that are presented in Figure 3 are used in a binned maximum likelihood fit to extract the constraints on the top-Higgs FCNC processes. The observed (expected) 95% CL limits on the top-quark FCNC branching fractions in the multilepton final states $\mathcal{B}(\text{t} \to \text{hc}) < 1.6 \times 10^{-3}$ (1.5×10^{-3}) and $\mathcal{B}(\text{t} \to \text{hu}) < 1.9 \times 10^{-3}$ (1.5×10^{-3}) are obtained. Multilepton searches provide an excellent sensitivity to the top-Higgs FCNC couplings; however, the existing results use only a partial data set, and further updates on these studies are anticipated in the future.

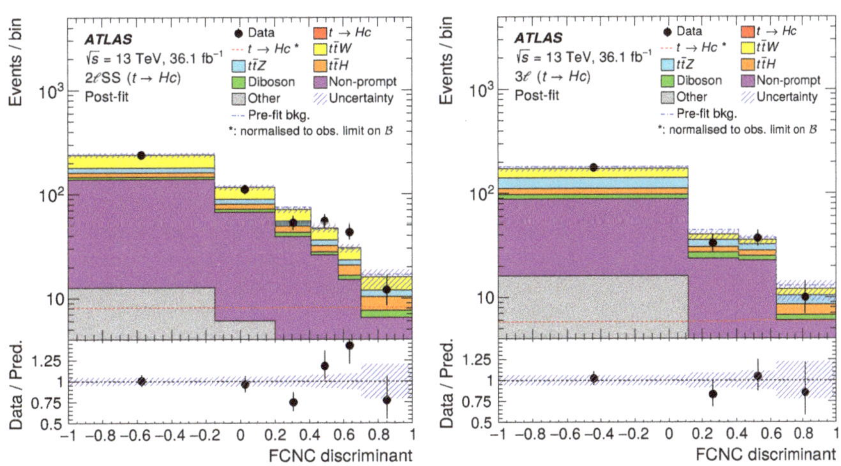

Figure 3. Distributions of the BDT discriminants using the top-Higgs FCNC signal selection in multilepton search channels in the same-sign dilepton (**left**) and trilepton (**right**) final states [58]. The presented BDT discriminant was optimised for the case of the t $\to$ hc FCNC decays.

A dedicated study of the top-Higgs FCNC effects in the final states with one or two hadronically decaying τ leptons was recently performed by ATLAS with 139 fb^{-1} of 13 TeV data [65]. The analysis strategy is similar to the one used in the previous analysis [57], with an increased number of kinematic regions sensitive to the signal production, in order to account for the single-top production channel for the top-Higgs FCNC. The dominant background in this search is associated with the presence of nonprompt τ leptons, estimated from the data. Other backgrounds are predicted by simulation. The obtained constraints on the branching fractions are $\mathcal{B}(\mathrm{t} \to \mathrm{hc}) < 9.4 \times 10^{-4}$ (4.8×10^{-4}) and $\mathcal{B}(\mathrm{t} \to \mathrm{hu}) < 6.9 \times 10^{-4}$ (3.5×10^{-4}).

2.3. $\mathrm{h} \to \mathrm{b\bar{b}}$

The Higgs boson decays to a pair of b quarks with the largest branching fraction of $\simeq$58% [66]. A considerable amount of background events is associated with the $\mathrm{t\bar{t}}$ production with additional hadronic jets. The analysis of this channel is systematically limited with the dominant contributions to the total uncertainty arising from the application of the heavy flavour jet identification techniques, as well as the modelling uncertainties relevant to the predictions of the top-quark production processes with additional jets. One of the important handles for suppressing background processes is the kinematic event reconstruction involving top quarks and additional jets. The assignment of the reconstructed final-state objects to the initial hard-process particles is performed using the MVA methods.

The top FCNC search in the $\mathrm{h} \to \mathrm{b\bar{b}}$ channel is performed in the final states with one isolated lepton and additional jets. The total integrated luminosity used in the ATLAS analysis corresponds to 36 fb^{-1} of 13 TeV data [57]. The CMS results use 101 fb^{-1} of data [64], additionally combined with the previously published result from the analysis of 36 fb^{-1} data [62]. The ATLAS analysis focuses on the study of the event topology with at least four jets in the final state, mainly relevant to the top-quark FCNC decays. The corresponding CMS analysis additionally includes the signal top-quark production mode of the signal events, and therefore, the requirement on the minimum number of reconstructed jets is set to a lower value. At least three b-tagged jets are required to be present in the event. In both analyses, the selected events are classified based on the number of jets and b-tagged jets. The dominant background contributions correspond to the top-quark pair production in association with light-flavour jets in the event categories with two b-tagged jets, while the associated production of top-quark pairs with heavy-flavour jets ($\mathrm{t\bar{t}b\bar{b}}$ and $\mathrm{t\bar{t}c\bar{c}}$) represents the dominant background in the case of the higher number of b-tagged jets. The theoretical predictions for these processes are subject to relatively large uncertainties due to the renormalisation and the factorisation scale variations arising from the different energy scales of the top-quark mass and the jet transverse momentum involved in the generation process, as well as the inclusion of heavy quark masses in the calculations [67]. The experimental uncertainties in the measurement of the production cross sections of these processes reach $\simeq$10–20% [68–71]. The background processes are further suppressed by using the discriminants that exploit the kinematic information of the selected reconstructed objects, defining the probability of an event to correspond to the signal process hypothesis. As shown in Figure 4, in the ATLAS analysis, this is performed by constructing the likelihood (LH) discriminant, while the BDT approach is used in the case of the CMS search. The binned maximum likelihood fits are performed to the data based on the described discriminants to extract the limits on the FCNC contributions, resulting in the observed (expected) 95% CL constraints on the top-quark FCNC branching fractions of $\mathcal{B}(\mathrm{t} \to \mathrm{hc}) < 4.2 \times 10^{-3}$ (4.0×10^{-3}) and $\mathcal{B}(\mathrm{t} \to \mathrm{hu}) < 5.2 \times 10^{-3}$ (4.9×10^{-3}). The resultant constraints in the CMS analysis are $\mathcal{B}(\mathrm{t} \to \mathrm{hc}) < 9.4 \times 10^{-4}$ (8.6×10^{-4}) and $\mathcal{B}(\mathrm{t} \to \mathrm{hu}) < 7.9 \times 10^{-4}$ (1.1×10^{-3}). The differences in the sensitivities in the published results by the two experiments are mainly due to the different size of the analysed data sample, as well as to the inclusion of the single top-quark production mode for the top-Higgs FCNC process in the case of the CMS analysis. A combination of the results obtained from the analyses of different Higgs boson decay channels, $\mathrm{h} \to \gamma\gamma$, $\mathrm{h} \to \mathrm{WW/ZZ}$, $\mathrm{h} \to \tau\tau$, and $\mathrm{h} \to \mathrm{b\bar{b}}$, was performed at ATLAS using 36 fb^{-1} of data, corresponding to the limits of $\mathcal{B}(\mathrm{t} \to \mathrm{hc}) < 1.1 \times 10^{-3}$ (8.3×10^{-4}) and $\mathcal{B}(\mathrm{t} \to \mathrm{hu}) < 1.1 \times 10^{-3}$ (8.3×10^{-4}) [57]. The ATLAS constraints on the top-Higgs FCNC interactions are competitive to the ones obtained in the analysis of the $h \to \gamma\gamma$ channel at CMS [63], which already uses all the available recorded data at 13 TeV.

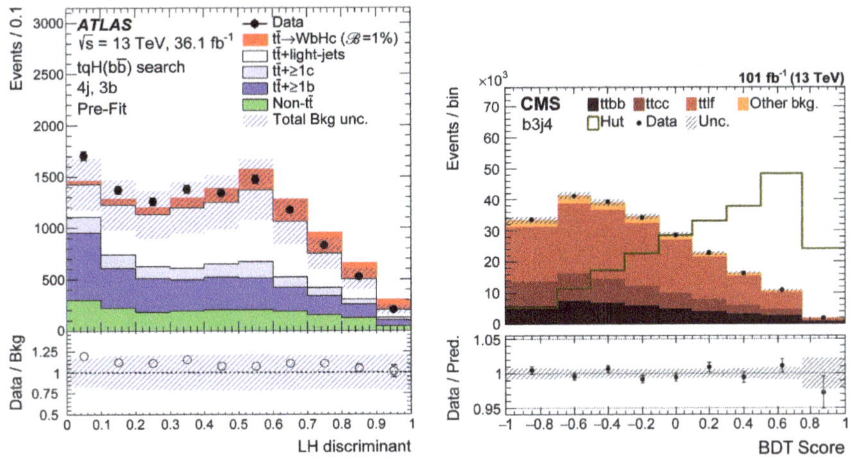

Figure 4. Distributions of the discriminants used in the analyses of the $h \to b\bar{b}$ channel at (**left**) ATLAS [57] and (**right**) CMS [64]. Selected events correspond to the final states with four reconstructed jets, three of which are identified as associated with heavy-flavour hadron decays. The pre-fit and post-fit results are shown for ATLAS and CMS, respectively.

2.4. Indirect Searches

The top-Higgs FCNC interactions can be indirectly constrained from the studies of the SM processes that can potentially include FCNC loop-level contributions involving top quarks. The relevant processes include the hadron electric dipole moments [72,73], $Z \to c\bar{c}$ decays [74], and $D^0 - \overline{D^0}$ mixing [75]. The indirect limits are competitive with the current direct constraints obtained at the LHC [76].

3. Global Approach to the FCNC Searches

A broad range of experimental searches for new physics phenomena have been using the κ-framework to parameterise the potential deviations from the SM predictions [77]. This framework defines a set of scaling factors for production cross sections and decay widths as a function of the new physics model parameters. While the κ-framework has proved to be very successful in theoretical interpretations of a large number of experimental results, it does not represent an ultimate approach to providing a complete systematic description of various new physics effects.

Given the absence of any strong evidence of new physics, the natural assumption is that new particles are much heavier than the SM particles, and its direct production at present is not achievable within the LHC energy range. The potentially induced new physics effects at the electroweak scale can be parameterised with a general effective field theory (EFT) approach that includes additional high-energy dimensional operators in the extended Lagrangian of the SM (SMEFT). The rich phenomenology of the SMEFT includes 59 independent operators, assuming baryon number conservation [78,79]. A full categorisation of EFT operators relevant to the top-quark sector and its interplay with other SM processes is summarised in refs. [80–82]. Several of these operators are relevant to the FCNC processes with top quarks. The Wilson coefficients (WCs) of the respective EFT operators can be constrained from the measured production cross sections, as well as from the study of the shapes of various kinematic variables. An EFT analysis of experimental observables represents a general approach to study potential deviations from the SM predictions that can be used to set constraints on various BSM models. The top-quark

FCNC EFT couplings comprise several dimension-six operators, which are discussed in refs. [81–85]. The FCNC EFT contributions can also interfere at higher orders [83,84,86].

The potential FCNC EFT effects with top quarks were probed in the experimental studies of top-gluon and top-photon FCNC processes at the LHC [56,87]. Additionally, several of the obtained experimental constraints on the FCNC top-quark decay branching fractions and strength κ-modifiers, which were described in Section 2, can be directly translated into the corresponding limits on the relevant WCs. The first direct measurement of the constraints on the EFT WCs relevant to the top-Higgs FCNC interactions was recently performed in ref. [65]. The re-interpretations of various experimental results that are sensitive to the top-Higgs EFT operators are also available [80,84].

4. New Scalar Bosons

Several extensions of the SM can induce sizable FCNC effects that can be experimentally probed at the LHC. There are two possible ways to introduce the top-quark FCNC in a BSM model. The first possibility is to increase the number of fermions, modifying the CKM matrix structure, escaping the GIM suppression. This approach is usually referred to as the minimal flavour violation (MFV). The second option is to involve new heavy particles in the loops of the higher-order diagrams, increasing the probability of FCNC transitions. The study of top-Higgs FCNC effects appears to be rather promising in various simple extensions of the SM, where the additional neutral scalar particles can potentially mix with the SM Higgs boson.

A dedicated estimate of the BSM-enhanced branching fractions of $t \to u(c)h$ decays shows the maximal values reaching $10^{-3} - 10^{-4}$ in some BSM models [19,88]. Such high event rates are being probed at the LHC using recorded data with the typical constraints set at the level of $\simeq 10^{-3}$. However, the maximal branching fractions are not necessarily associated with the most favourable parameter space of a BSM model and can potentially involve additional model tuning.

The SM Higgs boson can have its additional partners in various BSM scenarios. The two-Higgs doublet model (2HDM) is one of the simplest extensions of the SM that introduces two Higgs doublets with five scalar particles: h^0, CP-odd A^0, CP-even H^0 ($m_H > m_h$), and $H^{\pm}$, where h^0 is the lightest CP-even SM-like Higgs boson [89–91]. The 2HDM contains seven parameters, with only two of them relevant at the leading order (LO), usually defined as $\cos(\beta - \alpha)$ and $\tan(\beta)$. The former parameter is related to the couplings of a scalar particle to vector bosons, while the latter represents the ratio of the vacuum expectation values of the heavy and the SM Higgs bosons.

There are four (I, II, III, and IV) types of 2HDM. The 2HDM-I and 2HDM-II do not include FCNC processes at the tree level due to the requirement of flavour conservation via the presence of a Z_2 symmetry. In these two types of 2HDM, all the fermions couple to the same Higgs boson. In the 2HDM-III without an imposed discrete symmetry, the fermions can couple to both Higgs doublets, and the tree-level FCNC transitions involving the top-charm FCNC couplings with the Higgs boson can be significantly enhanced [92]. A combined fit of various results from the direct and indirect experimental searches favours the alignment limit $\cos(\beta - \alpha) \to 0$ with $\cos(\beta - \alpha) < 0.1 - 0.4$, with some additional dependence on $\tan(\beta)$, the mass of the scalar boson, as well as the type of the model [93–105]. The alignment scenario corresponds to the case, when h^0 and the SM Higgs boson share the same couplings.

In the aligned two-Higgs doublet model (A2HDM), it is assumed that both Yukawa matrices are aligned in the flavour space to avoid the FCNC at the tree level [96]. The enhanced one-loop-induced $t \to ch$ decays can occur in such models [97]. Special extensions of the 2HDM models can incorporate the FCNC at the tree level, such as the top-quark 2HDM (T2HDM) [98]. In this model, it is assumed that the top quark is the only elementary fermion that couples to the non-SM Higgs doublet to generate its large mass in a natural way, therefore allowing the top-Higgs FCNC due to a small $\cos(\beta - \alpha)$ admixture of the exotic neutral Higgs boson. The study of the $t \to ch$ decays represents a promising channel to probe the T2HDM and 2HDM-III at the LHC [99–102]. There are also the so-called BGL modifications of the 2HDM, where the tree-level top-Higgs FCNC transitions can be associated with either up- or down-type quarks, preserving the structure of the CKM matrix [103–105].

Additional Higgs doublets can naturally appear in the context of supersymmetric (SUSY) theories. The minimal supersymmetric standard model (MSSM) is the simplest extension of the SM representing the 2HDM-II with additional supersymmetric particle content [106–108]. At the tree level, this model contains two non-SM parameters: the mass of the CP-odd Higgs boson, m_A, and $\tan(\beta)$. An effective MSSM with the lightest CP-even SM Higgs boson is referred to as the hMSSM, where the properties of the SM Higgs boson define the remaining masses and couplings of the MSSM [91,109,110]. This approximation of MSSM is only completely valid at the moderate values of $\tan(\beta)$. The recent LHC experimental searches generally disfavour small values of m_A below $\simeq$600 GeV within the hMSSM [93–105]. The predicted top-Higgs FCNC rates in a general MSSM can reach 10^{-7} [111], while in the case of the R-parity violation (RPV) in a general SUSY model, these transitions can be enhanced to 10^{-5} [112].

An extended MSSM with baryon (B) and lepton (L) numbers as local symmetries, broken near the electroweak scale, is known as BLMSSM [113–116]. This model can incorporate an enhancement of $t \to ch$ rates at one loop [117]. The next-to-minimal supersymmetric standard model (NMSSM) represents an extension of the MSSM that naturally generates the mass parameter μ in the Higgs superpotential at the electroweak scale and resolves the so-called μ-problem [118,119]. The new neutral scalars considered in the MSSM theories can potentially mix with the SM Higgs boson and therefore generate top-Higgs FCNC at the tree level.

The addition of the exotic vector-like quark to the CKM matrix provides a way to escape the GIM mechanism. The top-Higgs FCNC transitions can be enhanced to 10^{-5} in the quark singlet (QS) [120] and alternative left-right models (ALRM) [121]. Similar enhancements can be achieved in the Littlest Higgs Model with T-parity (LHT) induced by interactions with the new T-odd gauge bosons and fermions [122]. The presence of the Kaluza–Klein fermion states in the Randall–Sundrum (RS) models with warped extra dimensions can produce sizable FCNC effects of the same order [123–125].

The new light neutral scalar singlets (S) are present in various supersymmetric extensions of the SM, including the NMSSM and the composite Higgs models (CHM) [126–130], with $t \to cS$ tree-level FCNC decays [131]. In such extensions, these scalars are considered as Nambu–Goldstone bosons (pNGBs) with the Higgs boson, representing a bound state of new strongly interacting dynamics. The large mass of the top quark can be generated through the mixing of elementary fermions with a composite operator of a high scaling dimension [132]. In CHM, the SM elementary particles can be seen as composite states that mix with its heavy partners. This model provides a promising explanation of the mass hierarchy of the SM by introducing a new physics scale and the idea of compositeness of the SM particles. The $t \to cS$ decays, with $m_S < m_t - m_c$, are expected to strongly dominate over the $t \to ch$ transitions in the CHM, providing a new window to constrain the new physics models via the top-quark FCNC searches with neutral light scalars. These processes are not yet studied experimentally. The predicted rates of the $t \to cS$ decays can be probed down to 10^{-5} with the existing LHC data [129,130].

In addition to the top-quark FCNC decays with a Higgs boson, one can also search for FCNC decays of the heavy neutral scalars (H) predicted in many BSM models. In a general 2HDM model, as well as in its extensions, such as the T2HDM, the probability of the $H \to t\bar{c}$ decay is proportional to $\sin(\beta - \alpha)$, while the probabilities of the $t \to ch$ decays are proportional to $\cos(\beta - \alpha)$ [133–137]. This represents an important complementarity of the top FCNC searches in the top-quark and heavy neutral scalar decays. At high energies, one of the dominant decays of the heavy scalars is the production of two top quarks, if the mass of the scalar particle exceeds the doubled mass of the top quark. However, in the heavy scalar mass range of 175 and 350 GeV, the $H \to t\bar{c}$ decays are associated with the largest branching fraction in the model parameter space, favoured by the current experimental constraints [138]. A study of the $H \to t\bar{c}$ decays is a promising way to search for heavy neutral scalar particles at the LHC [133–137]. The dominant production mode for heavy scalars is expected to be the gluon–gluon fusion process; however, in the context of the "flavorful" 2HDM (F2HDM), that removes the 2HDM-intrinsic Z_2 discrete symmetry and additionally modifies the structure of the Yukawa matrices [139–141], the dominant channel is the single top-associated production with a heavy neutral scalar ($pp \to tH \to tt\bar{c}$), resulting in the presence of two same-sign top quarks in the final state [142]. Similar final states are relevant to the top-quark FCNC searches within the T2HDM [133–137]. The searches for the heavy scalar FCNC decays with top quarks are also proposed within the Froggatt–Nelsen singlet model (FNSM), mostly relevant for the HL-LHC data analysis [143].

5. Future Perspectives

The LHC has accumulated about 25 fb^{-1} of proton–proton collision data at 7 and 8 TeV, as well as nearly 140 fb^{-1} of data at 13 TeV. The latest studies from the LHC on the top-quark FCNC processes therefore focus on the analysis of the 13 TeV data. The next round of the data taking at the LHC is planned for 2022, where it is expected that the total accumulated statistics will be doubled, reaching 300 fb^{-1} by the end of the LHC project. The future experiments at the High-Luminosity LHC (HL-LHC) are planned to bring almost an order of magnitude larger data set of 3 ab^{-1} due to a significant increase in the instantaneous luminosity of the colliding proton beams up to 10^{35} cm^{-2} s^{-1}, representing a 5 to 7 times higher luminosity with respect to its nominal value. The projected sensitivities of the top-quark FCNC searches, following the preliminary estimates of the expected performances of the upgraded ATLAS and CMS detectors at the HL-LHC, indicate a significant improvement in the constraints on the branching fractions of the top-quark FCNC decays after the analysis of the full LHC statistics, reaching an order of magnitude [144].

There are several major international projects under consideration in the high-energy physics domain, defining an evolving strategy for this field for many years to come. The Large Hadron electron Collider (LHeC) is proposed as an extension of the LHC project, re-using the existing proton accelerator complex and combining it with a new electron accelerator for the production of 60 GeV electron beams for the study of the deep inelastic scattering at high energies [145–148]. The planned experiments at the LHeC are mostly sensitive to the top-γ/Z FCNC couplings, and the projected limits are expected to be comparable to the corresponding sensitivities at the HL-LHC [149–151]. The study of the top-Higgs FCNC interactions appears to be less promising at the LHeC, the expected sensitivity of which has been already surpassed by the latest LHC results [152].

The electron–proton collisions are considered as part of the Future Circular Collider (FCC) project, involving several experiments targeting different types of high-energy collisions. The FCC-eh machine will collide a 60 GeV electron with 50 TeV proton beams, produced by the FCC accelerator [153,154]. Due to the increased energy of the proton beams, relatively similar sensitivities for the top-Higgs FCNC couplings are expected to the ones of the HL-LHC. Most of the improvements are anticipated for the top-γ/Z FCNC couplings [144,155]. The planned experiments at the high-precision electron–positron collider (FCC-ee) will also be very sensitive to the top-γ/Z FCNC couplings [156]. The dominant sensitivity to the top-Higgs FCNC processes at the FCC-ee is mainly associated with the top-quark decay channels. The FCC-hh machine with proton–proton collisions at $\sqrt{s}$ = 100 TeV will allow increased sensitivity to all relevant top-quark FCNC couplings, probing the $t \to ch$ decay branching fractions down to $\simeq 10^{-5}$ [157–161]. The High-Energy LHC (HE-LHC) project will adopt the FCC-hh technology to use proton–proton collisions at $\sqrt{s}$ = 27 TeV. The HE-LHC is viewed as capable to improve the HL-LHC limits on the top-quark FCNC couplings by an order of magnitude [162,163]. The linear electron–position colliders, such as the ILC and CLIC, are also associated with good prospects for the top-quark FCNC studies [164–168]. However, the projected ILC/CLIC sensitivities for the top-Higgs FCNC interactions are not expected to reach the sensitivity level of the corresponding studies at the HL-LHC.

A summary of the described experimental results and future projections is presented in Figure 5. The analysis of the $\simeq$140 fb^{-1} of the LHC 13 TeV data allows reaching the 95% confidence level limits of the order of 10^{-4} and 10^{-3} for the $t \to uh$ and $t \to ch$ decay branching fractions, respectively. These experimental limits are obtained from the analysis of the the Higgs boson decays to photon pairs, and therefore, the presented results are expected to be further improved when combined with the results obtained in the analysis of other Higgs boson's decay modes. The illustrated sensitivities for future colliders are also obtained in the analyses of specific channels. The comparison with the ultimate sensitivities is expected to be more complete, once the LHC and the future projection results become available for all the relevant decay channels of the Higgs boson. Based on the considered projections, the best expected sensitivity of $\simeq 10^{-5}$ is associated with the experiments at the FCC-hh.

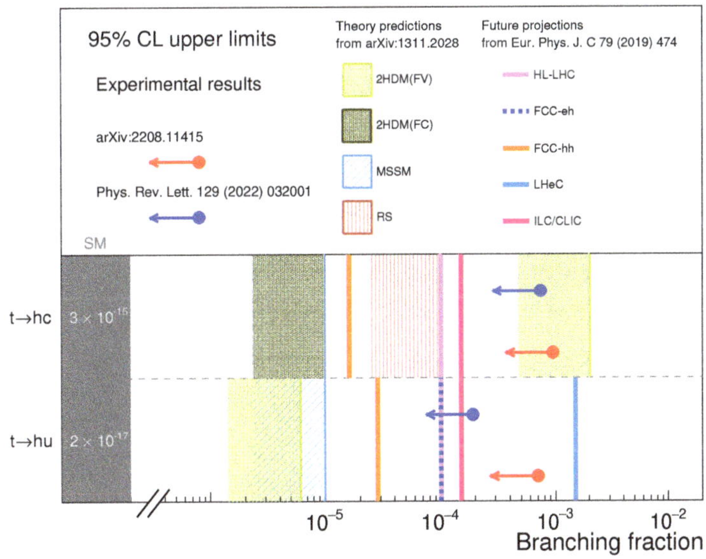

Figure 5. Summary of the best experimental constraints to date on the top-Higgs FCNC processes at the LHC [63,65], including sensitivity projections for future colliders. The results are also compared to various BSM predictions that correspond to the maximal expected branching fractions in a given model. Adapted from ref. [144].

The described sensitivities of the future experiments mainly correspond to the studies of the top-Higgs FCNC couplings, with only a few projections available for some of the top-quark flavour-changing neutral scalar processes. While the searches for new scalars via the top-quark FCNC appear to be highly relevant for the HL-LHC, as well as its successors, these processes are not yet explored to the full extent with the existing LHC data.

6. Summary

The sensitivity of the LHC experiments has reached the level of being able to rule out several BSM models assuming the maximal branching fractions of the top-quark FCNC decays. The study of the top-quark FCNC processes involving the neutral scalar bosons is an excellent probe of the new physics effects in a number of BSM scenarios, including additional Higgs doublets and scalar singlets through partial compositeness. In some models, the production of new scalars can be significantly enhanced by the flavour-changing neutral scalar couplings, and therefore, these studies represent a very promising direction to look for additional heavy and light partners of the discovered Higgs boson. Beyond the LHC, the upcoming experiments at the HL-LHC and FCC are expected to come with even better sensitivities to probe the top-quark anomalous couplings with the new scalars. The analysis of the LHC and the future collider data will remain the only way to directly probe the top-quark flavour-changing neutral scalar interactions in the next decades.

Funding: This research received no external funding.

Informed Consent Statement: Not applicable.

Data Availability Statement: Not applicable.

Acknowledgments: NC is supported by the FCT—Fundação para a Ciências e a Tecnologia, Portugal, under project CERN/FIS-PAR/0032/2021.

Conflicts of Interest: The authors declare no conflict of interest.

References

1. Kobayashi, M.; Maskawa, T. CP violation in the renormalizable theory of weak interaction. *Prog. Theor. Phys.* **1973**, *49*, 652. [CrossRef]
2. Glashow, S.L.; Iliopoulos, J.; Maiani, L. Weak interactions with lepton-hadron symmetry. *Phys. Rev. D* **1970**, *2*, 1285. [CrossRef]
3. Cabibbo, N. Unitary symmetry and leptonic decays. *Phys. Rev. Lett.* **1963**, *10*, 531. [CrossRef]
4. Maiani, L. The GIM Mechanism: Origin, predictions and recent uses. *arXiv* **2013**, arXiv:1303.6154.
5. Augustin, J.-E. et al. [SLAC Collaboration]. Discovery of a narrow resonance in e^+e^- annihilation. *Phys. Rev. Lett.* **1974**, *33*, 1406. [CrossRef]
6. Aubert, J.J. et al. [E598 Collaboration]. Experimental observation of a heavy particle. *J. Phys. Rev. Lett.* **1974**, *33*, 1404. [CrossRef]
7. Herb, S.W. et al. [E288 Collaboration]. Observation of a dimuon resonance at 9.5 GeV in 400 GeV proton-nucleus collisions. *Phys. Rev. Lett.* **1977**, *39*, 252. [CrossRef]
8. Abe, F. et al. [CDF Collaboration]. Observation of top quark production in $\bar{p}p$ collisions. *Phys. Rev. Lett.* **1995**, *74*, 2626. [CrossRef]
9. Abachi, S. et al. [D0 Collaboration]. Search for high mass top quark production in $p\bar{p}$ collisions at $\sqrt{s}$ = 1.8 TeV. *Phys. Rev. Lett.* **1995**, *74*, 2422. [CrossRef]
10. de Gouvêa, A.; Vogel, P. Lepton flavor and number conservation, and physics beyond the standard model. *Prog. Part. Nucl. Phys.* **2013**, *71*, 75. [CrossRef]
11. Celi, F.; Nicolo, D. Lepton flavour violation experiments. *Adv. High Energy Phys.* **2014**, *2014*, 282915. [CrossRef]
12. Calibbi, L.; Signorelli, G. Charged lepton flavour violation: An experimental and theoretical introduction. *Riv. Nuovo C.* **2018**, *41*, 71. [CrossRef]
13. Aubert, B. et al. [BaBar Collaboration]. Measurements of the $B \to X_s\gamma$ branching fraction and photon spectrum from a sum of exclusive final states. *Phys. Rev. D* **2005**, *72*, 052004. [CrossRef]
14. Nishimura, K. et al. [Belle Collaboration]. First measurement of inclusive $B \to X_s\eta$ decays. *Phys. Rev. Lett.* **2010**, *105*, 191803. [CrossRef] [PubMed]
15. Mohapatra, D. et al. [Belle Collaboration]. Observation of $b \to d\gamma$ and determination of V_{td}/V_{ts}. *Phys. Rev. Lett.* **2006**, *96*, 221601. [CrossRef]
16. Lin, S.-W. et al. [Belle Collaboration]. Observation of B decays to two kaons. *Phys. Rev. Lett.* **2007**, *98*, 181804. [CrossRef]
17. Lees, J.P. et al. [BaBar Collaboration]. Searches for rare or forbidden semileptonic charm decays. *Phys. Rev. D* **2011**, *84*, 072006. [CrossRef]
18. Batley, J.R. et al. [NA48/2 Collaboration]. Precise measurement of the $K^\pm \to \pi^\pm e^+e^-$ decay. *Phys. Lett. B* **2009**, *677*, 246. [CrossRef]
19. Aguilar-Saavedra, J.A. Top flavour-changing neutral interactions: Theoretical expectations and experimental detection. *Acta Phys. Polon. B* **2004**, *35*, 2695.
20. Zhang, J.J.; Li, C.S.; Gao, J.; Zhang, H.; Li, Z.; Yuan, C.-P.; Yuan, T.-C. Next-to-leading-order QCD corrections to the top-quark decay via model-independent flavor-changing neutral-current couplings. *Phys. Rev. Lett.* **2009**, *102*, 072001. [CrossRef]
21. Zhang, C.; Maltoni, F. Top-quark decay into Higgs boson and a light quark at next-to-leading order in QCD. *Phys. Rev. D* **2013**, *88*, 054005. [CrossRef]
22. Forslund, M.; Kidonakis, N. Soft-gluon corrections for single top quark production in association with electroweak bosons. *arXiv* **2019**, arXiv:1909.02619.
23. Guzzi, M.; Kidonakis, N. tZ' production at hadron colliders. *Eur. Phys. J. C* **2020**, *80*, 467. [CrossRef]
24. Aaij, R. et al. [LHCb Collaboration]. Search for lepton-universality violation in $B^+ \to K^+\ell^+\ell^-$ decays. *Phys. Rev. Lett.* **2019**, *122*, 191801. [CrossRef]

25. Aaij, R. et al. [LHCb Collaboration]. Test of lepton universality with $B^0 \to K^{*0}\ell^+\ell^-$ decays. *JHEP* **2017**, *8*, 55. [CrossRef]
26. Aaij, R. et al. [LHCb Collaboration]. Differential branching fractions and isospin asymmetries of $B \to K^{(*)}\ell^+\ell^-$ decays. *JHEP* **2014**, *6*, 133. [CrossRef]
27. Aaij, R. et al. [LHCb Collaboration]. Measurements of the S-wave fraction in $B^0 \to K^+\pi\mu^+\mu^-$ decays and the $B^0 \to K^*(892)^0\mu^+\mu^-$ differential branching fraction. *JHEP* **2016**, *11*, 47. Erratum: *JHEP* **2017**, *4*, 142. [CrossRef]
28. Aaij, R. et al. [LHCb Collaboration]. Angular analysis and differential branching fraction of the decay $B_s^0 \to \phi\mu^+\mu^-$. *JHEP* **2015**, *9*, 179. [CrossRef]
29. D'Amico, G.; Nardecchia, M.; Panci, P.; Sannino, F.; Strumia, A.; Torre, R.; Urbano, A. Flavour anomalies after the R_{K^*} measurement. *JHEP* **2017**, *9*, 10. [CrossRef]
30. Capdevila, B.; Crivellin, A.; Descotes-Genon, S.; Matias, J.; Virto, J. Patterns of new physics in $b \to s\ell^+\ell^-$ transitions in the light of recent data. *JHEP* **2018**, *1*, 93. [CrossRef]
31. Arbey, A.; Hurth, T.; Mahmoudi, F.; Neshatpour, S. Hadronic and new physics contributions to $b \to s$ transitions. *Phys. Rev. D* **2018**, *98*, 095027. [CrossRef]
32. Kim, T. J.; Ko, P.; Li, J.; Park, J.; Wu, P. Correlation between $R_{D^{(*)}}$ and top quark FCNC decays in leptoquark models. *JHEP* **2019**, *7*, 25. [CrossRef]
33. Bansal, S.; Capdevilla, R.M.; Kolda, C. On the Minimal Flavor Violating Leptoquark explanation of the $R_{D^{(*)}}$ anomaly. *Phys. Rev. D* **2019**, *99*, 035047. [CrossRef]
34. Achard, P. et al. [L3 Collaboration]. Search for single top production at LEP. *Phys. Lett. B* **2002**, *549*, 290. [CrossRef]
35. Abbiendi, G. et al. [OPAL Collaboration]. Search for single top quark production at LEP2. *Phys. Lett. B* **2001**, *521*, 181. [CrossRef]
36. Barate, R. et al. [ALEPH Collaboration]. Search for single top production in e^+e^- collisions at $\sqrt{s}$ = 189 GeV - 202 GeV. *Phys. Lett. B* **2000**, *494*, 33. [CrossRef]
37. Abdallah, J. et al. [DELPHI Collaboration]. Search for single top production via FCNC at LEP at $\sqrt{s}$ = 189 GeV - 208 GeV. *Phys. Lett. B* **2005**, *590*, 21. [CrossRef]
38. Aktas, A. et al. [H1 Collaboration]. Search for single top quark production in *ep* collisions at HERA. *Eur. Phys. J. C* **2004**, *33*, 9. [CrossRef]
39. Chekanov, S. et al. [ZEUS Collaboration]. Search for single-top production in ep collisions at HERA. *Phys. Lett. B* **2003**, *559*, 153–170. [CrossRef]
40. Abramowicz, H. et al. [ZEUS Collaboration]. Search for single-top production in ep collisions at HERA. *Phys. Lett. B* **2012**, *708*, 27. [CrossRef]
41. Aaron, F.D. et al. [H1 Collaboration]. Search for single top quark production at HERA. *Phys. Lett. B* **2009**, *678*, 450. [CrossRef]
42. Heister, A. et al. [ALEPH Collaboration]. Search for single top production in e^+e^- collisions at $\sqrt{s}$ up to 209 GeV. *Phys. Lett. B* **2002**, *543*, 173. [CrossRef]
43. Abazov, V. et al. [D0 Collaboration]. Search for flavor changing neutral currents via quark-gluon couplings in single top quark production using 2.3 fb^{-1} of $p\bar{p}$ collisions. *Phys. Lett. B* **2010**, *693*, 81. [CrossRef]
44. Abazov, V. et al. [D0 Collaboration]. Search for production of single top quarks via *tcg* and *tug* flavor-changing-neutral-current couplings. *Phys. Rev. Lett.* **2007**, *99*, 191802. [CrossRef]
45. Aaltonen, T. et al. [CDF Collaboration]. Search for the flavor-changing neutral-current decay $t \to Zq$ in $p\bar{p}$ collisions at $\sqrt{s}$ = 1.96 TeV. *Phys. Rev. Lett.* **2008**, *101*, 192002. [CrossRef]
46. Aaltonen, T. et al. [CDF Collaboration]. Search for top-quark production via flavor-changing neutral currents in W+1 jet events at CDF. *Phys. Rev. Lett.* **2009**, *102*, 151801. [CrossRef]
47. Sirunyan, A. et al. [CMS Collaboration]. Search for associated production of a Z boson with a single top quark and for tZ flavour-changing interactions in *pp* collisions at $\sqrt{s}$ = 8 TeV. *JHEP* **2017**, *7*, 003. [CrossRef]
48. Khachatryan, V. et al. [CMS Collaboration]. Search for anomalous Wtb couplings and flavour-changing neutral currents in t-channel single top quark production in *pp* collisions at $\sqrt{s}$ = 7 and 8 TeV. *JHEP* **2017**, *2*, 28. [CrossRef]
49. Khachatryan, V. et al. [CMS Collaboration]. Search for anomalous single top quark production in association with a photon in *pp* collisions at $\sqrt{s}$ = 8 TeV. *JHEP* **2016**, *4*, 35. [CrossRef]
50. Chatrchyan, S. et al. [CMS Collaboration]. Search for flavor changing neutral currents in top quark decays in *pp* collisions at $\sqrt{s}$ = 7 TeV. *Phys. Lett. B* **2013**, *718*, 1252. [CrossRef]
51. Chatrchyan, S. et al. [CMS Collaboration]. Search for flavor-changing neutral currents in top-quark decays $t \to Zq$ in *pp* collisions at $\sqrt{s}$ = 8 TeV. *Phys. Rev. Lett.* **2014**, *112*, 171802. [CrossRef] [PubMed]
52. Aaboud, M. et al. [ATLAS Collaboration]. Search for flavour-changing neutral current top-quark decays $t \to qZ$ in proton-proton collisions at $\sqrt{s}$ = 13 TeV with the ATLAS detector. *JHEP* **2018**, *7*, 176. [CrossRef]
53. Aad, G. et al. [ATLAS Collaboration]. Search for single top-quark production via flavour changing neutral currents at 8 TeV with the ATLAS detector. *Eur. Phys. J. C* **2016**, *76*, 55. [CrossRef] [PubMed]
54. Aad, G. et al. [ATLAS Collaboration]. Search for flavour-changing neutral current top-quark decays to qZ in pp collision data collected with the ATLAS detector at $\sqrt{s}$ = 8 TeV. *Eur. Phys. J. C* **2016**, *76*, 12. [CrossRef] [PubMed]
55. Aad, G. et al. [ATLAS Collaboration]. A search for flavour changing neutral currents in top-quark decays in *pp* collision data collected with the ATLAS detector at $\sqrt{s}$ = 7 TeV. *Eur. Phys. J. C* **2012**, *9*, 139. [CrossRef]

56. Aad, G. et al. [ATLAS Collaboration]. Search for flavour-changing neutral currents in processes with one top quark and a photon using 81 fb^{-1} of pp collisions at $\sqrt{s}$ = 13 TeV with the ATLAS experiment. *Phys. Lett. B* **2019**, *800*, 135082. [CrossRef]
57. Aaboud, M. et al. [ATLAS Collaboration]. Search for top-quark decays $t \to Hq$ with 36 fb^{-1} of pp collision data at $sqrts$ = 13 TeV with the ATLAS detector. *JHEP* **2019**, *5*, 123. [CrossRef]
58. Aaboud, M. et al. [ATLAS Collaboration]. Search for flavor-changing neutral currents in top quark decays $t \to Hc$ and $t \to Hu$ in multilepton final states in proton-proton collisions at $sqrts$ = 13 TeV with the ATLAS detector. *Phys. Rev. D* **2018**, *98*, 032002. [CrossRef]
59. Aaboud, M. et al. [ATLAS Collaboration]. Search for top quark decay $t \to qH$, with $H \to \gamma\gamma$, in $sqrts$ = 13 TeV pp collisions using the ATLAS detector *JHEP* **2017**, *10*, 129. [CrossRef]
60. Aad, G. et al. [ATLAS Collaboration]. Search for flavour-changing neutral current top quark decays $t \to Hq$ in pp collisions at $sqrts$ = 8 TeV with the ATLAS detector. *JHEP* **2015**, *12*, 61. [CrossRef]
61. Khachatryan, V. et al. [CMS Collaboration]. Search for top quark decays via Higgs-boson-mediated flavor-changing neutral currents in pp collisions at $sqrts$ = 8 TeV. *JHEP* **2017**, *2*, 79. [CrossRef]
62. Sirunyan, A. et al. [CMS Collaboration]. Search for the flavor-changing neutral current interactions of the top quark and the Higgs boson which decays into a pair of b quarks at $sqrts$ = 13 TeV. *JHEP* **2018**, *6*, 102. [CrossRef]
63. Tumasyan, A. et al. [CMS Collaboration]. Search for flavor-changing neutral current interactions of the top quark and Higgs boson in final states with two photons in proton-proton collisions at $\sqrt{s}$ = 13 TeV. *Phys. Rev. Lett.* **2022**, *129*, 032001. [CrossRef]
64. Tumasyan, A. et al. [CMS Collaboration]. Search for flavor-changing neutral current interactions of the top quark and the Higgs boson decaying to a bottom quark-antiquark pair at $\sqrt{s}$ = 13 TeV. *JHEP* **2022**, *2*, 169. [CrossRef]
65. Tumasyan, A. et al. [ATLAS Collaboration]. Search for flavour-changing neutral current interactions of the top quark and the Higgs boson in events with a pair of τ-leptons in pp collisions at $\sqrt{s}$ = 13 TeV with the ATLAS detector. *arXiv* **2022**, arXiv:2208.11415.
66. LHC Higgs Cross Section Working Group. *Handbook of LHC Higgs Cross Sections: 4. Deciphering the Nature of the Higgs Sector*; CERN Yellow Reports: Geneva, Switzerland, 2017. [CrossRef]
67. Bevilacqua, G.; Worek, M. On the ratio of $t\bar{t}b\bar{b}$ and $t\bar{t}jj$ cross sections at the CERN Large Hadron Collider. *JHEP* **2014**, *7*, 135. [CrossRef]
68. Aaboud, M. et al. [ATLAS Collaboration]. Measurements of fiducial and differential cross-sections of $t\bar{t}$ production with additional heavy-flavour jets in proton-proton collisions at $\sqrt{s}$ = 13 TeV with the ATLAS detector. *JHEP* **2019**, *4*, 46. [CrossRef]
69. Sirunyan, A.M. et al. [CMS Collaboration]. Measurement of the cross section for $t\bar{t}$ production with additional jets and b jets in pp collisions at $\sqrt{s}$ = 13 TeV. *JHEP* **2020**, *7*, 125. [CrossRef]
70. Sirunyan, A.M. et al. [CMS Collaboration]. Measurement of the $t\bar{t}b\bar{b}$ production cross section in the all-jet final state in pp collisions at $\sqrt{s}$ = 13 TeV. *Phys. Lett. B* **2020**, *803*, 135285. [CrossRef]
71. Sirunyan, A.M. et al. [CMS Collaboration]. First measurement of the cross section for top quark pair production with additional charm jets using dileptonic final states in pp collisions at $\sqrt{s}$ = 13 TeV. *Phys. Lett. B* **2021**, *820*, 136565. [CrossRef]
72. Harnik, R.; Kopp, J.; Zupan, J. Flavor violating Higgs decays. *JHEP* **2013**, *3*, 26. [CrossRef]
73. Gorbahn, M.; Haisch, U. Searching for $t \to c(u)h$ dipole moments. *JHEP* **2014**, *6*, 33. [CrossRef]
74. Larios, F.; Martinez, R.; Perez, M.A. Constraints on top quark FCNC from electroweak precision measurements. *Phys. Rev. D* **2005**, *72*, 057504. [CrossRef]
75. Fernandez, A.; Pagliarone, C.; Ramirez-Zavaleta, F.; Toscano, J.J. Higgs mediated double flavor violating top decays in effective theories. *J. Phys. G* **2010**, *37*, 085007. [CrossRef]
76. Hesari, H.; Khanpour, H.; Najafabadi, M. Direct and indirect searches for top-Higgs FCNC couplings. *Phys. Rev. D* **2015**, *92*, 113012. [CrossRef]
77. David, A. et al. [LHC Higgs Cross Section Working Group]. LHC HXSWG interim recommendations to explore the coupling structure of a Higgs-like particle. *arXiv* **2012**, arXiv:1209.0040.
78. Buchmuller, W.; Wyler, D. Effective lagrangian analysis of new interactions and flavour conservation. *Nucl. Phys. B* **1986**, *268*, 621. [CrossRef]
79. Grzadkowski, B.; Iskrzyński, M.; Misiak, M.; Rosiek, J. Dimension-six terms in the Standard Model Lagrangian. *JHEP* **2010**, *10*, 85. [CrossRef]
80. Aguilar-Saavedra, J.A.; Degrande, C.; Durieux, G.; Maltoni, F.; Vryonidou, E.; Zhang, C.; Barducci, D.; Brivio, I.; Cirigliano, V.; Dekens, W.; et al. Interpreting top-quark LHC measurements in the standard-model effective field theory. *arXiv* **2018**, arXiv:1802.07237.
81. Aguilar-Saavedra, J.A. A minimal set of top anomalous couplings. *Nucl. Phys. B* **2009**, *812*, 181. [CrossRef]
82. Aguilar-Saavedra, J.A. A minimal set of top-Higgs anomalous couplings. *Nucl. Phys. B* **2009**, *821*, 215. [CrossRef]
83. Degrande, C.; Maltoni, F.; Wang, J.; Zhang, C. Automatic computations at next-to-leading order in QCD for top-quark flavor-changing neutral processes. *Phys. Rev. D* **2015**, *91*, 034024. [CrossRef]
84. Durieux, G.; Maltoni, F.; Zhang, C. Global approach to top-quark flavor-changing interactions. *Phys. Rev. D* **2015**, *91*, 074017. [CrossRef]
85. Afik, Y.; Bar-Shalom, S.; Soni, A.; Wudka, J. New flavor physics in di- and tri-lepton events from single-top at the LHC and beyond. *Phys. Rev. D* **2021**, *103*, 075031. [CrossRef]

86. Barros, M.; Castro, N.F.; Erdmann, J.; Geßner, G.; Kröninger, K.; La Cagnina, S.; Peixoto, A. Study of interference effects in the search for flavour-changing neutral current interactions involving the top quark and a photon or a Z boson at the LHC. *Eur. Phys. J. Plus* **2020**, *135*, 339. [CrossRef]
87. Sirunyan, A.M. et al. [CMS Collaboration]. Search for new physics in top quark production in dilepton final states in proton-proton collisions at $\sqrt{s} = 13$ TeV. *Eur. Phys. J. C* **2019**, *79*, 886. [CrossRef]
88. Bardhan, D.; Bhattacharyya, G.; Ghosh, D.; Patra, M.; Raychaudhuri, S. A detailed analysis of flavour-changing decays of top quarks as a probe of new physics at the LHC. *Phys. Rev. D* **2016**, *94*, 015026. [CrossRef]
89. Branco, G.C.; Ferreira, P.M.; Lavoura, L.; Rebelo, M.N.; Sher, M.; Silva J.P. Theory and phenomenology of two-Higgs-doublet-models. *Phys. Rept.* **2012**, *516*, 1. [CrossRef]
90. Gunion, J.F.; Haber, H.E. CP-conserving two-Higgs-doublet-model: The approach of the decoupling limit. *Phys. Rev. D* **2003**, *67*, 075019. [CrossRef]
91. Maiani, L.; Polosa, A.D.; Riquer, V. Bounds to the Higgs sector masses in minimal supersymmetry from LHC data. *Phys. Lett. B* **2013**, *724*, 274. [CrossRef]
92. Hou, W.-S. Tree level $t \to ch$ or $h \to t\bar{c}$ decays. *Phys. Lett. B* **1992**, *296*, 179. [CrossRef]
93. Sirunyan, A. et al. [CMS Collaboration]. Combined measurement of Higgs boson couplings in proton-proton collisions at $\sqrt{s} = 13$ TeV. *Eur. Phys. J. C* **2019**, *79*, 421. [CrossRef] [PubMed]
94. Aaboud, M. et al. [ATLAS Collaboration]. Combined measurements of Higgs boson production and decay using up to 80 fb^{-1} of proton-proton collision data at $\sqrt{s} = 13$ TeV collected with the ATLAS experiment. *Phys. Rev. D* **2020**, *101*, 012002. [CrossRef]
95. Aaboud, M. et al. [ATLAS Collaboration]. Search for heavy resonances decaying into WW in $e\nu\mu\nu$ final state in pp collisions at $\sqrt{s} = 13$ TeV with the ATLAS detector. *Eur. Phys. J. C* **2018**, *78*, 24. [CrossRef] [PubMed]
96. Pich, A.; Tuzon, P. Yukawa alignment in the two-higgs-doublet model. *Phys. Rev. D* **2009**, *80*, 091702. [CrossRef]
97. Abbas, G.; Celis, A.; Li, X.-Q.; Lu, J.; Pich, A. Flavour-changing top decays in the aligned two-Higgs-doublet model. *JHEP* **2015**, *6*, 005. [CrossRef]
98. Das, A. K.; Kao, C. A two Higgs doublet model for the top quark. *Phys. Lett. B* **1996**, *372*, 106. [CrossRef]
99. Kao, C.; Cheng, H.-I.; Hou, W.-S.; Sayre, J. Top decays with flavor changing neutral Higgs interactions at the LHC. *Phys. Lett. B* **2012**, *716*, 225. [CrossRef]
100. Chen, K.-F.; Hou, W.-S.; Kao, C.; Kohda, M. When the Higgs meets the Top: Search for $t \to ch^0$ at the LHC. *Phys. Lett. B* **2013**, *725*, 378. [CrossRef]
101. Jain, R.; Kao, C. Charming top decays with flavor changing neutral Higgs boson and WW at hadron colliders. *Phys. Rev. D* **2019**, *99*, 055036. [CrossRef]
102. Baum, I.; Eilam, G.; Bar-Shalom, S. Scalar flavor changing neutral currents and rare top quark decays in a two Higgs doublet model "for the top quark". *Phys. Rev. D* **2008**, *77*, 113008. [CrossRef]
103. Branco, G.; Grimus, W.; Lavoura, L. Relating the scalar flavor changing neutral couplings to the CKM matrix. *Phys. Lett. B* **1996**, *380*, 119. [CrossRef]
104. Botella, F.; Branco, G.; Rebelo, M. Minimal flavour violation and multi-Higgs models. *Phys. Lett. B* **2010**, *687*, 194. [CrossRef]
105. Botella, F.; Branco, G.; Nebot, M.; Rebelo, M. Flavour-changing Higgs couplings in a class of two Higgs doublet models. *Eur. Phys. J. C* **2016**, *76*, 161. [CrossRef]
106. Djouadi, A. The anatomy of electroweak symmetry breaking Tome II: The Higgs bosons in the Minimal Supersymmetric Model. *Phys. Rept.* **2008**, *459*, 1. [CrossRef]
107. Fayet, P. Supersymmetry and weak, electromagnetic, and strong interactions. *Phys. Lett. B* **1976**, *64*, 159. [CrossRef]
108. Fayet, P. Spontaneously broken supersymmetric theories of weak, electromagnetic, and strong interactions. *Phys. Lett. B* **1977**, *69*, 489. [CrossRef]
109. Djouadi, A.; Maiani, L.; Moreau, G.; Polosa, A.; Quevillon, J.; Riquer, V. The post-Higgs MSSM scenario: Habemus MSSM? *Eur. Phys. J. C* **2013**, *73*, 2650. [CrossRef]
110. Djouadi, A.; Maiani, L.; Polosa, A.; Quevillon, J.; Riquer, V. Fully covering the MSSM Higgs sector at the LHC. *JHEP* **2015**, *6*, 168. [CrossRef]
111. Dedes, A.; Paraskevas, M.; Rosiek, J.; Suxho, K.; Tamvakis, K. Rare top-quark decays to Higgs boson in MSSM. *JHEP* **2014**, *11*, 137. [CrossRef]
112. Eilam, G.; Gemintern, A.; Han, T.; Yang, J. M.; Zhang, X. Top-quark rare decay $t \to ch$ in R-parity-violating SUSY. *Phys. Lett. B* **2001**, *510*, 227. [CrossRef]
113. Perez, P.F.; Wise, M.B. Baryon and lepton number as local gauge symmetries. *Phys. Rev. D* **2010**, *82*, 011901. [CrossRef]
114. Perez, P.F.; Wise, M.B. Low energy supersymmetry with baryon and lepton number gauged. *Phys. Rev. D* **2011**, *84*, 055015. [CrossRef]
115. Perez, P.F.; Wise, M.B. Breaking local baryon and lepton number at the TeV scale. *JHEP* **2011**, *8*, 68. [CrossRef]
116. Amond, J.M.; Perez, P.F.; Formal, B.; Spinner, S. Higgs boson decays, baryon number violation, and supersymmetry at the LHC. *Phys. Rev. D* **2012**, *85*, 115024. [CrossRef]
117. Gao, T.-J.; Feng, T.-F.; Sun, F.; Zhang, H.-B.; Zhao, S.-M. Top quark decay to a 125 GeV Higgs in BLMSSM. *Chin. Phys. C* **2015**, *39*, 073101. [CrossRef]

118. Ellwanger, U.; Hugonie, C.; Teixeira, A. M. The Next-to-Minimal Supersymmetric Standard Model. *Phys. Rept.* **2010**, *496*, 1. [CrossRef]
119. Kim, J.E.; Nilles, H.P. The μ-problem and the strong CP-problem. *Phys. Lett. B* **1984**, *138*, 150. [CrossRef]
120. Del Aguila, F.; Aguilar-Saavedra, J.A.; Miquel, R. Constraints on top couplings in models with exotic quarks. *Phys. Rev. Lett.* **1999**, *82*, 1628. [CrossRef]
121. Gaitan, R.; Miranda, O.; Gabral-Rosetti, L. Rare top quark and Higgs boson decays in alternative left-right symmetric models. *Phys. Rev. D* **2005**, *72*, 034018. [CrossRef]
122. Yang, B.; Liu, N.; Han, J. Top quark flavour-changing neutral-current decay to a 125 GeV Higgs boson in the littlest Higgs model with T parity. *Phys. Rev. D* **2014**, *89*, 034020. [CrossRef]
123. Azatov, A.; Toharia, M.; Zhu, L. Higgs mediated flavor changing neutral currents in warped extra dimensions. *Phys. Rev. D* **2009**, *80*, 035016. [CrossRef]
124. Casagrande, S.; Goertz, F.; Haisch, U.; Neubert, M.; Pfoh, T. The custodial Randall-Sundrum model: From precision tests to Higgs physics. *JHEP* **2010**, *9*, 14. [CrossRef]
125. Dey, U.K.; Jha, T. Rare top decays in Minimal and Non-minimal Universal Extra Dimension. *Phys. Rev. D* **2016**, *94*, 056011. [CrossRef]
126. Dimopoulos, S.; Preskill, J. Massless composites with massive constituents. *Nucl. Phys. B* **1982**, *199*, 206. [CrossRef]
127. Kaplan, D.B.; Georgi, H. SU(2) x U(1) breaking by vacuum misalignment. *Phys. Lett. B* **1984**, *136*, 183. [CrossRef]
128. Kaplan, D.B.; Georgi, H.; Dimopoulos, S. Composite Higgs scalars. *Phys. Lett. B* **1984**, *136*, 187. [CrossRef]
129. Cacciapaglia, G.; Ferretti, G.; Flacke, T.; Serodio, H. Light scalars in composite Higgs models. *Front. Phys.* **2019**, *7*, 22. [CrossRef]
130. Castro, N.; Chala, M.; Peixoto, A.; Ramos, M. Novel flavour-changing neutral currents in the top quark sector. *JHEP* **2020**, *10*, 38. [CrossRef]
131. Banerjee, S.; Chala, M.; Spannowsky, M. Top quark FCNCs in extended Higgs sectors. *Eur. Phys. J. C* **2018**, *78*, 683. [CrossRef]
132. Kaplan, D. Flavor at SSC energies: A new mechanism for dynamically generated fermion masses. *Nucl. Phys. B* **1991**, *365*, 259. [CrossRef]
133. Altunkaynak, B.; Hou, W.-S.; Kao, C.; Kohda, M.; McCoy, B. Flavor changing heavy Higgs interactions at the LHC. *Phys. Lett. B* **2015**, *751*, 135. [CrossRef]
134. Hou, W.-S.; Modak, T. Probing top changing neutral Higgs couplings at colliders. *Mod. Phys. Lett. A* **2021**, *36*, 2130006. [CrossRef]
135. Kohda, M.; Modak, T.; Hou, W.-S. Searching for new scalar bosons via triple-top signature in $cg \to tS^0 \to tt\bar{t}$. *Phys. Lett. B* **2018**, *776*, 379. [CrossRef]
136. Hou, W.-S.; Kohda, M.; Modak, T. Constraining a lighter exotic scalar via same-sign top. *Phys. Lett. B* **2018**, *786*, 212. [CrossRef]
137. Hou, W.-S.; Hsu T.-H.; Modak, T. Constraining $t \to u$ flavor changing neutral Higgs coupling at the LHC. *Phys. Rev. D* **2020**, *102*, 055006. [CrossRef]
138. Gori, S.; Grojean, C.; Juste, A.; Paul, A. Heavy Higgs searches: Flavour matters. *JHEP* **2018**, *1*, 108. [CrossRef]
139. Knapen, S.; Robinson, D.J. Disentangling mass and mixing hierarchies. *Phys. Rev. Lett.* **2015**, *115*, 161803. [CrossRef]
140. Altmannshofer, W.; Gori, S.; Robinson, D.J.; Tuckler, D. The Flavor-locked Flavorful Two Higgs Doublet Model. *JHEP* **2018**, *3*, 129. [CrossRef]
141. Altmannshofer, W.; Eby, J.; Gori, S.; Lotito, M.; Martone, M.; Tuckler, D. Collider signatures of flavorful Higgs bosons. *Phys. Rev. D* **2016**, *94*, 115032. [CrossRef]
142. Altmannshofer, W.; Maddock, B.; Tuckler, D. Rare top decays as probes of flavorful Higgs bosons. *Phys. Rev. D* **2019**, *100*, 015003. [CrossRef]
143. Arroyo-Urena, M.A.; Fernandez-Tellez, A.; Tavares-Velasco, G. Flavor changing Flavon decay $\phi \to tc(\phi = H_F, A_F)$ at the high-luminosity large hadron collider. *arXiv* **2019**, arXiv:1906.07821.
144. FCC Collaboration. FCC Physics Opportunities: Future Circular Collider Conceptual Design Report Volume 1. *Eur. Phys. J. C* **2019**, *79*, 474. [CrossRef]
145. Fernandez, J.L.A. et al. [LHeC Collaboration]. A Large Hadron Electron Collider at CERN: Report on the physics and design concepts for machine and detector. *J. Phys. G Nucl. Part. Phys.* **2012**, *39*, 075001. [CrossRef]
146. Bruening, O.; Klein, M. The Large Hadron Electron Collider. *Mod. Phys. Lett. A* **2013**, *28*, 1330011. [CrossRef]
147. Klein, M. *From My Vast Repertoire . . . : Guido Altarelli's Legacy*; Forte, S., Aharon, L.; Ridolfi, G., Eds.; World Scientific Publishing: Singapore, 2018. [CrossRef]
148. Klein, M. Deep inelastic scattering at the energy frontier. *Ann. Der Phys.* **2016**, *528*, 138. [CrossRef]
149. Turk Cakir, I.; Yilmaz, A.; Denizli, H.; Senol, A.; Karadeniz, H.; Cakir, O. Probing the anomalous FCNC couplings at Large Hadron Electron Collider. *Adv. High Energy Phys.* **2017**, *2017*, 1572053. [CrossRef]
150. Behera, S.; Islam, R.; Kumar, M.; Poulose, P.; Rahaman, R. Fingerprinting the top quark FCNC via anomalous Ztq couplings at the LHeC. *Phys. Rev. D* **2019**, *100*, 015006. [CrossRef]
151. Cakir, O.; Yilmaz, A.; Turk Cakir, I.; Senol, A.; Denizli, H. Probing top quark FCNC $tq\gamma$ and tqZ couplings at future electron-proton colliders. *Nucl. Phys. B* **2019**, *944*, 114640. [CrossRef]
152. Wang, X.; Sun, H.; Luo, X. Searches for the anomalous FCNC top-Higgs couplings with polarized electron beam at the LHeC. *Adv. High Energy Phys.* **2017**, *2017*, 4693213. [CrossRef]

153. Bordry, F.; Benedikt, M.; Bruning, O.; Jowett, J.; Rossi, L.; Schulte, D.; Stapnes, S.; Zimmermann, F. Machine parameters and projected luminosity performance of proposed future colliders at CERN. *arXiv* **2018**, arXiv:1810.13022.
154. Bruning, O.; Jowett, J.; Klein, M.; Pellegrini, D.; Schulte, D.; Zimmermann, F. *Future Circular Collider Study FCC-he Baseline Parameters*; Technical Report; CERN: Meyrin, Switzerland, 2018.
155. Behera, S.; Poulose, P. Tracing the anomalous tqg and $tq\gamma$ flavor changing interactions at the FCC-he. *arXiv* **2020**, arXiv:2007.02236.
156. FCC Collaboration. FCC-ee: The Lepton Collider: Future Circular Collider Conceptual Design Report Volume 2. *Eur. Phys. J. ST* **2019**, *228*, 261. [CrossRef]
157. FCC Collaboration. FCC-hh: The Hadron Collider: Future Circular Collider Conceptual Design Report Volume 3. *Eur. Phys. J. ST* **2019**, *228*, 755. [CrossRef]
158. Mandrik, P. et al. [the FCC Study Group]. Prospect for top quark FCNC searches at the FCC-hh. *J. Phys. Conf. Ser.* **2019**, *1390*, 012044. [CrossRef]
159. Papaefstathiou, A.; Tetlalmatzi-Xolocotzi G. Rare top quark decays at a 100 TeV proton-proton collider: $t \to bWZ$ and $t \to hc$. *Eur. Phys. J. C* **2018**, *78*, 214. [CrossRef]
160. Oyulmaz, K.Y.; Senol, A.; Denizli, H.; Cakir, O. Top quark anomalous FCNC production via tqg couplings at FCC-hh. *Phys. Rev. D* **2019**, *99*, 115023. [CrossRef]
161. Oyulmaz, K.Y.; Senol, A.; Denizli, H.; Yilmaz, A.; Turk Cakir, I.; Cakir, O. Probing anomalous $tq\gamma$ and tqg couplings via single top production in association with photon at FCC-hh. *Eur. Phys. J. C* **2019**, *79*, 83. [CrossRef]
162. Abada, A. et al. [FCC Collaboration]. Future Circular Collider Study: Vol. 4 The High-Energy LHC (HE-LHC). *Eur. Phys. J. Spec. Top.* **2019**, *228*, 1109. [CrossRef]
163. Liu, Y.-B.; Moretti, S. Probing the top-Higgs boson FCNC couplings via the $h \to \gamma\gamma$ channel at the HE-LHC and FCC-hh. *Phys. Rev. D* **2020**, *101*, 075029. [CrossRef]
164. Moortgat-Pick, G.; Abe, T.; Alexander, G.; Ananthanarayan, B.; Babich, A.A.; Bharadwaj, V.; Barber, D.; Bartl, A.; Brachmann, A.; Chen, S. The role of polarized positions and electrons in revealing fundamental interactions at the Linear Collider. *Phys. Rept.* **2008**, *460*, 131. [CrossRef]
165. Zarnecki, A. Sensitivity of CLIC at 380 GeV to the top FCNC decay $t \to cH$. *J. Phys. Conf. Ser.* **2017**, *873*, 012049. [CrossRef]
166. Zarnecki, A. Top-quark physics at the first CLIC stage. *arXiv* **2018**, arXiv:1810.05487.
167. Abramowicz, H. et al. [CLICdp Collaboration]. Top-quark physics at the CLIC electron-positron linear collider. *JHEP* **2019**, *11*, 003. [CrossRef]
168. Melic, B.; Patra, M. Exploring the top-Higgs FCNC couplings at polarized linear colliders with top spin observables. *JHEP* **2017**, *01*, 048. [CrossRef]

universe

MDPI

Review

Tests of Charge–Parity Symmetry and Lepton Flavor Conservation in the Top Quark Sector

Kai-Feng Chen [1,*,†] and Reza Goldouzian [2,†]

1 Department of Physics, National Taiwan University, Taipei 10617, Taiwan
2 Department of Physics 225, Nieuwland Science Hall University of Notre Dame, Notre Dame, IN 46556-5670, USA
* Correspondence: kai-feng.chen@cern.ch
† These authors contributed equally to this work.

Abstract: The Standard Model (SM) of particle physics is the most general renormalizable theory which is built on a few general principles and fundamental symmetries with the given particle content. However, multiple symmetries are not built into the model and are simply consequences of renormalizabilty, gauge invariance, and particle content of the theory. It is crucial to test the validity of these types of symmetries and related conservation laws experimentally. The CERN LHC provides the highest sensitivity for testing the SM symmetries at high energy scales involving heavy particles such as the top quark. In this article, we are going to review the recent experimental searches of charge–parity and charged-lepton flavor violation in the top quark sector.

Keywords: charge–parity violation; charged-lepton flavor violation; top quark

Citation: Chen, K.-F.; Goldouzian, R. Tests of Charge–Parity Symmetry and Lepton Flavor Conservation in the Top Quark Sector. *Universe* **2023**, *9*, 62. https://doi.org/10.3390/universe9020062

Academic Editors: Efe Yazgan, Pedro Ferreira da Silva and Jinmin Yang

Received: 16 November 2022
Revised: 9 January 2023
Accepted: 9 January 2023
Published: 20 January 2023

1. Introduction

The Standard Model (SM) of particle physics is an extremely successful theory that has been extensively verified against experimental results. However, it can not explain several fundamental aspects of nature such as neutrino masses, dark matter, and baryon asymmetry in the universe. The SM Lagrangian is built based on a few principles including gauge symmetries and renormalizability. While the gauge symmetry is considered as fundamental rules in the SM Lagrangian, there are some symmetries in the SM that come from the structure imposed by the fundamental rules, called global symmetries. These symmetries can be exact or approximate and are very powerful tools for testing the SM validity. To explain the SM open issues, many models beyond-the-SM (BSM) have been proposed. These BSM models usually introduce new particles and interactions which lead to violation of the SM global symmetries. Therefore, testing the SM global symmetries is also considered as a unique window for probing BSM physics scenarios.

The combined transformation of parity (P) and charge-conjugation (C) operations was believed to be exact symmetry, until the discovery of CP violation in the neutral Kaon system in 1964 [1]. A model proposed by Kobayashi and Maskawa [2], denoted as the KM model, pointed out that an irreducible complex phase in the quark mixing matrix, known as the Cabibbo–Kobayashi–Maskawa (CKM) matrix, would enable CP violation in a natural manner, if there are at least three generations of quarks. With the discovery of the charm, bottom, and top quarks, together with the discovery of CP violation in the B-meson system [3,4], the KM model has been confirmed and became one of the key ingredients of the modern SM. However, even with the big success of the KM model, the known CP asymmetry in the SM is still far from enough to form a matter-dominant universe, hence it is necessary to look for additional sources of CP violation experimentally.

Top quark is playing an important role in the B-meson mixing, which is one of the main processes that CP violation can actually occur in the bottom sector. Unlike the bottom quark, the top quark itself has a rather monotonic flavor structure. Top always decays to bottom and W-boson since $|V_{tb}|$ is much larger than $|V_{td}|$ and $|V_{ts}|$. As top is much heavier than any other quark, the GIM mechanism [5] is very effective and the flavor-changing neutral current processes are very suppressed. The CP violation effect vanishes in the limit of $m_d, m_s \approx 0$; no other interfered process generates a non-zero strong phase to have measurable CP asymmetries, which require the presence of both strong and weak phases.

However, if there are contributions from new physics, the situation can be very different. Any hint of CP violation in the top sector will be a smoking gun signal for new physics [6]. With more than 10^8 pairs of top quarks produced at the LHC, the experiments have a very good chance to probe sub-percent effects already. However, as the over-dominant $t \to bW$ decay, the usual measurements which involve different decays as introduced for strange and bottom quarks cannot be repeated for top. Alternative observables have to be introduced instead.

In the SM with massless neutrinos, the mixing of neutrino flavors is forbidden. Consequently, the flavor of charged leptons is conserved and couplings of leptons to gauge bosons are lepton flavor universal (LFU). The lepton flavor conservation in the neutral-lepton sector was found to be violated after the discovery of neutrino oscillations. Neutrino oscillations could also give rise to charged-lepton flavor violating (CLFV) processes. Because of the smallness of the neutrino masses, these processes are highly suppressed and are far below experimental sensitivity. Although the SM predictions for the CLFV processes are below experimental sensitivity, many theoretical scenarios beyond the SM, such as the two-Higgs doublet model [7], the minimal supersymmetric model [8], and the inverse seesaw model [9], predict detectable CLFV rates. Any evidence for such rare processes would therefore serve as a clear signature of physics beyond the SM.

The hints of LFU violations have been reported in semileptonic B decays, and the experimental evidence has risen over the past few years [10,11]. It is known that models accommodating violation of lepton universality generally also lead to observable effects in lepton flavor violation [12]. Models that can describe these small deviations also predict measurable observables in the top quark sector [13]. For example, certain leptoquark models that can accommodate the observed deviation in the B sector would imply branching fractions of $t \to \ell\ell' c \approx 10^{-6}$, with ℓ and ℓ' representing different-flavor charged leptons. Searching for CLFV processes related to the top quark would be complimentary to the searches in the B meson sector [14].

In this article, we discuss the measurements that have been carried out by ATLAS and CMS experiments in the context of CP violation in the top quark sector. In addition, we review the results of the first search for the top quark CLFV interactions performed by the CMS collaboration.

2. Search for CP Violation in Single Top Events

ATLAS has presented a note on the search for CP violation in the lepton plus jets decay of single top quarks [15]. The top quarks are produced via t-channel where the top quarks are highly polarized, and it is possible to define CP-violating sensitive observables with the angles derived by the top decay products. The top quarks are exclusively decaying to W boson and bottom quark, and the Wtb vertex in the general effective operator framework can be expressed as [16,17]

$$\mathcal{L}_{Wtb} = -\frac{g}{\sqrt{2}}\bar{b}\gamma^{\mu}(V_L P_L + V_R P_R)tW_{\mu}^{-} - \frac{g}{\sqrt{2}}\bar{b}\frac{i\sigma^{\mu\nu}q_{\nu}}{m_W}(g_L P_L + g_R P_R)tW_{\mu}^{-} + \text{h.c.}, \quad (1)$$

where the weak coupling constant is denoted by g; m_W and q_ν are the mass and four momentum of the W boson. The left (right) handed projection operators, vector, and tensor coupling are given by P_L (P_R), V_L (V_R), and g_L (g_R), respectively. In the SM, the V_L is simply the CKM element V_{tb}, and the anomalous couplings, V_R, g_L, and g_R all vanish to zero.

In order to probe these couplings, it is possible to construct angular asymmetry observables based on the decay products of the top quark, such as a forward-backward asymmetry,

$$A_{FB} = \frac{N_{\text{evt}}(\cos\theta > 0) - N_{\text{evt}}(\cos\theta < 0)}{N_{\text{evt}}(\cos\theta > 0) + N_{\text{evt}}(\cos\theta < 0)}, \tag{2}$$

where N_{evt} is the number of events with the criterion $\cos\theta > 0$ or $\cos\theta < 0$, and θ is an angle formed by the direction of top decay daughters. A typical choice of the angle (denoted as θ^*) could be the angle between the direction of the lepton from the W decay in the W boson rest frame and direction of the W boson in the top quark rest frame. However, the asymmetry defined with θ^* is not really sensitive to the anomalous couplings related to CP violating complex phases.

As suggested in Refs. [18,19], in the t-channel single top production, the top quarks are mostly polarisation in the direction of the spectator quark; hence, by defining a new reference direction, $\vec{N} = \vec{s_t} \times \vec{q}$, where $\vec{s_t}$ is the direction of the spectator quark, and $\vec{q}$ is the momentum of W boson, all in the rest frame of the top quark. A new angle θ^N can be defined by the angle between the lepton in the W boson rest frame and the new direction $\vec{N}$. The forward-backward asymmetry defined by θ^N, denoted as A^N_{FB}, could provide information of the anomalous coupling g_R, in particular for the imaginary part which is sensitive to the CP violation. In the SM, the coupling g_R is very close to zero, $(-7.17 - 1.23i) \times 10^{-3}$ [20].

ATLAS has performed a measurement of A^N_{FB} and provided a bound on the imaginary part of g_R based on the data collected at $\sqrt{s} = 7$ TeV, corresponding to an integrated luminosity of 4.66 fb^{-1}. The data events are required to have at least an electron or a muon at the trigger. At the analysis level, an offline reconstructed electron or muon, significant transverse missing energy E_T^{miss}, and two jets (one of the jets must be tagged as originated from b-quark) are required for the signal candidates. The candidate electron should have a transverse momentum p_T greater than 25 GeV, and within $|\eta| < 2.47$ (while the transition region $1.37 < |\eta| < 1.52$ is excluded). For the candidate muon, the minimal p_T is also 25 GeV and should be within the region $|\eta| < 2.5$. Jets are required to have a minimal p_T greater than 30 GeV; E_T^{miss} and the transverse mass of the reconstructed W boson should be above 30 GeV as well.

The top quark is reconstructed from the b-jet and the W boson, where the W boson is reconstructed with the charged lepton and neutrino momenta. The transverse momentum of neutrino is assumed to be the same as E_T^{miss}, while the longitudinal component is solved by the constraint of W boson mass. In the case of two possible solutions, the solution with smaller longitudinal neutrino momentum is selected. Based on the reconstructed momenta of top quark and W boson, the decay angle θ^N can be derived. The distribution of $\cos\theta^N$ includes the detector effects and has to be unfolded back to the parton level. The forward-backward asymmetry A^N_{FB} is then computed based on the unfolded distribution of $\cos\theta^N$:

$$A^N_{FB} = 0.031 \pm 0.065\,(\text{stat.})^{+0.029}_{-0.031}\,(\text{syst.}). \tag{3}$$

The major systematic uncertainties considered in the measurement of A^N_{FB} are: signal t-channel single top modeling, $t\bar{t}$ modeling, background normalization, jet energy scale and resolution, lepton selection and trigger. The effects of rest systematic sources are all smaller than 0.005.

The relation between A^N_{FB} and g_R can be approximated by $A^N_{FB} = 0.64P\Im(g_R)$ [21] if the values of g_R are small (assuming $V_L = 1$, $V_R = 0$, and $g_L = 0$). Assuming a value of $P = 0.9$ derived from the knowledge of top quark polarization, the limit on the value of $\Im(g_R)$ is obtained to be $[-0.20, 0.30]$ at the 95% confidence level. The measured A^N_{FB} and the derived limit on $\Im(g_R)$ are both consistent with the predictions from the SM, and no significant CP-violating effect is observed in the t-channel single top production.

3. Measurements of T-Odd Triple-Product Observables

A triple-product correlation is constructed with three directional vectors and takes the form of $\vec{v_1} \cdot (\vec{v_2} \times \vec{v_3})$. The vectors could be particle momenta or spins. The combined triple product is odd under time-reversal (T) transformation; under the CPT theorem, a T-odd observable is also a CP-odd observable. In the $pp \to t\bar{t} \to b\bar{b}W^+W^-$ production and decay chain, there are multiple momentum vectors which can be reconstructed experimentally, but how to select useful combinations of triple-products is a non-trivial task. The authors of Refs. [22,23] examined a series of triple-product observables and studied the contributions from a chromoelectric dipole moment (CEDM) of the top quark. The magnetic and electric couplings between the top quarks and gluons can be expressed as

$$\mathcal{L} = \frac{g_s}{2}\bar{t}T^a\sigma^{\mu\nu}(a_t^g + i\gamma_5 d_t^g)tG^a_{\mu\nu}, \tag{4}$$

where the strong coupling constant and the gluon field strength tensor are denoted by g_s and $G^a_{\mu\nu}$, respectively; the parameters a_t^g and d_t^g are for the chromomagnetic and chromoelectric dipole moments. The term d_t^g can be further written as $d_t^g = \frac{\sqrt{2}v}{\Lambda^2}\Im(d_{tG})$, where Λ is the scale of the BSM phenomena, $v \approx 246$ GeV is the vacuum expectation value, and d_{tG} is the CEDM parameter. CMS has performed several studies along this proposal, including dilepton channel [24] and lepton plus jets decays [25,26].

3.1. Search in the Dilepton Channel

The analysis in the dilepton channel exploits the top-pair production, with both W bosons decaying leptonically to define two CP-odd correlations:

$$\mathcal{O}_1 = \epsilon(p_t, p_{\bar{t}}, p_{\ell^+}, p_{\ell^-}), \tag{5}$$

$$\mathcal{O}_3 = \epsilon(p_b, p_{\bar{b}}, p_{\ell^+}, p_{\ell^-}), \tag{6}$$

where p_t ($p_{\bar{t}}$), p_b ($p_{\bar{b}}$), p_{ℓ^-} (p_{ℓ^+}) are the four momenta of top (anti-top) quark, bottom (anti-bottom) quark, and lepton (anti-lepton), respectively, and ϵ is the Levi–Civita tensor. The CP violation can be tested by the measurements of the asymmetries,

$$A_{\mathcal{O}_i} = \frac{N(\mathcal{O}_i > 0) - N(\mathcal{O}_i < 0)}{N(\mathcal{O}_i > 0) + N(\mathcal{O}_i < 0)}. \tag{7}$$

These two asymmetries are the observables with the highest sensitivity and linear to d_{tG}.

The data used in the dilepton analysis are from pp collisions at 13 TeV correspond to an integrated luminosity of 35.9 fb^{-1}. The events are required to pass the single or dilepton trigger conditions, and then categorized according to the combination of lepton flavors, e^+e^-, $\mu^+\mu^-$, or $e^\pm\mu^\mp$. Electron candidates are reconstructed with CMS tracker and calorimeter information, must have a $p_T > 25$ (20) GeV for the leading (sub-leading) candidate, and are within the volume $|\eta| < 2.4$. Muon candidates are reconstructed with a combination of tracker and muon system information and fulfill the same criteria on the transverse momentum and pseudorapidity. The jets are reconstructed using the anti-k_T algorithm with a distance parameter $R = 0.4$, and are required to have $p_T > 30$ GeV and $|\eta| < 2.4$. If a jet candidate is too close ($\Delta R = \sqrt{\Delta\eta^2 + \Delta\phi^2} < 0.4$) to a lepton candidate, the jet itself is removed.

An event is required to have two charged leptons, and at least two jets. One of the selected jets must be tagged as originating from bottom quark. The events with additional lepton of $p_T > 20$ GeV are discarded. The invariant mass of dilepton pairs must be greater than 20 GeV to suppress contributions from low-mass resonance decay and the Drell–Yan process. The same-flavor dilepton events with an invariant mass in the region between 76 and 106 GeV are also rejected to suppress the contributions from Z boson decays; however, these events in the Z mass region are still used to normalize the Drell–Yan background contribution in the analysis.

The four momenta of quark and anti-quark used in the construction of observables are resolved by a reconstruction based on kinematic information: the transverse missing momentum is assumed to originate from the neutrinos, and the mass of reconstructed W boson and top quark should be equal to 80.4 GeV and 172.5 GeV, respectively. Detector resolution effects are taken into account through a smearing of the measured energies and directions of the reconstructed objects. The efficiency of this reconstruction procedure is around 90%; events without solutions for the neutrino momenta are excluded.

The observables are computed using the resolved four momenta of quark and anti-quark. A maximum likelihood fit is introduced to extract the asymmetries from the observables. The effects of systematic uncertainties are estimated by varying the nominal inputs and samples by the uncertainties, and repeating the full measurements. Most of the uncertainty sources are naturally cancelled in the asymmetries. The remaining major uncertainties are from limited simulated background samples, jet energy resolution and scale, top modeling such as color reconnection. None of the uncertainty source results have a variation larger than 0.003 on the asymmetries. The resulting asymmetries, which are summarized in Table 1, are found to be consistent with zero and used to derive a measurement on the CEDM parameter d_{tG}. The measured asymmetries and the coefficients of CEDM term are consistent with the expectation from the SM.

Table 1. The measured asymmetries $A_{\mathcal{O}_1}$ and $A_{\mathcal{O}_3}$, and the CEDM parameter d_{tG}.

Observable	Asymmetry ($\times 10^{-3}$)	d_{tG}
$\mathcal{O}_1$	$2.4 \pm 2.8(\text{stat}) \pm 2.8(\text{syst})$	$0.10 \pm 0.12(\text{stat}) \pm 0.12(\text{syst})$
$\mathcal{O}_3$	$0.4 \pm 2.8(\text{stat}) \pm 2.2(\text{syst})$	$0.00 \pm 0.13(\text{stat}) \pm 0.10(\text{syst})$

3.2. Analysis in the Lepton Plus Jets Channel

The lepton plus jets analysis exploits a different set of observables recommended in Ref. [23]:

$$\mathcal{O}_3 = Q_\ell \epsilon(p_b, p_{\bar{b}}, p_\ell, p_{j_1}) \propto Q_\ell \vec{p}^{\,*}_b \cdot (\vec{p}^{\,*}_\ell \times \vec{p}^{\,*}_{j_1}), \tag{8}$$

$$\mathcal{O}_6 = Q_\ell \epsilon(P, p_b - p_{\bar{b}}, p_\ell, p_{j_1}) \propto Q_\ell (\vec{p}_b - \vec{p}_{\bar{b}}) \cdot (\vec{p}_\ell \times \vec{p}_{j_1}), \tag{9}$$

$$\mathcal{O}_{12} = q \cdot (p_b - p_{\bar{b}}) \epsilon(P, q, p_b, p_{\bar{b}}) \propto (\vec{p}_b - \vec{p}_{\bar{b}})_z \cdot (\vec{p}_b \times \vec{p}_{\bar{b}})_z, \tag{10}$$

$$\mathcal{O}_{14} = \epsilon(P, p_b + p_{\bar{b}}, p_\ell, p_{j_1}) \propto (\vec{p}_b + \vec{p}_{\bar{b}}) \cdot (\vec{p}_\ell \times \vec{p}_{j_1}). \tag{11}$$

The symbol $*$ indicates the momenta at the $b\bar{b}$ center-of-mass frame; the z subscript indicates the projection along the beam axis; P (q) is the sum (difference) of the four-momenta of the protons in the pp collision; $p_j 1$ represents the momentum of the jet with highest transverse momentum; Q_ℓ is the lepton charge. The presence of CP violation will result in a nonzero asymmetry defined in Equation (7).

The study uses the data collected at $\sqrt{s} = 13$ TeV corresponding to an integrated luminosity of 138 fb^{-1}. Trigger requirement includes the presence of an isolated lepton with a p_T above 24–35 GeV. In the offline analysis, the electron candidates are required to have $p_T > 38$ GeV and within $|\eta| < 2.4$ (excluding the gap between barrel and endcap calorimeter, $1.44 < |\eta| < 1.57$). Muon candidates are required to have a transverse momentum above 30 GeV and within $|\eta| < 2.4$. The flavor of lepton (electron or muon) is used to categorizing the events too.

Jets are clustered using the anti-k_T algorithm with a distance parameter of 0.4, with a minimal p_T threshold of 30 GeV. Jets should be separated from the selected lepton candidate by an angular distance of $\Delta R > 0.4$. Jets from the hadronization of bottom quarks are identified using a deep-learning neural network based algorithm.

The events are required to have a reconstructed lepton, and at least four reconstructed jets; two of the jets should be tagged as bottom quarks. The association of top quark and anti-top quark with the final state jets and leptons are resolved with a χ^2 algorithm that introduces the masses of top quark and W boson as constrained conditions: $\chi^2 = [(m_{jjb} - m_t)/\sigma_t]^2 + [(m_{jj} - m_W)/\sigma_W]^2$, where m_{jjb} is the invariant mass of three jets (two non-b-tagged and one b-tagged); m_t, σ_t, m_W and σ_W are the mass of top quark, resolution of reconstructed top quark mass, mass of W boson, and resolution of reconstructed W boson mass, respectively. The object assignment is decided by selecting the permutation with the lowest χ^2 score. By imposing the requirements of $\chi^2 < 20$ and $m_{lb} < 150$ GeV, the fraction of correctly assigned b jets is around 74% with an efficiency of 65% and a purity of $t\bar{t}$ events is 95%.

The measured asymmetry can be diluted by the detector and mis-reconstruction. In the analysis, the effect is parametrized with a dilution factor D, and the raw asymmetry (denoted as A'_{CP}) and the ideal A_{CP} are related as a multiplicative correction $A'_{CP} = DA_{CP}$. The values of D and the associated systematic uncertainties have been determined from simulations: $D(\mathcal{O}_3) = 0.46^{+0.01}_{-0.02}$, $D(\mathcal{O}_6) = 0.44^{+0.01}_{-0.02}$, $D(\mathcal{O}_12) = 0.74^{+0.01}_{-0.02}$, and $D(\mathcal{O}_14) = 0.60 \pm 0.01$. The observables $\mathcal{O}_3$ and $\mathcal{O}_6$ require distinguishing bottom and anti-bottom quarks, resulting in a lower chance of correct reconstruction and smaller D value. Nevertheless, as A_{CP} and A'_{CP} are linearly dependent, a non-zero A'_{CP} is already evidence of CP violation in the top sector; hence, the raw asymmetry A'_{CP} is considered as the primary result of the analysis.

The asymmetries are computed with the signal yields with positive or negative values of $\mathcal{O}_i$. These yields are determined with an extended maximum likelihood fit to the invariant mass distributions of lepton and b-tagged jet. Systematic uncertainties are also largely cancelled in the asymmetry measurements. The detector effects are studied using an event-mixing method, which is mixing the momentum of the b-tagged jet and the highest p_T light-flavor jet across different events. Other experimental and theoretical systematic uncertainties are all found to be tiny in this analysis, resulting in a total systematic uncertainty smaller than 0.001 on the asymmetries. The measured asymmetries can be used to provide a constraint on the CEDM contributions. The detector effects are first removed by dividing the dilution factors. The constraints on CEDM parameter are computed from the corrected asymmetries and then combined with the correlation among the CP observables taking into account. The resulting asymmetries as well as the derived CEDM parameters d_{tG} are summarized in Table 2; the parameter d_{tG} is measured to be $0.04 \pm 0.10 \pm 0.07$ with a combination of the results from the four observables. The measured asymmetries on the CP observables are consistent with the expectation from the SM, and show no hints for CP-violating effects.

Table 2. The measured raw A'_{CP}, the corrected asymmetry A_{CP}, and the derived CEDM parameter d_{tG}. The first uncertainty is statistical, and the second is systematic.

Observable	A'_{CP} (%)	A_{CP} (%)	d_{tG}
$\mathcal{O}_3$	$-0.05 \pm 0.09^{+0.04}_{-0.07}$	$-0.10 \pm 0.20 \pm 0.14$	$+0.04 \pm 0.11 \pm 0.07$
$\mathcal{O}_6$	$-0.13 \pm 0.09^{+0.05}_{-0.07}$	$-0.30 \pm 0.21 \pm 0.16$	$+0.25 \pm 0.20 \pm 0.15$
$\mathcal{O}_{12}$	$+0.09 \pm 0.09^{+0.03}_{-0.05}$	$+0.12 \pm 0.13 \pm 0.07$	$+0.45 \pm 0.47 \pm 0.27$
$\mathcal{O}_{14}$	$-0.17 \pm 0.09^{+0.09}_{-0.02}$	$-0.29 \pm 0.16 \pm 0.14$	$-0.81 \pm 0.48 \pm 0.44$

4. Search for CLFV in Top Quark Production and Decay

A search for CLFV in both top quark associated production, and decay is presented by the CMS Collaboration [27]. The effective field theory approach is followed for parametrizing the CLFV effects. The top related operators are categorized based on their Lorentz structure to vector-, scalar- and tensor-like operators as the following:

$$O_{\text{vector}} = (\bar{l}_a\gamma^{\mu}l_b)(\bar{q}_c\gamma_{\mu}q_d) + (\bar{l}_a\gamma^{\mu}l_b)(\bar{u}_c\gamma_{\mu}u_d) + (\bar{e}_a\gamma^{\mu}e_b)(\bar{q}_c\gamma_{\mu}q_d) + (\bar{e}_a\gamma^{\mu}e_b)(\bar{u}_c\gamma_{\mu}u_d), \tag{12}$$

$$O_{\text{scalar}} = (\bar{l}_a e_b)\ \varepsilon\ (\bar{q}_c u_d) + \text{h.c}, \tag{13}$$

$$O_{\text{tensor}} = (\bar{l}_a\sigma^{\mu\nu}e_b)\ \varepsilon\ (\bar{q}_c\sigma_{\mu\nu}u_d) + \text{h.c}, \tag{14}$$

where $a \neq b$ are lepton-flavor indices, c and d are quark-flavor indices, q and l represent left-handed fermion doublets, u and e the right-handed fermion singlets, τ^I the Pauli matrices, $\varepsilon \equiv i\tau^2$ is the antisymmetric $SU(2)$ tensor, $\sigma^{\mu\nu} = \frac{i}{2}[\gamma^{\mu}, \gamma^{\nu}]$, and γ^{μ} the Dirac matrices. Three Wilson coefficients C_{vector}, C_{scalar}, and C_{tensor} are probed individually in this analysis. The CLFV interactions contribute to the single top production and decay of the top quark in top quark–antiquark pair production ($t\bar{t}$). Due to the larger cross section of the CLFV process in the production mode compared to the decay mode and more distinctive kinematic distributions of the production mode with respect to the SM background, the production mode plays a leading role in the sensitivity of the search. This analysis presents results of the first search for "eμtu" and "eμtc" CLFV interactions in the eμ final state.

The analysis is based on pp collisions collected by the CMS detector at the LHC at a center-of-mass energy of 13 TeV, corresponding to an integrated luminosity of 138 fb^{-1}. Events with one oppositely charged electron–muon pair in the final state, along with at least one jet identified as originating from a bottom quark (b-tagged jet) are selected. The leading (sub-leading) lepton p_T should be greater than 25(20) GeV and to lie within $|\eta| < 2.4$. Selected events are required to have at least one b-tagged jet with $p_T > 20\,\text{GeV}$ and $|\eta| < 2.4$. Selected background events dominated with SM $t\bar{t}$ events ($\approx$90%), followed by single top quark production in association with a W boson (tW) ($\approx$10%). The contributions from the SM background processes are estimated using the simulated events. Events are categorized further based on number of b-tagged jets to signal and $t\bar{t}$ control regions by requiring exactly one and greater than 1 b-tagged jets, respectively. A boosted decision tree (BDT) is trained based on the distinctive features of the signal process in the production mode with respect to the main backgrounds to maximize the sensitivity of the search.

Various sources of systematic uncertainty from modeling of the detector response and theoretical modeling of the signal and background processes are considered in this search. The final BDT distribution in the signal region $t\bar{t}$ control regions fitted simultaneously using a binned likelihood function to test for the presence of signal events. All the systematic uncertainties are treated as nuisance parameters in the fit. The fit results are consistent with the SM prediction and show no evidence for the presence of the CLFV signal. Therefore, upper limits are set on the signal cross sections at a 95% confidence level using the modified frequentist CLs method. Upper limits on the cross section of the CLFV processes are translated to the upper limits on the Wilson coefficients and related branching fractions of the top quark $\mathcal{B}(t \to e\mu q)$, q = u (c) quark. Limits obtained for vector-, scalar-, and tensor-like interactions are summarized in Table 3. In Figure 1, the results for two-dimensional limits on CLFV Wilson coefficients and branching fractions are displayed.

Table 3. Expected and observed 95% CL upper limits on the CLFV Wilson coefficients and top quark CLFV branching fractions.

Vertex	Int. Type	$C_{e\mu tq}/\Lambda^2$ [TeV^{-2}] Exp	Obs	$\mathcal{B}(10^{-6})$ Exp	Obs
eμtu	Vector	0.12	0.12	0.14	0.13
	Scalar	0.23	0.24	0.06	0.07
	Tensor	0.07	0.06	0.27	0.25

Table 3. *Cont.*

Vertex	Int. Type	$C_{e\mu tq}/\Lambda^2$ [TeV^{-2}] Exp	Obs	$\mathcal{B}(10^{-6})$ Exp	Obs
eμtc	Vector	0.39	0.37	1.49	1.31
	Scalar	0.87	0.86	0.91	0.89
	Tensor	0.24	0.21	3.16	2.59

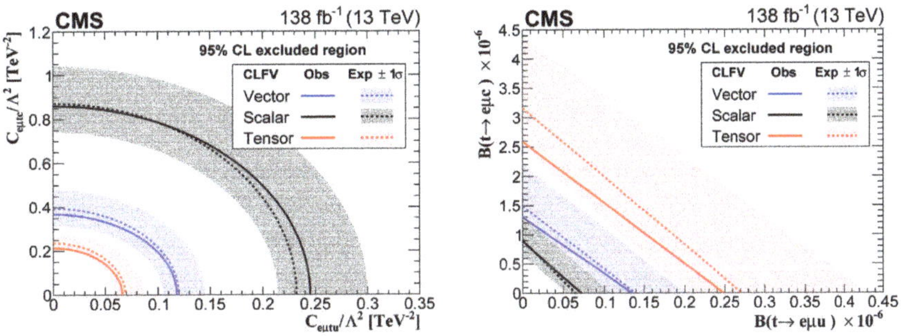

Figure 1. The observed 95% CL exclusion limits on the eμtc; of the eμtu Wilson coefficient (**left**) and $\mathcal{B}(t \to e\mu c)$ as a function of $\mathcal{B}(t \to e\mu u)$ (**right**) for the vector-, scalar-, and tensor-like CLFV interactions. The hatched bands indicate the regions containing 68% of the distribution of limits expected under the background only hypothesis.

5. Conclusions and Outlook

In this article, several studies on the tests of charge–parity (CP) symmetry and lepton flavor conservation have been reviewed. ATLAS collaboration has performed a measurement of a CP-violating forward-backward asymmetry in t-channel single top quark events, where top quarks are expected to be highly polarized. This measurement is consistent with null forward-backward asymmetry, and CP is conserved. The result has been used to derive a constraint on the anomalous coupling under the effective operator framework. CMS collaboration has looked for CP asymmetries with top quark pair productions in dilepton and lepton plus jets' final states. The CP asymmetries are tested with the CP-odd triple-product observables, constructed using the four-momenta reconstructed from with the final-state particles. The asymmetries of the proposed observables are measured to be consistent with zero and then converted to the measurements of the chromoelectric dipole moment contribution to the top quark. A search for charged-lepton flavor violation (CLFV) has been carried out by CMS in top quark production and decay as well. Events with oppositely charged electron–muon pairs are selected in the study, and no significant excess over the background is observed. Limits are set on the decay branching fractions as well as the strength of four-fermion effective operators in the framework of an effective field theory approach.

Although none of the existing experimental searches to date shows evidence of symmetry breaking effects in the top sector, the studies should be still pursued. As the Standard Model of particle physics failed to explain several important aspects of the nature, such as baryogenesis of the universe and the origin of neutrino masses, substantial tests of these fundamental symmetries and the conservation laws governing them are the key methods to explore the open issues in the particle physics. As the top quark does play an important role in the Standard Model, it is mandatory to refine the existing measurements and look for other possibilities in the future. In particular, the upcoming high-luminosity LHC project will provide more than a factor of twenty statistics of top quarks, compared to the currently accumulated datasets during LHC Run-2. Precision of existing analyses can be further improved by one or two orders of magnitudes. More observables can be further examined with the data produced at future LHC or with the planned high-energy e^+e^- colliders.

Author Contributions: Writing—review and editing, K.-F.C. and R.G. All authors have read and agreed to the published version of the manuscript.

Funding: This research received no external funding.

Institutional Review Board Statement: Not applicable.

Informed Consent Statement: Not applicable.

Data Availability Statement: No new data were created or analyzed in this study. Data sharing is not applicable to this article.

Conflicts of Interest: The authors declare no conflict of interest.

References

1. Christenson, J.H.; Cronin, J.W.; Fitch, V.L.; Turlay, R. Evidence for the 2π Decay of the K_2^0 Meson. *Phys. Rev. Lett.* **1964**, *13*, 138–140. [CrossRef]
2. Kobayashi, M.; Maskawa, T. CP Violation in the Renormalizable Theory of Weak Interaction. *Prog. Theor. Phys.* **1973**, *49*, 652–657. [CrossRef]
3. Abe, K.; Abe, R.; Abe, T.; Adachi, I.; Ahn, B.S.; Aihara, H.; Akatsu, M.; Asano, Y.; Aso, T.; Aushev, T.; et al. Observation of mixing induced CP violation in the neutral B meson system. *Phys. Rev. D* **2002**, *66*, 032007. [CrossRef]
4. Aubert, B. Improved measurement of the CP-violating asymmetry amplitude $\sin 2\beta$. *Frascati Phys. Ser.* **2002**, *27*, 515–538.
5. Glashow, S.L.; Iliopoulos, J.; Maiani, L. Weak Interactions with Lepton-Hadron Symmetry. *Phys. Rev. D* **1970**, *2*, 1285–1292. [CrossRef]
6. Atwood, D.; Bar-Shalom, S.; Eilam, G.; Soni, A. CP violation in top physics. *Phys. Rept.* **2001**, *347*, 1–222. [CrossRef]
7. Diaz-Cruz, J.L.; Toscano, J.J. Lepton flavor violating decays of Higgs bosons beyond the standard model. *Phys. Rev. D* **2000**, *62*, 116005. [CrossRef]
8. Crivellin, A.; Fabisiewicz, Z.; Materkowska, W.; Nierste, U.; Pokorski, S.; Rosiek, J. Lepton flavour violation in the MSSM: Exact diagonalization vs. mass expansion. *J. High Energy Phys.* **2018**, *6*, 003. [CrossRef]
9. Malinsky, M.; Ohlsson, T.; Xing, Z.Z.; Zhang, H. Non-unitary neutrino mixing and CP violation in the minimal inverse seesaw model. *Phys. Lett. B* **2009**, *679*, 242–248. [CrossRef]
10. Glattauer, R.; Schw, A.C.; Abdesselam, A.; Adachi, I.; Adamczyk, K.; Aihara, H.; Said, S.A.; Asner, D.M.; Aushev, T.; Ayad, R. Measurement of the decay $B \to D\ell\nu_\ell$ in fully reconstructed events and determination of the Cabibbo-Kobayashi-Maskawa matrix element $|V_{cb}|$. *Phys. Rev. D* **2016**, *93*, 032006. [CrossRef]
11. LHCb Collaboration; Aaij, R. Test of lepton universality in beauty-quark decays. *Nat. Phys.* **2022**, *18*, 277–282. [CrossRef]
12. Glashow, S.L.; Guadagnoli, D.; Lane, K. Lepton Flavor Violation in B Decays? *Phys. Rev. Lett.* **2015**, *114*, 091801. [CrossRef] [PubMed]
13. Bißmann, S.; Grunwald, C.; Hiller, G.; Kröninger, K. Top and Beauty synergies in SMEFT-fits at present and future colliders. *J. High Energy Phys.* **2021**, *6*, 10. [CrossRef]
14. Kim, T.J.; Ko, P.; Li, J.; Park, J.; Wu, P. Correlation between $R_{D^{(*)}}$ and top quark FCNC decays in leptoquark models. *J. High Energy Phys.* **2019**, *7*, 025. [CrossRef]
15. ATLAS Collaboration. *Search for $\mathcal{CP}$ Violation in Single Top Quark Events in pp Collisions at $\sqrt{s}$ = 7 TeV with the ATLAS Detector*; ATLAS-CONF-2013-032; ATLAS Experiment, CERN: Geneva, Switzerland, 2013.
16. Aguilar-Saavedra, J.A. A Minimal set of top anomalous couplings. *Nucl. Phys. B* **2009**, *812*, 181–204. [CrossRef]
17. Aguilar-Saavedra, J.A. A Minimal set of top-Higgs anomalous couplings. *Nucl. Phys. B* **2009**, *821*, 215–227. [CrossRef]
18. Mahlon, G.; Parke, S.J. Single top quark production at the LHC: Understanding spin. *Phys. Lett. B* **2000**, *476*, 323–330. [CrossRef]
19. Schwienhorst, R.; Yuan, C.P.; Mueller, C.; Cao, Q.H. Single top quark production and decay in the t-channel at next-to-leading order at the LHC. *Phys. Rev. D* **2011**, *83*, 034019. [CrossRef]
20. Gonzalez-Sprinberg, G.A.; Martinez, R.; Vidal, J. Top quark tensor couplings. *J. High Energy Phys.* **2011**, *7*, 094. [CrossRef]
21. Aguilar-Saavedra, J.A.; Bernabeu, J. W polarisation beyond helicity fractions in top quark decays. *Nucl. Phys. B* **2010**, *840*, 349–378. [CrossRef]
22. Gupta, S.K.; Mete, A.S.; Valencia, G. CP violating anomalous top-quark couplings at the LHC. *Phys. Rev. D* **2009**, *80*, 034013. [CrossRef]
23. Hayreter, A.; Valencia, G. T-odd correlations from the top-quark chromoelectric dipole moment in lepton plus jets top-pair events. *Phys. Rev. D* **2016**, *93*, 014020. [CrossRef]
24. CMS Collaboration. *Search for CP Violating Top Quark Couplings in pp Collisions at $\sqrt{s}$ = 13 TeV*; Report Number CMS-PAS-TOP-18-007, CMS-PAS-TOP-18-007; CERN: Geneva, Switzerland, 2022.

25. The CMS Collaboration; Khachatryan, V.; Sirunyan, A.M.; Nayak, A.; Bargassa, P.; Beirão Da Cruz E Silva, C.; Calpas, B.; Di Francesco, A.; Faccioli, P.; Ferreira, Parracho, P.G.; et al. Search for CP violation in $t\bar{t}$ production and decay in proton-proton collisions at $\sqrt{s} = 8$ TeV. *J. High Energy Phys.* **2017**, *3*, 101. [CrossRef]
26. CMS Collaboration. Search for CP violation using $\mathrm{t\bar{t}}$ events in the lepton+jets channel in pp collisions at $\sqrt{s}$ = 13 TeV. *arXiv* **2022**, arXiv:2205.02314.
27. CMS Collaboration; Tumasyan, A. Search for charged-lepton flavor violation in top quark production and decay in pp collisions at $\sqrt{s}$ = 13 TeV. *J. High Energy Phys.* **2022**, *6*, 82. [CrossRef]

universe

MDPI

Essay

On Extra Top Yukawa Couplings of a Second Higgs Doublet

George Wei-Shu Hou

Department of Physics, National Taiwan University, Taipei 10617, Taiwan; wshou@phys.ntu.edu.tw

Abstract: A very likely New Physics in plain sight, but that the community does not see, is a second Higgs doublet that has a second set of Yukawa couplings. The extra *tt* and *tc* couplings can each drive baryogenesis, with O(1) Higgs quartic couplings providing a first order electroweak phase transition. A natural cancellation mechanism can tame electron EDM, if extra *ee*, *tt* couplings "know" the known fermion mass and mixing hierarchies. Colliding *c* with *g* produces tH/A, bH^+ via extra *tc* coupling, and together with extra *tt* coupling give *ttc*(bar), *ttt*(bar), and *btb*(bar) signatures at the LHC. Extra *tu* coupling can also be probed, but more definitive would be the B to $\mu\nu$ and $\tau\nu$ decay rate ratio. Myriad extra Yukawa couplings can make an impact on flavor physics and CP violation, including on muon g-2. The opening to the prelude of a new physics Higgs and flavor era may unfold before us.

Keywords: second Higgs doublet; Yukawa couplings; baryogenesis; Higgs quartic couplings; phase transition; electric dipole moment (EDM); mass and mixing hierarchy; LHC; flavor physics; *CP* violation; muon g-2

Citation: Hou, G.W.-S. On Extra Top Yukawa Couplings of a Second Higgs Doublet. *Universe* **2022**, *8*, 475. https://doi.org/10.3390/universe8090475

Academic Editors: Pedro Ferreira da Silva and Freya Blekman

Received: 25 April 2022
Accepted: 31 August 2022
Published: 9 September 2022

1. Introduction: Historical Development of Extra Top Yukawa Couplings

Mass is central to physics, energy alone is not enough. Without mass, an electron will not bind to a nucleus to form an atom, nor planets to its sun to form solar systems. For elementary particles with no known structure, mass is dynamically generated by coupling to the Higgs field. For the SU(2) gauge theory of the weak interactions of the standard model (SM), Weinberg introduced [1] a complex Higgs doublet to spontaneously generate the vacuum expectation value $v \cong 246$ GeV, giving mass to vector bosons via the Higgs mechanism [2–4]: $2m_V = gv$, with g the gauge coupling. As left-handed fermions come in weak doublets while right-handed fermions in singlets, Weinberg introduced [1] the Yukawa interaction Lagrangian by balancing the gauge charge of the left–right fermion bilinear with the Higgs doublet. Thus, charged fermion mass generation is analogous to the Higgs mechanism: $\sqrt{2}m_f = \lambda_f v$, with λ_f the Yukawa coupling. This linear relation of Higgs couplings vs mass was spectacularly demonstrated recently [5] by ATLAS and CMS, from $\lambda_t \cong 1$ for top, down to λ_μ for the muon [6] that is over a thousand times weaker! Yukawa couplings are now measured dynamically.

With Yukawa couplings now "real", they are also "complex". The chiral nature of fermions in the above discussion means that the Yukawa couplings involved in fermion mass generation are complex. Kobayashi and Maskawa demonstrated [7] that it takes three generations of u- and d-type quarks to yield a single *CP* violating (CPV) phase; all laboratory-verified CPV effects so far [5] can be accounted for by this unique phase.

With $\lambda_t \cong 1$, the strongest coupling we know, we turn to the possibility of extra top Yukawa couplings. To explain the absence of antimatter in the Universe is one of the greatest challenges facing particle physics, which provides the biggest motivation for exploring extra top Yukawa couplings. The fourth generation (4G) naturally comes to mind [8], which, besides $\lambda_{t'}$, $\lambda_{b'}$ themselves, there are their mixings with the top quark. Indeed, around 2010 there was some hope [9] that 4G may provide enough CPV for electroweak baryogenesis (EWBG [10]), although a sufficient first order phase transition seemed lacking. However, except discovering [5] the expected SM-like $h(125)$, no new physics has emerged at the LHC, and 4G is also absent. Furthermore, $h(125)$ production through gluon–gluon fusion is consistent with only the top quark in the triangle loop, rather than having t, t', b' all contributing [8], which would have given an order of magnitude larger cross section.

Next would be vector-like quarks (VLQ), where left- and right-handed quarks have the same gauge charge, but can mix with SM quarks. For example, when the top remained unseen in the early 1990s, a left–right singlet Q was invoked [11] to hide the top via $t \to cH$ (H here is SM Higgs boson) by t–Q mixing, i.e., extra top Yukawa couplings. It was refuted, however, that t–Q mixing cannot generate sufficiently large c–t mixing [12], as it would still be suppressed by m_c/m_Q. There is further the issue of m_Q scale: be it singlet or doublet, the gauge invariant m_Q is arbitrary, making its plausibility dubious, as we have never seen one [5]. Since the VLQ representations do not follow the SM pattern, they are seen as more exotic than 4G. However, there were still some interest in VLQs, e.g., in the LEP2 era [13].

With $h(125)$ discovery at the LHC, however, at the same time that it spelled the demise of 4G [14,15], VLQ became in vogue. For example, in the context of "composite Higgs" models [16–20], to protect the lightness of $h(125)$, one "engineers" a cancellation of top-loop corrections by additional VLQ loops. This bears analogy to SUSY where there are "scalar top partners" (stop), so these VLQs are also called vector-like "top partners". The cancellation mechanism points m_Q to the TeV scale. Before long, one had handbooks and hunter's guides [21,22]. However, a decade after $h(125)$ discovery, experimental bounds from various singlet/doublet T and B VLQ searches have reached the TeV scale, but they still have not been seen [5]. Are VLQ top partners going the way of SUSY particles? The pursuit certainly continues.

In this article, we comment on another type of extra top Yukawa couplings, those that ought to be present if a second Higgs doublet (2HDM) exists [23] in *Nature*. Like 4G, the existence of a second Higgs doublet is a conservative extension of SM. The "traditional" approach, however, is to enforce the Natural Flavor Conservation (NFC) condition of Glashow and Weinberg [24], that only one doublet can generate fermion masses, which forbids the extra Yukawa couplings altogether by fiat! Though clearly ad hoc, the NFC condition is usually implemented via a Z_2 symmetry, hence come in two types [23]: in Model I, both u- and d-type quarks receive mass from the same doublet; for Model II, they receive mass from separate doublets. The latter arises automatically in SUSY, making 2HDM II the most popular. Dropping NFC, however, there is a third type, 2HDM III [25], one that possesses extra Yukawa couplings. Such couplings were used to stress $t \to ch$ [25] as a decay mode to watch, where h is some Higgs boson lighter than top.

So how does one address Glashow's concern of flavor changing neutral Higgs (FCNH) couplings? As pointed out by Cheng and Sher [26], by capitalizing on fermion mass and especially mixing hierarchies that emerged after the NFC paper—and never predicted by SM— *Nature* has spoken her mind! It was further clarified [25] that one need not assume a specific mass-mixing "Ansatz", as in Ref. [26], but the known mass-mixing hierarchies may suffice to hide the effect of the second doublet. Just before the top finally emerged, it was noted that $t \to ch$ for some light h boson could actually still hide the top below M_W [27], but *Nature* soon unveiled the surprisingly heavy top at Fermilab.

Fast forward to LHC, the lightness of $h(125)$ itself indeed immediately prompted [25] the experimental pursuit of $t \to ch$, first done by ATLAS [28], and the clarification [29] that, aside from the extra FCNH coupling ρ_{tc}, there is a mixing factor $\cos(\beta - \alpha)$ in SUSY notation, and later called plainly $\cos\gamma$ in 2HDM III. Surprisingly, the emergent "alignment" phenomenon [30], that h so resembles the SM Higgs boson, helped clarify the situation further: the mixing angle $c_\gamma \equiv \cos\gamma$ of h with the exotic CP-even Higgs, H, is small, which can explain the absence [5] of $t \to ch$ so far. The emerging picture is that the mass-giving doublet Φ, to which h belongs, does not mix much with the exotic second doublet Φ' that is not involved in generating v. The absence so far in searches [5] for the H, pseudoscalar A and H^+ bosons from Φ' means the Φ and Φ' doublets are somewhat separate in mass scale. We shall argue that one need not send the exotic Higgs bosons to multi-TeV, as in the minds of some SUSY advocates, but they should be sub-TeV in mass for reasons [31] we shall discuss. Such a mass range should be fully explored at the LHC in any case.

Reflecting our times, with no new physics found, it is common to assume that the scale Λ of beyond SM (BSM) physics is high, hence a popular approach is [32] effective field theory (EFT), i.e., expanding in powers of $1/\Lambda$ with higher dimension operators. We do not take this approach, however, as we do not think dimension-4 operators are exhausted yet.

Thus, this article proceeds as follows. We promote 2HDM III and call it the general 2HDM (g2HDM), where an extra Higgs doublet exists but without NFC, and let *Nature* speak her design. Descending down this "Road Not Taken" by most, we present in Section 2 two sets of dimension-4 dynamical operators, namely the extra Higgs quartic couplings and the extra Yukawa couplings. In Section 3, we assert that the extra top Yukawa couplings ρ_{tt} and ρ_{tc} can separately drive EWBG, while the current bound on the electron electric dipole moment (eEDM) can be satisfied with finite ρ_{ee}, where intriguingly its ratio with ρ_{tt} must reflect the ratio of λ_e/λ_t with specific phase correlation. That is, the two types of Yukawa matrices "know" each other. For the sake of EWBG, we argue that the extra Higgs bosons ought to be sub-TeV in mass, hence ripe for the LHC to explore. In Section 4, we recount the main production processes at the LHC, all with ρ_{tc} as crux, but also comment on accessing ρ_{tu}. Turning to flavor, we explore in Section 5 the special ratio of $\mathcal{B}(B \to \mu\nu)/\mathcal{B}(B \to \tau\nu)$, that if it is found to deviate from 0.0045 by Belle II, it not only would rule out both SM and 2HDM II, but point to $\rho_{tu} \neq 0$ in g2HDM, which would provide impetus for collider study. We then generalize and make clear that the pursuit of extra Higgs bosons with extra Yukawa couplings (not just top-related) is an experimental question, altogether involving over 60 new parameters. A new era on Higgs and flavor could unfold before us, and we offer the prospects and conclusions in Section 6.

This article is written in essay style, giving a narrative on the physics but avoiding formulas as much as possible. A companion article with considerably more detail on the new Higgs/flavor era can be found in Ref. [33].

2. Two Sets of Dimension-4 Dynamical Operators

The general 2HDM, or g2HDM, has an extra Higgs doublet that possesses extra Yukawa couplings, i.e., without the NFC condition imposed. It thus provides two sets of dimension-4 operators to be scrutinized at the LHC and by flavor experiments: extra Higgs quartics, and extra Yukawa couplings. This is in contrast with the present EFT trend [32], that there is a high scale Λ for BSM operators, or else one has particles with weaker than SM couplings. Actually, g2HDM does involve weaker couplings, in general, but is not part of the dark sector, with coupling strengths to be determined by experiment.

Generalizing the familiar Higgs potential of SM, $V(\Phi) = \mu^2|\Phi|^2 + \lambda|\Phi|^4$ with $\mu^2 < 0$, the Higgs potential of g2HDM, i.e., with no Z_2 symmetry imposed, is [31,34]

$$V(\Phi, \Phi') = \mu_{11}^2|\Phi|^2 + \mu_{22}^2|\Phi'|^2 - \left(\mu_{12}^2\Phi^\dagger\Phi' + \text{h.c.}\right) + \frac{1}{2}\eta_1|\Phi|^4 + \frac{1}{2}\eta_2|\Phi'|^4 + \eta_3|\Phi|^2|\Phi'|^2 + \eta_4|\Phi^\dagger\Phi'|^2 + \left[\frac{1}{2}\eta_5(\Phi^\dagger\Phi')^2 + \left(\eta_6|\Phi|^2 + \eta_7|\Phi'|^2\right)\Phi^\dagger\Phi' + \text{h.c.}\right], \quad (1)$$

in notation of Ref. [31], with Φ the mass-giving doublet that is responsible for v, and Φ' the exotic doublet with $\langle\Phi'\rangle = 0$, i.e., $\mu_{22}^2 > 0$. This natural separation is called the Higgs basis [35]. With $\mu_{11}^2 < 0$, v is generated in the usual way, except for a slight change in convention [31]. A second minimization condition $\mu_{12}^2 = \eta_6 v^2/2$ eliminates μ_{12}^2, and η_6 is the sole parameter for Φ-Φ' mixing. Note that η_6, η_7 would be absent under usual Z_2 to impose NFC. Thus, g2HDM has two inertial mass parameters $\mu_{11}^2 < 0$ (transferred to v) and $\mu_{22}^2 > 0$, plus seven quartics η_i. It is more intuitive than 2HDM I and II, where both μ_{11}^2, $\mu_{22}^2 < 0$ (transferred to v_1, v_2), with μ_{12}^2 playing the dual role of inertial mass and Φ-Φ' mixing, while η_6 and η_7 are absent. In SUSY, μ_{12}^2 is now often considered at several TeV^2.

The extra Yukawa couplings for charged fermions are [34,36,37]

$$-\frac{1}{\sqrt{2}}\sum_{f=u,d,\ell}\bar{f}_i\Big[\big(-\lambda_i^f\delta_{ij}\,s_\gamma+\rho_{ij}^f\,c_\gamma\big)\,h+\big(\lambda_i^f\delta_{ij}\,c_\gamma+\rho_{ij}^f\,s_\gamma\big)\,H-i\,\mathrm{sgn}(Q_f)\,\rho_{ij}^f\,A\Big]R\,f_j$$
$$-\bar{u}_i\Big[(V\rho^d)_{ij}\,R-(\rho^{u\dagger}V)_{ij}\,L\Big]d_j\,H^+-\bar{\nu}_i\,\rho_{ij}^\ell\,R\,\ell_j\,H^++\text{h.c.},\qquad (2)$$

in notation of Ref. [37], where i and j are summed over generations, $L,R=(1\mp\gamma_5)/2$ are projections, and V the Cabibbo–Kobayashi–Maskawa matrix; the lepton matrix is taken as unity due to vanishing neutrino mass. The h–H mixing angle c_γ is known to be small [30], but the value in g2HDM would be harder to extract than 2HDM II. The smallness of c_γ means that h largely arises from Φ, while Φ' gives exotic Higgs bosons H, A, and H^+. As stressed in the Introduction, the fermion mass hierarchies $m_1^2 \ll m_2^2 \ll m_3^2$ across generations plus $m_b^2 \ll m_t^2$, together with the mixing hierarchy $|V_{ub}|^2 \ll |V_{cb}|^2 \ll |V_{us}|^2 \ll |V_{tb}|^2 \cong 1$, if reflected also in the ρ_{ij}^f Yukawa matrices, may be *Nature*'s mechanism in hiding the exotic bosons so far. None of these relations were predicted, and constitute together the flavor enigma, with the non-flavor $c_\gamma^2 \ll 1$ thrown in as bonus. For example, the bound on $t\to ch$ is now below 10^{-3}, but this can be largely absorbed by c_γ^2, rather than small ρ_{tc}^2.

Equation (1) has 8 parameters (including c_γ) besides v, and Equation (2) has $9\times 3\times 2=54$ parameters, but the latter extend from similar SM parameters, which are already plenty. Linking to CPV in the Heavens (baryon asymmetry of the Universe) and on Earth (electron EDM constraint), one finds profound implications, as discussed in the next section:

- CPV for EWBG calls for $\mathcal{O}(1)$ extra top Yukawa couplings, while first order phase transition calls for $\mathcal{O}(1)$ Higgs quartics. The latter, in turn, suggests sub-TeV exotic Higgs masses, as we shall see.
- For the electron EDM constraint, the diagonal extra electron Yukawa coupling ρ_{ee} needs to correlate with extra top Yukawa coupling ρ_{tt} that echoes the known Yukawa coupling pattern.

3. Driving EWBG and Facing eEDM: Extra tt, tc and ee Couplings

A main reason to pursue extra Yukawa couplings is for *CP* violation: Kobayashi–Maskawa phase [7] can account for all CPV measured so far on Earth, but the CPV needed for baryogenesis is tremendously larger. As a Belle member, we keenly recall the time when the detector was under construction, there was the sense that, even if we demonstrated the Kobayashi–Maskawa phase, the baryon asymmetry of the Universe was far out of reach. That agony was only uplifted when the Belle direct CPV result [38] pointed to a possible fourth generation solution for baryogenesis [9]. However, with 4G now out of favor, as there are no clear hints of new physics whatsoever, the extra Yukawa couplings for each type of charged fermions in g2HDM should be kept and scrutinized experimentally.

In this vein, it was found [39] that the product

$$\lambda_t\,\mathrm{Im}\,\rho_{tt},\qquad (3)$$

can robustly account for EWBG: with $\lambda_t \simeq 1$ measured, a best guess for $|\rho_{tt}|$ is also $\mathcal{O}(1)$ and applicable to the imaginary part, making the product—interfering Φ and Φ'—rather transparent and more convincing than the previous 4G work [9]. Interestingly, if ρ_{tt} turns out accidentally small, then ρ_{tc} could kick in as a back-up option. However, it would need to be [39] close to 1 with near maximal CPV phase, hence is less robust than via ρ_{tt}, where strength at 0.1 is more than enough. Note that ρ_{bb} has also been employed [40,41] for EWBG, but whether the estimate of scattering off the expanding bubble wall is trustworthy enough has been questioned [42] for such light quarks. Furthermore, the strength of ρ_{bb}, by analogy with λ_b, is likely much weaker, and strongly constrained [36] by $b\to s\gamma$.

One prerequisite (of Sakharov) for EWBG [10] is a first order electroweak phase transition, which is possible if [43] quartic couplings η_i are also $\mathcal{O}(1)$. It was argued [31] that this implies H, A, and H^+ are sub-TeV in mass, or else perturbation theory starts to fail for exotic Higgs scatterings, which can be tested in principle at the LHC. Although it is not in itself a failure, one loses control of predictions. Similarly, μ_{22}^2/v^2 should [31] also be $\mathcal{O}(1)$, otherwise a large inertial mass would damp the needed scattering off the bubble wall of our very early Universe, hence quench EWBG.

Worthy of mention [31] is the η_1 coupling, the analog of $\lambda\Phi^4$ of SM. It could be only slightly less than 1 in strength, with $\mathcal{O}(1)$ η_6 coupling (h–H mixing) helping to push m_h down to the observed value by level repulsion. Thus, one may discover considerably larger hh production than SM expectation and harbinger [44] first order phase transition.

It is therefore interesting that, for the sake of EWBG, the exotic H, A, and H^+ bosons would be sub-TeV in mass, and the dimension-4 operators of Equations (1) and (2) are just right for the LHC to probe, which we shall turn to in the next section. However, before that, one needs to face the challenge of low energy precision measurements such as electron EDM.

Although $\lambda_t \,\mathrm{Im}\,\rho_{tt}$ being $\mathcal{O}(1)$ can achieve the lofty goal of baryogenesis, it does "expose" one to the current frontier of eEDM, d_e. The ACME experiment already pushed the frontier in 2014 [45], so in our EWBG work, we made [39] a simplified estimate by turning off ρ_{ee}, the extra diagonal electron Yukawa coupling. The point was to make a projection for the ACME upgrade to check. To our surprise, the ACME2018 update excluded [46] our "prediction" altogether, and we were forced to put ρ_{ee} back for a more complete study. However, the study turned out fruitful [47]: we found a (still) simplified ansatz for a cancellation mechanism, if *Nature* follows the pattern that (r depends on loop functions)

$$\frac{\mathrm{Im}\,\rho_{ff}}{\mathrm{Im}\,\rho_{tt}} = r\,\frac{\lambda_f}{\lambda_t}, \quad \frac{\mathrm{Re}\,\rho_{ff}}{\mathrm{Re}\,\rho_{tt}} = -r\,\frac{\lambda_f}{\lambda_t}, \tag{4}$$

i.e., extra Yukawa couplings echo the known mass-mixing hierarchy pattern of SM, with particular phase correlation! Even values that are two orders of magnitude below [47] ACME2018 bound can be entertained. However, we think the current bound of $\sim 1 \times 10^{-29}\, e\,\mathrm{cm}$ [46] should be scrutinized carefully, hopefully by several experiments using different approaches. A discovery at, or not far below, this bound would be exciting.

Equation (4) confirms our implicit assumption that the ρ^f matrices of Equation (2) "know" the mass and mixing hierarchies that *Nature* has revealed through SM Yukawa couplings long ago. We remark that the ρ_{tc} mechanism for baryogenesis evades eEDM constraint altogether [47], but the ρ_{tt} mechanism is more robust.

4. Crux of Production at Hadron Colliders: Extra tc Coupling

Gluon–gluon fusion production of H and A proceed via sizable ρ_{tt}, with H, A subsequently decaying to $t\bar{t}$, $t\bar{c}$ and also $\tau\mu$, $\tau\tau$. The $t\bar{t}$ final state is of interest, interfering with the enormous QCD production of $t\bar{t}$ pairs, with associated difficulty in analysis due to the rise-dip "signal" [48]. Experimental studies are ongoing; for example, a CMS study with partial Run 2 data reported some global excess [49] of 1.9σ at 400 GeV, with local excess higher. We await experimental progress. The $t\bar{c}$ final state is naively simple, but the catch is the unknown $t\bar{c}$ mass resolution. For H, A decaying to lepton pairs, if one takes $\rho_{\tau\mu}$, $\rho_{\tau\tau} \sim \lambda_\tau$, one suffers from branching ratio suppression as $t\bar{c}$, $t\bar{t}$ thresholds [50] turn on.

Thus, we advocate exotic Higgs production in association with a top or a bottom quark, with three main production processes: a gluon excites a charm quark, which emits H/A (H^+) via basically the ρ_{tc} coupling and turn into top (bottom). At the parton level, the processes are (we refer to Ref. [51] for a brief review):

1. $cg \to tH/tA \to tt\bar{c}$: Same-Sign Top plus c-jet [52];
2. $cg \to tH/tA \to tt\bar{t}$: Triple-Top [52];

3. $cg \to bH^+ \to bt\bar{b}$: Single-Top plus two b-jets [53].

4.1. Top-Associated Neutral Higgs Production

The production relies on the not well-constrained ρ_{tc} coupling. If ρ_{tt} vanishes or $m_{H,A} < 2m_t$, the H, A decay only to $t\bar{c}$ ($t\bar{t}c$ signature suffers high backgrounds). For finite ρ_{tt} and above $t\bar{t}$ threshold, the $tt\bar{t}$ final state would mutually dilute $tt\bar{c}$ cross-section.

Thus, top-associated H and A production lead to the two signatures of $tt\bar{c}$ and $tt\bar{t}$. For the former "Same-Sign-Top plus jet" production, the hard (charm) jet provides additional signature for discriminating against background. For the latter "Triple-Top" production, the cross-section can be several hundred times larger [51,52] than SM expectation [54], providing exquisite signature for detailed study at the high luminosity LHC (HL-LHC).

4.2. Bottom-Associated Charged Higgs Production

A simple but subtle extension of the above is $cg \to bH^+ \to bt\bar{b}$, i.e., the H^+ boson decays to $t\bar{b}$, while production is through H^+ emission from $c \to b$ transition. Here lies the subtlety: if one thinks in 2HDM II mindset, the process would be suppressed by V_{cb}. However, following Equation (2) carefully and working out the $V\rho^d$ or $\rho^{u\dagger}V$ products, one finds [53] that the $\bar{c}bH^+$ vertex is not V_{cb}-suppressed in g2HDM, hence receives V_{tb}/V_{cb} enhancement over 2HDM II! The $\bar{c}bH^+$ vertex is on equal footing with $\bar{t}bH^+$ vertex, which governs $H^+ \to t\bar{b}$ decay, that an undiluted ρ_{tc} coupling also governs $cg \to bH^+$ production in g2HDM. Thus, not only $cg \to bH^+$ is on equal footing as tH/tA associated production, it is favored in phase space by only requiring a light accompanying b quark instead of top.

One may think single top [5] production should provide stringent constraint. However, once again the two hard b and $\bar{b}$ jets are unusual for single top production, and discriminating against QCD-produced $b\bar{b}$ should be straightforward. Our collider study of the "Single Top plus two b-jets" signature did not spot [53] particularly worrisome backgrounds.

The $t\bar{t}t\bar{t}$ ($4t$) production at the LHC has a SM cross-section at 12 fb^{-1} level [54,55], larger than triple-top in SM and has been pursued [5] by both ATLAS and CMS. Lacking a similar pursuit for triple-top, we utilized [56] the $4t$ results to constrain ρ_{tc}–ρ_{tt} parameter space. However, we believe a genuine effort on triple-top should be pushed as the signature is quite different, and with possibility of discovering a much larger cross-section than $4t$, which may in turn provide extra backgrounds to some approaches of studying $4t$.

4.3. Enter the tu Coupling

So far we have considered ρ_{tt}, ρ_{tc} couplings but not ρ_{tu}. This is prudent, given that mass-mixing hierarchies imply ρ_{tu} should be considerably weaker. However, with two u quarks per proton, one should be careful. For instance, in the latest $t \to ch$ search by CMS [57] with $h \to \gamma\gamma$, one does incorporate single-top production off a valence u quark, and the constraint on $t \to uh$ is indeed more stringent than on $t \to ch$.

We have done a collider study [58] for $4t$ feed down to $ug \to tH/tA$, keeping ρ_{tu} but not ρ_{tc}. As in the $t \to qh$ study, we find stronger constraint on ρ_{tu} than for the analogous $cg \to tH/tA$ study, but certainly not as stringent as by $\sqrt{m_u/m_c}$, as implied by Cheng–Sher ansatz [26]. A dedicated search [58] comparing significance of positively versus negatively charged same-sign dilepton events could be fruitful, especially at the HL-LHC. Note that keeping both ρ_{tu} and ρ_{tc} in such a study may not be fruitful, since there is no good tool in separating c from u jets. This brings us to the next topic: a unique probe of ρ_{tu} through the ratio of $B \to \mu\nu$ and $B \to \tau\nu$ decay rates.

5. Turning to Flavor: Ratio of B to Muon+Neutrino vs. Tau+Neutrino

The subtlety of $cg \to bH^+$ being enhanced by V_{tb}/V_{cb} [53] in g2HDM compared with 2HDM II was originally uncovered through the study of $B \to \mu\nu$ vs $\tau\nu$ rate ratio. As we noted earlier in a flavor physics and CP violation (FPCP) review [59], the ratio

$$\frac{\mathcal{B}(B \to \mu\nu)}{\mathcal{B}(B \to \tau\nu)} = \frac{m_\mu^2 (m_B^2 - m_\mu^2)}{m_\tau^2 (m_B^2 - m_\tau^2)} \cong 0.0045, \tag{5}$$

holds for both SM and 2HDM II, because of the m_ℓ-independent correction factor for the latter [60] that we uncovered long ago. However, thanks to a referee remark, we stressed [59] that the SM ratio need not hold for g2HDM because more Yukawa couplings enter, hence provides a very interesting probe for Belle II. It turned out [61] rather interesting.

For $B \to \ell\nu$ decay, $\bar{u}$ annihilates the b quark, the decay constant f_B accounts for the disappearance of the meson, and the virtual H^- turns into $\ell\bar{\nu}_{\ell'}$ in g2HDM; this ℓ' index is the first subtlety [61]. As $\nu_{\ell'}$ goes undetected, according to Equation (2), the ℓ' flavor should be summed over, bringing in the extra lepton Yukawa couplings $\rho_{\ell\ell'}$ in g2HDM! That is, $\rho_{\mu\tau}$ for $B \to \mu\nu_\tau$, and $\rho_{\tau\mu}$ for $B \to \tau\nu_\mu$, besides the diagonal $\rho_{\mu\mu}$, $\rho_{\tau\tau}$, respectively.

The quark side is even more subtle [61], for both $V\rho^d$ and $\rho^{u\dagger}V$ in Equation (2). The former gives $\sum_i \rho_{ib} V_{ui} = \rho_{bb} V_{ub} + \rho_{sb} V_{us} + \rho_{db} V_{ud} \cong \rho_{bb} V_{ub}$, with ρ_{sb} and ρ_{db} constrained severely at tree level by B_s and B_d mixings. The smallness of $\rho_{bb} \sim \lambda_b$ and V_{ub} suppression means this term, the one that is operative for 2HDM II [60] by simple replacement, is small compared with $\rho^{u\dagger}V$ in g2HDM. This latter sum gives $\sum_i \rho_{iu}^* V_{ib} = \rho_{tu}^* V_{tb} + \rho_{cu}^* V_{cb} + \rho_{uu}^* V_{ub} \cong \rho_{tu}^* V_{tb}$, as ρ_{cu} is constrained by D^0 mixing and ρ_{uu} is suppressed by mass-mixing hierarchy, with both terms further CKM-suppressed. Since ρ_{tu} is poorly known experimentally while receiving $|V_{tb}/V_{ub}| \sim 300$ enhancement (!) compared to $\rho_{bb} V_{ub}$, this makes $B \to \ell\nu$ very interesting for Belle and Belle II [61]. Let us not analyze further, but note that $B \to \mu\nu$ can still deviate from SM expectation (but f_B-dependent), while $B \to \tau\nu$ is expected to be SM-like, which is the case observed [5], taking ratio cancels common factors, such as f_B.

The subtleties on the neutrino and/or the quark side were missed by earlier studies, such as Refs. [62,63], but could become definitive for g2HDM in the near future. Additionally, it was through clarifying the $V\rho^d$ and $\rho^{u\dagger}V$ products that led subsequently [53] to the $cg \to bH^+ \to bt\bar{b}$ process, where production is V_{tb}/V_{cb} enhanced in amplitude compared with 2HDM II thinking.

The current Belle result [64] for $\mathcal{B}(B \to \mu\nu) = (5.3 \pm 2.0 \pm 0.9) \times 10^{-7}$ is consistent with SM but with errors still sizable, so this ratio can be measured in the near future when Belle II accumulates just a factor of two or three times Belle data. If any deviation from 0.0045 is observed, it would not only be BSM, but beyond 2HDM II. We would then know $\rho_{tu} \neq 0$ in g2HDM, which should stimulate LHC studies. It furthermore points to finite $\rho_{\tau\mu}$ ($\rho_{\mu\tau}$), which can also be studied at the LHC via $gg \to H, A \to \tau\mu$ [50,65].

Put simply, Equation (5) probes the extra Yukawa coupling product $\rho_{tu}\rho_{\tau\mu}$ in g2HDM, receiving V_{tb}/V_{ub} enhancement in amplitude and thereby provides a sophisticated probe [61] of g2HDM through subtle H^+ effects. Any deviation from the SM value of 0.0045 would also rule against [59] 2HDM II that is automatic in SUSY.

We started with extra top Yukawa couplings because they are likely the strongest and least constrained, and can drive [39,47] baryogenesis. However, by now it should be clear that the issue of extra Yukawa couplings in 2HDM— if an extra scalar doublet actually exists—is an experimental question. The NFC condition of Glashow and Weinberg [24] is plainly ad hoc, dated, and ought to be retired, as it is best left for experiment to arbitrate.

We have barely touched upon the issue of flavor. A more detailed survey finds [66] that $\mu \to e\gamma$, $\mu N \to eN$ conversion and $\tau \to \mu\gamma$ are rather interesting via ρ_{tt} enhancement through the two-loop mechanism [67]. Furthermore, $\bar{e}\mu\bar{q}q$ operators may turn $\mu N \to eN$ into sophisticated probes of diagonal ρ_{qq} couplings by utilizing many different nuclei, and with help from nuclear physics in evaluating nuclear matrix elements. For rare B

decays [66], on one hand g2HDM effects quite often hide themselves and may be hard to probe, but especially $B \to \mu\nu$, $\tau\nu$ as we have advocated [61], and $B_{s,d} \to \mu\mu$ that are hotly pursued at the LHC [5], these modes with SM expectations can probe g2HDM effects through interference, and appear rather promising [66].

For the "B anomalies", the effects are so large that they cannot arise from, but can coexist [66] with, g2HDM. As to the recently confirmed [68] muon $g-2$ anomaly, it can [69] be explained by a known [70–75] one-loop mechanism in g2HDM, if $\rho_{\tau\mu} \sim \rho_{\mu\tau} \sim 20$ times larger than $\lambda_\tau \sim 0.01$. What was surprising was that [69] CMS $gg \to H, A \to \tau\mu$ [65] search provided more stringent bound than $\tau \to \mu\gamma$ by Belle [76]. Given the mass-mixing hierarchy, we do not particularly favor such large $\rho_{\tau\mu}$ strength, but there should be no doubt that *Nature* "reigns" over all things. If the one-loop mechanism is behind muon $g-2$, it would be a great boon [77] to muon physics, making the aforementioned leptonic processes far more interesting, including perhaps [78] the observation of μEDM in the not too distant future! We would see a renaissance of muon-related physics.

6. Prospects and Conclusions

In addition to the broadened impact on the flavor front as discussed in the previous section, on LHC and future colliders side, our three main processes of $cg \to tt\bar{c}$, $tt\bar{t}$ [52], $bt\bar{b}$ [53] in Section 4 are only starting points. If there exists a second Higgs doublet that is sub-TeV in mass, the myriad extra couplings promise rich phenomena for future collider studies, which we refer to Ref. [51] for a little more discussion. For the one-loop mechanism behind muon $g-2$ in g2HDM, one could have spectacular [69] signatures, such as $pp \to b\tau\mu W$ or $btcW$, with $\tau\mu$ or tc descending from a neutral H, A scalar, while bW does not come from top but arises from a combination of bH^+ production and H^+ weak decay. Our sub-TeV exotic Higgs masses were argued based on EWBG, but we do not yet know the actual spectrum. Once we learn the spectrum at the LHC, the extra dimension-4 couplings of Equations (1) and (2) should lead to very rich phenomena that await us at the LHC, and at future colliders.

One remark we would like to make is in regards the extra ρ^d Yukawa matrix. The K^0, B_d^0, and B_s^0 systems are the most sensitive probes of FPCP that we have, in particular, to the FCNH ρ_{ds}, ρ_{db} and ρ_{sb} couplings: we could have observed spectacular "BSM" effects in meson mixings and rare decays since long ago. The fact that all three systems behave according to SM should have implications. It is our conjecture that *Nature* somehow deactivated this sector, that the ρ^d matrix is close to diagonal; otherwise there would be arbitrary tuning space with ρ_{tt} loop effects. This enhances our doubt that ρ_{bb} is behind baryogenesis, inasmuch as it can carry a CPV phase. Perhaps it traces back to *Nature*'s choice of $\lambda_b \ll \lambda_t$, where both couplings are now experimentally measured, and the astonishing hierarchy confirmed.

With $\mathcal{O}(1)$ Higgs quartics, and with possibly $\mathcal{O}(1)$ extra top Yukawa couplings ρ_{tc} and ρ_{tt}, together with 50 more (likely weaker) flavor parameters in the form of extra Yukawa couplings, we may be just at the opening to the "prelude" of a new Higgs and flavor era that could start to unfold before us. We have dubbed this prospect "the Decadal Mission" [33].

In conclusion, *Nature* may, or may not, have g2HDM in store for us, but we must walk the walk to probe these sub-TeV extra Higgs bosons and carry out this "mission" towards unveiling a possible new physics Higgs and Flavor era.

Funding: This research is funded by MOST 110-2639-M-002-002-ASP of Taiwan, and also NTU 111L104019, 111L894801.

Acknowledgments: I have enjoyed collaborative works with my able collaborators as listed in the References. I thank Paoti Chang, Kai-Feng Chen, and Chi-Jen David Lin for helping to realize the ASP Project.

Conflicts of Interest: The author declares no conflicts of interest.

References

1. Weinberg, S. A Model of Leptons. *Phys. Rev. Lett.* **1967**, *19*, 1264. [CrossRef]
2. Higgs, P.W. Broken symmetries, massless particles and gauge fields. *Phys. Lett.* **1964**, *12*, 132. [CrossRef]
3. Brout, R.; Englert, F. Broken Symmetry and the Mass of Gauge Vector Mesons. *Phys. Rev. Lett.* **1964**, *13*, 321.
4. Higgs, P.W. Broken Symmetries and the Masses of Gauge Bosons. *Phys. Rev. Lett.* **1964**, *13*, 508. [CrossRef]
5. Zyla, P.A.; Barnett, R.M.; Beringer, J.; Dahl, O.; Dwyer, D.A.; Groom, D.E.; Lin, C.-J.; Lugovsky, K.S.; Pianori, E.; Robinson, D.J.; et al. Review of particle physics. *Prog. Theor. Exp. Phys.* **2020**, *2020*, 083C01.
6. Sirunyan, A.M.; Tumasyan, A.; Adam, W.; Bergauer, T.; Dragicevic, M.; Erö, J.; Escalante Del Valle, A.; Frühwirth, R.; Jeitler, M.; Krammer, N.; et al. Evidence for Higgs Boson Decay A Pair Muons. *J. High Energy Phys.* **2021**, *1*, 148. [CrossRef]
7. Kobayashi, M.; Maskawa, T. CP Violation in the Renormalizable Theory of Weak Interaction. *Prog. Theor. Phys.* **1973**, *49*, 652. [CrossRef]
8. Holdom, B.; Hou, W.S.; Hurth, T.; Mangano, M.L.; Sultansoy, S.; Unel, G. Four Statements about the Fourth Generation. *PMC Phys. A* **2009**, *3*, 4. [CrossRef]
9. Hou, W.S. Source of CP Violation for the Baryon Asymmetry of the Universe. *Chin. J. Phys.* **2009**, *47*, 134. [CrossRef]
10. Morrissey, D.E.; Ramsey-Musolf, M.J. Electroweak baryogenesis. *New J. Phys.* **2012**, *14*, 125003. [CrossRef]
11. Mukhopadhyaya, B.; Nandi, S. Evading the top mass bound at the Tevatron: New signals for the top. *Phys. Rev. Lett.* **1991**, *66*, 285. [CrossRef]
12. Hou, W.S. The Top quark cannot evade the tevatron mass bound via mixing with singlet quarks. *Phys. Rev. Lett.* **1992**, *69*, 3587. [CrossRef] [PubMed]
13. del Aguila, F.; Aguilar-Saavedra, J.A.; Miquel, R. Constraints on Top Couplings in Models with Exotic Quarks. *Phys. Rev. Lett.* **1999**, *82*, 1628. [CrossRef]
14. Djouadi, A.; Lenz, A. Sealing the fate of a fourth generation of fermions. *Phys. Lett. B* **2012**, *715*, 310. [CrossRef]
15. Eberhardt, O.; Herbert, G.; Lacker, H.; Lenz, A.; Menzel, A.; Nierste, U.; Wiebusch, M. Impact of a Higgs boson at a mass of 126 GeV on the standard model with three and four fermion generations. *Phys. Rev. Lett.* **2012**, *109*, 241802. [CrossRef] [PubMed]
16. Dugan, M.J.; Georgi, H.; Kaplan, D.B. Anatomy of a composite higgs model. *Nucl. Phys. B* **1985**, *254*, 299. [CrossRef]
17. Kaplan, D.B. Flavor at SSC energies: A new mechanism for dynamically generatedfermion masses. *Nucl. Phys. B* **1991**, *365*, 259. [CrossRef]
18. Contino, R.; Da Rold, L.; Pomarol, A. Light custodians in natural composite Higgs models. *Phys. Rev. D* **2007**, *75*, 55014. [CrossRef]
19. Contino, R.; Kramer, T.; Son, M.; Sundrum, R. Warped/composite phenomenology simplified. *J. High Energy Phys.* **2007**, *5*, 074. [CrossRef]
20. Matsedonskyi, O.; Panico, G.; Wulzer, A. Light Top Partners for a Light Composite Higgs. *J. High Energy Phys.* **2013**, *1*, 164. [CrossRef]
21. Aguilar-Saavedra, J.A.; Benbrik, R.; Heinemeyer, S.; Pérez-Victoria, M. Handbook of vectorlike quarks: Mixing and single production. *Phys. Rev. D* **2013**, *88*, 094010. [CrossRef]
22. De Simone, A.; Matsedonskyi, O.; Rattazzi, R.; Wulzer, A. A First Top Partner Hunter's Guide. *J. High Energy Phys.* **2013**, *4*, 004. [CrossRef]
23. Branco, G.C.; Ferreira, P.M.; Lavoura, L.; Rebelo, M.N.; Sher, M.; Silva, J.P. Theory and phenomenology of two-Higgs-doublet models. *Phys. Rept.* **2012**, *516*, 1. [CrossRef]
24. Glashow, S.L.; Weinberg, S. Natural Conservation Laws for Neutral Currents. *Phys. Rev. D* **1977**, *15*, 1958. [CrossRef]
25. Hou, W.S. Tree level $t \to ch$ or $h \to t\bar{c}$ decays. *Phys. Lett. B* **1992**, *296*, 179. [CrossRef]
26. Cheng, T.P.; Sher, M. Mass Matrix Ansatz and Flavor Nonconservation in Models with Multiple Higgs Doublets. *Phys. Rev. D* **1987**, *35*, 3484. [CrossRef]
27. Hou, W.S. Is the top quark really heavier than the W boson? *Phys. Rev. Lett.* **1994**, *72*, 3945. [CrossRef]
28. Aad, G.; Abbott, B.; Abdallah, J.; Abdel Khalek, S.; Abdinov, O.; Aben, R.; Abi, B.; Abolins, M.; AbouZeid, O.S.; Abramowicz, H.; et al. Search Top Quark Decays $t \to QH$ $H \to \gamma\gamma$ Using ATLAS Detect. *J. High Energy Phys.* **2014**, *6*, 008. [CrossRef]
29. Chen, K.F.; Hou, W.S.; Kao, C.; Kohda, M. When the Higgs meets the Top: Search for $t \to ch^0$ at the LHC *Phys. Lett. B* **2013**, *725*, 378. [CrossRef]
30. Aad, G.; Abbott, B.; Abdallah, J.; Abdinov, O.; Abeloos, B.; Aben, R.; AbouZeid, O.S.; Abraham, N.L.; Abramowicz, H.; Abreu, H.; et al. Measurements of the Higgs boson production and decay rates and constraints on its couplings from a combined ATLAS and CMS analysis of the LHC pp collision data at $\sqrt{s} = 7$ 8 TeV. *J. High Energy Phys.* **2016**, *8*, 045. [CrossRef]
31. Hou, W.S.; Kikuchi, M. Approximate Alignment in Two Higgs Doublet Model with Extra Yukawa Couplings. *EPL* **2018**, *123*, 11001. [CrossRef]
32. Pomarol, A. The SM EFT & new physics. In Proceedings of the LHCP2021, Paris, France, 7–12 June 2021.
33. Hou, W.S. Decadal Mission for the New Physics Higgs/Flavor Era. *Chin. J. Phys.* **2022**, *77*, 432–451. [CrossRef]
34. Davidson, S.; Haber, H.E. Basis-independent methods for the two-Higgs-doublet model. *Phys. Rev. D* **2005**, *72*, 035004. [CrossRef]
35. Botella, F.J.; Silva, J.P. Jarlskog-like invariants for theories with scalars and fermions. *Phys. Rev. D* **1995**, *51*, 3870. [CrossRef]
36. Altunkaynak, B.; Hou, W.S.; Kao, C.; Kohda, M.; McCoy, B. Flavor Changing Heavy Higgs Interactions at the LHC. *Phys. Lett. B* **2015**, *751*, 135. [CrossRef]

37. Hou, W.S.; Modak, T. Prospects for tZH and tZh production at the LHC. *Phys. Rev. D* **2020**, *101*, 035007. [CrossRef]
38. Lin, S.W.; Unno, Y.; Hou, W.S.; Chang, P.; Adachi, I.; Aihara, H.; Akai, K.; Arinstein, K.; Aulchenko, V.; Aushev, T.; et al. Difference in direct charge-parity violation between charged and neutral B meson decays. *Nature* **2008**, *452*, 332–335. [PubMed]
39. Fuyuto, K.; Hou, W.S.; Senaha, E. Electroweak baryogenesis driven by extra top Yukawa couplings. *Phys. Lett. B* **2018**, *776*, 402. [CrossRef]
40. Modak, T.; Senaha, E. Electroweak baryogenesis via bottom transport. *Phys. Rev. D* **2019**, *99*, 115022. [CrossRef]
41. Modak, T.; Senaha, E. Probing Electroweak Baryogenesis induced by extra bottom Yukawa coupling via EDMs and collider signatures. *J. High Energy Phys.* **2020**, *11*, 025. [CrossRef]
42. Cline, J.M.; Laurent, B. Electroweak baryogenesis from light fermion sources: A critical study. *Phys. Rev. D* **2021**, *104*, 083507. [CrossRef]
43. Kanemura, S.; Okada, Y.; Senaha, E. Electroweak baryogenesis and quantum corrections to the triple Higgs boson coupling. *Phys. Lett. B* **2005**, *606*, 361. [CrossRef]
44. Reichert, M.; Eichhorn, A.; Gies, H.; Pawlowski, J.M.; Plehn, T.; Scherer, M.M. Probing baryogenesis through the Higgs boson self-coupling. *Phys. Rev. D* **2018**, *97*, 075008. [CrossRef]
45. Baron, J.; Campbell, W.C.; DeMille, D.; Doyle, J.M.; Gabrielse, G.; Gurevich, Y.V.; Hess, P.W.; Hutzler, N.R.; Kirilov, E.; Kozyryev, I.; et al. Order of Magnitude Smaller Limit on the Electric Dipole Moment of the Electron. *Science* **2014**, *343*, 269.
46. Andreev, V.; Ang, D.G.; DeMille, D.; Doyle, J.M.; Gabrielse, G.; Haefner, J.; Hutzler, N.R.; Lasner, Z.; Meisenhelder, C.; O'Leary, B.R.; et al. Improved limit on the electric dipole moment of the electron. *Nature* **2018**, *7727*, 355.
47. Fuyuto, K.; Hou, W.S.; Senaha, E. Cancellation mechanism for the electron electric dipole moment connected with the baryon asymmetry of the Universe. *Phys. Rev. D* **2020**, *101*, 011901. [CrossRef]
48. Carena, M.; Liu, Z. Challenges and opportunities for heavy scalar searches in the $t\bar{t}$ channel at the LHC. *J. High Energy Phys.* **2016**, *11*, 159. [CrossRef]
49. Sirunyan, A.M.; Tumasyan, A.; Adam, W.; Ambrogi, F.; Bergauer, T.; Brandstetter, J.; Dragicevic, M.; Erö, J.; Escalante Del Valle, A.; Flechl, M.; et al. Search Heavy Higgs Bosons Decaying A Top Quark Pair Proton-Proton Collisions $\sqrt{s} = 13$ TeV. *J. High Energy Phys.* **2020**, *4*, 171.
50. Hou, W.S.; Jain, R.; Kao, C.; Kohda, M.; McCoy, B.; Soni, A. Flavor Changing Heavy Higgs Interactions with Leptons at Hadron Colliders. *Phys. Lett. B* **2019**, *795*, 371. [CrossRef]
51. Hou, W.S.; Modak, T. Probing Top Changing Neutral Higgs Couplings at Colliders. *Mod. Phys. Lett. A* **2021**, *36*, 21300064. [CrossRef]
52. Kohda, M.; Modak, T.; Hou, W.S. Searching for new scalar bosons via triple-top signature in $cg \to tS^0 \to tt\bar{t}$. *Phys. Lett. B* **2018**, *776*, 379. [CrossRef]
53. Ghosh, D.K.; Hou, W.S.; Modak, T. Sub-TeV H^+ Boson Production as Probe of Extra Top Yukawa Couplings. *Phys. Rev. Lett.* **2020**, *1256*, 221801. [CrossRef] [PubMed]
54. Barger, V.; Keung, W.Y.; Yencho, B. Triple-Top Signal of New Physics at the LHC. *Phys. Lett. B* **2010**, *687*, 70. [CrossRef]
55. Frederix, R.; Pagani, D.; Zaro, M. Large NLO corrections in $t\bar{t}W^{\pm}$ and $t\bar{t}t\bar{t}$ hadroproduction from supposedly subleading EW contributions. *J. High Energy Phys.* **2018**, *2*, 031. [CrossRef]
56. Hou, W.S.; Kohda, M.; Modak, T. Implications of Four-Top and Top-Pair Studies on Triple-Top Production. *Phys. Lett. B* **2019**, *798*, 134953. [CrossRef]
57. CMS Collaboration. Search for flavor-changing neutral current interactions of the top quark and Higgs boson in final states with two photons in proton-proton collisions at $\sqrt{s} = 13$ TeV. *arXiv* **2021**, arXiv:2111.02219.
58. Hou, W.S.; Hsu, T.H.; Modak, T. Constraining the $t \to u$ flavor changing neutral Higgs coupling at the LHC. *Phys. Rev. D* **2020**, *102*, 055006. [CrossRef]
59. Chang, P.; Chen, K.F.; Hou, W.S. Flavor Physics and CP Violation. *Prog. Part. Nucl. Phys.* **2017**, *97*, 261. [CrossRef]
60. Hou, W.S. Enhanced charged Higgs boson effects in $B^- \to \tau\bar{\nu}$, $\mu\bar{\nu}$ and $b \to \tau\bar{\nu} + X$. *Phys. Rev. D* **1993**, *48*, 2342. [CrossRef]
61. Hou, W.S.; Kohda, M.; Modak, T.; Wong, G.G. Enhanced $B \to \mu\bar{\nu}$ decay at tree level as probe of extra Yukawa couplings. *Phys. Lett. B* **2020**, *800*, 135105. [CrossRef]
62. Crivellin, A.; Kokulu, A.; Greub, C. Flavor-phenomenology of two-Higgs-doublet models with generic Yukawa structure. *Phys. Rev. D* **2013**, *87*, 094031. [CrossRef]
63. Chen, C.H.; Nomura, T. Charged Higgs boson contribution to $B_q^- \to \ell\bar{\nu}$ and $\bar{B} \to (P,V)\ell\bar{\nu}$ in a generic two-Higgs doublet model. *Phys. Rev. D* **2018**, *98*, 095007. [CrossRef]
64. Prim, M.T.; Bernlochner, F.U.; Goldenzweig, P.; Heck, M.; Adachi, I.; Adamczyk, K.; Aihara, H.; Al Said, S.; Asner, D.M.; Atmacan, H.; et al. Search $B^+ \to \mu^+ \nu_\mu$ $B^+ \to \mu^+ N$ Incl. Tagging. *Phys. Rev. D* **2020**, *101*, 032007. [CrossRef]
65. Sirunyan, A.M.; Tumasyan, A.; Adam, W.; Ambrogi, F.; Bergauer, T.; Brandstetter, J.; Dragicevic, M.; Erö, J.; Escalante Del Valle, A.; Flechl, M.; et al. Search Lepton Flavour Violating Decays A Neutral Heavy Higgs Boson $\mu\tau$ Eτ Proton-Proton Collisions $\sqrt{s} =$ 13 TeV. *J. High Energy Phys.* **2020**, *3*, 103.
66. Hou, W.S.; Kumar, G. Muon Flavor Violation in Two Higgs Doublet Model with Extra Yukawa Couplings. *Phys. Rev. D* **2020**, *102*, 115017. [CrossRef]
67. Chang, D.; Hou, W.S.; Keung, W.Y. Two loop contributions of flavor changing neutral Higgs bosons to $\mu \to e\gamma$. *Phys. Rev. D* **1993**, *48*, 217. [CrossRef]

68. Abi, B.; Albahri, T.; Al-Kilani, S.; Allspach, D.; Alonzi, L.P.; Anastasi, A.; Anisenkov, A.; Azfar, F.; Badgley, K.; Baessler, S.; et al. Measurement of the positive muon anomalous magnetic moment to 0.46 ppm. *Phys. Rev. Lett.* **2021**, *126*, 141801. [CrossRef]
69. Hou, W.S.; Jain, R.; Kao, C.; Kumar, G.; Modak, T. Collider Prospects for Muon $g-2$ in General Two Higgs Doublet Model. *Phys. Rev. D* **2021**, *104*, 075036. [CrossRef]
70. Assamagan, K.A.; Deandrea, A.; Delsart, P.A. Search for the lepton flavor violating decay $A^0/H^0 \to \tau^{\pm}\mu^{\mp}$ at hadron colliders. *Phys. Rev. D* **2003**, *67*, 035001. [CrossRef]
71. Davidson, S.; Grenier, G.J. Lepton flavour violating Higgs and $\tau \to \mu\gamma$. *Phys. Rev. D* **2010**, *81*, 095016. [CrossRef]
72. Omura, Y.; Senaha, E.; Tobe, K. Lepton-flavor-violating Higgs decay $h \to \mu\tau$ and muon anomalous magnetic moment in a general two Higgs doublet model. *J. High Energy Phys.* **2015**, *5*, 028. [CrossRef]
73. Omura, Y.; Senaha, E.; Tobe, K. τ- and μ-physics in a general two Higgs doublet model with $\mu-\tau$ flavor violation. *Phys. Rev. D* **2016**, *94*, 055019. [CrossRef]
74. Iguro, S.; Tobe, K. $R(D^{(*)})$ in a general two Higgs doublet model. *Nucl. Phys. B* **2017**, *925*, 560. [CrossRef]
75. Iguro, S.; Omura, Y.; Takeuchi, M. Testing the 2HDM explanation of the muon $g-2$ anomaly at the LHC. *J. High Energy Phys.* **2019**, *11*, 130. [CrossRef]
76. Uno, K.; Hayasaka, K.; Inami, K.; Adachi, I.; Aihara, H.; Asner, D.M.; Atmacan, H.; Aushev, T.; Ayad, R.; Babu, V.; et al. Search Lepton-Flavor Tau-Lepton Decays $\ell\gamma$ Belle. *J. High Energy Phys.* **2021**, *10*, 19.
77. Hou, W.S.; Kumar, G. Charged lepton flavor violation in light of muon $g-2$. *Eur. Phys. J. C* **2021**, *81*, 1132. [CrossRef]
78. Hou, W.S.; Kumar, G.; Teunissen, S. Charged Lepton EDM with Extra Yukawa Couplings. *J. High Energy Phys.* **2022**, *1*, 092. [CrossRef]

universe

MDPI

Review

Dark Matter Searches with Top Quarks

J. Katharina Behr *,† and Alexander Grohsjean *,†

Deutsches Elektronen-Synchrotron DESY, Notkestr. 85, 22607 Hamburg, Germany
* Correspondence: katharina.behr@desy.de (J.K.B.); alexander.grohjsean@desy.de (A.G.)
† These authors contributed equally to this work.

Abstract: Collider signatures with top quarks provide sensitive probes of dark matter (DM) production at the Large Hadron Collider (LHC). In this article, we review the results of DM searches in final states with top quarks conducted by the ATLAS and CMS Collaborations at the LHC, including the most recent results on the full LHC Run 2 dataset. We highlight the complementarity of DM searches in final states with top quarks with searches in other final states in the framework of various simplified models of DM. A reinterpretation of a DM search with top quarks in the context of an effective field theory description of scalar dark energy is also discussed. Finally, we give an outlook on the potential of DM searches with top quarks in LHC Run 3, at the high-luminosity LHC, and possible future colliders. In this context, we highlight new benchmark models that could be probed by existing and future searches as well as those that predict still-uncovered signatures of anomalous top-quark production and decays at the LHC.

Keywords: top quark; dark matter; WIMP; LHC

Citation: Behr, J.K.; Grohsjean, A. Dark Matter Searches with Top Quarks. *Universe* **2023**, *9*, 16. https://doi.org/10.3390/universe9010016

Academic Editors: Efe Yazgan, Pedro Ferreira da Silva and Jinmin Yang

Received: 18 November 2022
Revised: 16 December 2022
Accepted: 20 December 2022
Published: 27 December 2022

1. Introduction

The particle nature of dark matter (DM) is one of the major puzzles in modern particle physics, despite long-standing evidence for its existence. As early as 1884, Lord Kelvin realized that the mass of the Milky Way derived from the velocity dispersion of the stars orbiting its center is very different from the mass of the visible stars. He considered the majority of stars in our galaxy to be dark bodies. One hundred and forty years later, overwhelming astronomical and cosmological evidence has been accumulated for the existence of dark matter (DM) across different scales, ranging from the rotational velocity of stars in ultra-faint galaxies over gravitational lensing effects to precision measurements of the cosmic microwave background [1–5].

It is well established that 85% of the matter in our Universe consists of DM. The dominant part of DM must be stable with a lifetime much longer than the age of the Universe. The fact that DM was already produced in the early Universe may provide a clue to nongravitational interactions. At the same time, the feature that DM must form cosmological structures consistent with current observations allows setting a limit on the strength of DM interactions with SM particles and with itself. It is clear that none of the Standard Model particles is consistent with all of these observations.

One of the highly-motivated theory paradigms for DM is the so-called WIMP (weakly interacting massive particle) paradigm, also known as the WIMP miracle [6]. Assuming DM to be produced via the freeze-out mechanism, one can achieve the observed relic density when the DM mass is close to the electroweak scale and when the DM coupling to Standard Model particles is on the order of the weak interaction. Consequently, DM particles could be produced and studied at the Large Hadron Collider (LHC) [7]. Alternative strategies to search for DM include *indirect detection* searches, which target recoils of DM particles from the local DM halo in sensitive underground detectors, as well as *indirect detection* searches, which aim to detect the products of DM interactions, such as DM annihilation into SM particles, e.g., gamma rays, within the visible universe. Recent reviews of results from direct- and indirect-detection searches can be found in Refs. [8,9], respectively.

An advantage of DM searches at colliders, besides the fact that DM would be produced under controlled experimental conditions, lies in the fact that they could provide access to the particles mediating the interactions between DM and the Standard Model. A DM mediator produced in proton–proton (pp) collisions could decay to DM particles. Such *invisible decays* could only be inferred via the presence of missing transverse momentum, $p_{\mathrm{T}}^{\mathrm{miss}}$, in the detector. However, a DM mediator decaying back into SM particles (*visible decays*) would provide direct access to its properties. DM searches at the LHC explore both avenues. To detect the invisible decays of a mediator, it is mandatory to produce the mediator in association with SM particles. In this review article, we will focus on the associated production with top quarks and, more generally, on the role of top quarks in the quest for DM. Best suited to study DM in top-quark channels are the two general-purpose detectors ATLAS [10] and CMS [11]. Recent reviews of DM searches at the LHC, including those without top quarks, can be found, e.g., in Refs. [12,13].

Discovered in 1995 at the Fermilab Tevatron collider [14,15], the top quark is the heaviest of all known elementary particles. In the case of a DM mediator with Yukawa-like couplings, the top quark would be ideal for discovery. Moreover, the top quark would allow for a first characterization of the dark sector. Due to its short lifetime, the top quark fully transmits its spin information to the decay particles. In turn, this allows inferring the spin of the mediator for both the associated production of top quarks and DM as well as for the decay of a mediator to a top-quark pair.

Another major unknown in the physics of our universe, besides the particle nature of DM, is the origin of its accelerating expansion [16,17], which is usually attributed to the presence of a yet-unknown repulsive force, referred to as dark energy (DE). If DE is a scalar field, it may be possible to produce it at the LHC. Similar to DM, DE would escape the detector unnoticed. DM searches with top quarks could be sensitive to DE production, as shown in Sections 2.4 and 4.4 of this review.

This article is structured as follows. After a detailed discussion of the underlying DM models in Section 2, we focus on the experimental signatures of DM searches involving top quarks at LHC in Section 3. In Section 4, current highlights and results from DM searches at LHC are summarized. We conclude with a discussion of uncovered signatures and models, followed by an outlook on prospects for discovering DM at future colliders in Sections 5 and 6.

2. Models with BSM Signatures Involving Top Quarks

Collider searches for DM are usually interpreted in the context of so-called *simplified models*, which contain a minimal set of new particles and couplings. Most of these models contain only a single Dirac DM particle and a single mediator particle. They are characterized by a minimal set of free parameters, namely, the masses of the DM and mediator particles and the couplings of the mediator to the SM and dark sector. Simplified models provide a convenient framework to compare searches in different final states and among different experiments. In the following, the simplified models used for the interpretation of DM searches involving top quarks are described. Additionally, an effective field theory (EFT) description of scalar DE is introduced.

2.1. Vector and Axial-Vector Mediators

2.1.1. Flavor-Conserving Interaction

A mediator with flavor-universal couplings to the SM quarks and leptons, respectively, is predicted in a simplified model that describes a flavor-conserving interaction between a fermionic WIMP DM particle χ and the SM fermions [18]. It is based on a simple extension of the SM by a new $U(1)$ gauge symmetry under which χ, as well as some of the SM fermions, are charged, thus allowing the mediator to couple to the SM sector. The interaction described by this gauge group is mediated by the s-channel exchange of a new, electrically neutral spin-1 particle Z' with either vector or axial-vector couplings to the DM and SM fields. It is referred to as *vector mediator* or *axial-vector mediator* in the following.

The model contains five free parameters [18]: the masses of the mediator, $m_{Z'}$, and the DM particle, m_χ, as well as the quark-flavor universal coupling g_q of the mediator to quarks, the lepton-flavor universal coupling g_ℓ of the mediator to leptons, and the coupling g_χ of the mediator to DM.

The mediator can decay either invisibly into a $\chi\bar{\chi}$ pair or visibly into a fermion–anti-fermion $f\bar{f}$ pair, as illustrated schematically by the left and right diagrams, respectively, in Figure 1. The former process can be detected as a $p_{\mathrm{T}}^{\mathrm{miss}} + X$ signature in the presence of initial-state radiation (ISR), where X can be a gluon, photon, or vector boson, depending on the type of ISR, while the latter process results in a resonant enhancement in the invariant mass spectrum of the $f\bar{f}$ pair.

Constraints on this model are derived in various parameter planes, including the $(m_{Z'}, m_\chi)$ plane for fixed couplings g_q, g_ℓ, g_χ [19] and as upper limits on g_q as a function of $m_{Z'}$, as shown in Section 4.1.1.

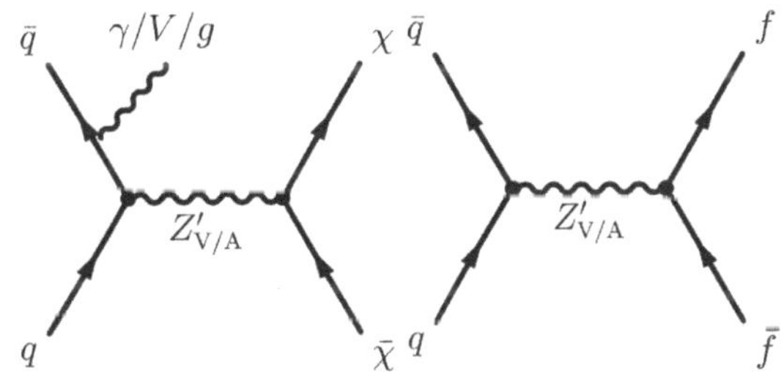

Figure 1. Schematic representation of the dominant production and decay modes of the simplified model with an s-channel vector or axial-vector mediator Z' [19].

2.1.2. Flavor-Changing Interaction

DM signatures with top quarks are predicted in simplified models containing a vector mediator Z'_{VFC} with a flavor-changing coupling V_{ut} to the top and up-quark. This type of model, referred to as *VFC model* in the following, is motivated, for example, by scenarios with DM in a hidden sector that only interacts with the SM sector via a flavor-changing coupling of a Z' boson [20,21]. The dominant production and decay modes of the VFC model are shown in Figure 2. The mediator can be produced on-shell in association with a single top or anti-top (left diagram) and decay either invisibly into DM or visibly into a top and up-quark. The former decay results in a $p_{\mathrm{T}}^{\mathrm{miss}} + t$ signature, often referred to as *mono-top*. The latter decay yields a characteristic final state with two top quarks (tt) or two anti-top quarks $\bar{t}\bar{t}$ (same-sign tt). This signature can be easily distinguished from the more abundant $t\bar{t}$ production via SM processes by the sign of the lepton charges in fully leptonic decays. Similar $tt/\bar{t}\bar{t}$ final states arise from the other two diagrams in Figure 2, which represent the t-channel exchange of the Z'_{VFC} mediator.

The VFC model is fully characterized by four free parameters: the mass of the mediator, $m_{Z'_{\mathrm{VFC}}}$, the mass of the DM particle, m_χ, the coupling of the mediator to DM, g_χ, and the flavor-changing coupling, g_{ut} [22]. The DM mass has no significant impact on the collider phenomenology of the VFC model, if $2m_\chi < m_{Z'_{\mathrm{VFC}}}$, and is fixed to a value of 1 GeV for existing collider searches [19]. Constraints on the VFC model are accordingly derived in several parameter planes involving the remaining free parameters (or dependent parameters): $m_{Z'_{\mathrm{VFC}}}$, g_{ut}, and the invisible branching ratio $\mathcal{BR}(\chi\bar{\chi})$ of the mediator.

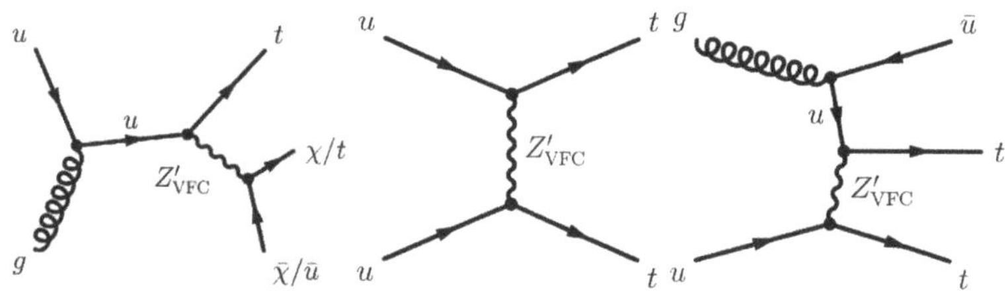

Figure 2. Schematic representation of the dominant production and decay modes of the VFC model [19].

2.2. Scalar and Pseudoscalar Mediators

A preferred coupling of DM to top quarks is predicted in simplified models containing a spin-0 mediator with Yukawa-like couplings to SM fermions. The mediator can be either a scalar (ϕ) or pseudoscalar (a). These models can be straightforwardly embedded in ultraviolet (UV) complete theories with extended Higgs sectors, such as two-Higgs-doublet models (2HDMs; see also Section 2.3). Assuming Yukawa-like couplings allows this class of models to satisfy strong constraints from flavor precision measurements. The dynamics of flavor violation are completely determined by the structure of the ordinary fermion Yukawa couplings, which is referred to as *minimal flavor violation (MFV)* [23].

The simplified models described in this section can be broadly categorized into models with a color-neutral and a color-charged interaction. An overview of the models falling into each category can be found in Ref. [19] and references therein. Two representative benchmark models used by the ATLAS and CMS collaborations are presented in the following.

2.2.1. Color-Neutral Interaction

A color-neutral interaction between an SM and a DM particle is described by a simplified model with a neutral, scalar, or pseudoscalar mediator [18,24] with Yukawa-like couplings to the SM fermions. The model has four free parameters: the mass of the DM particle, m_χ, the mass of the mediator, $m_{\phi/a}$, the coupling of the mediator to DM, g_χ, and the coupling of the mediator to SM fermions. The latter is parameterized by a flavor-universal coupling constant $g_q \equiv g_u = g_d = g_\ell$, which modifies the SM-like Yukawa coupling of the mediator to fermions [24], thus satisfying the requirements of MFV. It should be noted that couplings to leptons are explicitly included in the model, but in practice, the related signatures play no significant role in the parameter space accessible to collider searches [18]. Couplings to vector bosons W, Z are not included in this simplified model [24]. The Yukawa-like couplings imply that the mediator is mostly produced via loop-induced gluon fusion via a heavy-quark dominated loop or in association with heavy-flavor quarks, mostly top quarks. Additionally, visible decays of the mediator preferentially result in heavy quarks. The dominant production and decay modes of the mediator with heavy-flavor quarks in the final state are shown in Figure 3. These are (from left to right):

- Visible decay of a mediator produced via gluon-fusion to heavy-flavor quarks, resulting in a resonant $t\bar{t}$ or $b\bar{b}$ signal.
- Associated production of a mediator that decays either visibly or invisibly with heavy-flavor quarks, leading to a $p_{\text{T}}^{\text{miss}} + t\bar{t}/b\bar{b}$ signature in the case of invisible mediator decay or characteristic fully visible $t\bar{t}t\bar{t}$, $t\bar{t}b\bar{b}$, $b\bar{b}b\bar{b}$ signatures.
- Associated production of an invisibly decaying mediator with a top quark and a light (d, u, s, c) quark, leading to a $p_{\text{T}}^{\text{miss}} + tj$ signature.
- Associated production of an invisibly decaying mediator with a top quark and a W boson, resulting in a $p_{\text{T}}^{\text{miss}} + tW$ signature.

Additional signatures not shown here include $p_{\text{T}}^{\text{miss}}$ + jet and $p_{\text{T}}^{\text{miss}} + V/h$ production.

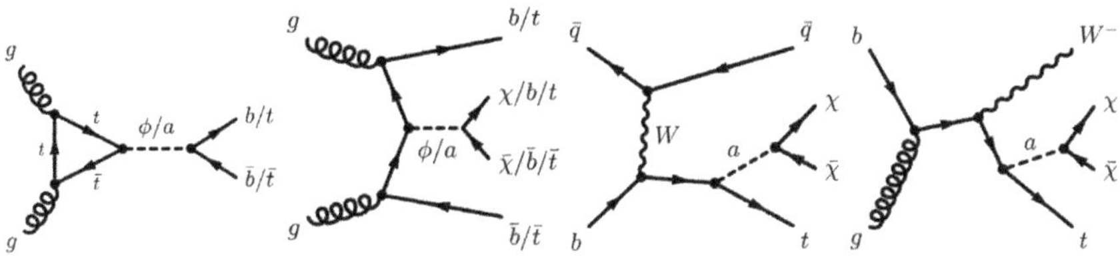

Figure 3. Schematic representation of the dominant production and decay modes with heavy-flavor quarks in the final state in the simplified model with a scalar (ϕ) or pseudoscalar (a) mediator [19].

It should be noted that, while the Yukawa-like coupling structure implies a greater importance of signatures involving top quarks rather than bottom quarks in the final state, signatures involving bottom quarks are still relevant as some UV completions of this simplified model involve a parameter modifying the relative importance of the couplings to up- and down-type quarks. In these UV completions, signatures involving bottom quarks can be more sensitive than signatures involving top quarks if the couplings to up-type quarks are suppressed.

2.2.2. Color-Charged Interaction

A color-charged interaction between the SM quarks and DM is described in a class of simplified models containing a scalar, color-triplet mediator particle. This type of simplified models is inspired by the minimal supersymmetric standard model (MSSM) [25,26] with first- and second-generation squarks and neutralino DM [19]. The mediator couplings to quarks and DM in the simplified models, however, can differ from those of the MSSM, leading to additional production diagrams.

Different models of color-charged mediators, differing by the mediator couplings to quarks, have been probed at the LHC. These include a model with preferred couplings of the mediator to the first and second quark generation, a model with preferred mediator couplings to bottom quarks, and a model with preferred mediator couplings to top quarks. Only the latter will be discussed in this review. The concrete realization of this model is documented in Ref. [21]. It contains a new $SU(2)_L$ singlet field that couples to right-handed quarks. The mediator corresponding to this field is produced from a down-type quark–anti-quark pair and decays to a top quark and a DM particle, as illustrated in Figure 4. This model can be related to the MSSM if an additional R-parity violating interaction of the top squark with the down-type quarks is assumed [19]. The free parameters of this model are the mass of the DM particle, m_χ, the mass of the mediator, m_{η_t}, the t-DM coupling strength of the mediator, λ_t, and the coupling strength of the mediator to down-type quarks, g_{ds}.

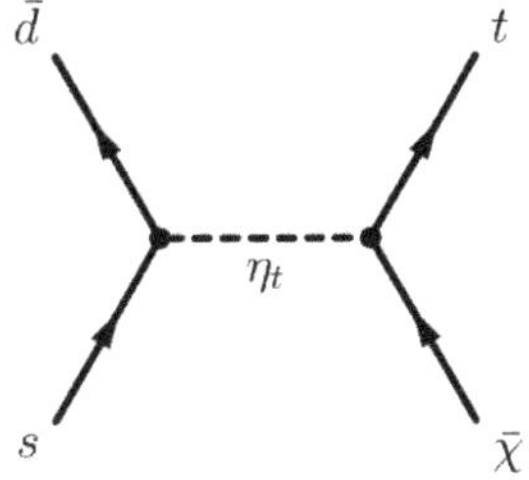

Figure 4. Schematic representation of p_T^{miss} + t production via a color-changing scalar mediator η_t [19].

2.3. Extended Higgs Sectors

Extended Higgs sectors are predicted by a range of BSM theories, such as supersymmetry [27], certain classes of axion models [28], or theories predicting additional sources of CP violation in the Higgs sector to explain the observed baryon asymmetry in the universe [29,30]. Extension of the SM Higgs sector by a second complex SU(2) doublet, referred to as two-Higgs-doublet models (2HDMs), are among the simplest and most studied models with an extended Higgs sector, historically due to their strong motivation from supersymmetry. In the past years, 2HDMs have also received considerable attention from the DM community as a means of embedding the simplified, mediator-based, models described in the previous sections in the context of a UV-complete and renormalizable framework with a broader collider phenomenology. Models of DM based on a 2HDM with a vector [31], pseudoscalar [32,33], and scalar [34] mediator have been proposed. Concrete realizations of the former two have been used as benchmark models by the LHC experiments. Models with vector mediators are not discussed in this review as final states with top quarks do not play a dominant role in their phenomenology. Models with a pseudoscalar mediator, on the other hand, feature a rich phenomenology involving relevant signatures with top quarks due to the Yukawa-type coupling of the mediator to SM fermions. Pseudoscalar mediators are also particularly interesting to study at the LHC as they are not strongly constrained by direct-detection experiments because the DM–nucleon scattering cross-section pseudoscalar couplings are strongly suppressed at tree-level by the momentum transfer in the nonrelativistic limit [35]. A concrete realization of a 2HDM with a pseudoscalar mediator that is used as a benchmark model by the LHC experiments is described in Section 2.3.

2HDM with a Pseudoscalar Mediator

A 2HDM with a pseudoscalar mediator a [32], referred to as 2HDM + a in the following, is a more complex simplified model that embeds the phenomenology of the simplified models with a color-neutral pseudoscalar mediator (Section 2.2.1) in more complete model with a second complex SU(2) doublet. The 2HDM in this model has a CP-conserving potential with a softly broken $\mathbb{Z}_2$ symmetry [36]. Its Higgs sector contains five Higgs bosons: two scalars, h and H, a pseudoscalar, A, and two charged Higgs bosons $H^{\pm}$. The alignment limit is assumed, meaning that one of the two scalars of the model is identified with the 125 GeV Higgs boson discovered in 2012. Furthermore, the Yukawa structure of the 2HDM is of type-II [37], meaning that couplings of the additional Higgs bosons to top quarks are preferred over those to other fermions at low values of the ratio of the two vacuum expectation values, $\tan\beta$, one of the model parameters with the biggest impact on the collider phenomenology of the model. The pseudoscalar mediator a mixes with the pseudoscalar A of the 2HDM with mixing angle θ.

The phenomenology of the 2HDM + a is fully defined by 14 free parameters, making it considerably more complex than the simplified models described in the previous sections. These parameters are as follows: the masses m_h, m_H, and m_A of the neutral Higgs bosons; the masses $m_{H^{\pm}}$ of the charged Higgs bosons; the mass m_a of the mediator; the mass m_χ of the DM particle; the coupling y_χ between DM and the mediator; the three quartic couplings λ_{P1}, λ_{P2}, λ_3 of the mediator to the SU(2) fields; the vacuum expectation value (VEV) v of the electroweak sector; the ratio $\tan\beta = \frac{v_2}{v_1}$ of the VEVs of the two Higgs fields; the mixing angle α between the two scalar Higgs bosons h and H; and the mixing angle θ between the pseudoscalar Higgs boson A and the mediator a.

The choice of the alignment limit ($\cos(\beta - \alpha)$=0) implies that $m_h = 125$ GeV and $v = 246$ GeV. The DM–mediator coupling is set to unity ($y_\chi = 1.0$) without significant impact on the phenomenology of the model. The setting $\lambda_3 = 3$ is chosen to ensure the stability of the Higgs potential in the mass ranges of interest of the heavy Higgs bosons [19]. Furthermore, the choice $\lambda_{P1} = \lambda_{P2} = \lambda_3 = 3$ maximizes the trilinear couplings between the CP-even and CP-odd neutral states [19]. Finally, the choice $m_A = m_H = m_{H^{\pm}}$ ensures compatibility of the model predictions with flavor constraints [32] and additionally simplifies the phenomenology of the model [19].

With these constraints, the remaining 2HMD+a parameter space can be described by the following five parameters: m_A, m_a, m_χ, $\sin\theta$, and $\tan\beta$. Representative benchmark scans of this parameter space have been defined by the LHC Dark Matter Working Group [38] with the aim to highlight different aspects of the phenomenology of this benchmark model and the interplay between searches targeting different signal processes across this parameter space. Additional benchmark scans are defined in Ref. [39].

The 2HDM + a predicts a rich phenomenology with a diverse range of final states. The dominant processes leading to final states with top quarks are shown in Figure 5, along with the leading diagrams for the resonant production of an invisibly decaying mediator with a Higgs or Z boson, leading to $p_{\mathrm{T}}^{\mathrm{miss}} + h$ and $p_{\mathrm{T}}^{\mathrm{miss}} + Z$ final states, respectively, which are among the most sensitive probes of the 2HDM + a. A full overview of the phenomenology of the 2HDM + a can be found in Refs. [32,38].

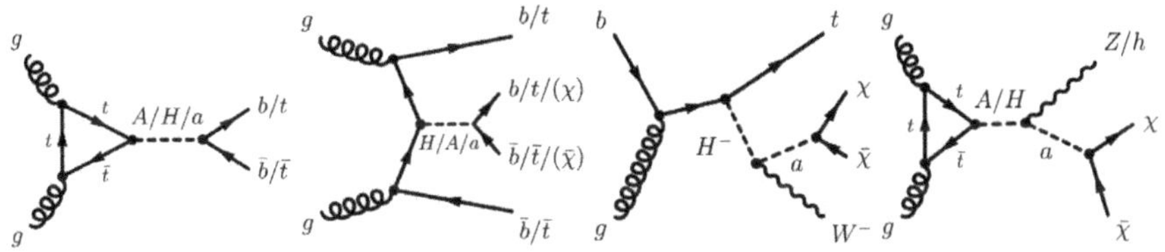

Figure 5. Schematic representation of relevant production and decay modes with top quarks leading to either top quarks in the final state or $p_{\mathrm{T}}^{\mathrm{miss}} + h/Z$ signatures. From left to right: resonant production of a neutral scalar or pseudoscalar particle $H/A/a$ decaying to $t\bar{t}$ or $b\bar{b}$; associated production with $b\bar{b}$ or $t\bar{t}$ of a single $H/A/a$ decaying either visibly to heavy flavor or invisibly to DM; associated production of a top quark and a charged Higgs boson decaying to a W boson and an invisibly decaying mediator a; resonant A/H production with subsequent decay to a Z/h boson and an invisibly decaying mediator a [19].

2.4. EFT Model of Scalar Dark Energy

Searches for DM signatures involving top quarks provide a powerful tool to probe models of scalar DE. The first reinterpretation of DM searches in the context of DE, which relied on the analysis of 36 fb^{-1} of LHC Run 2 data [19], used an EFT implementation [40] of the Horndeski theories [41] to describe DE production at the LHC [19]. The latter introduce a new scalar field, ϕ_{DE}, corresponding to DE, that couples to gravity.

The EFT model contains two classes of operators: operators that are invariant under a shift symmetry $\phi_{\mathrm{DE}} \to \phi_{\mathrm{DE+constant}}$ and operators that break this symmetry. The former contain only derivative couplings of the DE field to SM fermions as direct Yukawa-type interactions break the shift symmetry. The latter induce direct couplings of the DE field to the SM fermions, such as Yukawa-type interactions, and are subject to tight experimental constraints [42].

Only shift-symmetric operators of the EFT model have been considered for the DE reinterpretation of LHC DM searches [19]. The model under consideration contains nine such operators, $\mathcal{O}_i^{(d)}$, where d denotes the dimensionality of the operator. This leads to nine possible terms in the Lagrangian, each suppressed by powers of a characteristic energy scale M_i^{d-4}, according to the operator's dimensionality:

$$\mathcal{L} = \mathcal{L}_{\mathrm{SM}} + \sum_{i=1}^{9} c_i \mathcal{L}_i = \mathcal{L}_{\mathrm{SM}} + \sum_{i=1}^{9} \frac{c_i}{M_i^{d-4}} \mathcal{O}_i^{(d)},$$

where the c_i denote the Wilson coefficients.

Only the phenomenology of the two leading, i.e., least suppressed, terms has been considered by the LHC experiments so far. These are of dimension eight and can be expressed in terms of the conformal anomaly, T_ν^ν ($= m\bar{\psi}\psi$ for a Dirac field), and the energy-momentum tensor of the SM Lagrangian $T^{\mu\nu}$ as follows:

$$\mathcal{L}_1 = \frac{\partial_\mu \phi_{\mathrm{DE}} \partial^\mu \phi_{\mathrm{DE}}}{M_1^4} T^\nu_\nu$$
$$\mathcal{L}_2 = \frac{\partial_\mu \phi_{\mathrm{DE}} \partial_\nu \phi_{\mathrm{DE}}}{M_2^4} T^{\mu\nu}.$$

The coupling described by the first term, $\mathcal{L}_1$, is proportional to the mass of the SM fermions to which the DE field couples, thus making collider signatures involving top quarks a sensitive probe of DE. A schematic representation of DE production at the LHC via this operator is shown in Figure 6. It describes the radiation of a pair of DE particles off a final-state top quark from SM $t\bar{t}$ production, leading to a $p_{\mathrm{T}}^{\mathrm{miss}}$ + $t\bar{t}$ signature.

The second operator, $\mathcal{L}_2$, involves derivatives of the SM fields and is therefore proportional to their momenta. Final states involving high-momentum intermediate states, of which a DE pair is radiated off, provide the best sensitivity to this operator. At a hadron collider such as the LHC, the most likely high-momentum intermediate state particles are hadronically interacting particles, such as gluons, leading to characteristic $p_{\mathrm{T}}^{\mathrm{miss}}$ + jet signatures as the smoking-gun signatures for DE production.

Constraints on the EFT model of DE have been derived using searches for both $p_{\mathrm{T}}^{\mathrm{miss}}$ + $t\bar{t}$ ($\mathcal{L}_1$ term) and $p_{\mathrm{T}}^{\mathrm{miss}}$ + jet signatures [19] ($\mathcal{L}_2$ term). Only the former are discussed in this review.

It should be noted that additional signatures, such as $p_{\mathrm{T}}^{\mathrm{miss}}$ + t production, are predicted based on the sub-leading operators. The exploration of these additional signatures and possible reinterpretations of further DM searches in the context of DE is left to future work.

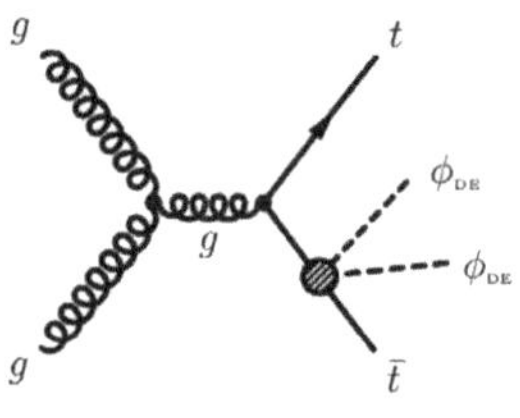

Figure 6. Schematic representation of the leading process of DE production in association with a $t\bar{t}$ pair in an EFT model of scalar DE via the operator $\mathcal{L}_1$ [19].

3. Experimental Signatures

Searches for DM in pp collisions involving single or multiple top quarks can be broadly split into two categories: searches for large $p_{\mathrm{T}}^{\mathrm{miss}}$ and searches for a DM mediator decaying into SM particles. Both classes rely on different analysis techniques. Common to all searches is a detailed exploration of the top-quark decay. Due to the almost diagonal structure of the CKM matrix and, in particular, V_{tb} being close to one, the top quark decays almost 100% of the time into a bottom quark and a W boson. The W boson itself decays with about 30% probability into a charged lepton, i.e., an electron, muon, or tau, and the corresponding neutrino, or into two quarks otherwise. Similar to DM particles, neutrinos can only be inferred from missing transverse momentum in the detector. Events with two top quarks or with a single top quark and a W boson are typically categorized into three orthogonal channels based on the lepton ($\ell = e, \mu$, including decays via τ leptons, i.e., $\tau \to e, \tau \to \mu$) multiplicity in the final state. The 0-lepton (0ℓ) final states arise in events in which both W bosons decay hadronically; 1-lepton (1ℓ) final states arise in events in which one W boson decays hadronically, the other leptonically; 2-lepton (2ℓ) final states arise if both W bosons decay leptonically.

When top quarks recoil against significant $p_{\mathrm{T}}^{\mathrm{miss}}$ or result from the decay of a very heavy resonance, top quarks are highly Lorentz-boosted and their decay products become highly collimated. In the case of hadronic top-quark decays, this means that the particle showers from the three final-state quarks can no longer be reconstructed as three separate small-radius (small-R) jets (*resolved decay*) but instead merge into a single large-radius (large-R) jet with characteristic substructure (*merged decay*). Merged top-quark decays are identified using dedicated *top tagging* algorithms.

3.1. Final States with Invisible Decays

3.1.1. $p_{\mathrm{T}}^{\mathrm{miss}}$ + t

Searches for the production of large $p_{\mathrm{T}}^{\mathrm{miss}}$ in association with a single top quark were conducted by both the ATLAS [43] and CMS [44] collaborations.

The ATLAS Collaboration performed a $p_{\mathrm{T}}^{\mathrm{miss}}$ + t search targeting merged hadronic top-quark decays using 139 fb^{-1} of $\sqrt{s} = 13$ TeV pp collision data [43]. Events are required to have $p_{\mathrm{T}}^{\mathrm{miss}}$ >250 GeV and contain at least one large-R (anti-k_t [45] $R = 1.0$) jet with transverse momentum $350 < p_T < 2500$ GeV and mass $40 < m < 600$ GeV. Additionally, the selected jet must be identified as a top-quark candidate via a dedicated top-tagging algorithm [46], which relies on a deep neural net (DNN) that uses jet kinematics and substructure variables as input [46,47]. The working point for the top tagging algorithm chosen for this analysis corresponds to a 50% top tagging efficiency.

Dedicated signal regions targeting resonant DM production via a color-charged scalar mediator (Section 2.2.2) and nonresonant DM production via a vector mediator with a V_{ut} coupling (Section 2.1.2) are defined based on the output score of XGBoost classifiers [48] that are trained on several event observables. Control regions are defined to constrain the dominant backgrounds from $t\bar{t}$ and V + jets production.

A similar search was performed by the CMS Collaboration [44]. Different from the ATLAS analysis, the result is based on data recorded in 2016 that only correspond to an integrated luminosity of 36 fb^{-1}. To identify the hadronically decaying top quark, CA15 jets were used. CA15 jets are clustered from particle flow candidates using the Cambridge–Aachen algorithm [45] with a distance parameter of 1.5. The CA15 jets must have a transverse momentum $p_T > 250$ GeV, $|\eta| < 2.4$ and an invariant mass of 110 GeV < m < 210 GeV. Furthermore, several substructure observables, such as the N-subjettiness [49] or so-called energy-correlation functions [50,51], are combined in a boosted decision tree (BDT) [52] to distinguish top-quark jets from the hadronization products of single light quarks or gluons. At 50% signal efficiency, the BDT background acceptance is 4.7%. The dominant backgrounds from $t\bar{t}$ and single vector bosons (Z, W, γ) are constrainedfrom dedicated control regions. The signal is probed in distributions of missing transverse energy $p_{\mathrm{T}}^{\mathrm{miss}}$ considering two signal regions which correspond to a BDT output between 0.1 and 0.45 and above 0.45, respectively.

The summary plots for the benchmark model with a color-charged scalar mediator in Section 4.2.2, which show the interplay between the $p_{\mathrm{T}}^{\mathrm{miss}}$ + t and same-sign tt (Section 3.2.1) searches, are based on an earlier search of the ATLAS Collaboration using 36 fb^{-1} of $\sqrt{s} = 13$ TeV pp collision [53]. This analysis statistically combines the results from two orthogonal channels, targeting semi-leptonic and hadronic top-quark decays, respectively.

3.1.2. $p_{\mathrm{T}}^{\mathrm{miss}}$ + tW and $p_{\mathrm{T}}^{\mathrm{miss}}$ + tj

Similar to the $p_{\mathrm{T}}^{\mathrm{miss}}$ + t searches described in Section 3.1.1, searches for $p_{\mathrm{T}}^{\mathrm{miss}}$ + tW target events with single top quarks produced in association with large $p_{\mathrm{T}}^{\mathrm{miss}}$ additionally require the existence of a second visible object. This can be either a W boson or a hadronic jet. The resulting signatures are referred to as $p_{\mathrm{T}}^{\mathrm{miss}}$ + tW and $p_{\mathrm{T}}^{\mathrm{miss}}$ + tj, respectively. It should be noted that searches in these final states are not orthogonal to the $p_{\mathrm{T}}^{\mathrm{miss}}$ + t searches discussed in Section 3.1.1, as the latter do not veto the presence of additional visible objects in the event, and hence implicitly include $p_{\mathrm{T}}^{\mathrm{miss}}$ + tj and $p_{\mathrm{T}}^{\mathrm{miss}}$ + tW signatures.

While $p_{\mathrm{T}}^{\mathrm{miss}}$ + t searches are traditionally used to constrain resonant DM production via a color-charged scalar mediator, and nonresonant DM production via a vector mediator with a flavor-violating V_{ut} coupling, as explained in Section 3.1.1, $p_{\mathrm{T}}^{\mathrm{miss}}$ + tW searches in particular are used to probe the 2HDM + a (Section 2.3) and, more recently, also simplified models with a scalar or pseudoscalar mediator (Section 2.2.1).

Simplified models with a scalar or pseudoscalar mediator predict both $p_{\mathrm{T}}^{\mathrm{miss}}$ + tW and $p_{\mathrm{T}}^{\mathrm{miss}}$ + tj production, as illustrated by the two right-most diagrams in Figure 3. The corresponding signal cross-sections are, up to mediator masses of 200 GeV, smaller than those of the dominant $p_{\mathrm{T}}^{\mathrm{miss}}$ + $t\bar{t}$ production mode discussed in Section 3.1.3. Therefore, $p_{\mathrm{T}}^{\mathrm{miss}}$ + tW and $p_{\mathrm{T}}^{\mathrm{miss}}$ + tj searches have not been used to constrain these simplified models by the ATLAS Collaboration. However, with the increased sensitivity of recent searches, single-top-associated production becomes more and more relevant, and a first search including $p_{\mathrm{T}}^{\mathrm{miss}}$ + tW and $p_{\mathrm{T}}^{\mathrm{miss}}$ + $t\bar{t}$ signatures was performed by the CMS Collaboration [54], as further discussed in Section 3.1.4.

$p_{\mathrm{T}}^{\mathrm{miss}}$ + tW and $p_{\mathrm{T}}^{\mathrm{miss}}$ + tj production is also predicted in the 2HDM + a. Compared to simplified models with a single (pseudo-)scalar mediator, this model contains additional production modes, illustrated, for example, by the third diagram in Figure 5, which lead to higher predicted signal cross-sections for $p_{\mathrm{T}}^{\mathrm{miss}}$ + tW and $p_{\mathrm{T}}^{\mathrm{miss}}$ + tj production. A search for $p_{\mathrm{T}}^{\mathrm{miss}}$ + tW and $p_{\mathrm{T}}^{\mathrm{miss}}$ + tj signatures, optimized specifically for 2HDM + a signal processes, was conducted by the ATLAS Collaboration [55] using 139 fb^{-1} of $\sqrt{s} = 13$ TeV pp collision data. The search considers events with one or two leptons (e,μ), at least one b-tagged jet, and significant $p_{\mathrm{T}}^{\mathrm{miss}}$ in three orthogonal categories. Two of them target $p_{\mathrm{T}}^{\mathrm{miss}}$ + tW production in final states with one or two leptons, while the third channel targets $p_{\mathrm{T}}^{\mathrm{miss}}$ + tj production in final states with exactly one lepton. The search was extended in the context of a preliminary analysis of the same dataset [56] to include events with highly energetic W boson decays in final states with zero leptons or one lepton. These provide additional sensitivity for large masses of the charged Higgs bosons. The newly added zero- and improved one-lepton channels are statistically combined with the two-lepton channel of Ref. [55].

3.1.3. $p_{\mathrm{T}}^{\mathrm{miss}}$ + $t\bar{t}$

Searches for DM or DE production in association with a $t\bar{t}$ pair target final states characterized by sizeable $p_{\mathrm{T}}^{\mathrm{miss}}$ and the presence of the $t\bar{t}$ decay products.

The CMS Collaboration released a search for DM in association with a $t\bar{t}$ pair using 137 fb^{-1} of data recorded at $\sqrt{s}$ = 13 TeV between 2016 and 2018 [57]. The analysis combines previous searches in final states with 0 [58], 1 [59], or 2 [60] leptons. While the primary target of the analyses is pair production of the supersymmetric partner of the top quark (*stop*), a reinterpretation of the combined result in a simplified DM model with scalar mediators is provided.

A central feature of the analysis in the 0-lepton channel is an advanced jet-tagging algorithm identifying hadronically decaying top quarks and W bosons with low and high Lorentz-boost. For the highly Lorentz-boosted regime, the DeepAK8 algorithm [61] is used, whereas in the resolved regime the DeepResolved algorithm [59] is explored, to tag top quarks in the intermediate transverse momentum range from 150 to 450 GeV. The analysis includes a total of 183 nonoverlapping signal regions. The contribution of each SM background process is estimated through measurements of event rates in dedicated background control samples that are translated to predicted event counts in the corresponding signal region with the aid of MC simulation.

The key requirements in the 1-lepton channel are exactly one lepton and $p_{\mathrm{T}}^{\mathrm{miss}} > 250$ GeV. Moreover, the transverse mass computed from the lepton and the missing momentum is required to be larger than 150 GeV to reduce the dominant background from SM $t\bar{t}$ and W + jets production, for which the transverse mass has a natural cutoff at the mass of the W boson. The SM production of dileptonic $t\bar{t}$ events, where one of the leptons is lost, is the largest remaining background. It is estimated through a set of dedicated control regions and reduced by using the modified topness variable [59]. The 1-lepton channel also exploits the jet tagging algorithms used in the 0-lepton channel, to identify hadronic top-quark decays. In order to enhance the sensitivity to different signal scenarios, including the case of small missing transverse momentum, events are categorized into a total of 39 nonoverlapping signal regions.

The search in the 2-lepton channel explores orthogonal signal regions based on the flavor of the leptons and three characteristic observables: the so-called missing transverse momentum significance [62] and two specific definitions of the stransverse mass [60,63]. The p_T^{miss} significance is given by the ratio of the p_T^{miss} over its resolution and it is particularly powerful to suppress events where detector effects and misreconstruction of particles from pileup interactions are the main source of missing transverse momentum. The key feature of the stransverse mass using leptons (lepton and b-quark jets) is that it retains a kinematic endpoint at the W-boson (top-quark) mass for SM background events from the leptonic decays of two W bosons (top quarks). The dominant backgrounds arise from $t\bar{t}$ and $t\bar{t}+Z$ production as well as single-top-quark production in the Wt channel. After a veto of the Z-boson mass window, i.e., $|m_{\ell\ell} - m_Z| > 15$ GeV, Drell–Yan production represents only a minor source of background.

A similar search using 139 fb^{-1} of LHC data was released by the ATLAS Collaboration exploring the 0-lepton [64], 1-lepton [65], and 2-lepton [66] channels separately. All three final states were combined afterwards into a single result [67]. In this context, the (0ℓ) channel search was further optimized through an improved selection of triggers targeting b-jets. Searches for p_T^{miss} + tW (Section 3.1.2) production were not included in this combination as their datasets are not orthogonal to those in the p_T^{miss} + $t\bar{t}$ by construction. Including them in a statistical combination is left to future publications. While, by now, the p_T^{miss} + $t\bar{t}$ searches discussed above have been interpreted in simplified models with a scalar or pseudoscalar mediator only (see Section 4.2.1), earlier searches, based on smaller datasets, have already been used to constrain a 2HDM with a pseudoscalar mediator (Section 4.3.1) and a model of scalar DE (Section 2.4).

3.1.4. p_T^{miss} + tW, p_T^{miss} + tj and p_T^{miss} + $t\bar{t}$

A first result exploring topologies of single top quark and top-quark pair associated production was released by the CMS Collaboration [54]. The analysis uses 36 fb^{-1} of data recorded in 2016 at 13 TeV and combines multiple selection categories in final states with 0 or 1 lepton. In the 1-lepton channel, dominant background is suppressed using a similar strategy as the one discussed in Section 3.1.3, while in the 0-lepton channel, dominant background is reduced by a cut on the missing transverse energy, the ratio of the leading jet transverse momentum over the total hadronic transverse energy in the event, and the minimum opening angle between the missing transverse energy and the two leading jets. To enhance the sensitivity to single top quark associated production, events are separated according to the number of identified b-quark jets. Events with a single b-tagged jet are further split into events with a central or forward jet. The categorization in terms of forward jets allows a further enhancement of t/$\bar{t}$+DM t-channel events. This production mode leads to final states with one top quark and an additional jet, which tends to be in the forward region of the detector, while the additionally produced b-quark is typically low in transverse momentum and therefore not reconstructed. A key observation of this search is the p_T^{miss} spectrum explored in a combined fit to different orthogonal signal regions. Overall, data are found to be in good agreement with the expected SM background. Due to the combination of single top quark and $t\bar{t}$ associated production, this analysis was able to derive the most stringent limits from LHC data on spin-0 mediators at that time.

3.2. *Final States without Invisible Decays*

3.2.1. Same-Sign tt

Events with a same-sign tt pair are identified via the leptonic decays of the W bosons from the two top quarks. They are required to contain two same-sign charged leptons, at least one b-jet, and significant p_T^{miss} from the two neutrinos resulting from the leptonic W boson decays.

A search in same-sign tt events was conducted by the ATLAS Collaboration, using 36 fb^{-1} of $\sqrt{s} = 13$ TeV data [22]. The signal region of this search is defined by requiring the presence of two positively charged leptons (e,μ) and at least one b-jet. Additionally, the scalar sum of the transverse momenta of all selected objects in the event, H_T, is required to be significant ($H_T > 750$ GeV). Further requirements on the $p_{\mathrm{T}}^{\mathrm{miss}}$ and the angular separation of the two leptons are imposed. The signal region is split into three orthogonal channels based on the lepton flavor (ee, $e\mu$, $\mu\mu$). The main backgrounds of this search are estimated using MC simulation, while the subdominant background from fake leptons is estimated using data-driven techniques.

3.2.2. $t\bar{t}$

A search for resonant $t\bar{t}$ production in the 0ℓ channel was conducted by the ATLAS Collaboration using 139 fb^{-1} of $\sqrt{s} = 13$ TeV data [68]. This search targets heavy vector and axial-vector resonances (including DM mediators) with masses > 1.4 TeV, resulting in two merged top-quark decays. Merged top-quark decays are identified using a deep neural net (DNN)-based top tagger trained on the distributions of various characteristic jet and jet substructure variables to distinguish top-quark from light-quark and gluon initiated jets. SM $t\bar{t}$ production constitutes the main, irreducible background to this search, followed by strong multijet production. The background spectrum is derived from data by fitting a smoothly falling function to the reconstructed $m_{t\bar{t}}$ distribution, similar to the approach classically chosen in di-jet resonance searches.

A larger range of resonance masses was probed by a search for resonant $t\bar{t}$ production in the 1ℓ channel, conducted by the ATLAS Collaboration on 36 fb^{-1} of $\sqrt{s} = 13$ TeV data [69]. This search targets both *merged* and *resolved* hadronic top-quark decays and is sensitive to resonance masses just above the $t\bar{t}$ kinematic threshold ($>2m_{\mathrm{top}}$). The main, irreducible background from SM $t\bar{t}$ production, as well as most other, smaller backgrounds, are estimated using MC simulation. Data-driven corrections are applied to the MC simulation of the W + jets background. The small background from strong multijet production is estimated with a fully data-driven approach.

A first search for heavy spin-1 resonances combining final states with 0, 1, and 2 leptons was performed by the CMS Collaboration using data recorded at $\sqrt{s} = 13$ TeV and corresponding to a total integrated luminosity of 35.9 fb^{-1} [70]. The analysis utilizes reconstruction techniques that are optimized for top quarks with high Lorentz-boosts, which requires the use of nonisolated leptons partially overlapping with b-quark jets and jet substructure techniques for top-quark tagging. Except for the QCD multijet background in the 0-lepton channel, the shapes of all backgrounds are estimated from MC simulation. The signal strength is extracted from the distributions of the reconstructed invariant mass of the top-quark pair for the 0- and 1-lepton channels and from the sum of missing transverse energy and the transverse momenta of all jets and leptons in the 2-lepton channel.

Interference effects between the resonant signal and background processes are not taken into account in the searches discussed above as they are irrelevant for spin-1 and spin-2 particles. However, this is not true for scalar and pseudoscalar resonances, such as additional heavy Higgs bosons, which are produced from gg initial states via heavy quark loops. The process $gg \to A/H \to t\bar{t}$ interferes strongly with the irreducible background from SM $t\bar{t}$ production, which is dominated by $gg \to t\bar{t}$. Interference effects significantly distort the resonance lineshape from a Breit–Wigner peak to a characteristic peak-dip or even more complicated structures. The treatment of these effects is nontrivial and requires dedicated analysis methods, in particular in the statistical analysis. Searches for heavy scalars and pseudoscalars were conducted by both the ATLAS [71] and CMS collaborations [72] in the 1ℓ and $1\ell + 2\ell$ channels, respectively. These searches are sensitive to the production of scalar and pseudoscalar DM mediators; however, due to the strong model-dependence of the interference patterns, no dedicated interpretation of these results in the context of DM models exists to date. An approximate reinterpretation of the results in Ref. [71] in the context of the 2HDM + a (Section 2.3) can be found in Ref. [32].

3.2.3. $t\bar{t}t\bar{t}$

Final states with four top quarks ($t\bar{t}t\bar{t}$) can arise from nonresonant processes predicted in the SM but are also predicted in BSM models allowing for the associated production of a heavy BSM resonance, which subsequently decays to $t\bar{t}$, with a $t\bar{t}$ pair. Four-top final states are particularly relevant in searches for heavy scalars and pseudoscalars, as the signal-background interference is negligible for associated production with $t\bar{t}$ compared to loop-induced production from gg initial states (Section 3.2.2). It should be noted, though, that the production cross-section for associated production is significantly lower than for loop-induced production.

Four-top final states are characterized by a high object multiplicity. Orthogonal signal regions can be defined based on the multiplicity of leptons (e, μ) in the final state, which corresponds to the number of top quarks with a leptonically decaying W boson.

The ATLAS Collaboration recently found evidence (4.3σ observed, 2.4σ expected significance) for four-top-quark production in a search focusing on the multilepton final state conducted on 139 fb^{-1} of $\sqrt{s} = 13$ TeV pp collision data [73]. The result is consistent with the SM prediction for four-top production within 1.7σ. A subsequent dedicated search for BSM four-top production on the same dataset specifically targets $t\bar{t}$-associated production of heavy scalar or pseudoscalar Higgs bosons A/H decaying to $t\bar{t}$ ($t\bar{t}\ A/H \rightarrow t\bar{t}t\bar{t}$) [74]. It is based on and extends the analysis strategy of Ref. [73] to increase the sensitivity to A/H production. In both the SM and BSM searches, events are required to contain either a same-sign lepton pair or at least three leptons. A multivariate discriminant based on a boosted decision tree (BDT) is used to separate between SM four-top production and other background processes, using event-level information such as jet and b-jet multiplicity as well as additional kinematic variables. The BSM search relies on a second BDT to subsequently distinguish between BSM and SM four-top production. This second BDT is parameterized as a function of the mass of the heavy Higgs boson by introducing the mass as a labeled input in the training [75]. The main, irreducible backgrounds arise from associated production of a $t\bar{t}$ pair with a boson and additional jets ($t\bar{t}$ + W + jets, $t\bar{t}$ + Z + jets, $t\bar{t}$ + h + jets). They are estimated using MC simulations with additional data-driven corrections applied in the case of $t\bar{t}$ + W + jets production. Smaller, reducible backgrounds arise mostly from $t\bar{t}$ + jets and tW + jets production with misidentified charge or fake/nonprompt leptons. These smaller backgrounds are estimated from data using dedicated control regions. No significant excess of events over the SM prediction is observed in the BSM four-top search and the results are interpreted in the context of a type-II 2HDM. No dedicated interpretation in the context of DM models has been performed. The constraints on the type-II 2HDM with $m_A = m_H$, however, indicate that this search can improve upon the current four-top constraints on the 2HDM + a parameter space included in the latest 2HDM + a summary plots of Ref. [76] (Section 4.3.1), which are based on a search in the single-lepton channel using 36 fb^{-1} of $\sqrt{s} = 13$ TeV data [77].

The CMS Collaboration reported an observed (expected) significance for $t\bar{t}t\bar{t}$ of 2.6σ (2.7σ) in the multilepton channel using 137 fb^{-1} of $\sqrt{s} = 13$ TeV pp collision data [78]. The search relies on a new multivariate classifier to maximize the sensitivity to the SM $t\bar{t}t\bar{t}$ signal. As in the equivalent ATLAS search, the main backgrounds from $t\bar{t}$ + boson + jets production are estimated using MC simulations. Data-driven corrections are applied in the cases of $t\bar{t}$ + W + jets and $t\bar{t}$ + Z + jets production. Backgrounds arising from charge misidentification or fake/nonprompt leptons are estimated from data. This result has been used to constrain scalar and pseudoscalar production in 2HDMs as well as in the simplified DM model with a scalar or pseudoscalar mediator (Section 2.2.1). No dedicated interpretation for the 2HDM + a is available, although the constraints on type-II 2HDMs suggest that the search will also constrain the 2HDM + a parameter space.

The searches described above have been optimized for nonresonant $t\bar{t}t\bar{t}$ production and/or production of heavy scalar or pseudoscalar resonances, including resonance masses below 1 TeV. An additional search targeting top-philic vector and axial-vector (Z') resonances with masses > 1 TeV was conducted by the ATLAS Collaboration. The preliminary result relies on 139 fb^{-1} of $\sqrt{s} = 13$ TeV data [79]. Unlike other searches in the $t\bar{t}t\bar{t}$ final state, this search was designed to reconstruct the BSM resonance explicitly from a pair of reclustered jets identified as merged top quarks. The results can, in principle, be used to constrain purely top-philic vector or axial-vector mediators to which classic $t\bar{t}$ resonance searches, which assume Z' production from light-quark or gluon initial states (Section 3.2.2), may not be sensitive. A dedicated interpretation of this search in the context of DM models is left to future work.

3.2.4. $tbH^{\pm}(tb)$

Final states with two top and two bottom quarks are sensitive to the associated production of a charged Higgs boson $H^{\pm}$ with a top and a bottom quark (tb) and its subsequent decay to tb.

The ATLAS Collaboration published a search for $tbH^{\pm}(tb)$ production using 139 fb^{-1} of $\sqrt{s} = 13$ TeV data [80]. It targets charged Higgs boson masses in the range 0.2–2.0 TeV. Events are required to contain exactly one electron or muon to suppress the large backgrounds from strong multi(-b)-jet production. The selected events are further classified according to the number of reconstructed jets and the number of b-jets among them. A neural network is used to enhance the separation between signal and background. The dominant background for this search is composed of $t\bar{t}$ jets events as well as single-top production in the Wt channel [81]. The backgrounds are modeled using MC simulations with additional data-driven corrections derived in a dedicated control region.

A search for charged Higgs bosons decaying into a top and a bottom quark in the 0-lepton final state was performed by the CMS Collaboration using proton–proton collision at $\sqrt{s}$ = 13 TeV from 2016 [82]. Two different scenarios were studied: the associated production with a top and bottom quark and the s-channel production of a charged Higgs. The results were combined with a search in final states with one or two leptons [83]. For production in association with a top quark, upper limits at the 95% confidence level on the charged Higgs production cross-section and branching fraction of 9.25 to 0.005 pb were obtained for charged Higgs masses in the range of 0.2 to 3 TeV. While there is no DM interpretation of the result by the CMS Collaboration, the result from ATLAS was interpreted in a 2HDM + a scenario, as further detailed in Section 4.3.1.

4. Results

4.1. Vector and Axial-Vector Mediators

4.1.1. Flavor-Conserving Interaction

Strong constraints on visible decays of the axial-vector (Figure 7) or vector (Figure 8) mediator $m_{Z'}$ are obtained from a variety of resonance and related searches that probe mediator masses in the range between 50 GeV [84] and 5000 GeV [76].

The latest constraints on axial-vector mediators released by the ATLAS Collaboration and based on data from pp collisions at $\sqrt{s} = 13$ TeV are shown in Figure 7. The coupling of the mediator to leptons is set to zero ($g_{\ell} = 0$), while the coupling to DM is set to unity ($g_{\chi} = 1.0$) and the DM mass is taken to be 10 TeV to kinematically suppress invisible mediator decays and highlight the interplay of constraints on visible mediator decays.

In the high mediator mass range, the main sensitivity comes from two searches for di-jet resonances, referred to as *di-jet* and *di-jet angular*. The former aims to identify local resonant enhancements in the di-jet invariant mass spectrum and targets narrow mediator widths. The latter, for which no results on the full LHC Run 2 dataset are available, relies on the di-jet angular separation to identify broader mediator widths that cannot be probed by the search in the invariant mass spectrum. Neither of the searches imposes quark-flavor specific selection requirements, and hence are sensitive to all possible hadronic decays of the mediator.

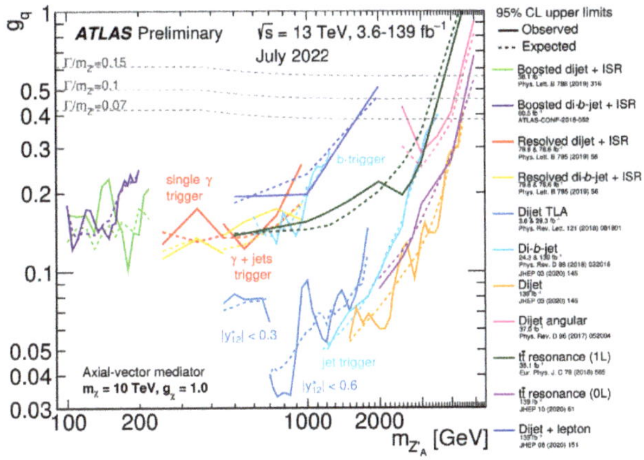

Figure 7. Upper limits at 95% CL on the coupling g_q of the mediator to quarks in a simplified model with a vector or axial-vector mediator obtained from different types of resonance searches using data from pp collisions at $\sqrt{s} = 13$ TeV. The DM mass is $m_\chi = 10$ TeV and its coupling to the mediator is $g_\chi = 1$ [76].

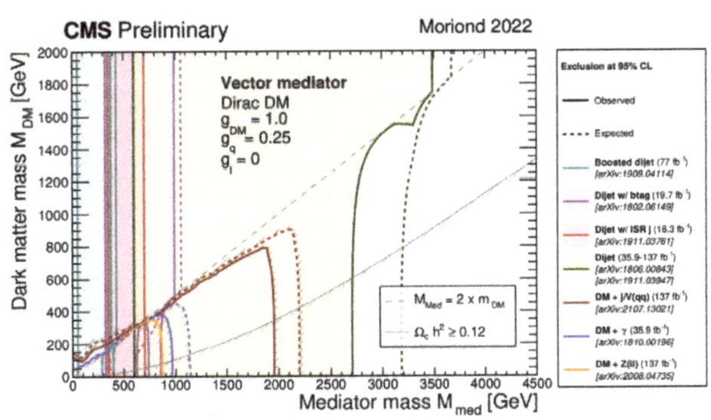

Figure 8. 95% CL observed and expected exclusion regions on vector mediators in the DM–mediator mass plane from searches with visible and invisible final states released by the CMS Collaboration [84]. Exclusions are computed for a leptophobic scenario with $g_l = 0$, a universal quark coupling of $g_q = 0.25$, and a DM coupling of $g_{\rm DM} = 1.0$.

Searches for $t\bar{t}$ resonances, which rely on top-quark identification algorithms to specifically identify the decays of the mediator to top quarks, have a slightly lower expected sensitivity to the coupling g_q than di-jet searches, although the observed limit is stronger than that from the di-jet search in some small regions of the mediator mass where the di-jet observed limit fluctuates upward. The use of top-quark identification allows for a stronger suppression of SM backgrounds compared to di-jet and also di-b-jet searches, in particular the background from strong multijet production. This effect partially compensates the disadvantage of probing only roughly $\frac{1}{6}$ of the hadronic mediator decays.

In Figure 8, constraints on vector mediators in the plane of the DM and the mediator mass from the CMS Collaboration [84] are shown. Different from Figure 7, results from visible and invisible decays are summarized. While searches with invisible final states are only possible when the mediator mass is about twice the DM mass, the sensitivity of searches for visible decays only depends on the DM mass through the width of the mediator. When the decay channel to DM particles opens up, the width of the mediator increases and resonant searches become less sensitive. The best sensitivity to vector mediators from $p_{\rm T}^{miss}$ searches is provided by DM searches with initial state radiation either from a gluon/quark jet or from the hadronic decay of a vector boson [85]. Searches with visible final states achieve best sensitivity down to 50 GeV when looking for a large radius jet that recoils against the mediator [86]. At high mass, the strongest constraints are obtained from di-jet searches [87]. The searches discussed in Section 3.2.2 probing vector mediators decaying into $t\bar{t}$ are not shown, as no dedicated interpretation of these results were performed in models of DM by the CMS Collaboration. However, the interpretation of the searches in generic vector particle models show comparable sensitivity between the results released by the ATLAS and CMS collaborations.

The collider constraints on simplified models with a vector or axial-vector mediator can be translated into limits on the spin-dependent DM–proton or DM–neutron and spin-independent DM–nucleon scattering cross-sections as a function of the DM mass to allow for a comparison with results from direct-detection experiments [88]. It should be noted that such a sensitivity comparison is highly dependent on the choice of model parameters, such as the mediator coupling type, as illustrated in the following. In Figure 9, constraints on the spin-independent DM–nucleon scattering cross-section (left) and the spin-dependent DM–neutron scattering cross-section (right) are shown for the case of a vector mediator and an axial-vector mediator, respectively. In the former case, the sensitivity from direct-detection experiments supersedes that from collider searches by several orders of magnitude for DM masses above 10 GeV. For smaller DM masses, collider experiments dominate the sensitivity as the sensitivity of direct-detection experiments is limited by the very low energy recoils that such low-mass DM particles would induce. In the case of an axial-vector mediator, collider experiments dominate the sensitivity across the studied DM mass range. A detailed discussion of the interplay between collider and direct-detection experiments in simplified models with a vector or axial-vector mediator can be found in Ref. [19].

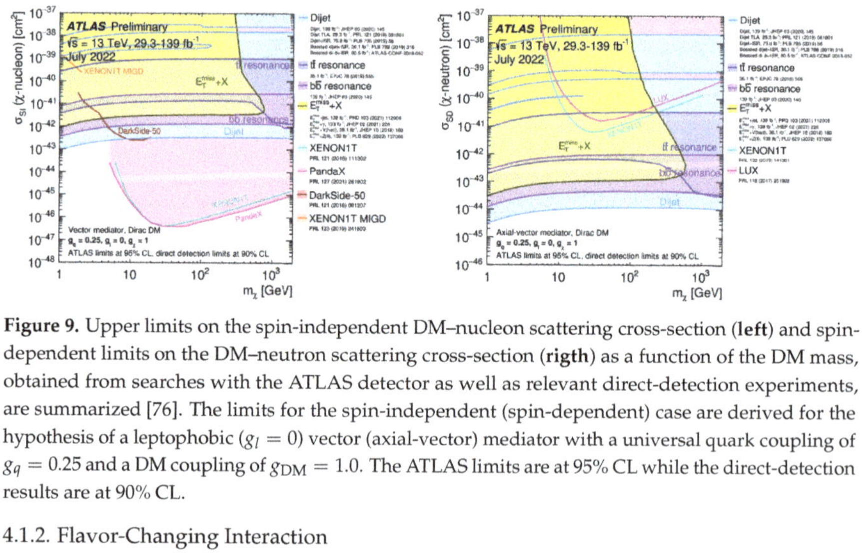

Figure 9. Upper limits on the spin-independent DM–nucleon scattering cross-section (**left**) and spin-dependent limits on the DM–neutron scattering cross-section (**rigth**) as a function of the DM mass, obtained from searches with the ATLAS detector as well as relevant direct-detection experiments, are summarized [76]. The limits for the spin-independent (spin-dependent) case are derived for the hypothesis of a leptophobic ($g_l = 0$) vector (axial-vector) mediator with a universal quark coupling of $g_q = 0.25$ and a DM coupling of $g_{\rm DM} = 1.0$. The ATLAS limits are at 95% CL while the direct-detection results are at 90% CL.

4.1.2. Flavor-Changing Interaction

The strongest constraints on the VFC model are obtained from searches targeting same-sign tt and $p_{\rm T}^{\rm miss} + t$ production on 36 fb^{-1} of pp collision data [19]. Results for two representative parameter planes are shown in Figure 10.

The left plot of Figure 10 shows a scan in the mediator mass versus the flavor-changing coupling g_{ut} while fixing the remaining two parameters at $m_\chi = 1$ GeV and $g_\chi = 1$. The $p_{\rm T}^{\rm miss} + t$ search provides stronger constraints on g_{ut} at lower mediator masses, excluding g_{ut} down to 0.07 at 1 TeV, while the same-sign tt search is more sensitive for mediator masses > 1.6 TeV, still excluding $g_{ut} > 0.3$ at 3 TeV. Mediator masses below 1 TeV were probed by the CMS Collaboration at $\sqrt{s}$ = 13 TeV and are shown in Figure 11. The $p_{\rm T}^{\rm miss} + t$

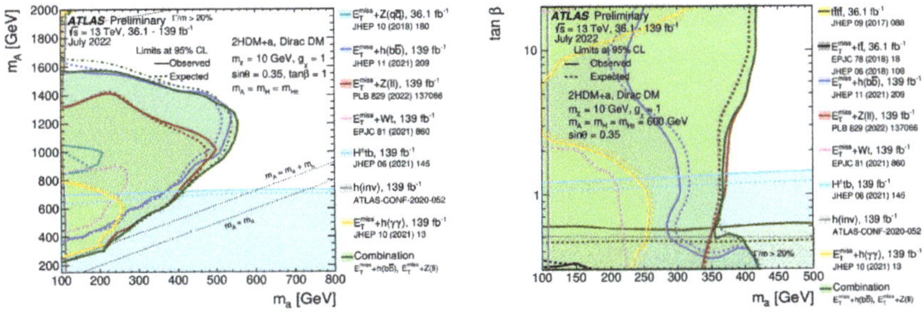

Figure 15. Regions in the 2HDM + a parameter space excluded at 95% CL by several individual searches targeting different signatures and a statistical combination of $p_{\mathrm{T}}^{\mathrm{miss}}$ +$Z(\ell\ell)$ and $p_{\mathrm{T}}^{\mathrm{miss}}$ +$h(b\bar{b})$ searches. The results are shown in the (m_a,m_A) plane (left) and the (m_a,tan β) plane (right). In the former case, tan $\beta = 1$, while in the latter case, $m_A = 600$ GeV. In both cases, the conditions $\sin\theta = 0.35$ and $m_A = m_H = m_{H^\pm}$ are imposed. All results are based on either the full 139 fb^1 of pp collision data at $\sqrt{s} = 13$ TeV or a subset of that dataset amounting to 36 fb^1 [76].

In the (m_a,tan β) plane with $m_A = m_H = m_{H^\pm} = 600$ GeV (right plot in Figure 15), the sensitivity is again dominated by the statistical combination of the $p_{\mathrm{T}}^{\mathrm{miss}}$ +$h(b\bar{t})$ and $p_{\mathrm{T}}^{\mathrm{miss}}$ +$Z(\ell\ell)$ searches and the search for $tbH^{\pm}(tb)$ production, which provide complementary constraints in this region of parameter space. Low values of tan β are fully excluded by the search for charged Higgs bosons decaying to tb. The constraints from the search targeting $t\bar{t}t\bar{t}$ production on 36 fb^{-1} of LHC Run 2 data are also shown. While they are notably weaker than the constraints from the charged-Higgs-boson search, which relies on the full Run 2 dataset amounting to 139^{-1}, the results from the search for $t\bar{t}A/H(t\bar{t})$ on 139 fb^{-1} of LHC Run 2 data [74] (Section 3.2.3) indicate that this final state may provide a comparable exclusion power to the charged-Higgs-boson search if reinterpreted in the context of this model.

Searches for $p_{\mathrm{T}}^{\mathrm{miss}}$ + $t\bar{t}$ production, which dominate the sensitivity to the simplified model with a color-neutral scalar or pseudoscalar mediator (Section 4.2.1), only weakly constrain the benchmark scenarios [38,39] probed at the LHC. It should, however, be noted that the $p_{\mathrm{T}}^{\mathrm{miss}}$ + $t\bar{t}$ constraints shown in Figure 15 are based on only 36 fb^{-1} of LHC Run 2 data and the sensitivity is mainly limited by low event rates. Hence, significantly stronger constraints are expected from a reinterpretation of searches using the full 139 fb^{-1} of LHC Run 2 data [67]. The sensitivity of the $p_{\mathrm{T}}^{\mathrm{miss}}$ + $t\bar{t}$ final state is expected to become comparable to that of searches in the $p_{\mathrm{T}}^{\mathrm{miss}}$ + h and $p_{\mathrm{T}}^{\mathrm{miss}}$ + Z final states for an integrated luminosity of 300 fb^{-1}, expected to be available after the end of LHC Run 3 (2022 2025) [32]. In this context, it should be noted that the cross-section for $p_{\mathrm{T}}^{\mathrm{miss}}$ + $t\bar{t}$ production is suppressed by $\sin\theta^2$, making this process more sensitive for large values of $\sin\theta$ [32]. Furthermore, for $m_a > 2 \cdot m_t$, visible mediator decays to $t\bar{t}$ are possible, reducing the invisible branching ratio $a \to \chi\chi$ and, hence, the sensitivity of the $p_{\mathrm{T}}^{\mathrm{miss}}$ + $t\bar{t}$ searches [32].

4.4. Scalar DE EFT Model

Searches in the $p_{\mathrm{T}}^{\mathrm{miss}}$ + $t\bar{t}$ final state were used to constrain the $\mathcal{L}_1$ operator in the EFT model of scalar DE (Section 2.4). Results from three independent analyses, each targeting a different $t\bar{t}$ decay mode (0-, 1-, 2-lepton channels), were used. No statistical combination was performed. Instead, the constraint from the analysis yielding the smallest CL$_s$ value for a given signal hypothesis was reinterpreted in the EFT model of DE. The strongest constraints arise from searches in the 0- and 1-lepton channels, with both contributing roughly equally.

The constraints are derived as a function of the effective coupling g_* associated with the UV completion of the EFT model and the effective mass scale M_1. It is assumed that the EFT is valid for momentum transfers $Q_{\mathrm{tr}} < g_* M$ [19]. For events failing this requirement, a conservative approach to correct the final limits based on the fraction of valid events, referred to as iterative rescaling [18], is applied.

The regions excluded at 95% CL are shown in Figure 16. Mass scales < 200 GeV are excluded for $g_* > \pi^2$. The sensitivity of the p_T^{miss} + $t\bar{t}$ signature to softer effective couplings g_* is limited by the EFT criterion as $t\bar{t}$-pair production typically involves large momentum transfers.

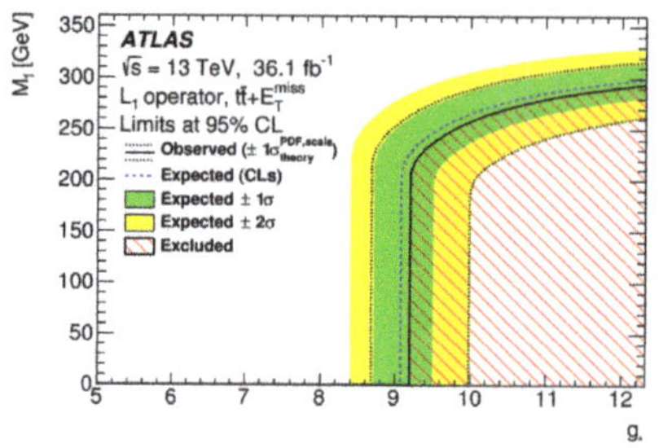

Figure 16. Regions in the plane of the effective coupling g_* associated with the UV completion of the EFT model and the effective mass scale M_1 for the $\mathcal{L}_\infty$ operator excluded at 95% CL by searches in the p_T^{miss} + $t\bar{t}$ final state [19].

5. Discussion

A variety of searches targeting top-quark production in association with DM or via visible decays of mediator particles were conducted by the ATLAS and CMS collaborations. No significant deviation from the SM prediction was observed; therefore, the results were used to constrain DM in a variety of simplified models as well as scalar DE described in an EFT model. Signatures involving top quarks often provide sensitivity in parameter regions not covered by other DM searches, underlining their importance as sensitive probes of DM at colliders. They provide a particularly relevant probe of models involving new particles with Yukawa-like interactions, which imply preferred couplings to top quarks.

It should be noted that many of the results and summary plots presented in this review are preliminary, as various searches on the full LHC Run 2 collision data are still ongoing. Furthermore, not all of the existing results have been interpreted in relevant benchmark models. Further results of DM searches with top quarks are expected to be released by both collaborations in the near future.

6. Outlook

6.1. LHC Run 3

The nonobservation of WIMP DM at the LHC and various direct-detection experiments to date has prompted the particle physics community to place a stronger focus on models and searches for non-WIMP DM as well as uncovered DM signatures at the LHC that can be probed during LHC Run 3 (2022–2025) and/or via reinterpretations of existing searches on LHC Run 2 data [89–91]. A few notable examples involving signatures with top quarks are given in the following.

6.1.1. ALPs

Axions and axion-like particles (ALPs) [92,93] have received increasing attention in recent years. A novel strategy to search for ALPs and, more generally, pseudo-Nambu–Goldstone bosons (pNGB) at the LHC was proposed in Ref. [94], focusing on nonresonant searches that would be sensitive to ALPs produced as off-shell s-channel mediators. It is motivated by the fact that the pNGB nature of the ALPs implies that their couplings to the SM are dominantly derivative, which leads to a cross-section enhancement for nonresonant ALPs production at center-of-mass energies $\hat{s} >> m_a$, where m_a denotes the mass of the ALP. The focus of recent studies is on constraining the ALP-boson (W, Z, h,

g, γ) coupling via nonresonant ZZ, $\gamma\gamma$, and gg [94], nonresonant ZZ and Zh [95], and nonresonant WW, $Z\gamma$ [96] production. The ALPs–fermion coupling can be predominantly probed via nonresonant $t\bar{t}$ production (illustrated by the left diagram in Figure 17) due to the Yukawa-like structure of the ALP–fermion couplings. No public results exist to date, but studies are ongoing.

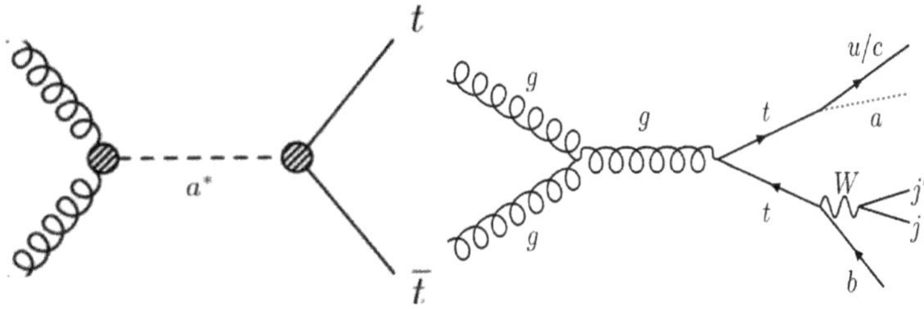

Figure 17. Schematic representation of nonresonant $t\bar{t}$ production via an off-shell s-channel ALP ((**left**), [97]) and SM $t\bar{t}$ production with subsequent decay of one of the top quarks to an up-type quark and a long-lived ALP ((**right**), [98]).

The ALPs–fermion coupling can also be probed in $p_{\mathrm{T}}^{\mathrm{miss}}$ + $t\bar{t}$ final states. These are sensitive to $t\bar{t}$-associated production of a single ALP with couplings to quarks derived from couplings to the bosonic sector and proportional to the fermion mass [99]. It should be noted that the $p_{\mathrm{T}}^{\mathrm{miss}}$ distribution predicted for this signal process is softer on average than that predicted by, e.g., stop production in supersymmetric models, emphasizing the importance of keeping the $p_{\mathrm{T}}^{\mathrm{miss}}$ threshold low in future searches.

Novel detector signatures involving exotic top-quark decays are predicted in models with flavor-violating ALPs [98], which are motivated by t-channel dark sector models [100] or Frogatt–Nielsen models of flavor [101]. These models predict flavor-violating decays of the top quark to an up-type quark and an ALP, with the ALP decaying predominantly to hadrons, either promptly or with a long lifetime. Precision measurements of single-top-quark production can constrain the parameter space of such models for prompt ALPs decays to jets and detector-stable ALPs. Displaced detector signatures are predicted for nonprompt ALPs decays within the detector volume. A novel search was proposed [98], focusing on exotic top-quark decays from SM $t\bar{t}$ production (right diagram in Figure 17), where one of the top quarks decays into an up-type quark and an ALP, which in turn decays into a displaced narrow jet within the calorimeter volume. This and other signatures involving long-lived particles (LLP) in top-quark decays have not yet been probed in dedicated searches at the LHC. They remain an exciting prospect for the analysis of LHC Run 3 data within the currently fast-growing field of LLPs searches at the LHC, a field that benefits in particular from novel trigger and reconstruction algorithms deployed by the ATLAS and CMS experiments for Run 3 data-taking.

6.1.2. Composite Pseudo-Nambu–Goldstone Bosons

Signatures with top quarks can also be used to probe still-viable WIMP models in which WIMP DM is made up of composite pNGBs [102]. In these models, both the SM Higgs boson and DM emerge from a TeV-scale strongly-coupled sector as pNGBs, and the SM–DM interaction is provided by higher-dimensional derivative couplings with the Higgs fields, which leads to a strong suppression of the DM scattering rates against SM particles. Thus, these models evade the strong constraints from direct-detection experiments, making collider searches particularly relevant. The pNGB DM contains additional interactions with the SM sector, besides the derivative Higgs portal, with preferential couplings to third-generation fermions being well motivated [102]. If couplings to top quarks are preferred over couplings to bottom quarks, e.g., in the case of Yukawa-type couplings, pNGB models can be probed at the LHC via associated production of pNGB DM with $t\bar{t}$ or a single top quark, i.e., in $p_{\mathrm{T}}^{\mathrm{miss}}$ + $t\bar{t}$ or $p_{\mathrm{T}}^{\mathrm{miss}}$ + t + X final states. Two possible production modes of pNGB leading to p_T^{miss} + tW final states via the Higgs portal and direct DM–top interactions are shown in Figure 18. Searches in these final states are complementary to searches for invisible Higgs boson decays in vector boson fusion (VBF) production as they are sensitive to pNGB interactions with fermions not accessible via the latter. Reinterpretations of existing p_T^{miss} + $t\bar{t}$ and p_T^{miss} + tW searches, as well as possible optimizations of future searches for pNGB production, could be interesting to explore during LHC Run 3.

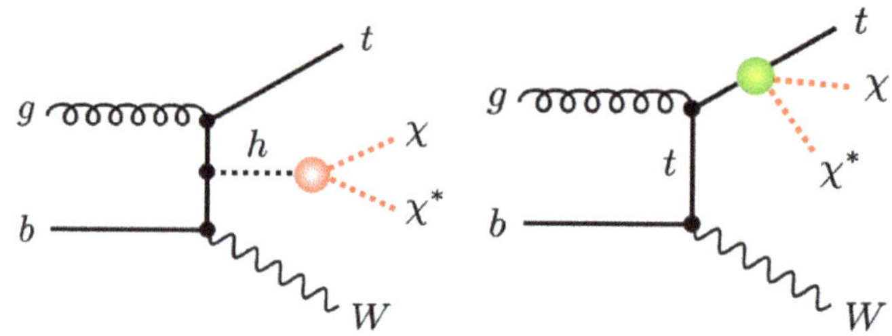

Figure 18. Schematic representation of p_T^{miss} + tW production via DM–Higgs operators (left) and DM–top operators in an EFT of composite pNGBs [102].

6.1.3. Dark Mesons

Final states with multiple top quarks are predicted in models with a strongly coupled dark sector consisting of composite particles that carry electroweak, but no color, charges [103]. These models not only address the hierarchy problem but can also provide a DM candidate in the form of a composite meson whose decays are suppressed via an automatic accidental symmetry.

The most promising target for collider searches is the dark meson sector, consisting of dark vector mesons ρ_D and dark pions π_D [103]. Signatures with multiple top or bottom quarks are predicted if a pair of dark pions with gaugephobic couplings to the SM is produced from the decay of a resonantly produced ρ_D ($pp \to \rho_D \to \pi_D\pi_D$). The dark pions then decay predominantly into third-generation fermions, with decays to $t\bar{t}$ (tb) dominating the branching fraction for π_D^0 ($\pi_D^{\pm}$) if the pion mass is above the $t\bar{t}$ (tb) production threshold. Depending on the charge of the intermediate ρ_D, different final states involving third-generation quarks are possible: $b\bar{b}t\bar{b}$, $t\bar{t}b\bar{b}$, $t\bar{t}t\bar{b}$.

Existing searches in multitop final states only weakly constrain the parameter space of these models [103]. This is due to the fact that small masses of the ρ_D and π_D are still viable, which means that the SM fermions in the final state tend to be rather soft. In searches at $\sqrt{s} = 13$ TeV, in particular, higher thresholds are imposed on the energy/momenta of the final-state objects or their vector sum. In order to probe dark pions, or more generically strongly-coupled-like models, dedicated searches targeting final states with a high multiplicity of low-momentum objects compatible with the decays of one or several low-momentum top quarks are needed.

6.2. HL-LHC and HE-LHC

The physics potential for DM searches involving top quarks during the high-luminosity phase of the LHC (HL-LHC, starting 2028) and the perspectives for a possible future high-energy LHC (HE-LHC) were studied in the context of a 2019 CERN Yellow Report [104]. The final HL-LHC dataset is expected to amount to an integrated luminosity of 3000 fb^{-1} at a center-of-mass energy $\sqrt{s} = 14$ TeV. The HE-LHC scenario relies on the assumption of a possible further upgrade of the LHC to a 27 TeV pp collider with a final integrated luminosity of 15,000 fb^{-1}.

Sensitivity studies were performed for the p_T^{miss} + $t\bar{t}$, p_T^{miss} + tW, p_T^{miss} + t, $t\bar{t}$, and $t\bar{t}t\bar{t}$ signatures within various benchmark models, including simplified models with a scalar or pseudoscalar mediator (Section 2.2.1), simplified models with a vector mediator with a flavor-changing coupling to the top and up-quark (Section 2.1.2), and the 2HDM + a

(Section 2.3). These studies were mostly based on the analysis tools and strategies used for the analysis of the partial LHC Run 2 dataset (2015–2016). They do not include further improvements, such as new machine-learning-based tools or background estimation strategies, implemented for the later analyses of the full LHC Run 2 dataset. A full review of the results of these sensitivity studies across the different final states and models is beyond the scope of this article, but a few general observations can be made. Overall, both the increase in integrated luminosity (HL-LHC) and center-of-mass energy (HE-LHC) lead to a significant sensitivity increase across the different final states. For example, the mass range for a (pseudo-)scalar mediator expected to be excluded by p_T^{miss} + $t\bar{t}$ searches in the simplified model of Section 2.2.1 with $g = g_q = g_\chi = 1.0$ (compare Figure 13) is expected to increase by a factor of two for the HL-LHC compared to the expected sensitivity for LHC Run 3, and by another factor of two for the HE-LHC compared to the HL-LHC.

The sensitivity of most of the searches is dominated by the systematic uncertainties on the main (often irreducible) background processes, for example, $t\bar{t} + V$ in the case of p_T^{miss} + $t\bar{t}$ searches. In $t\bar{t}$ final states, these typically arise from two sources: firstly, uncertainties related to reconstructed objects, such as the energy scale for hadronic jets, and, secondly, uncertainties arising from the modeling of SM processes, such as missing higher-order corrections. These uncertainties can vary between a few percent and a few tens of percent, depending on the process and kinematic region. The former are expected to decrease with increasing integrated luminosity as the statistical uncertainties on the measurements from which they are derived are reduced accordingly. A further reduction of these uncertainties can be expected due to the development of better and more refined calibration methods. The latter can be reduced significantly through profiling in a likelihood fit to data if appropriate, background-enriched control regions are defined. Improved theoretical predictions, for example, for differential cross-sections at higher orders in perturbation theory, can also significantly boost the sensitivity of many searches.

In the case of the HE-LHC, in addition to the improvements due to the larger integrated luminosity, the larger center-of-mass energy provides access to mediator masses beyond the kinematic reach of the (HL-)LHC and to processeswith small signal cross-sections.

6.3. FCC-hh

Similar considerations to those for the HE-LHC apply to the case of a potential future hadron collider operating at center-of-mass energies beyond that of the LHC and HE-LHC. The most prominent example is that of the FCC-hh, the Future Circular Collider, in its operation mode as a hadron collider with a center-of-mass energy of $\sqrt{s} = 100$ TeV [105]. Few dedicated studies regarding the sensitivity of DM searches with top quarks at the FCC-hh exist. For example, in Ref. [106], the sensitivity of the 2-lepton p_T^{miss} + $t\bar{t}$ final state to Higgs portal models and their extensions is discussed. In general, a significant increase in the accessible mass range of both mediators and DM particles is expected, as well as a significant increase in the sensitivity to smaller DM–SM couplings, rendering detector signatures involving decays of long-lived particles away from the interaction point that is highly relevant. Moreover, top quarks appearing in the final states of FCC-hh collision can be extremely boosted, underlining the need for high-resolution detectors to identify very collimated decays, as well as the use of advanced pattern recognition methods for top-quark tagging. A particularly interesting observation is the fact that associated production of a single Higgs boson with $t\bar{t}$ becomes the dominant Higgs boson production mode at Higgs boson transverse momenta of 1-2 TeV and above, a kinematic regime that would be well-populated at the FCC-hh [107]. According to initial studies [107], searches for invisible Higgs boson decays in this production mode would feature a very low background contamination ($S/B \sim 1$) and, hence, provide excellent sensitivity to Higgs portal models with small couplings. The corresponding final state would be p_T^{miss} + $t\bar{t}$ with highly boosted top quarks.

6.4. Future e^+e^- Colliders

No studies of DM searches with top quarks exist for future e^+e^- colliders, such as the International Linear Collider (ILC) [108], the Compact Linear Collider (CLIC) [109,110], the Future Circular Collider FCC-ee [111,112], and the Circular Electron Positron Collider (CEPC) [113,114]. This can be mostly attributed to the fact that these machines are primarily designed for Higgs boson and top-quark precision measurements, rather than a broad range of BSM (including DM) searches, and that their foreseen center-of-mass energies are, in many cases, below or close to the $t\bar{t}$ production threshold. For example, operation modes at $\sqrt{s} = 240$ GeV (250 GeV), i.e., around the maximum of the Zh production cross-section, are foreseen for the FCC-ee and the CEPC (ILC). Additional operation modes in the range 350–365 GeV (FCC-ee, CEPC) and 380 GeV (CLIC) are foreseen for top-quark precision measurements. Higher center-of-mass energies of 1 TeV (ILC) and 1–3 TeV could be possible for the linear e^+e^- machines to allow for wider range of BSM searches. Hence, direct DM production in association with at least one top quark, leading to $p_{\mathrm{T}}^{\mathrm{miss}} + t\bar{t}$ and $p_{\mathrm{T}}^{\mathrm{miss}} + t + X$ final states, while in principle possible, is trivially limited by the available center-of-mass energy. Nevertheless, the foreseen precision scans of the $t\bar{t}$ production threshold at the FCC-ee could, in principle, be sensitive to anomalous resonant or nonresonant $t\bar{t}$ production linked with DM or DM mediators as well as anomalous top-quark decays. Further studies are needed to understand the prospects for DM searches with top quarks at future e^+e^- colliders.

6.5. Conclusions

Collider signatures with top quarks provide sensitive probes of DM predicted by a wide range of models, and possibly even to DE signatures. Searches targeting top-quark production in association with DM or via visible decays of mediator particles have been performed by the ATLAS and CMS Collaborations, with many searches on the full LHC Run 2 collision data still ongoing. As shown in this review, DM searches involving top quarks often provide sensitivity in parameter regions not covered by other DM searches, underlining their importance as sensitive probes of DM at colliders. The upcoming LHC Run 3 opens up further opportunities to improve upon existing results or to explore new signatures, for example, involving top quarks in association with long-lived particle signatures.

Author Contributions: All authors equally contributed to all the parts of the manuscript. All authors have read and agreed to the published version of the manuscript.

Funding: This review article was produced without external funding.

Data Availability Statement: Not applicable.

Acknowledgments: K.B. thanks the Helmholtz Association for the support through the "Young Investigator Group" initiative. The authors acknowledge funding by the Deutsche Forschungsgemeinschaft (DFG, German Research Foundation) under Germany's Excellence Strategy—EXC 2121 "Quantum Universe"—390833306.

Conflicts of Interest: The authors declare no conflict of interest.

Abbreviations

The following abbreviations are used in this manuscript:

DM	Dark matter
DE	Dark energy
EFT	Effective field theory

References

1. Trimble, V. Existence and Nature of Dark Matter in the Universe. *Annu. Rev. Astron. Astrophys.* **1987**, *25*, 425. [CrossRef]
2. Bertone, G.; Hooper, D.; Silk, J. Particle dark matter: evidence, candidates and constraints. *Phys. Rep.* **2005**, *405*, 279. [CrossRef]

3. Feng, J.L. Dark Matter Candidates from Particle Physics and Methods of Detection. *Annu. Rev. Astron. Astrophys.* **2010**, *48*, 495. [CrossRef]
4. Hinshaw, G.; Larson, D.; Komatsu, E.; Spergel, D. N.; Bennett, C. L.; Dunkley, J. M.; Nolta, R.; Halpern, M.; Hill, R. S.; Odegard, N. Nine-Year Wilkinson Microwave Anisotropy Probe (WMAP) Observations: Cosmological Parameter Results. *Astrophys. J. Suppl.* **2013**, *208*, 19. [CrossRef]
5. Planck Collaboration. Planck 2018 results. I. Overview and the cosmological legacy of Planck. *Astron. Astrophys.* **2020**, *641*, A1. [CrossRef]
6. Steigman, G.; Turner, M.S. Cosmological constraints on the properties of weakly interacting massive particles. *Nucl. Phys. B* **1985**, *253*, 375. [CrossRef]
7. Evans, L.; Bryant, P. LHC Machine. *JINST* **2008**, *3*, S08001. [CrossRef]
8. Billard, J.; Boulay, M.; Cebrián, S.; Covi, L.; Fiorillo, G.; Green, A.; Kopp, J.; Majorovits, B.; Palladino, K.; Petricca, F. Direct detection of dark matter—APPEC committee report. *Rept. Prog. Phys.* **2022**, *85*, 056201. [CrossRef]
9. Heros, P.d.l. Status, Challenges and Directions in Indirect Dark Matter Searches. *Symmetry* **2020**, *12*, 1648. [CrossRef]
10. ATLAS Collaboration. The ATLAS Experiment at the CERN Large Hadron Collider. *JINST* **2008**, *3*, S08003.
11. CMS Collaboration. The CMS experiment at the CERN LHC. *JINST* **2008**, *3*, S08004.
12. Buchmueller, O.; Doglioni, C.; Wang, L.T. Search for dark matter at colliders. *Nat. Phys.* **2017**, *13*, 217–223. [CrossRef]
13. Boveia, A.; Doglioni, C. Dark Matter Searches at Colliders. *Ann. Rev. Nucl. Part. Sci.* **2018**, *68*, 429. [CrossRef]
14. CDF Collaboration. Observation of top quark production in ppbar collisions with the Collider-Detector at Fermilab. *Phys. Rev. Lett.* **1995**, *74*, 2626. [CrossRef] [PubMed]
15. D0 Collaboration. Observation of the Top Quark. *Phys. Rev. Lett.* **1995**, *74*, 2632. [CrossRef] [PubMed]
16. Riess, A.G.; Filippenko, A.V.; Challis, P.; Clocchiatti, A.; Diercks, A.; Garnavich, P.M.; Gilliland, R.L.; Hogan, C.J.; Jha, S.; Kirshner, R.P.; Leibundgut, B.R.U.N.O. [Supernova Search Team]. Observational evidence from supernovae for an accelerating universe and a cosmological constant. *Astron. J.* **1998**, *116*, 1009. [CrossRef]
17. Perlmutter, S.; Aldering, G.; Goldhaber, G.; Knop, R.A.; Nugent, P.; Castro, P.G.; Deustua, S.; Fabbro, S.; Goobar, A.; Groom, D.E.; Hook, I.M. [Supernova Cosmology Project]. Measurements of Ω and Λ from 42 high redshift supernovae. *Astrophys. J.* **1999**, *517*, 565. [CrossRef]
18. ATLAS/CMS Dark Matter Forum. Dark Matter benchmark models for early LHC Run-2 Searches: Report of the ATLAS/CMS Dark Matter Forum. *Phys. Dark Univ.* **2020**, *27*, 100371. [CrossRef]
19. ATLAS Collaboration. Constraints on mediator-based dark matter and scalar dark energy models using $\sqrt{s} = 13$ TeV pp collision data collected by the ATLAS detector. *JHEP* **2019**, *5*, 142.
20. Kamenik, J.F.; Zupan, J. Discovering Dark Matter Through Flavor Violation at the LHC. *Phys. Rev. D* **2011**, *84*, 111502. [CrossRef]
21. Boucheneb, I.; Cacciapaglia, G.; Deandrea, A.; Fuks, B. Revisiting monotop production at the LHC. *JHEP* **2015**, *1*, 17. [CrossRef]
22. ATLAS Collaboration, Search for new phenomena in events with same-charge leptons and b-jets in pp collisions at $\sqrt{s} = 13$ TeV with the ATLAS detector. *JHEP* **2018**, *12*, 39.
23. D'Ambrosio, G.; Giudice, G.F.; Isidori, G.; Strumia, A. Minimal flavor violation: An Effective field theory approach. *Nucl. Phys. B* **2002**, *645*, 155. [CrossRef]
24. Buckley, M.R.; Feld, D.; Goncalves, D. Scalar Simplified Models for Dark Matter. *Phys. Rev. D* **2015**, *91*, 015017. [CrossRef]
25. Fayet, P. Supersymmetry and weak, electromagnetic and strong interactions. *Phys. Lett. B* **1976**, *64*, 159. [CrossRef]
26. Fayet, P. Spontaneously broken supersymmetric theories of weak, electromagnetic and strong interactions. *Phys. Lett. B* **1977**, *69*, 489 [CrossRef]
27. Djouadi, A. The Higgs sector of supersymmetric theories and the implications for high-energy colliders. *Eur. Phys. J. C* **2009**, *59*, 389. [CrossRef]
28. Particle Data Group. The Review of Particle Physics. *Prog. Theor. Exp. Phys.* **2020**, *2020*, 083C01. [CrossRef]
29. Carena, M.; Ellis, J.; Lee, J.S.; Pilaftsis, A.; Wagner, C.E. CP Violation in Heavy MSSM Higgs Scenarios. *JHEP* **2016**, *2*, 123. [CrossRef]
30. Fuchs, E.; Weiglein, G. Impact of CP-violating interference effects on MSSM Higgs searches. *Eur. Phys. J. C* **2018**, *78*, 87. [CrossRef]
31. Berlin, A.; Lin, T.; Wang, L.T. Mono-Higgs Detection of Dark Matter at the LHC. *JHEP* **2014**, *6*, 078. [CrossRef]
32. Bauer, M.; Haisch, U.; Kahlhoefer, F. Simplified dark matter models with two Higgs doublets: I. Pseudoscalar mediators. *JHEP* **2017**, *5*, 138. [CrossRef]
33. Goncalves, D.; Machado, P.A. N.; No, J.M. Simplified Models for Dark Matter Face their Consistent Completions. *Phys. Rev. D* **2017**, *95*, 055027. [CrossRef]
34. Bell, N.F.; Busoni, G.; Sanderson, I.W. Self-consistent Dark Matter Simplified Models with an s-channel scalar mediator. *JCAP* **2017**, *3*, 15. [CrossRef]
35. Abe, T.; Fujiwara, M. ; Hisano, J. Loop corrections to dark matter direct detection in a pseudoscalar mediator dark matter model. *JHEP* **2019**, *2*, 28. [CrossRef]
36. Gunion, J.F.; Haber, H. E. The CP conserving two Higgs doublet model: The Approach to the decoupling limit. *Phys. Rev. D* **2003**, *67*, 075019. [CrossRef]
37. Branco, G.C.; Ferreira, P.M.; Lavoura, L.; Rebelo, M.N.; Sher, M.; Silva, J.P. Theory and phenomenology of two-Higgs-doublet models. *Phys. Rept.* **2012**, *516*, 1. [CrossRef]

38. LHC Dark Matter Working Group. LHC Dark Matter Working Group: Next-generation spin-0 dark matter models. *Phys. Dark Univ.* **2020**, *27*, 100351. [CrossRef]
39. ATLAS Collaboration. ATLAS-CONF-2021-036. Available online: https://atlas.web.cern.ch/Atlas/GROUPS/PHYSICS/CONFNOTES/ATLAS-CONF-2021-036 (accessed on 10 October 2022).
40. Brax, P.; Burrage,C.; Englert,C.; Spannowsky, M. LHC Signatures Of Scalar Dark Energy. *Phys. Rev. D* **2016**, *94*, 084054. [CrossRef]
41. Horndeski, G.W. Second-order scalar-tensor field equations in a four-dimensional space. *Int. J. Theor. Phys.* **1974**, *10*, 363. [CrossRef]
42. Joyce, A.; Jain, B.; Khoury,J. ; Trodden, M. Beyond the Cosmological Standard Model. *Phys. Rept.* **2015**, *568*, 1. [CrossRef]
43. ATLAS Collaboration. ATLAS-CONF-2022-036. Available online: https://atlas.web.cern.ch/Atlas/GROUPS/PHYSICS/CONFNOTES/ATLAS-CONF-2022-036/ (accessed on 10 October 2022).
44. CMS Collaboration. Search for dark matter in events with energetic, hadronically decaying top quarks and missing transverse momentum at $\sqrt{s} = 13$ TeV. *JHEP* **2018**, *6*, 027.
45. Cacciari, M.; Salam, G.P.; Soyez, G. The anti-k_t jet clustering algorithm. *JHEP* **2008**, *4*, 63. [CrossRef]
46. ATLAS Collaboration. Performance of top-quark and *W*-boson tagging with ATLAS in Run 2 of the LHC. *Eur. Phys. J. C* **2019**, *79*, 375. [CrossRef]
47. ATLAS Collaboration. ATL-PHYS-PUB-2020-017. Available online: https://atlas.web.cern.ch/Atlas/GROUPS/PHYSICS/PUBNOTES/ATL-PHYS-PUB-2020-017/ (accessed on 10 October 2022).
48. Chen, T.; Guestrin,C. XGBoost: A Scalable Tree Boosting System Proceedings of the 22nd ACM SIGKDD International Conference on Knowledge Discovery and Data Mining. *KDD* **2016**, *16*, 785.
49. Thaler, J.; Tilburg, K.V. Identifying Boosted Objects with N-subjettiness. *JHEP* **2011**, *3*, 15. [CrossRef]
50. Larkoski, A.J.; Salam ,G. P.;Thaler, J. Energy Correlation Functions for Jet Substructure. *JHEP* **2013**, *6*, 108. [CrossRef]
51. Moult, I.; Necib,L.; Thaler, J. New Angles on Energy Correlation Functions. *JHEP* **2016**, *12*, 153. [CrossRef]
52. Hoecker, A.; Speckmayer, P.; Stelzer, J.; Therhaag, J.; Toerne, E.; Voss, H.; Backes, M.; Carli, T.; Cohen, O.; Christov, A.; et al. TMVA—Toolkit for Multivariate Data Analysis. *arXiv* **2007**, arXiv:0703039.
53. ATLAS Collaboration. Search for large missing transverse momentum in association with one top-quark in proton-proton collisions at $\sqrt{s} = 13$ TeV with the ATLAS detector. *JHEP* **2019**, *5*, 41.
54. CMS Collaboration. Search for dark matter produced in association with a single top quark or a top quark pair in proton-proton collisions at $\sqrt{s} = 13$ TeV. *JHEP* **2019**, *3*, 141.
55. ATLAS Collaboration. Search for dark matter produced in association with a single top quark in $\sqrt{s} = 13$ TeV *pp* collisions with the ATLAS detector. *Eur. Phys. J. C* **2021**, *81*, 860.
56. ATLAS Collaboration. ATLAS-CONF-2022-012. Available online: https://atlas.web.cern.ch/Atlas/GROUPS/PHYSICS/CONFNOTES/ATLAS-CONF-2022-012/ (accessed on 10 October 2022).
57. CMS Collaboration. Combined searches for the production of supersymmetric top quark partners in proton-proton collisions at $\sqrt{s}$= 13 TeV. *Eur. Phys. J. C* **2021**, *81*, 970. [CrossRef]
58. CMS Collaboration. Search for top squark production in fully-hadronic final states in proton-proton collisions at s = 13 TeV. *Phys. Rev. D* **2021**, *104*, 052001. [CrossRef]
59. CMS Collaboration. Search for direct top squark pair production in events with one lepton, jets, and missing transverse momentum at 13 TeV with the CMS experiment. *JHEP* **2020**, *5*, 32.
60. CMS Collaboration. Search for top squark pair production using dilepton final states in pp collision data collected at $\sqrt{s} = 13$ TeV. *Eur. Phys. J. C* **2021**, *81*, 3. [CrossRef]
61. CMS Collaboration. Identification of heavy, energetic, hadronically decaying particles using machine-learning techniques. *JINST* **2020**, *15*, P06005. [CrossRef]
62. CMS Collaboration. Performance of missing transverse momentum reconstruction in proton-proton collisions at $\sqrt{s} = 13$ TeV using the CMS detector. *JINST* **2019**, *14*, P07004. [CrossRef]
63. CMS Collaboration. Search for top squarks and dark matter particles in opposite-charge dilepton final states at $\sqrt{s} = 13$ TeV. *Phys. Rev. D* **2018**, *97*, 032009. [CrossRef]
64. ATLAS Collaboration. Search for a scalar partner of the top quark in the all-hadronic $t\bar{t}$ plus missing transverse momentum final state at $\sqrt{s} = 13$ TeV with the ATLAS detector. *Eur. Phys. J. C* **2020**, *80*, 737.
65. ATLAS Collaboration. Search for new phenomena with top quark pairs in final states with one lepton, jets, and missing transverse momentum in *pp* collisions at $\sqrt{s} = 13$ TeV with the ATLAS detector. *JHEP* **2021**, *4*, 174.
66. ATLAS Collaboration. Search for new phenomena in events with two opposite-charge leptons, jets and missing transverse momentum in pp collisions at $\sqrt{s} = 13$ TeV with the ATLAS detector. *JHEP* **2021**, *4*, 165.
67. ATLAS Collaboration. Constraints on spin-0 dark matter mediators and invisible Higgs decays using ATLAS 13 TeV *pp* collision data with two top quarks and missing transverse momentum in the final state. *arXiv* **2022**, arXiv:2211.05426.
68. ATLAS Collaboration. Search for $t\bar{t}$ resonances in fully hadronic final states in *pp* collisions at $\sqrt{s} = 13$ TeV with the ATLAS detector. *JHEP* **2020**, *10*, 061.
69. ATLAS Collaboration. Search for heavy particles decaying into top-quark pairs using lepton-plus-jets events in proton–proton collisions at $\sqrt{s} = 13$ TeV with the ATLAS detector. *Eur. Phys. J. C* **2018**, *78*, 565. [CrossRef] [PubMed]
70. CMS Collaboration. Search for resonant $t\bar{t}$ production in proton-proton collisions at $\sqrt{s} = 13$ TeV. *JHEP* **2019**, *4*, 31.

71. ATLAS Collaboration. Search for Heavy Higgs Bosons A/H Decaying to a Top Quark Pair in pp Collisions at $\sqrt{s} = 8$ TeV with the ATLAS Detector. *Phys. Rev. Lett.* **2017**, *119*, 191803. [CrossRef]
72. CMS Collaboration. Search for heavy Higgs bosons decaying to a top quark pair in proton-proton collisions at $\sqrt{s} = 13$ TeV. *JHEP* **2020**, *4*, 171.
73. ATLAS Collaboration. Evidence for $t\bar{t}t\bar{t}$ production in the multilepton final state in proton–proton collisions at $\sqrt{s} = 13$ TeV with the ATLAS detector. *Eur. Phys. J. C* **2020**, *80*, 1085.
74. ATLAS Collaboration. Search for $t\bar{t}H/A \to t\bar{t}t\bar{t}$ production in the multilepton final state in proton-proton collisions at $\sqrt{s} = 13$ TeV with the ATLAS detector. *arXiv* **2022**, arXiv:2211.01136.
75. Baldi, P.; Cranmer, K.; Faucett, T.; Sadowski, P.; Whiteson, D. Parameterized neural networks for high-energy physics. *Eur. Phys. J. C* **2016**, *76*, 235. [CrossRef]
76. ATLAS Collaboration. ATL-PHYS-PUB-2022-036. Available online: https://atlas.web.cern.ch/Atlas/GROUPS/PHYSICS/PUBNOTES/ATL-PHYS-PUB-2022-036/ (accessed on 10 October 2022).
77. ATLAS Collaboration. Search for new phenomena in a lepton plus high jet multiplicity final state with the ATLAS experiment using $\sqrt{s} = 13$ TeV proton-proton collision data. *JHEP* **2017**, *9*, 088.
78. CMS Collaboration. Search for production of four top quarks in final states with same-sign or multiple leptons in proton-proton collisions at $\sqrt{s} = 13$ TeV. *Eur. Phys. J. C* **2020**, *80*, 75. [CrossRef]
79. ATLAS Collaboration. ATLAS-CONF-2021-048. Available online: https://atlas.web.cern.ch/Atlas/GROUPS/PHYSICS/CONFNOTES/ATLAS-CONF-2021-048/ (accessed on 10 October 2022).
80. ATLAS Collaboration. Search for charged Higgs bosons decaying into a top quark and a bottom quark at $\sqrt{s} = 13$ TeV with the ATLAS detector. *JHEP* **2021**, *6*, 145.
81. ATLAS Collaboration. ATLAS-CONF-2022-008. Available online: https://atlas.web.cern.ch/Atlas/GROUPS/PHYSICS/CONFNOTES/ATLAS-CONF-2022-008/ (accessed on 10 October 2022).
82. CMS Collaboration. Search for charged Higgs bosons decaying into a top and a bottom quark in the all-jet final state of pp collisions at s $\sqrt{s}$ =13 TeV. *JHEP* **2020**, *7*, 126.
83. CMS Collaboration. Search for a charged Higgs boson decaying into top and bottom quarks in events with electrons or muons in proton-proton collisions at **s** $\sqrt{s}$= 13 TeV. *JHEP* **2020**, *1*, 096.
84. CMS Collaboration. CMS Exotica Summary Plots for 13 TeV Data. Available online: https://twiki.cern.ch/twiki/bin/view/CMSPublic/SummaryPlotsEXO13TeV#DM_summary plots
85. CMS Collaboration. Search for new particles in events with energetic jets and large missing transverse momentum in proton-proton collisions at $\sqrt{s}$ = 13 TeV. *JHEP* **2021**, *11*, 153.
86. CMS Collaboration. Search for low mass vector resonances decaying into quark-antiquark pairs in proton-proton collisions at $\sqrt{s}$ = 13 TeV. *Phys. Rev. D* **2019**, *100*, 11.
87. CMS Collaboration. Search for high mass dijet resonances with a new background prediction method in proton-proton collisions at $\sqrt{s}$ = 13 TeV. *JHEP* **2020**, *5*, 033.
88. Boveia, A.; Buchmueller, O.; Busoni, G.; D'Eramo, F.; De Roeck, A.; De Simone, A.; Doglioni, C.; Dolan, M.J.; Genest, M.-H.; Hahn, K.; et al. Recommendations on presenting LHC searches for missing transverse energy signals using simplified ss-channel models of dark matter. *Phys. Dark Univ.* **2020**, *27*, 100365. [CrossRef]
89. Haisch, U.; Polesello, G. Searching for production of dark matter in association with top quarks at the LHC. *J. High Energy Phys.* **2019**, *2019*, 29. [CrossRef]
90. Pani, P.; Polesello, G. Dark matter production in association with a single top-quark at the LHC in a two-Higgs-doublet model with a pseudoscalar mediator. *Phys. Dark Univ.* **2018**, *21*, 8. [CrossRef]
91. Pinna, D.; Zucchetta, A.; Buckley, M.R.; Canelli, F. Single top quarks and dark matter. *Phys. Rev. D* **2017**, *96*, 035031. [CrossRef]
92. Gavela, M.B.; Parga, N. Ward identities and the physical interpretations of anomalies in stochastic quantization. *Phys. Lett. B* **1986**, *169*, 73. [CrossRef]
93. Choi, K.; Kang, K.; Kim, J.E. Effects of η' in low-energy axion physics. *Phys. Lett. B* **1986**, *181*, 145–149. [CrossRef]
94. Gavela, M.B.; No, J.M.; Sanz, V.; de Trocóniz, J.F. Nonresonant Searches for Axionlike Particles at the LHC. *Phys. Rev. Lett.* **2020**, *124*, 051802. [CrossRef]
95. CMS Collaboration. Search for heavy resonances decaying to ZZ or ZW and axion-like particles mediating nonresonant ZZ or ZH production at $\sqrt{s}$ = 13 TeV. *JHEP* **2022**, *4*, 87.
96. Carra, S.; Goumarre, V.; Gupta, R.; Heim, S.; Heinemann, B.; Küchler, J.; Meloni, F.; Quilez, P.; Yap, Y.C. Constraining off-shell production of axionlike particles with $Z\gamma$ and WW differential cross-section measurements. *Phys. Rev. D* **2021**, *104*, 092005. [CrossRef]
97. Bonilla, J. Available online: https://ep-news.web.cern.ch/content/non-resonant-searches-open-new-avenues-hunt-axion-particles (accessed on 10 October 2022).
98. Carmona, A.; Elahi, F.; Scherb, C.; Schwaller, P. The ALPs from the top: searching for long lived axion-like particles from exotic top decays. *J. High Energy Phys.* **2022**, *2022*, 122. [CrossRef]
99. Brivio, I.; Gavela, M.B.; Merlo, L.; Mimasu, K.; No, J.M.; Del Rey, R.; Sanz, V. ALPs Effective Field Theory and Collider Signatures. *Eur. Phys. J. C* **2017**, *77*, 572. [CrossRef]
100. Renner, S.; Schwaller, P. A flavoured dark sector. *J. High Energy Phys.* **2018**, *2018*, 052. [CrossRef]

101. Froggatt, C.; Nielsen, H.B. Hierarchy of Quark Masses, Cabibbo Angles and CP Violation. *Nucl. Phys. B* **1979**, *147*, 277. [CrossRef]
102. Haisch, U.; Polesello, G.; Schulte, S. Searching for pseudo Nambu-Goldstone boson dark matter production in association with top quarks. *JHEP* **2021**, *9*, 206. [CrossRef]
103. Kribs, G.D.; Martin, A.; Ostdiek, B.; Tong, T. Dark Mesons at the LHC. *JHEP* **2019**, *7*, 133. [CrossRef]
104. Dainese, A.; Mangano, M.; Meyer, A.B; Nisati, A.; Salam, G.; Vesterinen, M.A. Report on the Physics at the HL-LHC, and Perspectives for the HE-LHC. CERN-2019-007. Available online: https://cds.cern.ch/record/2703572 (accessed on 10 October 2022).
105. Benedikt, M.; Capeans G.M. ; Cerutti, F.; Goddard, B. ; Gutleber, J.; Jimenez, J.M.; Mangano, M.; Mertens, V.; Osborne, J.A.; Otto, T. FCC-hh: The Hadron Collider: Future Circular Collider Conceptual Design Report Volume 3. Future Circular Collider. CERN-ACC-2018-0058. Available online: https://cds.cern.ch/record/2651300 (accessed on 10 October 2022).
106. Dutta, B.; Kamon, T.; Ko, P.; Li, J. Prospects for discovery and spin discrimination of dark matter in Higgs portal DM models and their extensions at 100 TeV *pp* collider. *Eur. Phys. J. C* **2018**, *78*, 595. [CrossRef]
107. Harris, P. Prospects for Discovery and Spin Discrimination of Dark Matter in Higgs Portal DM Models and Their Extensions at 100 TeV *pp* Collider. Talk at FCC Week 2017. Available online: https://indico.cern.ch/event/556692/contributions/2592531/attachments/1469109/2272523/PCH_FCC_DM_v2.pdf (accessed on 10 October 2022).
108. Behnke, T.; James, E. B.; Brian, F.; Juan, F.; Mike, H.; James, M.E.P.; Michael P.; Marcel, S.; Nicholas, W.; Hitoshi, Y. The International Linear Collider Technical Design Report—Volume 1: Executive Summary. *arXiv* **2013**, arXiv:1306.6327.
109. Aicheler, M.; Burrows, P.; Draper, M.; Garvey, T.; Lebrun, P.; Peach, K.; Phinney, N.; Schmickler, H.; Schulte, D.; Toge, N. A Multi-TeV Linear Collider Based on CLIC Technology: CLIC Conceptual Design Report. CERN Yellow Reports: Monographs 2012. Available online: https://cds.cern.ch/record/1500095 (accessed on 10 October 2022).
110. Linssen, L.; Miyamoto, A.; Stanitzki, M.; Weerts, H. Physics and Detectors at CLIC: CLIC Conceptual Design Report. *arXiv* **2012**, arXiv:1202.5940.
111. The FCC Collaboration. FCC Physics Opportunities. *Eur. Phys. J. C* **2019**, *79*, 474. [CrossRef]
112. The FCC Collaboration. FCC—The lepton collider. *Eur. Phys. J. S. T.* **2019**, *228*, 261.
113. CEPC Study Group. CEPC Conceptual Design Report: Volume 1—Accelerator. *arXiv* **2018**, arXiv:1809.00285.
114. CEPC Study Group. CEPC Conceptual Design Report: Volume 2—Physics & Detector. *arXiv* **2018**, arXiv:1811.10545.

universe

MDPI

Communication

Prospects for Probing Axionlike Particles at a Future Hadron Collider through Top Quark Production

Yasaman Hosseini and Mojtaba Mohammadi Najafabadi *

School of Particles and Accelerators, Institute for Research in Fundamental Sciences (IPM), Tehran P.O. Box 19395-5531, Iran; yasaman.hosseini@cern.ch
* Correspondence: mojtaba@cern.ch

Abstract: Axionlike particles (ALPs) emerge from spontaneously broken global symmetries in high energy extensions of the Standard Model (SM). This causes ALPs to be among the objectives of future experiments that intend to search for new physics beyond the SM. We discuss the reach of future pp collider FCC-hh in probing the ALP model parameters through top quark pair production associated with ALP ($t\bar{t}$ + ALP) in a model-independent approach. The search is performed in the semi-leptonic decay mode of $t\bar{t}$ and the analysis is performed using a parametric simulation of the detector response for a projected integrated luminosity of 30 ab^{-1}. It is shown that $t\bar{t}$ + ALP production at the FCC-hh is a promising channel with significant sensitivity to probe the ALP coupling with gluons. The ALP coupling with gluons obtained from HL-LHC and other experiments is presented for comparison.

Keywords: top quark; models beyond the standard model

Citation: Hosseini, Y.; Mohammadi Najafabadi, M. Prospects for Probing Axionlike Particles at a Future Hadron Collider through Top Quark Production. *Universe* **2022**, *8*, 301. https://doi.org/10.3390/universe8060301

Academic Editor: Jinmin Yang

Received: 29 March 2022
Accepted: 23 May 2022
Published: 26 May 2022

1. Introduction

Axionlike particles (ALPs) are pseudo-Goldstone bosons that can appear from the spontaneous breaking of some global symmetries at energy scales well above the electroweak scale. In recent years, there has been much interest in ALPs because of their various notable aspects. ALPs possess many applications based on their masses and couplings in the parameter space. ALPs can solve the strong CP problem [1] and they are appropriate candidates for non-thermal Dark Matter (DM) [2]. ALPs can play a vital role in baryogenesis, giving an explanation for the observed imbalance in matter and antimatter [3], and are able to explain the neutrino mass problem through an ALP–neutrino interaction that causes neutrinos to earn mass [4]. Furthermore, ALPs can address the muon anomalous magnetic dipole moment [5] and the excess observed in the rare K meson studies reported by the KOTO experiment [6].

ALPs are mostly probed in a model-independent effective field theory (EFT) framework. The strength of ALPs' couplings to SM fields is proportional to the inverse of $U(1)$ spontaneous symmetry breaking scale f_a, which is much higher than the electroweak symmetry breaking scale of the SM. Thus far, a remarkable region of the ALP parameter space in terms of its mass and couplings has been probed or will be studied by cosmological observations, low-energy experiments, and collider searches [7–14].

Very light ALPs with masses below the electron pair mass ($m_a < 2m_e$) are only allowed to decay into a pair of photons. Based on the ALPs' masses and couplings, heavier ones are allowed to decay into hadrons and charged leptons. The decay rates of light ALPs are usually very small, such that they can travel a long distance before they decay. Long-lived ALPs appear as invisible particles at colliders; therefore, they appear as missing energy in the detectors since they decay outside the detector environment. There are several proposals for searches at collider experiments to probe long-lived ALPs via mono-jet, mono-V ($V = \gamma, W, Z$), and jet + γ [15–22]. Searches for ALPs via exotic Higgs decays $H \to Z + a$ and $H \to a + a$ with ALP decays to diphoton and dilepton at the LHC have provided remarkable sensitivities in a vast region of parameter space [10,11,23–25]. There are searches for ALPs through the production of dijet in association with an ALP and jet+ALP at the LHC and FCC-hh, which can be found, for instance, in Refs. [15,21]. It has been shown that the dijet+ALP channel using multivariate analysis provides strong sensitivity to the ALP coupling with gluons. Although the bounds on the ALP coupling with gluons from dijet+ALP and jet+ALP [21] are very strong, it is worth performing complementary searches through $t\bar{t} + ALP$. Furthermore, the structure of the fermionic ALP couplings is specific as it consists of the Yukawa matrices; as a result, the ALP is expected to couple more strongly to third-generation quarks. This makes $t\bar{t} + ALP$ an important channel by which to explore the ALP model.

In this paper, we propose a search for strong and fermionic couplings of ALPs through the associated production of an ALP with a pair of $t\bar{t}$ in proton–proton collisions at the future circular collider (FCC-hh) [26] at a center-of-mass energy of 100 TeV. In particular, the focus is on a region of the parameter space in which an ALP does not decay inside the detector and manifest as missing energy.

This paper is organized as follows. In Section 2, an introduction to the ALP model is presented. Section 3 is dedicated to presenting the details of the search for the ALP model using $t\bar{t} + ALP$. In Section 4, a summary of the results and discussion is given.

2. Effective Lagrangian for Axionlike Particles

The theoretical framework adopted throughout this work is a linear effective field theory where electroweak physics beyond the SM is expressed by a linear EFT expansion versus gauge-invariant operators ordered by their mass dimension. The model includes SM plus an ALP, where the scale of the new physics is the ALP decay constant f_a. The most general effective Lagrangian describing ALP interactions with SM fields up to dimension $D = 5$ operators has the following form [15]:

$$\begin{aligned} \mathcal{L}_{eff}^{D\leq 5} &= \mathcal{L}_{SM} + \frac{1}{2}(\partial^\mu a)(\partial_\mu a) - \frac{1}{2}m_a^2 a^2 + c_{a\Phi}\mathbf{O}_{a\Phi}^{\psi} \\ &- c_{gg}\frac{a}{f_a}G_{\mu\nu}^{A}\tilde{G}^{\mu\nu,A} - c_{WW}\frac{a}{f_a}W_{\mu\nu}^{A}\tilde{W}^{\mu\nu,A} - c_{BB}\frac{a}{f_a}B_{\mu\nu}\tilde{B}^{\mu\nu}, \end{aligned} \tag{1}$$

where

$$\mathbf{O}_{a\Phi}^{\psi} \equiv i(\bar{Q}_L\mathbf{Y}_U\tilde{\Phi}u_R - \bar{Q}_L\mathbf{Y}_D\Phi d_R - \bar{L}_L\mathbf{Y}_E\Phi e_R)\frac{a}{f_a} + \text{h.c.} \tag{2}$$

where e_R,d_R,u_R are $SU(2)_L$ singlets and L_L and Q_L are the $SU(2)_L$ doublets. The ALP EFT Lagrangian of Equation (1) is implemented in `FeynRules` [27] according to the notation of Ref. [15]. The obtained Universal FeynRules Output (UFO) model [28] (http://feynrules.irmp.ucl.ac.be/attachment/wiki/ALPsEFT/ALP_linear_UFO.tar.gz (accessed on 31 December 2021)) is embedded in `MadGraph5_aMC@NLO` [29] to compute the cross-sections and to generate the ALP signal events.

ALP Decays

According to the ALP interactions presented by the effective Lagrangian Equation (1), an ALP is allowed to decay into pairs of SM particles. For the MeV-scale ALPs, the decays into photons, charged leptons, and light hadrons are dominant. The diphoton decay mode is the most important one for light ALPs with mass $m_a < 2m_e = 1.022$ MeV. As m_a increases to $2m_e$ and above, the leptonic decay mode $a \to l^+l^-$ becomes accessible. The ALP hadronic decay modes appear when $m_a > m_\pi$ and arise from the ALP decays $a \to gg$ and $a \to q\bar{q}$. The triple pion decay modes $a \to \pi^+\pi^-\pi^0$ and $a \to \pi^0\pi^0\pi^0$ are the main hadronic modes for $m_a < 1$ GeV. Other ALP hadronic decay modes such as $a \to \pi^0\gamma\gamma$ and $a \to \pi^+\pi^-\gamma$ are suppressed with respect to $\pi^+\pi^-\pi^0$ and $3\pi^0$ due to the presence of powers of the fine structure constant [17].

One should note that a fraction of ALPs decay inside the detector environment and consequently they do not appear as missing energy. The decay length of ALP $\mathcal{L}_a$ is proportional to $\sqrt{\gamma^2 - 1}/\Gamma_a$, where Γ_a and γ are the total width and ALP Lorentz factor, respectively. The ALP decay probability in the detector volume is proportional to $e^{-\mathcal{L}_{\rm det}/\mathcal{L}_a}$. $\mathcal{L}_{\rm det}$ is the transverse distance of the detector component from the collision point. In this study, the probability that the ALP does not decay inside the detector and escapes detection is considered event-by-event. The total width of the ALP is obtained from Ref. [30], where the chiral perturbation theory and vector meson dominance model have been used in width calculations. For instance, the decay probability for an ALP with $m_a = 10$ MeV and $|\vec{p}_a|$ = 242 GeV is 0.0053, while the decay probability for an ALP with $m_a = 70$ MeV and $|\vec{p}_a|$ = 387 GeV is 0.999.

3. ALP Production Associated with a Pair of Top Quarks

Top quark pair production in association with an ALP in proton–proton collisions at a center-of-mass energy of 100 TeV is used to probe the parameter space of the ALP model. As indicated previously, the focus is on the ALPs that do not decay within the detector volume and are not detected by the detectors appearing as missing momentum. Figure 1 depicts the representative Feynman diagrams for $t\bar{t} + ALP$ in proton–proton collisions.

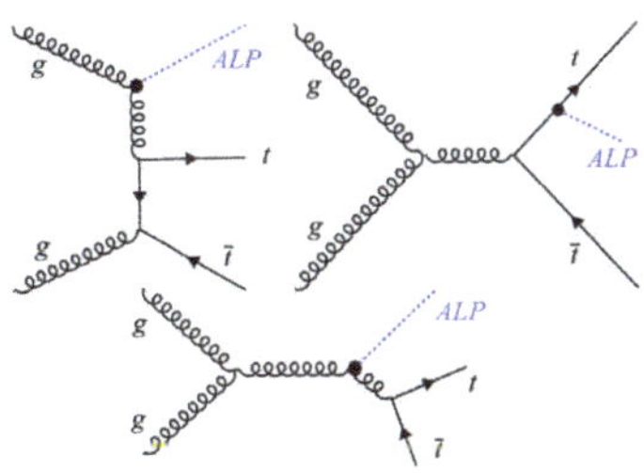

Figure 1. Representative leading order Feynman diagrams for production of a pair of top quarks with an ALP in pp collisions.

At the production level, this process is sensitive to c_{gg} and $c_{a\Phi}$. Assuming one non-vanishing ALP coupling at a time, the cross-sections $\sigma(pp \to t\bar{t} + ALP)(c_{XX})$ at leading order (LO) read:

$$\sigma(c_{gg}) = 459.6\Big(\frac{c_{gg}}{f_a}\Big)^2 \text{pb},$$
$$\sigma(c_{a\Phi}) = 2.45\Big(\frac{c_{a\Phi}}{f_a}\Big)^2 \text{pb}, \tag{3}$$

where f_a is in units of TeV and the cross-sections are calculated using `MadGraph5_aMC@NLO` with the NNPDF23 [31] as the parton distribution functions (PDFs) of a proton. The cross-sections are obtained for the value of ALP mass $m_a = 1$ MeV and change up to 10% when m_a increases to 100 MeV. This is expected as, in this mass range, m_a is negligible in comparison to the typical energy scale of the process. The total cross-section of the SM $t\bar{t}$ production at leading order calculated by `MadGraph5_aMC@NLO` is 24,673.5 pb. From Equation (3), it is clear that there is more sensitivity to c_{gg} than $c_{a\Phi}$, which is due to the fact that c_{gg} appears in both initial and final states and the large gluon PDF. Since the $t\bar{t} + ALP$ rate has no significant sensitivity for $c_{a\Phi}$ coupling with respect to c_{gg}, a weaker bound on $c_{a\Phi}$ is expected.

As the ALP escapes detection, the $t\bar{t} + ALP$ can be probed through the $t\bar{t} + E_T^{\rm miss}$ signature. A similar signature has been studied by the CMS and ATLAS collaborations in Refs. [32,33] to explore simplified models for dark matter where a mediator exists that couples to both the SM particles and dark matter. These studies investigate the production of a fermionic dark matter through a color-neutral scalar or pseudo-scalar particle (ϕ) exchange, where the couplings between the new (pseudo)scalar and SM particles are Yukawa-like. Therefore, the mediator is expected to be produced mainly in association with heavy quarks or through loop-induced gluon–gluon fusion. The distinctive signature for dark matter in $t\bar{t}+\phi$ production followed by $\phi \to \chi\chi$, where χ is the dark matter field, is the emergence of a high missing transverse momentum recoiling against the $t\bar{t}$ system.

In this analysis, as in the past LHC search [20,32], the concentration is on the semi-leptonic $t\bar{t}$ decay channel and follows a similar selection. The final state consists of one charged lepton, four jets, and large missing transverse momentum. The main background sources to the signal arise from $t\bar{t}$, $W+jets$, $Z+jets$, single top production, and diboson. All background contributions are estimated from simulation. Both signal and background processes are generated using `MadGraph5_aMC@NLO` at leading order and passed through `Pythia` [34] to perform showering and hadronization. `Delphes 3.5.0` [35], and the FCC-hh detector card (https://github.com/delphes/delphes/blob/master/cards/FCC/FCChh.tcl (accessed on 31 December 2021)) is used for detector simulation. The jet finding is performed using `FastJet` [36] using an anti-k_t algorithm with a distance parameter of 0.4 [37], considering the particle-flow reconstruction approach as described in Ref. [35]. Several signal samples are generated with ALP masses from 1 MeV to 150 MeV and f_a is taken to be 1 TeV. Based on the final state, events are selected by applying the following requirements:

- Only one isolated charged lepton (e, μ) with $p_{\rm T} \geq 30$ GeV and $|\eta| \leq 2.5$. Events containing additional charged leptons with $p_{\rm T} \geq 10$ GeV that fulfill loose isolation criteria are discarded. Isolated leptons are chosen with the help of the isolation variable $I_{\rm Rel}$ according to the definition given in Ref. [35]. Similar to Ref. [32], $I_{\rm Rel}$ is required to be less than 0.15 for muons and 0.035 for electrons. For loose electrons (muons), $I_{\rm Rel}$ is required to be less than 0.126 (0.25).
- At least three jets with $p_{\rm T} \geq 30$ GeV and $|\eta| \leq 2.5$ from which one must be tagged as a b-jet. B-jet identification is based on a parametric approach that relies on Monte Carlo generator information. The probability for b-jet identification is according to the parameterization of the b-tagging efficiency available in the FCC-hh detector card. For a jet with $10 < p_T < 500$ GeV and $|\eta| < 2.5$, the b-tagging efficiency is taken to be 82% and misidentification rates are 15% and 1% for c-quark jets and light flavor jets, respectively.
- The magnitude of missing transverse momentum to be greater than 160 GeV.

For further reduction of $t\bar{t}$ and $W+jets$ backgrounds, the transverse mass $M_{\rm T} = \sqrt{2p_{\rm T,l}E_{\rm T}^{\rm miss}(1-\cos\Delta\phi(\vec{p}_{\rm T,l}, \vec{E}_{\rm T}^{\rm miss}))}$ has to be greater than 160 GeV. Moreover, the magnitude of the vector sum of all jets with $p_{\rm T} > 20$ GeV and $|\eta| < 5.0$, $\rm H_T$, is required to be larger than 120 GeV. To suppress the contribution of SM $t\bar{t}$ background, a lower cut value of 200 GeV is applied on the $\rm M_{T2}^{W}$ variable. The $\rm M_{T2}^{W}$ variable has been introduced in Ref. [38] in searches for supersymmetric partners of the top quark. To ensure the validity of the considered effective Lagrangian, it is required that its suppression scale f_a must be larger than the typical energy scale of the process. Therefore, in each event, the energy scale of the process $\sqrt{\hat{s}}$ has to be much less than f_a. In this work, the ALP appears as missing momentum and $\sqrt{\hat{s}}$ is not totally measurable. As a result, to provide the validity of the effective theory, f_a is compared to the magnitude of missing transverse momentum. The magnitude of missing transverse momentum is required to be less than f_a in each event. The signal efficiency after the cuts is found to be 12.7% for the case of $m_a = 1$ MeV. The total number of background events after the cuts corresponding to an integrated luminosity of 30 ab^{-1} is 2.12×10^7. The signal and background efficiencies after lepton and jet selection and the cuts on $M_{\rm T}$, $\rm M_{T2}^{W}$, and $\rm H_T$ are presented in Table 1.

Table 1. Efficiency of cuts for two signal cases with (c_{gg}/f_a = 0.1 TeV^{-1}, m_a = 1 MeV); ($c_{a\Phi}/f_a$ = 0.1 TeV^{-1}, m_a = 1 MeV), and for background processes after lepton and jet selection and applying cuts on M_T, MET, H_T, M_{T2}^W.

Cut	$c_{gg}/f_a = 0.1$ TeV^{-1}	$c_{a\Phi}/f_a = 0.1$ TeV^{-1}	$t\bar{t}$	Single Top	W+jets	Z+jets	Diboson
Lepton and jet selection, M_T, MET, H_T, M_{T2}^W	12.7%	3.7%	0.0077%	0.0044%	5.65×10^{-6}%	5.78×10^{-6}%	0.0046%

In order to constrain c_{XX}/f_a coupling, the first step is to set an upper limit on the signal cross-section. The expected upper 95% CL limit on the signal cross-section in the background-only hypothesis is obtained using the standard Bayesian approach [39]. Comparing the upper bound on the signal cross-section with the theoretical cross-section, the 95% CL upper limits on $|c_{XX}/f_a|$ are derived. The expected 95% CL upper bound on $|c_{gg}/f_a|$ for $m_a = 1$ MeV is found to be:

$$|\frac{c_{gg}}{f_a}| \leq 0.00446\ \text{TeV}^{-1} \ @\ 30\ \text{ab}^{-1}, \tag{4}$$

The prospect at HL-LHC for $m_a = 1$ MeV is [20]: $|\frac{c_{gg}}{f_a}| \leq 0.063$ TeV^{-1} @ 3000 fb^{-1}. Excluded regions in the $(|c_{gg}/f_a|, m_a)$ plane at 95% CL from $t\bar{t} + ALP$ are presented in Figure 2. The regions are corresponding to integrated luminosities of 3000 fb^{-1} for the LHC and 30 ab^{-1} for the FCC-hh at the center-of-mass energies of 14 and 100 TeV, respectively. For the case of non-vanishing $c_{a\Phi}$ coupling, using the related signal and background efficiency in Table 1, the upper bound on $|c_{a\Phi}/f_a|$ for m_a = 1 MeV is found to be 0.11 TeV^{-1}. This limit is two orders of magnitude looser than the one derived on $|c_{gg}/f_a|$, which is due to the weaker dependence of the signal cross-section on $c_{a\Phi}/f_a$ than c_{gg}/f_a. The analysis does not have sensitivity to $|c_{gg}/f_a|$ greater than approximately 10^{-3} since, in this region, the ALP will decay inside the detector, and this is in contrast to our assumption of ALP being long-lived and detected as missing energy. Moreover, for a heavier ALP, its decay length tends to zero and, consequently, it will decay inside the detector. It is notable that the limits are obtained considering only statistical uncertainty. In the case of including systematic uncertainties similar to Ref. [32], the upper limit on $|c_{gg}/f_a|$ is weakened by around 2.2%.

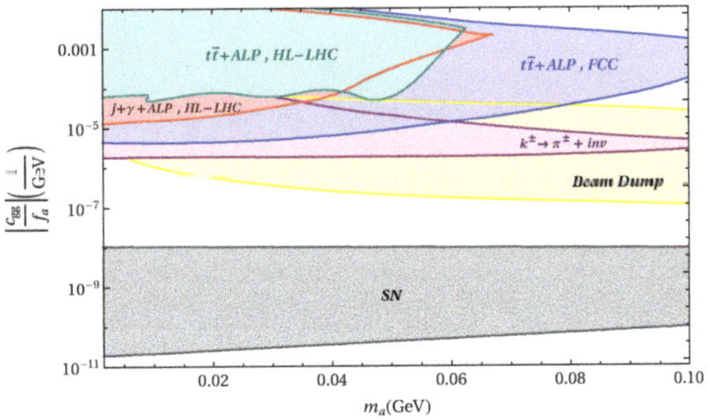

Figure 2. The expected excluded regions of the ALP model parameter space $(|c_{gg}/f_a|, m_a)$ at 95% CL obtained from $t\bar{t} + ALP$ and $j + \gamma + ALP$ channels are presented. The regions derived from $t\bar{t} + ALP$ and $j + \gamma + ALP$ processes at HL-LHC are corresponding to an integrated luminosity of 3 ab^{-1} and are adapted from Ref. [20]. The blue region shows the constraint obtained in the present analysis using $t\bar{t} + ALP$ process at FCC-hh at a center-of-mass energy of 100 TeV with an integrated luminosity of 30 ab^{-1}. The grey region denoted by SN presents the bound from supernova neutrino burst duration adapted from Ref. [7]. The region labeled by $K^{\pm} \to \pi^{\pm} + inv$ adapted from [8] (purple) and beam dump (yellow) present the constraints from Kaon decay and from the proton beam dump experiment CHARM adapted from Ref. [9].

4. Discussion

ALPs are CP odd scalar particles arising from spontaneously broken global $U(1)$ symmetries that can address some of the SM shortcomings, such as the strong CP problem, baryon asymmetry, neutrino mass, and dark matter. The potential of $t\bar{t} + ALP$ production to probe the parameter space of light ALPs at FCC-hh is studied. In general, light ALPs have a long lifetime and do not decay inside the detector, appearing as missing momentum in the final state. For the ALP mass $m_a = 1$ MeV, the obtained upper limit on ALP coupling with gluons $|c_{gg}/f_a|$ at FCC-hh is found to be 0.00446 TeV^{-1}. This bound corresponds to the ultimate integrated luminosity that the FCC-hh will eventually operate at based on the benchmark specifications. As seen in Figure 2, the limit on $|c_{gg}/f_a|$ varies slightly as the ALP mass increases. In order to compare the limits obtained in this analysis with those already derived at HL-LHC, the expected upper limits on $|c_{gg}/f_a|$ at 95% CL from $t\bar{t}$ + ALP and $j + \gamma + ALP$ are presented in Figure 2. A comparison shows that the constraints obtained from FCC-hh are stronger than the limits derived from $t\bar{t}$ + ALP and $j + \gamma + ALP$ analyses at HL-LHC by one to three orders of magnitude depending on the ALP mass. Results of Figure 2 indicate that the analysis of $t\bar{t}$ + ALP FCC-hh is able to span a large area in the ALP parameter space that is not accessible by $K^{\pm} \to \pi^{\pm} + inv$, SN, and beam dump experiments. It can be concluded that the $t\bar{t}$ + ALP production at FCC-hh provides an excellent solution in exploring the light ALP physics as a significant portion of the parameter space is accessible through this channel.

Author Contributions: All authors have equally contributed to all the parts of the manuscript. All authors have read and agreed to the published version of the manuscript.

Funding: This research received no external funding.

Data Availability Statement: Not applicable.

Acknowledgments: The authors are grateful to Yotam Soreq for the insightful comments and for providing data of Ref. [30]. The authors would also like to thank Pedro Ferreira da Silva and Efe Yazgan for inviting us to contribute to the open-access Special Issue, *Top Quark at the New Physics Frontier.*

Conflicts of Interest: The authors declare no conflict of interest.

References

1. Hooft, G. Symmetry Breaking Through Bell-Jackiw Anomalies. *Phys. Rev. Lett.* **1976**, *37*, 8–11. [CrossRef]
2. Dolan, M.J.; Kahlhoefer, F.; McCabe, C.; Schmidt-Hoberg, K. A taste of dark matter: Flavour constraints on pseudoscalar mediators. *J. High Energy Phys.* **2015**, *3*, 171; Erratum in *J. High Energy Phys.* **2015**, *7*, 103. [CrossRef]
3. Jeong, K.S.; Jung, T.H.; Shin, C.S. Adiabatic electroweak baryogenesis driven by an axionlike particle. *Phys. Rev. D* **2020**, *101*, 035009. [CrossRef]
4. Chen, C.S.; Tsai, L.H. Peccei-Quinn symmetry as the origin of Dirac Neutrino Masses. *Phys. Rev. D* **2013**, *88*, 055015. [CrossRef]
5. Chang, D.; Chang, W.F.; Chou, C.H.; Keung, W.Y. Large two loop contributions to g-2 from a generic pseudoscalar boson. *Phys. Rev. D* **2001**, *63*, 091301. [CrossRef]
6. Kitahara, T.; Okui, T.; Perez, G.; Soreq, Y.; Tobioka, K. New physics implications of recent search for $K_L \to \pi^0 \nu\bar{\nu}$ at KOTO. *Phys. Rev. Lett.* **2020**, *124*, 071801. [CrossRef]
7. Fukuda, H.; Harigaya, K.; Ibe, M.; Yanagida, T.T. Model of visible QCD axion. *Phys. Rev. D* **2015**, *92*, 015021. [CrossRef]
8. Izaguirre, E.; Lin, T.; Shuve, B. Searching for Axionlike Particles in Flavor-Changing Neutral Current Processes. *Phys. Rev. Lett.* **2017**, *118*, 111802. [CrossRef]
9. Döbrich, B.; Jaeckel, J.; Kahlhoefer, F.; Ringwald, A.; Schmidt-Hoberg, K. ALPtraum: ALP production in proton beam dump experiments. *J. High Energy Phys.* **2016**, *2*, 018. [CrossRef]
10. Chatrchyan, S.; Khachatryan, V.; Sirunyan, A.M.; Tumasyan, A.; Adam, W.; Aguilo, E. Search for a Non-Standard-Model Higgs Boson Decaying to a Pair of New Light Bosons in Four-Muon Final States. *Phys. Lett. B* **2013**, *726*, 564–586. [CrossRef]
11. Chatrchyan, S.; Khachatryan, V.; Sirunyan, A.M.; Tumasyan, A.; Adam, W.; Aguilo, E. Search for light bosons in decays of the 125 GeV Higgs boson in proton-proton collisions at $\sqrt{s} = 8$ TeV. *J. High Energy Phys.* **2017**, *10*, 76. [CrossRef]
12. Liu, J.; Dona, K.; Hoshino, G.; Knirck, S.; Kurinsky, N.; Malaker, M. Broadband Solenoidal Haloscope for Terahertz Axion Detection. *Phys. Rev. Lett.* **2022**, *128*, 131801. [CrossRef] [PubMed]
13. Aad, G.; Abbott, B.; Abbott, D.C.; Abed, A.; Abeling, K.; Abhayasinghe, D.K. Measurement of light-by-light scattering and search for axion-like particles with 2.2 nb^{-1} of Pb+Pb data with the ATLAS detector. *J. High Energy Phys.* **2021**, *11*, 50. [CrossRef]

14. Sirunyan, A.M.; Erbacher, R.; Carrillo Montoya, C.A.; Carvalho, W.; Górski, M.; Kotlinski, D. Evidence for light-by-light scattering and searches for axion-like particles in ultraperipheral PbPb collisions at $\sqrt{s_{NN}} = 5.02$ TeV. *Phys. Lett. B* **2019**, *797*, 134826. [CrossRef]
15. Brivio, I.; Gavela, M.B.; Merlo, L.; Mimasu, K.; No, J.M.; del Rey, R.; Sanz, V. ALPs Effective Field Theory and Collider Signatures. *Eur. Phys. J. C* **2017**, *77*, 572. [CrossRef] [PubMed]
16. Mimasu, K.; Sanz, V. ALPs at Colliders. *J. High Energy Phys.* **2015**, *6*, 173. [CrossRef]
17. Bauer, M.; Neubert, M.; Thamm, A. Collider Probes of Axion-Like Particles. *J. High Energy Phys.* **2017**, *12*, 44. [CrossRef]
18. Bonilla, J.; Brivio, I.; Machado-Rodríguez, J.; de Trocóniz, J.F. Nonresonant Searches for Axion-Like Particles in Vector Boson Scattering Processes at the LHC. *arXiv* **2022**, arXiv:2202.03450.
19. Bauer, M.; Neubert, M.; Renner, S.; Schnubel, M.; Thamm, A. Flavor probes of axion-like particles. *arXiv* **2021**, arXiv:2110.10698.
20. Ebadi, J.; Khatibi, S.; Mohammadi Najafabadi, M. New probes for axionlike particles at hadron colliders. *Phys. Rev. D* **2019**, *100*, 015016. [CrossRef]
21. Haghighat, G.; Haji Raissi, D.; Mohammadi Najafabadi, M. New collider searches for axionlike particles coupling to gluons. *Phys. Rev. D* **2020**, *102*, 115010. [CrossRef]
22. Bauer, M.; Heiles, M.; Neubert, M.; Thamm, A. Axion-Like Particles at Future Colliders. *Eur. Phys. J. C* **2019**, *79*, 74. [CrossRef]
23. Bauer, M.; Neubert, M.; Thamm, A. LHC as an Axion Factory: Probing an Axion Explanation for $(g-2)_\mu$ with Exotic Higgs Decays. *Phys. Rev. Lett.* **2017**, *119*, 031802. [CrossRef]
24. Aad, G.; Abbott, B.; Abbott, D.C.; Abud, A.A.; Abeling, K.; Abhayasinghe, D.K. Search for new phenomena in events with an energetic jet and missing transverse momentum in pp collisions at $\sqrt{s}$ =13 TeV with the ATLAS detector. *Phys. Rev. D* **2021**, *103*, 112006. [CrossRef]
25. Aad, G.; Abbott, B.; Abbott, D.; Abed, A.; Abeling, K.; Abhayasinghe, D. Search for Higgs boson decays into a pair of pseudoscalar particles in the $bb\mu\mu$ final state with the ATLAS detector in pp collisions at $\sqrt{s}$=13 TeV. *Phys. Rev. D* **2022**, *105*, 012006. [CrossRef]
26. FCC Collaboration. FCC-hh: The Hadron Collider: Future Circular Collider Conceptual Design Report Volume 3. *Eur. Phys. J. Spec. Top.* **2019**, *228*, 755–1107. [CrossRef]
27. Alloul, A.; Christensen, N.D.; Degrande, C.; Duhr, C.; Fuks, B. FeynRules 2.0—A complete toolbox for tree-level phenomenology. *Comput. Phys. Commun.* **2014**, *185*, 2250–2300. [CrossRef]
28. Degrande, C.; Duhr, C.; Fuks, B.; Grellscheid, D.; Mattelaer, O.; Reiter, T. UFO—The Universal FeynRules Output. *Comput. Phys. Commun.* **2012**, *183*, 1201–1214. [CrossRef]
29. Alwall, J.; Herquet, M.; Maltoni, F.; Mattelaer, O.; Stelzer, T. MadGraph 5: Going Beyond. *J. High Energy Phys.* **2011**, *6*, 128. [CrossRef]
30. Aloni, D.; Soreq, Y.; Williams, M. Coupling QCD-Scale Axionlike Particles to Gluons. *Phys. Rev. Lett.* **2019**, *123*, 031803. [CrossRef] [PubMed]
31. Ball, R.D.; Bertone, V.; Carrazza, S.; Deans, C.S.; Del Debbio, L.; Forte, S. Parton distributions with LHC data. *Nucl. Phys. B* **2013**, *867*, 244–289. [CrossRef]
32. Sirunyan, A.M.; Tumasyan, A.; Adam, W.; Asilar, E.; Bergauer, T.; Brstetter, J. Search for dark matter produced in association with heavy-flavor quark pairs in proton-proton collisions at $\sqrt{s} = 13$ TeV. *Eur. Phys. J. C* **2017**, *77*, 845. [CrossRef] [PubMed]
33. Aaboud, M.; Aad, G.; Abbott, B.; Abeloos, B.; Abidi, S.H.; AbouZeid, O.S. Search for dark matter produced in association with bottom or top quarks in $\sqrt{s} = 13$ TeV pp collisions with the ATLAS detector. *Eur. Phys. J. C* **2018**, *78*, 18. [CrossRef]
34. Sjostrand, T.; Mrenna, S.; Skands, P.Z. PYTHIA 6.4 Physics and Manual. *J. High Energy Phys.* **2006**, *5*, 26. [CrossRef]
35. de Favereau, J.; Delaere, C.; Demin, P.; Giammanco, A.; Lemaître, V.; Mertens, A.; Selvaggi, M. DELPHES 3, A modular framework for fast simulation of a generic collider experiment. *J. High Energy Phys.* **2014**, *2*, 57. [CrossRef]
36. Cacciari, M.; Salam, G.P.; Soyez, G. FastJet User Manual. *Eur. Phys. J. C* **2012**, *72*, 1896. [CrossRef]
37. Cacciari, M.; Salam, G.P.; Soyez, G. The anti-k_t jet clustering algorithm. *J. High Energy Phys.* **2008**, *4*, 63. [CrossRef]
38. Bai, Y.; Cheng, H.C.; Gallicchio, J.; Gu, J. Stop the Top Background of the Stop Search. *J. High Energy Phys.* **2012**, *7*, 110. [CrossRef]
39. Bertram, I.; Landsberg, G.L.; Linnemann, J.; Partridge, R.; Paterno, M.; Prosper, H.B. *A Recipe for the Construction of Confidence Limits*; Fermi National Accelerator Lab.: Batavia, IL, USA, 2000. [CrossRef]

MDPI
St. Alban-Anlage 66
4052 Basel
Switzerland
www.mdpi.com

Universe Editorial Office
E-mail: universe@mdpi.com
www.mdpi.com/journal/universe

www.ingramcontent.com/pod-product-compliance
Lightning Source LLC
Chambersburg PA
CBHW061838210926
51850CB00006B/917

9783725806133